Proceedings in Life Sciences

Molecular Genetics of the Bacteria-Plant Interaction

Edited by A. Pühler

With 154 Figures

Springer-Verlag
Berlin Heidelberg New York Tokyo 1983

Professor Dr. ALFRED PÜHLER
Biologie VI (Genetik), Universität Bielefeld
Universitätsstraße
4800 Bielefeld, FRG

ISBN 3-540-12798-4 Springer-Verlag Berlin Heidelberg New York Tokyo
ISBN 0-387-12798-4 Springer-Verlag New York Heidelberg Berlin Tokyo

Library of Congress Cataloging in Publication Data. Main entry under title: Molecular genetics of the bacteria-plant interaction. (Proceedings in life sciences) Includes index. 1. Bacterial genetics. 2. Microorganisms, Nitrogen-fixing. 3. Plant genetics. 4. Molecular genetics. 5. Symbiosis. I. Pühler, A. II. Title: Bacteria-plant interaction. III. Series. QH434.M64 1983 589.9'015 83-14710

Offsetprinting and bookbinding: Brühlsche Universitätsdruckerei, Giessen

2131/3130-543210

Preface

During the last decade it has become clear that bacteria-plant interactions play an important role in agricultural science. In particular, symbiotic nitrogen fixation, plant tumorigenesis, and plant pathogenic bacteria have been extensively investigated. Therefore, several of my colleagues suggested the importance of organizing a specialized symposium on the "Molecular genetics of the bacteria-plant interaction". This symposium was held in Germany, at the University of Bielefeld, in the autumn of 1982. The International Program Advisory Board, together with the National Organizing Committee, initially decided to concentrate on the following three topics:

1. *Rhizobium*-plant interaction
2. *Agrobacterium*-plant interaction
3. Plant pathogenic bacteria

In retrospect, these topics proved to be most popular: Nearly all important groups working in this field contributed to the Bielefeld Symposium. The papers presented are now published as the Proceedings. It will be obvious that this book represents the most comprehensive collection of papers describing the molecular genetics of the bacteria-plant interaction. Scientists, as well as students, can use this book to learn about the basic principles and the current ideas that have recently been developed in this growing research area.

The Symposium was financed by the Bundesministerium für Forschung und Technologie (Federal Ministry of Science and Technology). I express on behalf of the Organizing committee its sincere appreciation to this sponsor, without whom the Symposium would not have been possible. In addition, the help of several industrial companies is acknowledged.

Finally, I would like to thank the students and colleagues of my department who spent a great deal of their time organizing the Bielefeld Symposium. I especially thank the Symposium secretaries, Anneliese Hofmann and Susanna Malmivaara, for their help in coordinating the meeting.

Bielefeld, July 1983

A. PÜHLER

Contents

A. General Introduction

B. The *Rhizobium*-Plant Interaction

I. The *Rhizobium meliloti-Medicago sativa* System

IV. The Interaction of Different *Rhizobium* Species with Plants

C. The *Agrobacterium*-Plant Interaction

I. The Tumor-Inducing Plasmids of *Agrobacterium*

D. Plant Pathogenic Bacteria and Related Aspects

Contributors

You will find the addresses at the beginning of the respective contribution

A. General Introduction

The Historical Background of the Bacteria-Plant Interaction

W. HEUMANN [1]

As an introduction to this book let me give you a few remarks about the historical background of bacteria-plant interaction. It is one of the miseries of our time that it has lost the sense of appreciation of history. Science history is not so much an enumeration of facts, it is instead the fascinating story of the development of scientific ideas.

About 150 years ago chemists and physiologists were driven by the question: "What is the source of nitrogen which is available to green plants?" Great names were involved on this problem: Priestley and Ingen-Housz for instance; they supposed that all plants absorb free nitrogen from the air, de Saussure and Liebig on the other hand denied it. Boussingault was the first to show, in 1838, that peas and clover could take nitrogen from the air, while wheat could not. The nitrogen question for plant nutrition became one of the main topics in the many Agricultural Experimental Stations, in large private farms and in Universities as well. The special position of legumes in this connection was known for a long time, but not understood. In the second half of the nineteenth century many reports were published, suggesting that bacteria or fungi might be concerned with the assimilation of free nitrogen and that the root nodules of legumes might be induced by these microorganisms (Fig. 1). This was the state of affairs when Hermann Hellriegel (Fig. 2) reported his discovery, on Monday 20 Sept. 1886, 96 years ago. It was on the 59th session of the meeting of Deutsche Naturforscher und Ärzte in Berlin, in section 29 (Agricult. Experiments). Hellriegel's title was: "What nitrogen source is available to the plants?" In his talk he compared the nitrogen source of Gramineae and Papilionaceae (legumes): "Graminae are solely dependent on nitrate from soil, whereas to the Leguminosae a second source is available, besides nitrogen from soil. This is the free elemental nitrogen of the air. The so-called root nodules of leguminosae are directly responsible for the assimilation of free nitrogen. These can be induced by a small amount of cultural soil or prevented by the exclusion of microorganisms." This was all very clear-cut. Hellriegel published these results in an extensive report together with H. Wilfarth, a chemist at Bernburg, Thüringen, in the Beilagenheft zu der Zeitschrift des Vereins der Rübenzuckerindustrie des Deutschen Reiches (1888, p. 234). J.H. Gilbert from Rothamsted-Station chaired the session in Berlin. He confirmed the results of Hellriegel in England together with J.B. Lawes (1889); W.O. Atwater and C.D. Woods did the same in the U.S. (1891/92), T. Schloesing and E. Laurent in France (1890/92), and V. Alpe and A.E. Menozzi (1892) in Italy. In 6 years Hellriegel's results were accepted and confirmed in all Europe and the U.S.A.

1 Institut für Mikrobiologie und Biochemie, Lehrstuhl für Mikrobiologie, Universität Erlangen-Nürnberg, 8520 Erlangen, FRG

Molecular Genetics of the Bacteria-Plant Interaction
(ed. by A. Pühler)

Fig. 1. Nodulated *Faba* in 1542 edition of Fuch's herbal. (According to Fred, Baldwin, McCoy: Root nodule bacteria and leguminous plant, Madison 1932)

Fig. 2. Hermann Hellriegel 1831–1895. (According to Fred, Baldwin, McCoy: Root nodule bacteria and leguminous plant, Madison 1932)

On this basis of Hellriegel's discovery it took not more than about ten years to solve all important questions for the further understanding of the *Rhizobium*/Legume symbiosis. Let me summarize a few data. The famous Dutch bacteriologist M.W. Beyerinck isolated the bacteria from nodules, grew them in pure culture and called them *Bacillus radicicola* (1888) (Fig. 3). Later this name was changed into *Rhizobium,* a proposal by B. Frank (1889). A. Prazmowski was the first to fulfil Robert Koch's postulate for parasitic bacteria (1890): he induced nodules on pea plants with a pure culture of *Rhizobium leguminosarum* and re-isolated the microsymbiont from the nodules. All experiments to receive nitrogen fixation with freeliving *Rhizobia* failed or gave undecisive results. On the other hand the gain of fixed nitrogen by the nodules was very well established. It was the merit of F. Nobbe and L. Hiltner, the two *Rhizobium* experts of the Royal Saxonian Plantphysiological Experiment Station at Tharand near Dresden, to define this problem clearly. They concluded that the differentiation of the rodshaped bacteria into bacteroids inside the nodule might be the precondition to trigger the nitrogen fixation, Nobbe, F. und Hiltner, L.: „Wodurch werden die Knöllchen-besitzenden Leguminosen befähigt, den freien atmosphärischen Stickstoff für sich zu verwerten?" Landwirtsch. Vers. Stn. 42:459–478 (1893). Nobbe and Hiltner were also the first ones to recognize the biotechnological poten-

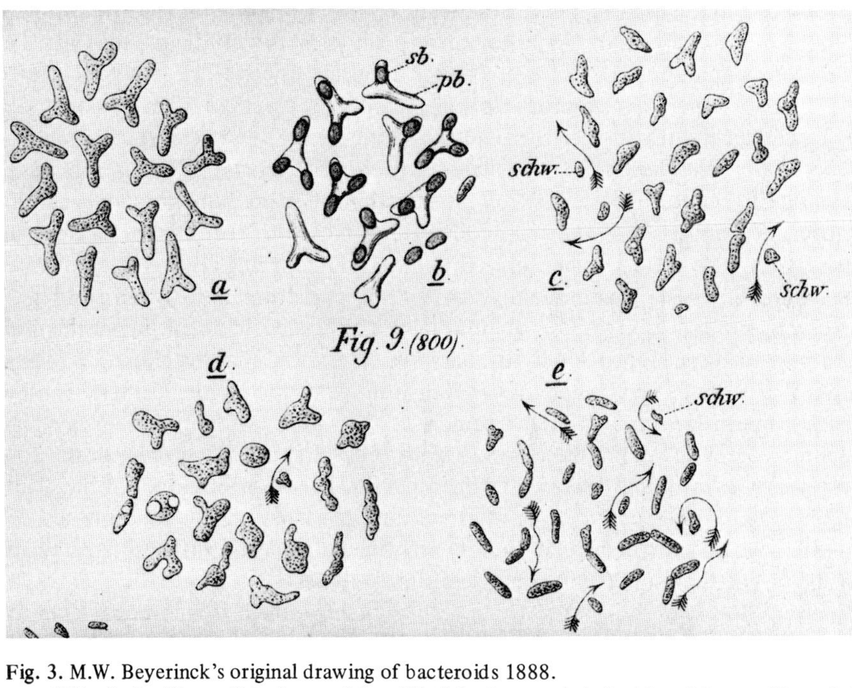

Fig. 3. M.W. Beyerinck's original drawing of bacteroids 1888.
"Fig. 9. *Bacillus radicicola* var. *fabae.* Die Schwärmer mit (schw) bezeichnet. *a* Normale Bacteroiden. *b* Bacteroiden mit Oeltropfen; *pb* primäres; *sb* secundäres „Bläschen". *c* Bacterien einer kleinen Colonie auf Fabastengelgelatine. *d* Bacteroiden einer sehr kleinen Colonie auf Fabastengelgelatine. *e* Bacterien in Fabastengeldecoct-Nährflüssigkeit"
(According to: Bot. Z. 1888, vol. 46, plate XI)

tialities of the *Rhizobium* system. They made a contract with the Farbwerke Hoechst to produce Nitragin, *Rhizobium* cultures to be delivered to the farmers to inoculate their farmland (Nobbe and Hiltner 1897). But the big business did not work out, Hoechst stopped the production of Nitragin 3 years later. The problems of host specificity and of inoculation substrate were not properly solved at that time. Later, especially in the U.S.A., the conditions for soil inoculation with *Rhizobium* were improved, so that even today, when properly applied, crop yields can be markedly increased by correct soil inoculation.

Before the end of the nineteenth century almost all fundamental aspects of the *Rhizobium*/Legume symbiosis were basically understood and a continuous production of papers turned around all these aspects without offering new ideas. The only later finding of real importance was in 1939 the discovery by Hideo Kubo that the red pigment of the nodules is hemoglobin which serves the oxygen transport for the energy supply of the fixation process (Kubo 1939).

At this point I will summarize a few data of the other important system of bacteria-plant interaction, also evolved by the Rhizobiaceae: the induction of plant tumors by *Agrobacterium tumefaciens.* The investigation of plant tumorgenesis started entirely independently from the problem of symbiotic *legume* nodulation. The recognition of the close taxonomic relationship between *Agrobacterium* and *Rhizobium* and therefore the obvious idea to compare the mechanisms of tumor induction and nodulation is a result of recent times. The pioneer work on the crown gall problem started with the American plant pathologist Erwin Frank Smith and the Danish veterinarian Carl Oluf Jensen, about 20 years after Hellriegel's *Rhizobium* work. In two papers E.F. Smith (1907a,b) published together with his coworker C.O. Townsend the result of having induced tumors with a pure culture of bacteria, *Agrobacterium tumefaciens,* in the plant *Paris daisy (Chrysanthemum frutescens).* Jensen (1910) in Europe started his investigation of plant tumors independently as models of comparison for animal tumors. He worked with beet tumors, in German known as Wurzelkropf der Rüben and showed that the inducing agent was *Agrobacterium tumefaciens.* The most prominent crown gall specialist who followed Smith and Jensen was, and still is, Armin C. Braun. Braun investigated in detail the mechanism of the plant cell transformation by the bacteria. He showed that the alteration of the plant cell is persistent and stable. After transformation, the inciting bacteria can be killed by thermal treatment of the tumor, at 46°–47°C, without influencing the autonomous growth of the transformed plant cells (Braun 1943, 1947). In the following years it was found that transformed plant cells are autotrophic for the plant hormones *auxin* and *cytokinin.* Further research has been mainly concentrated on the difference in growth regulation between transformed and differentiated plant cells and on the tumor-inducing principle of the bacteria.

If we follow the *Agrobacterium* and *Rhizobium* problem from around 1945 to the present time, it seems remarkable that the classical era of bacterial genetics from 1945 to 1960 a time which revolutionized the entire biology – has almost completely been ignored by the *Rhizobium* and *Agrobacterium* research. I think I was the only person who started conjugation experiments with *Rhizobia* in the 1950's (Heumann 1962). But more recently the Rhizobiaceae scene was changed dramatically by three main discoveries:

1. The discovery that nitrogenase reduces with high affinity acetylene to ethylene (Dilworth 1966, Schollhorn and Burris 1967).
2. The transfer of *Klebsiella* nif genes into *E. coli* by conjugation (Dixon and Postgate 1972).
3. The molecular characterization of the *Ti* plasmid of *A. tumefaciens,* followed by the demonstration of the transfer and stable integration of the T-DNA into the plant DNA, recent results in the *Agrobacterium* field, achieved almost simultaneously by three different groups: Nester and Chilton (1977) in Seattle, U.S.A., Van Montagu and Schell (1979), Gent, Belgium, and Schilperoort and Ledeboer (1978), Leiden, Holland.

With these discoveries a new scientific era has started in applied as well as in basic genetics. This era, as short as it is yet, has already produced so many fascinating results that a congress of all experts seems justifiable to define the state of research and to clarify the lines of new development.

If we regard the general speed of research-progress in the first 100 years of Rhizobiaceae research it seems relatively slow. Fortunately we are now in the good position that in the meantime molecular biology and gene technology have developed very powerful tools and methods, which will enable us to make an efficient approach to the next century of research in bacteria-plant interaction.

References

Alpe V, Menozzi A (1892) Studi e ricerche sulla questione dell' assimilazione dell' azoto per parte delle piante. Bol Nat Agrar (Roma) 14 (1):747–779

Atwater WO, Woods CD (1892) The fixation of free nitrogen by plants. Conn Agric Exp Stn Annu Rep (Storrs) 5:17–22

Beyerinck MW (1888) Die Bakterien der Papilionaceen-Knöllchen. Bot Z 46:725–735, 741–750, 757–771, 781–790, 797–804

Braun AC (1943) Studies on tumor inception in the crown-gall disease. Am J Bot 30:674–677

Braun AC (1947) Thermal studies on the factors responsible for tumor initiation in crown-gall. Am J Bot 34:234–240

Chilton MD, Drummond HJ, Merlo DJ, Sciaky D, Montoya AL, Gordon MP, Nester EW (1977) Stable incorporation of plasmid DNA into higher plant cells: the molecular basis of crown-gall tumor-genesis. Cell 11:263–271

Dilworth MT (1966) Acetylen reduction by nitrogen-fixing preparations from *Clostridium pasteureanum.* Biochim Biophys Acta 127:285–294

Dixon RA, Postgate JR (1972) Genetic transfer of nitrogen fixation from *Klebsiella pneumoniae* to *Escherichia coli.* Nature 237:102–104

Frank B (1889) Über die Pilzsymbiose der Leguminosen. Ber Dtsch Bot Ges 7:332–346

Hellriegel H, Wilfarth H (1888) Untersuchungen über die Stickstoffnahrung der Gramineen und Leguminosen. Beilagen Z Ver Rübenzucker Ind Dtsch Reiches, 234 pp

Heumann W (1962) Genetische Untersuchung sternbildender Bakterien. Z Vererbungsl 93:441–452

Jensen CO (1910) Von echten Geschwülsten bei Pflanzen. Rapp Conf Int Etude Cancer 2:243–254

Kubo H (1939) Über Hämoprotein aus den Wurzelknöllchen von Leguminosen. Acta Phytochim (Jpn) 11:195–200

Lawes JB, Gilbert JH (1891) The sources of the nitrogen of our leguminous crops. J R Agric Soc Engl 2, Ser 3:657–702

Ledeboer AM (1978) Large plasmids in *Rhizobiaceae.* I. Studies on the transcription of the tumor inducing plasmid from *Agrobacterium tumefaciens* in sterile crown gall tumor cells. II. Studies on large plasmids in different *Rhizobium* species. Thesis Univ Leiden, The Netherlands

Montagu M Van, Schell J (1979) The plasmid of *Agrobacterium tumefaciens.* In: Timmis KN, Pühler A (eds) Plasmids of medical environmental and commercial importance. Developments in genetics, vol I. Elsevier/North-Holland Biomedical Press, Amsterdam New York, p 494

Nobbe F, Hiltner L (1893) Wodurch werden die knöllchenbesitzenden Leguminosen befähigt, den freien atmosphärischen Stickstoff für sich zu verwerten? Landwirtsch Vers Stn 42:459–478

Nobbe F, Hiltner L (1897) Die Boden-Impfung für Leguminosen mit reinkultivierten Bakterien. Farbwerke vorm. Meister Lucius u. Brüning, Hoechst a. Main, p 32

Prazmowski A (1890) Die Wurzelknöllchen der Erbse. Landwirtsch Vers Stn 37:161–238

Schloesing T Jr, Laurent E (1890) Sur la fixation de l'azote gazense par les Léguminenses. C R Acad Sci (Paris) III:750–753

Schollhorn R, Burris RH (1967) Acetylene as a competitive inhibitor of N_2 fixation. Proc Natl Acad Sci USA 58:213–216

Smith EF, Townsend CO (1907a) A plant-tumor of bacterial origin. Science 25:671–673

Smith EF, Townsend CO (1907b) Ein Pflanzentumor bakteriellen Ursprungs. Zentralbl Bakteriol Parasitenkd Infektionskr Abt 2, 20:89–91

The *Rhizobium*-Plant Interaction

J.E. BERINGER [1]

Introduction

During the last 10 years we have seen major advances in our knowledge of *Rhizobium* genetics. There are now circular linkage maps of the chromosomes of four species of fast-growing rhizobia (see Beringer et al. 1982) and we know the location of a number of genes involved in nodule formation and nitrogen fixation. Some aspects of the interaction between *Rhizobium* and plants that need further study before we are able to exploit our genetic expertise to the full to develop improved strains of *Rhizobium* for inoculation purposes, will be discussed.

Events Prior to Infection

The rhizobia are a group of Gram-negative soil bacteria which can live for long periods in the soil in the absence of host plants. The extent to which rhizobia are adapted to life in soil is not clear, though it is important because of its relevance in helping us to understand why rhizobia nodulate leguminous plants and how long inoculant strains may survive in the soil. Unfortunately, it is not possible to measure the size of the population of rhizobia in a given soil. An estimate of the number of rhizobia capable of nodulating test plants can be obtained. However, counts obtained by finding the lowest dilution of a soil sample that will nodulate a particular host give no indication of the number of rhizobia capable of nodulating other hosts or those which may no longer carry nodulation genes. The shortcomings in our ability to enumerate rhizobia are important for two main reasons. The first is that rhizobia compete with each other in their interaction with roots. Thus a strain that is used as an inoculant will only form a reasonable proportion of nodules if it is very competitive or there are very few other rhizobia that will interfere with its interaction with the host plant. Usually indigenous rhizobia are very competitive and their presence provides an important limitation on the ease with which improved rhizobia can be used as inoculants (Brockwell et al. 1968, Bromfield and Jones 1980).

---

1 Soil Microbiology Department, Rothamsted Experimental Station, Harpenden, Hertfordshire, AL5 2JQ, UK

Molecular Genetics of the Bacteria-Plant Interaction
(ed. by A. Pühler)

The second reason why soil populations are important is that they provide a source of bacteria which may be able to exchange DNA with introduced inoculant strains. In other chapters in this book *Rhizobium* and *Agrobacterium* plasmids will be discussed and reference will be made to the transfer of these plasmids among members of the Rhizobiaceae. It is likely that a single legume plant will have about 10^8 rhozobia associated with its root system. Thus if plasmid transfer occurs as infrequently as 10^{-6} per day at least 100 new genotypes of rhizobia will be produced per day per plant. The great majority of "new " strains will be lost, but any that are at a selective advantage will be selected for and could become predominant. No data are available to show what has happened to introduced rhizobia at the level of maintenance of plasmids or even novel genes introduced within these strains, though there are reports of the decline in numbers of serologically identifiable or drug-resistant rhizobia which have been added to soils (Dudman and Brockwell 1968).

The loss of identifiable rhizobia after inoculation need not be related to genetic exchange. For example soils with a very low pH seldom contain rhizobia and when such soils are inoculated numbers of rhizobia decline rapidly (Parker et al. 1977). The decline is presumably due to the inability of rhizobia to survive at low pH, though loss of nodulation genes as a result of low pH conditions would give a similar decline in the numbers of bacteria nodulating test plants. Just how important physical factors such as pH, temperature and moisture stress may be on the maintenance of symbiotically-important genes needs to be assessed. Growth at elevated temperatures in vitro can be used to cure some *Rhizobium* and *Agrobacterium* strains of some of their plasmids.

Infection

For *Rhizobium* in the soil to form nodules on a plant a number of interactions must occur. The first is that the bacteria must gain access to sites that are susceptible to infection. For the majority of plants which are infected via roots this need not be a problem providing reasonable numbers of bacteria are present in the soil; however for plants such as *Sesbania* and *Aeschynomene,* which also form nodules on their stems, access to suitable sites can limit the number of nodules formed. Whether or not chemotaxis to root exudates plays a role in nodulation is debatable. Rhizobia have been shown to be chemotactic (Currier 1980, Götz et al. 1982, Gaworzewska and Carlile 1982) and non-motile mutants have been found to form nodules (Napoli and Albersheim 1980). However, it has yet to be shown whether chemotaxis has a role in increasing the competitive advantage of strains that are particularly well adapted to certain host plants. Motile strains are apparently more competitive than non-motile ones (Ames and Bergman 1981).

A most important aspect of infection is the recognition that is involved between the host plant and *Rhizobium* strain. We observe this mainly by the ability of a given combination of *Rhizobium* strain and host plant to form a nitrogen fixing symbiosis and this assessment has led to the formation of the well known cross-inoculation groups and species of *Rhizobium*. However, it is clear that there are many stages in nodule development at which recognition can occur. One extreme is the situation where no infection occurs; the opposite is when the nodule develops to the point where bacteroids are formed but there is no haemoglobin or nitrogen fixation. Between these extremes there are many points at which an infection can fail to develop. Hopefully an analysis of defective nodule formation using mutants and wild-type strains will enable us to identify the stages involved. A problem in studies of this nature is that infection, nodule formation, and nitrogen fixation are so strongly influenced by the environment and the genotype of the host and *Rhizobium.* For example in vitro *R. leguminosarum* nodulates *Phaseolus vulgaris* well, though there is no nitrogen fixation (Beynon et al. 1980). However, when *P. vulgaris* is sown in the field in soils containing no *R. phaseoli* no nodules are formed even though large numbers of *R. leguminosarum* are present which are able to nodulate peas *(Pisum sativum)* and *Vicia faba* very effectively.

The influence of environment and genotype × genotype interactions should be streassed because it is likely to affect nearly all observations of interactions between *Rhizobium* and plants. For example, Nutman identified five genes in clover *(Trifolium pratense)* that were involved in nodule formation and showed that the symbiotic response of four of these was *Rhizobium* strain dependent (Nutman 1969). A similar dependence on *Rhizobium* genotype is seen with other legumes (see Nutman 1969, Beringer and Johnston 1978). Pankhurst et al. (1974) showed that a riboflavine-requiring mutant of *R. trifolii* formed ineffective (Fix^-) nodules on some varieties of clover, while it was Fix^+ on others and demonstrated that the host difference was due to the availability of flavin within the host. These examples of genotype variation were obtained from the few reports where mutants were tested against a range of genotypes; something that is seldom done in genetic studies of *Rhizobium.*

Nitrogen Fixation

The agricultural significance of the *Rhizobium*/legume symbiosis is that nitrogen is fixed, and as a result nodulated plants require little soil nitrogen for growth. Because nitrogen fixation requires large amounts of energy, crop yields might be increased if *Rhizobium* strains or legume varieties were found that were more efficient in their use of photosynthate for ammonia production and assimilation. About the only direct genetic approach towards obtaining more efficient *Rhizobium* strains being pursued at present is the manipulation of hydrogen uptake genes (see Chapters by Brewin and by Cantrell, Haughland and Evans). The development of improved strains of *Rhizobium* for inoculation purposes will presumably be based upon the selection of strains which are more efficient at promoting plant growth. At present we are limited in the range of techniques available for assessing the effectiveness of symbiotic associations and thus are hampered in efforts to study the genetics of "efficiency". This problem is equally relevant to studies of the role of the plant, which is likely to be of major importance. Indeed it is relevant to ask why so much attention is paid to genetic studies of *Rhizobium* when it is likely that the potential for obtaining large improvements in the symbiosis may be determined largely by the plant.

The Future

Rhizobium and *Agrobacterium* are very closely related (see this vol.) and genetic studies of *Rhizobium* plasmids have undoubtedly benefitted from the work on *Agrobacterium* and presumably will continue to do so. Really dramatic advances in producing nitrogen fixing plant species by, for example, introducing nitrogen fixation genes into plants using *Agrobacterium* T-DNA are possible, but problems associated with the regulation of nitrogenase activity and the extreme oxygen sensitivity of the enzyme need to be overcome.

We can expect advances in our genetic unterstanding of the symbiosis, but studies of factors involved in the efficiency of nitrogen fixation in nodules and interactions between rhizobia in the soil will require major advances in methodology. Because the objective of genetic manipulation is the production if improved strains of *Rhizobium* for inoculum purposes, the assessment of genetic improvement will be in the field. This implies that considerable effort will need to be put into methods for assessing the value of strains; something that is too difficult to do on a large scale at present. Thus a major challenge to geneticists in the future is going to be the design of experiments to enable them to identify useful functions and to assess the role and value of these functions. The quality and range of genetic work with *Rhizobium* and *Agrobacterium* that is reported in this book suggests that improved strains of *Rhizobium* will be produced and that legume production worldwide will benefit from the genetic manipulation of these interesting and important bacteria, particularly if there is sufficient research on the host and inoculant use.

References

Ames P, Bergman K (1981) Competitive advantage provided by bacterial motility in the formation of nodules by *Rhizobium meliloti*. J Bacteriol 148:728–729

Beringer JE, Johnston AWB (1978) The genetics of the *Rhizobium*-legume symbiosis. In: Isotopes in biological dinitrogen fixation. Int Atom Energ Ag, Vienna, p 27–39

Beringer JE, Johnston AWB, Kondorosi A (1982) Genetic maps of *Rhizobium leguminosarum, R. meliloti, R. phaseoli* and *R. trifolii*. Genet Maps 2:130–131

Beynon JL, Beringer JE, Johnston AWB (1980) Plasmids and host range in *Rhizobium leguminosarum* and *Rhizobium phaseoli*. J Gen Microbiol 120:421–429

Brockwell J, Dudman WF, Gibson AH, Hely FW, Robinson AC (1968) An integrated programme for the improvement of legume inoculant strains. Transact 9th Int Congr Soil Sci, vol II, pp 103–114

Bromfield ESP, Jones GD (1980) Studies of double strain occupancy of nodules and the competitive ability of *Rhizobium trifolii* on red and white clover grown in soil agar. Ann Appl Biol 94:51–59

Currier WW (1980) Chemotaxis of a Birdsfoot Trefoil strain of *Rhizobium* to simple sugars. FEMS Microbiol Lett 8:43–46

Dudman WF, Brockwell J (1968) Ecological studies of root nodule bacteria introduced into field environments. 1. A survey of field performance of clover inoculants by gel immune diffusion serology. Aust J Agric Res 19:739–747

Gaworzewska ET, Carlile MJ (1982) Positive chemotaxis of *Rhizobium leguminosarum* and other bacteria towards root exudates from legumes and other plants. J Gen Microbiol 128:1179–1188

Götz R, Limmer N, Ober K, Schmitt R (1982) Motility and chemotaxis in two strains of *Rhizobium* with complex flagella. J Gen Microbiol 128:789–798

Napoli C, Albersheim P (1980) Infection and nodulation of clover by nonmotile *Rhizobium trifolii.* J Bacteriol 141:979–980

Nutman PS (1969) Genetics of symbiosis and nitrogen fixation in legumes. Proc R Soc London Ser B 156:122–137

Pankhurst CE, Schwinghamer EA, Thorne SW, Bergersen FJ (1974) The flavin content of clovers relative to symbiosis with a riboflavin-requiring mutant of *Rhizobium trifolii.* Plant Physiol 53:198–205

Parker CA, Trinick MJ, Chatel DL (1977) Rhizobia as soil and rhizosphere inhabitants. In: Hardy RWF, Gibson AH (eds) A treatise on dinitrogen fixation, Sect. IV. Wiley Interscience, New York

La Piste des Opines

J. TEMPÉ and A. PETIT [1]

La Piste des Opines – the opine trail – is the trail opened in 1956 by George Morel and Claude Lioret. Beginning with the discovery in crown gall tissue cultures of those specific compounds that we now call opines, this trail led to the description, at the molecular level, of the relations established in a crown gall tumour, between *Agrobacterium tumefaciens* and its host plant.

The reason we gave our paper this somewhat unexpected title is that one of our colleagues once compared the way we produced our contribution to the description of the crown gall system with hiking on a trail in a country where almost everybody drives a powerful motorcar on the freeway (P. Schaeffer, personal communication).

In this paper, we shall present our views on opine research and its recent developments. However, since this is seen from one of the viewpoints in that mark the opine trail it is quite possible that some workers in the field will not recognize the landscape of which they only have glimpses from the freeway.

Early Opine Research

As mentioned before, the first steps on the opine trail date back to 1956, when, at a meeting of the Société Française de Physiologie Végétale, G. Morel (1956) and, C. Lioret (1956), independently, anounced that they had found unknown specific compounds in crown gall tissue cultures. These preliminary communications were followed by the identification of lysopine (Biemann et al. 1960), octopine (Ménagé and Morel 1964) and octopinic acid (Ménagé and Morel 1965).

In the mid-sixties, the significance of the synthesis of these amino acid derivatives was far from being clear. The original claim of Lioret (1956, 1957) that lysopine was specific for crown gall tissue had been challenged (Seitz and Hochster 1964), and Lioret himself had concluded a paper on the physiological significance of the presence of lysopine in crown gall tissues with the following statement: «la lysopine paraît être un métabolite naturel existant également dans les tissus normaux» (Lysopine appears to be a natural metabolite also present in normal tissues) (Lioret 1966). However, in G. Morel's laboratory, we had no doubt that, in the plant kingdom, octopine was a specific marker for crown gall cells: plant extracts from crown gall

1 Group de Recherches sur les Intéractions entre Microorganismes et Plantes, Institut de Microbiologie, Bâtiment 409, Université de Paris-Sud, 91405 Orsay, France

Molecular Genetics of the Bacteria-Plant Interaction
(ed. by A. Pühler)

tissues were analysed by high resolution ion exchange chromatography and octopine was identified in the effluent by its retention time and a color reaction specific for guanidine derivatives such as arginine and octopine (Ménagé et al. 1966, Tempé et al. 1966). Our "guanidine analyzer" was accurate, reliable and sensitive. In addition to routine analyses, this equipment was also used for analysing extracts prepared from tissues labelled with radioactive precursors of octopine. Thus incubating crown gall tissues with ^{14}C-arginine always resulted in high incorporation of the radioactivity into octopine, wheras in similar experiments with normal tissues no radioactive octopine could be detected (Goldmann-Ménagé 1970). Thus we were confident that if octopine was at all present in normal tissues, its concentration was certainly below 0.1% of what we found it to be in crown gall tissues, about 2 μmol/g of dry weight.

At this time, G. Morel and A. Goldmann-Ménagé were devoting much work to the structural identification of another arginine derivative, which they later named nopaline. It was present in high amounts, up to 60 μmol/g of dry weight, in crown gall tissues from *Opuntia vulgaris* (in French: nopal). This crown gall tissue culture strain was exceptional in that it was the only one not to contain octopine (Ménagé and Morel 1966). At that time the current hypothesis was that this observation was reflecting the metabolic differences between crassulacean plants like *Opuntia* (Metzler 1977) and other plant species (Goldmann-Ménagé 1970, Morel 1970).

Having at hand the guanidine analyzer, J. Tempé was studying crown gall tissues from various origins, hoping to correlate octopine concentration with growth characteristics. The idea for these experiments came from a discussion he had had with A.C. Braun when visiting him, in 1967. The purpose of the planned experiments was to test Morel's hypothesis that octopine synthesis was a consequence of the metabolic changes that occur during tumour transformation. If that was true, the concentration of octopine in crown gall tissues could reflect the extent to which the tissues were transformed. A.C. Braun had offered to provide us with tissues with different growth patterns. The first sample we analyzed was the T37 teratoma line. This line which had been isolated from a tumour incited with strain T37, continuously produced buds which developed into leafy shoots. Braun had been able to recover normal plants from this line (Braun 1959). We were expecting to find a concentration gradient of octopine from the inner part of the callus to the outer parts, the shoots, and the leaves, and an overall lower concentration as compared with undifferentiated lines. Instead, we found a huge amount of nopaline.

Besides demonstrating that our hypotheses were wrong this result indicated that most probably the type of arginine derivative present in crown gall tumours was specified by the strain of *Agrobacterium tumefaciens* inoculated. This was readily demonstrated by analyzing extracts from tumours incited on various plants with either strain B6 or T37. All the tumours formed on plants, including crassulaceae, inoculated with B6 contained octopine, and all those on plants inoculated with T37 contained nopaline (Goldmann et al. 1968). This observation was taken as evidence that the bacterial genome was specifying the phenotype of crown gall cells.

These results were then extended to other *A. tumefaciens* strains which, therefore, could be characterized by the type of arginine derivative that was present in the tumours they had incited (Petit et al. 1970). In order to understand the significance of our observations, we were trying to correlate the data on the guanidine content of crown gall tumours, with bacterial phenotypes. After a few unsuccessful attempts to stimulate various strains to synthesize octopine or nopaline, we studied the bacterial degradation of these compounds. Since lysopine degradation by *Agrobacterium* had already been reported (Lejeune and Jubier 1968), it was not unexpected that our strains degraded octopine or nopaline. What was new was that opine synthesis in crown gall tumours and degradation by the bacteria showed exactly the same strain specificity: a strain that catabolized nopaline incited, on all susceptible plant species, tumours that contained nopaline. Similarly, a strain that catabolized octopine incited tumours that contained octopine. Therefore, it was possible, knowing the bacterial phenotype with respect to catabolism of octopine or nopaline, to predict the tumour phenotype. Thus we had been able to confirm that the bacterial genome dictates the phenotype of crown gall tumours and we had established a biochemical relation between *A. tumefaciens* strains and crown gall cells. We concluded from these observations that: «au cours de la transformation tumorale un sytème enzymatique propre à la bactérie est probablement transmis à la cellule végétale, ce qui suppose la présence d'une fraction du génome bactérien, douée de continuité génétique, dans les cellules transformées, ainsi que son expression dans ces cellules» (during the course of tumour transformation an enzymatic system is probably transferred from the bacterium to the plant cell, which supposes the presence and the expression in crown gall cells of a fraction of the bacterial genome, endowed with genetic continuity) (Petit et al. 1970).

A. tumefaciens strains were then classified according to their opine type (Table 1). We found that most of them were either of the octopine- or the nopaline-type. Among the few strains which did not fit into this classification, some appeared to be of mixed type, and this was probably an artifact, others did not catabolize octopine or nopaline and the tumours they incited did not contain either of these opines. These were later termed null-type strains (Sciaky et al. 1978).

We also studied the pathway of octopine degradation by the octopine strain B6: we found that the first product of this degradation was arginine, which was an indication that the pathway of octopine degradation was opposite to the synthetic pathway in tumours. This observation was the basis for our next working hypothesis: we proposed that during the tumour transformation, the degradative genes carried by the bacterial genome were transferred to the plant cell in which their expression resulted in opine synthesis. The main advantage of this hypothesis was that, if it were true, it should be quite easy to give proof of gene transfer taking place during crown gall induction: we anticipated that among bacterial mutants affected in opine degradation some should have lost the capability to trigger opine synthesis in tumours.

That crown gall transformation may involve DNA transfer from the bacterium to the plant cell had been proposed earlier by Braun (1947) and Klein (1954). In the 1960's, many workers thought they could prove this hypothesis by biochemical methods such as hybridization or immunochemistry (Schilperoort et al. 1967, Milo and Srivastava 1969, Quétier et al. 1969, Schilperoort et al. 1969, 1973). However the evidence contributed by such studies was later questioned (Bomhoff 1974, Dons 1975, Loening et al. 1975, Chilton et al. 1975) and serious doubts were cast on the validity of earlier results, including those concerning the specificity of opines as biochemical markers for crown gall cells (Johnson et al. 1974, Wendt-Gallitelli and

Table 1. Correlation between opine synthesis in crown gall tumours and opine degradation by *Agrobacterium tumefaciens*

Strain number	Octopine		Nopaline	
	Synthesis	Degradation	Synthesis	Degradation
A6	+	+	–	–
A281	–	–	–	–
ACH5	+	+	–	–
Anthemis	–	–	+	–
AT1	–	–	–	–
AT4	–	–	–	–
B2	+	+	–	–
B6	+	+	–	–
B78	–	–	+	+
B91	+	+	–	–
Bo542	–	–	–	–
CHR3	–	–	+	+
C58	–	–	+	+
Delft	–	–	+	+
Eu6 [a]	–	–	–	+
H5	–	–	+	+
H100	–	–	+	+
Incense cedar	–	–	+	+
K10	+	+	–	–
K15/73	–	–	–	–
A. rubi	–	–	+	+
R10	+	+	–	–
SP	+	+	–	–
S353	–	–	+	+
S379	+	+	–	–
T37	–	–	+	+
W1	+	+	–	–
IIBNV6 [b]		–		–
IIBV7	–	–	+	+
9P	–	–	+	+
42IV	–	–	+	+
61	+	+	–	–
71	–	–	+	+
181 [a]	–	–	–	+
1001	+	+	–	–
15955	+	+	–	–

[a] 181 and Eu6 are exceptional strains which degrade nopaline although the tumours they elicit do not contain any identified opine

[b] Avirulent strain

Dobrigkeit 1973). Braun himself, whose earlier work had constituted the first basis for the idea that crown gall tumour transformation could be the result of DNA transfer (Braun 1947), became an eager supporter of the idea that crown gall, and, generally, cancer, results from epigenetic changes, that is changes in the expression of genetic information, rather than changes in genetic information itself (Braun 1974).

Ti-Plasmids and Opines

In this rather grim perspective of opine research, the discovery of the Ti-plasmids (Zaenen et al. 1974) in J. Schell's laboratory, in Ghent, came as a breakthrough, casting a new light on the crown gall problems and providing new approaches, and new tools, to solve them. On the opine trail, as well as in the whole field of crown gall research, the Ti-plasmids were the touchstones on which all the previous evidence could be tested. Thus our results, which had been confirmed by Bomhoff in R. Schilperoort's laboratory (Bomhoff 1974) were quickly established on firm ground with the demonstration that genes involved in opine synthesis and opine degradation were located on the Ti-plasmids (Bomhoff et al. 1976, Montoya et al. 1977). The fact that the bacterial genes specifying the opine-type of the tumours were located on the same genetic element as the pathogenicity determinants constituted good support for the gene transfer hypothesis.

Ti-plasmid mutants affected in octopine catabolism were isolated. All of those which were still virulent induced the formation of tumours that contained octopine (Klapwijk et al. 1976, Montoya et al. 1977, Petit and Tempé 1978) disproving our working hypothesis according to which the same genes were operating in plant and in bacterial cells. The genetic approach we had taken to prove the gene transfer hypothesis had been unfruitful, however, this result did not invalidate the hypothesis.

Finally, hybridization experiments, performed by M.D. Chilton and her colleagues in Seattle, with restriction fragments of the Ti-plasmid as probes, contributed decisive evidence for the gene transfer hypothesis (Chilton et al. 1977). These results were soon confirmed (Schell 1978, Schell et al. 1979, Thomashow et al. 1980, Lemmers et al. 1980) and the existence of T-DNA (*T*ransferred *DNA*) was therefore definitively established. Insertion mutagenesis (Hernalsteens et al. 1980), transcription studies (Willmitzer et al. 1982a), in vitro translation (Schroeder et al. 1981, Murai and Kemp 1982), DNA cloning and sequencing (De Greve et al., submitted) have now proven beyond doubt that the enzymes responsible for octopine or nopaline synthesis in crown gall cells are coded for by Ti-plasmid genes which have become stably integrated into the plant cell genome, thus establishing the validity of the gene transfer hypothesis.

The Opine Concept

What was then the biological rationale for opine synthesis in crown gall cells and opine degradation by the bacteria? The new theory, we called it the opine concept, tried to fit together the data on opine synthesis and degradation with the earlier observations of A. Kerr (1969, 1971) on pathogenicity transfer between *Agrobacterium* strains in crown gall tumours and the discovery that Ti-plasmids were conjugative plasmids, the conjugative activity of which was inducible by opines (Petit et al. 1978b). This theory resulted from discussions involving mainly Jeff Schell, Allen Kerr and Jacques Tempé.

The Opine Concept (Petit et al. 1978a, Tempé et al. 1979, Guyon et al. 1980) proposes that, in the plant-bacterium interaction, opines act as chemical mediators of parasitism and that this role is essential. In other words, the main features of the Ti-plasmids are the opine related functions by the virtues of which the plasmids and their bacterial hosts manage to tap the plant metabolic pools for their own benefit. The opine synthesis functions serve the unique purpose to enslave the host plant to the production of opines which are specific substrates for the opine catabolic functions. The simultaneous expression of these complementary functions provides the energy that is required for the propagation of the Ti-plasmids.

The Genetic Colonization scheme formulated by J. Schell, M. van Montagu and their coworkers (Schell 1978, Schell et al. 1979) adds to this theory the precision that the mechanism by which this is achieved is the transfer of Ti plasmid genes from the bacterium into the plant genome.

Development of the Opine Concept

Since up to this point the only advances we had made on the opine trail resulted from entering the wrong way, it was not without scepticism that we set out to test the validity of the opine concept. Besides curiosity, one of the reasons for understanding this was that if the theory was to be disproven, we would rather help doing it.

Theoretically, the best way to assess the validity of the opine concept would be to perform ecological studies. If valid, the concept should allow to make predictions concerning the evolution of bacterial populations in the soil. For instance one could predict that a mutant strain that would have lost the opine catabolic functions would be less competitive than the isogenic wild-type strain in a situation where the Ti-plasmid pathogenic functions could be expressed. Certainly it is not difficult to conceive experiments along these lines. However, we felt that, being unfamiliar with those kinds of studies, it would probably be difficult to design experimental procedures in such a way that they give fool proof results. We, therefore, decided to take another approach and to look for opines in systems where they had not been described yet. We find this approach has, over the ecological one, the added advantage to be more dynamic, since if positive results are obtained, they will not only give support to the opine concept, but they will also advance our knowledge of other systems.

Opines are Present in "Null Type" Crown Gall Tumours

As mentioned before, besides the octopine- and nopaline-type tumours, and bacteria, which provided the experimental basis for the opine concept, so called null type (Sciaky et al. 1978) tumours and bacteria had been identified by the negative criterion that the tumours did not contain either octopine or nopaline and that the strains inciting those tumours did not degrade these opines.

We decided to work with strain Bo542 since it had been previously shown that this strain efficiently transfers its plasmid to recipients during in planta crosses (Sciaky et al. 1978). In such crosses, which were first described by A. Kerr (1969, 1971), opines could play a dual role, as conjugation inducers (Petit et al. 1978b) and as selective nutrients in providing conditions favoring transconjugants relative to recipients.

We looked, in tumours incited by Bo542 or it transconjugant A281 (Sciaky et al. 1978), for the presence of compounds which could play the role of opines. The requirements for these were that they should be tumour-specific and, at the same time, serve as catabolic substrates for the transconjugant A281 but not for the Ti-plasmid-less strain C58C1. Accordingly, opine containing tumour extracts should, when incorporated as sole nutrients into a culture medium, allow for better growth of the transconjugant strain than of C58C1. This differential growth effect should not be seen with extracts from normal tissues. Indeed, small differences in total growth were sometimes seen when crude tumour extracts were incorporated into mineral medium and inoculated with A281 or C58C1. However, this was not always the case and even when positive differences were observed, these were generally only a few percent of total growth. We then reasoned that if opines were present in our extracts, the exhausted medium to which they had been added and on which the Ti-plasmidless strain had been grown should be enriched for these. We therefore purified the organic components of the spent medium by ion exchange and we recycled

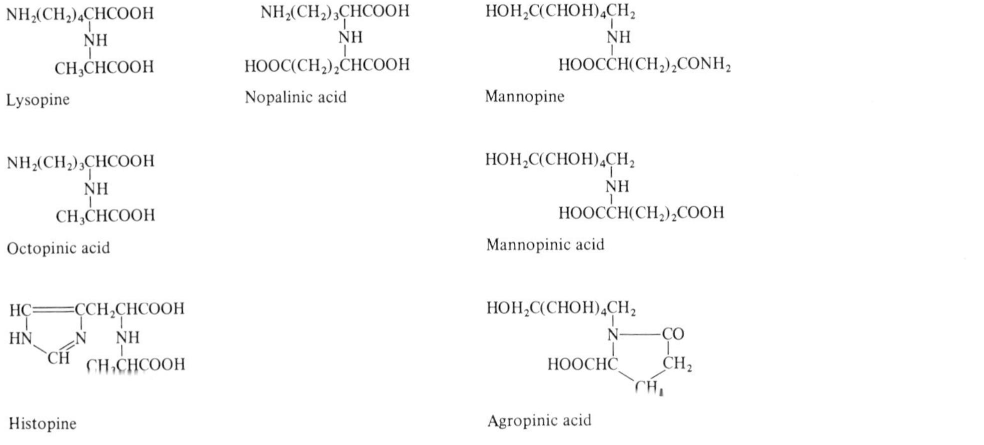

Chart 1. Chemical structures of opines. Source: Biemann et al. (1960), Ménagé and Morel (1964, 1965), Goldmann et al. (1969), Firmin and Fenwick (1977), Kemp (1977), Coxon et al. (1980), Tate et al. (1982), Petit et al. (1983)

it into fresh medium. Indeed we observed a very reproducible differential growth effect. Therefore the culture of C58C1 on the crude extract achieved an efficient purification of specific growth factors. In contrast, normal tissues, treated in the same way did not contain Ti-plasmid-specific growth substrates.

These results confirmed that, indeed, null type tumours contained opine activity (Guyon et al. 1980). Ion exchange chromatography of the amino acid fraction yielded enough purified material for chemical and biological characterization. The cryptic opine was shown to be an efficient growth substrate for both null-type and octopine-type strains, but not for nopaline-type strains. It was shown to be identical to agropine (Chart 1), a new opine that had been isolated from octopine-type tumours (Firmin and Fenwick 1978, Tate et al. 1982).

We have recently investigated in more detail these agropine-containing tumours. We have established that they generally contain three other opines: mannopine, mannopinic acid and agropinic acid (Chart 1), which share with agropine the same carbon-nitrogen backbone (Dahl et al., in preparation).

Agrocinopines and the Biological Control of Crown Gall

Another contribution to the extension of the opine concept came from A. Kerr's work on biocontrol of crown gall. Kerr and Htay (1974) have shown that it is possible to prevent crown gall by treating susceptible plants with a suspension of the non oncogenic *A. radiobacter* strain K84. The method was shown to work, very reproducibly, when the pathogen was a nopaline type strain. The reason for this biocontrol is the production by strain K84 of a microcin called agrocin 84 (Kerr and Htay 1974, Roberts et al. 1977). It was demonstrated that growth of the pathogen is inhibited by agrocin 84 and that the gene responsible for agrocin sensitivity is located on the nopaline Ti-plasmid (Engler et al. 1975). J.G. Ellis and P.J. Murphy reasoned that a suicide function such as agrocin 84 sensitivity would be unlikely to be maintained on a plasmid. With new opines being discovered in various kinds of tumours, the inference that agrocin 84 was a toxic substrate for uptake or catabolism of a yet undiscovered opine was rather straightforward. Indeed, Ellis and Murphy (1981) showed that nopaline tumours contain a new class of opines which they called agrocinopines and which appear to be the normal substrates for the agrocin sensitivity function carried by the nopaline Ti-plasmid. Agrocinopine A discovered in nopaline tumours was found to enhance agrocin 84 sensitivity in strains harboring a nopaline Ti-plasmid. This increase in sensitivity is due to induction of an active transport system which accepts agrocin 84 and agrocinopine A as substrates. The sensitivity of uninduced nopaline strains to agrocin 84 is due to the fact that the transport system is partially constitutive.

In tumours incited by the agropine type strains Bo542 and A281 another compound that interfered with agrocin 84 was found. This compound which has been called agrocinopine C is antagonistic to agrocin 84 toxicity for nopaline strains. In contrast, it induces agrocin sensitivity of the otherwise resistant strain A281. Strain Bo542, which was the donor strain for the Ti-plasmid of A281, is resistant to agrocin 84, even in the presence of agrocinopine C, although it must have the determinants for agrocin 84 sensitivity since these can be demonstrated in its transconjugant A281. We have found that Bo542 produces a microcin that is very similar, if not identical, with agrocin 84. Therefore it is likely that the immunity of this strain to agrocin 84 is correlated with the production of this molecule by this strain.

When the agrocin 84 producer strain *A. radiobacter* K 84 was studied, it was found that it could use nopaline as sole carbon and nitrogen source (Kerr and Roberts 1976), and that this catabolic function is carried by a plasmid. Strain 84 was isolated from soil in the vicinity of peach plants bearing nopaline type crown gall tumours. One may therefore wonder whether this strain is not a kind of "superparasite", which being unable to trigger opine synthesis, since it is non-pathogenic, is nevertheless able to get access to opines by controlling, via agrocin 84 production, the propagation of the pathogenic strain and therefore its utilisation of nopaline synthesized by the transformed plant cells. Although no experimental data have been presented to support this hypothesis we find it worth mentioning it here since it definitely fits with the opine concept. In this case also ecological studies could probably help assess the validity of the theory.

The Role of Opines in the Propagation of the Ti-Plasmids

The spread of a plasmid within a bacterial population can be the result of two main mechanisms:

1. A higher growth rate of the strain when the plasmid it harbors carries functions susceptible to confer it a better ability to grow. Thus bacteria harboring catabolic plasmids multiply at a faster rate in an environment where the substrate of the plasmid catabolic function constitutes a significant part of the nutrients available.
2. The conjugative activity of the plasmid which contributes to its propagation within the bacterial population.

Opines appear to be involved in both mechanisms, since their presence gives a selective advantage to bacteria carrying a Ti-plasmid and since they also induce the conjugative activity of the Ti-plasmids.

Obviously both these mechanisms could significantly contribute to the propagation of the Ti-plasmids within the *Agrobacterium* soil population, provided susceptible plants were present and carried tumours. Again no ecological data are available to evaluate the validity of these assumptions.

The idea that both the catabolic functions and the induction of the conjugative functions may be important for the propagation of the Ti-plasmids is supported by the fact that in all the crown gall tumours types studied, both catabolic opines and conjugation inducing opines have been characterized (Table 2).

In connection with the postulated role of opines in nature, it is interesting to note that the opines which induce conjugative activity are always found in low amounts relative to the opines which only have a catabolic role.

Table 2. Opine types of crown gall tumours and *A. tumefaciens* strains

Tumour opine type (example strains)	Catabolic opines	Conjugative opines
Octopine (A6, B6, R10. ACH5, 15955)	Octopine, lysopine, octopinic acid, histopine, agropine, agropinic acid, mannopine, mannopinic acid	Octopine
Nopaline (C58, T37, H100)	Nopaline, nopalinic acid, agrocinopine A [a]	Agrocinopine A
Agropine (Bo542), AT1)	Agropine, agropinic acid, mannopine, mannopinic acid, agrocinopine C [a]	Agrocinopine C

Source: Ellis et al. (1982), Petit et al. (1978b), Tempé et al. (1978)

[a] Agrocinopines B and D, which have been detected in extracts from respectively nopaline tumours and agropine tumours may be degradation products of agrocinopine A and C, they are therefore not mentioned here since it is not known whether they are really present in tumours (Ellis, person. commun.). Agropinic acid, which is also a degradation product of agropine and mannopine, appears to be formed both in vivo and in vitro, is therefore listed here

Extension of the Opine Concept to the Hairy Root System

Classically, in the genus *Agrobacterium* two main pathogenic species have been described. There are *A. tumefaciens* and *A. rhizogenes,* the latter being responsible for the hairy root disease which results in the formation on dicotyledonous plants of numerous secondary roots at the point of inoculation (Fig. 1). Involvement of plasmids in the hairy root disease has been suggested by Moore et al. (1979) and demonstrated by White and Nester (1980). These authors have shown that most (A. rhizogenes strains carry three plasmids. In two of the strains studied, the largest plasmid is a cointegrate of the two smaller ones. The plasmid of intermediate size has been found to carry the genes for pathogenicity. We have reached similar conclusions for other stains (Chilton et al. 1982).

One of the main reasons for chosing this system for further work on the opine concept was that, as had been shown for *A. tumefaciens,* pathogenicity of *A. rhizogenes* can be transferred to non-pathogenic *Agrobacteria* by in planta crosses (Moore et al. 1979), which, because of the analogy with the previous case, suggests that opines are involved in this system too.

The identification of opines in hairy root material was made easy by the knowledge that *A. rhizogenes* strains could utilize opines of the agropine family for their growth (Tepfer and Tempé 1981). Analysis of extracts from hairy root tissues incited by a collection of six strains revealed that these fell into two classes with respect to their opine content (Table 3) (Petit et al. 1983). The agropine-type hairy root is similar to the corresponding crown gall: it contains all four opines of the agropine family (Chart 1), and, in addition, an agrocinopine which appears to be identical to agrocinopine A. The mannopine-type hairy root has, so far, no equivalent in the crown gall system. The opines associated with this type are: mannopine, mannopinic

Fig. 1. Hairy root on *Kalanchoe tubiflora; left* inoculated with *A. rhizogenes* 15834, *right* inoculated with strain 8196. Photograph taken 7 months after inoculation

acid, agropinic acid and an agrocinopine which is similar to agrocinopine C. We have studied the catabolic properties of *A. rhizogenes* strains with respect to the opines of the agropine family (Table 3). The picture which emerges from this study is entirely consistent with the fact that these substances are opines and therefore supports the opine concept. However, there is one difference which we would like to discuss now. When we studied the catabolic properties of the agropine-type strains, we found that one of our three strains (strain HRI) could only catabolize agropine, although it did elicit the formation of roots that contained all four opines of the family. The reason for this behaviour was understood later, when we studied the properties of a collection of transconjugants of agropine strains. We found that when the selection had been performed on agropine, the transconjugants either carried the complete set of plasmids of the donor strain, or only the plasmid of intermediate size. Both types of transconjugants were virulent. When the selection was performed on mannopine or mannopinic acid we found that we, again, obtained two classes of transconjugants, one which contained all the plasmids of the donor, while the other one carried only the smaller plasmid. The latter transconjugants were found to be avirulent.

Table 3. Opine types of hairy root and *A. rhizogenes* strains

Strain designation	Opines in hairy root	Opines degraded by strain
I. Agropine type		
A4, 15834	Agropine, mannopine, mannopinic acid, agropinic acid, agrocinopine A [a]	Agropine, mannopine, mannopinic acid, agropinic acid
HRI	Agropine, mannopine, mannopinic acid, agropinic acid, agrocinopine? [b]	Agropine
II. Mannopine type		
TR7, TR101, 8196	Mannopine, mannopinic acid, agropinic acid, agrocinopine C [a]	Mannopine, mannopinic acid (agropinic acid) [c]

Source: Petit et al. (1983)

[a] Agropines have been detected on extracts from hairy root cultures by high voltage paper electrophoresis. They have been tentatively identified, by chemical tests and biological assay as agrocinopine A, in agropine-type roots, and as agrocinopine C, in mannopine-type roots. No degradation studies have been performed on hairy root agrocinopines

[b] Not analyzed for agrocinopines

[c] Strain 8196 grew poorly while TR7 and TR101 failed to grow on this substrate, however mutants that could utilize this substrate appeared after 3–4 weeks of incubation

The catabolic properties of the transconjugants were studied, and were shown to fall into three classes (Table 4). Transconjugants which had the same plasmid content as the donor could catabolize all the opines of the agropine family, transconjugants with only the intermediate plasmid catabolized only agropine, whereas transconjugants with the smaller plasmid catabolized mannopine, mannopinic acid and agropinic acid. These results are consistent with the idea that in *A. rhizogenes* strains of the agropine-type the catabolic functions are dissociated, some are borne on the virulence plasmid, the others on the smaller plasmid. Eckhardt analysis of the plasmid content of the wild type strain HRI which only catabolizes agropine, showed that this strain has only one plasmid of the same size as the virulence plasmid in other strains. The unusual behaviour of strain HRI can be therefore explained by the fact that it carries only the virulence plasmid.

Table 4. Catabolic properties of transconjugants from *Agrobacterium rhizogenes* strains

Strain	Plasmid content	Virulence	Utilisation of substrate			
			Agropine	Mannopine	Mannopinic acid	Agropinic
C58C1 (pAr15834abc)	a, b, c	+	+	+	+	+
C58C1 (pAr15834a)	a	–	–	+	+	+
C58C1 (pArA4a)	a	–	–	+	+	+
C58C1 (pArA4b)	b=pRiA4	+	+	–	–	–
C58C1 (pArA4abc)	a, b, c	+	+	+	+	+
C58C1 (pRiHRI)	pRiHRI	+	+	–	–	–
C58C1 (pAr8196b)	b=pRi8196	+	–	+	+	–

Source: Petit et al. (1983). + = Normal growth; – = Very thin growth

In mannopine strains we found the situation was simpler since all the opine related frunctions we have studied are carried by the virulence plasmid, which, as for agropine strains is of intermediate size.

We have not studied agrocinopine catabolism, and we have not studied whether agrocinopines induce the conjugative activity of the *A. rhizogenes* plasmids.

The fact that opines are present in all the systems we have studied provides good support for the opine concept. In hairy root cells as well as in crown gall cells T-DNA is present (Chilton et al. 1982, Willmitzer et al. 1982b, White et al. 1982, Spano et al. submitted) and thus, so far, the salient features of both systems are the same, except for the fact that in the agropine-type *A. rhizogenes* strains the catabolic functions are not all carried by the virulence plasmid.

Does the Opine Concept Apply to Other Plant-Bacteria Interactions?

Which are, among other systems of interaction between plants and bacteria, those for which the opine concept could conceivably operate? Since, according to this concept the biological rationale for opine synthesis is to create an ecological niche for the pathogen by providing a chemical environment selective for the strains carrying the pathogenic plasmids, we have chosen to study those interactions for which the involvement of plasmids is established, or at least suggested.

The choice of *Rhizobium meliloti* for these studies was rather obvious since:

- *Rhizobium sp.* and *Agrobacterium sp.* belong to the same family
- *R. meliloti* is taxonomically very close to *Agrobacterium* (De Ley et al. 1966).

We have looked for opine-like compounds in alfalfa nodules incited with various strains of *R. meliloti*. One of the difficulties encountered in this work was to obtain sufficient amounts of nodules. This problem was solved by growing the plants under aeroponic conditions (Tempé et al. 1982). We found that alfalfa nodules (Fig. 2) incited by *R. meliloti* L530 contain a compound which is never detected in the other parts of the plant and which can be catabolized by this strain and not by any of the other strains studied (Fig. 3). Biological purification of the nodule extract has been performed as described for agropine from formerly called null-type tumours (Guyon et al. 1980). The cationic fraction recovered after this step can be used as sole carbon and nitrogen source for strain L530, but not by other *R. meliloti* strains (Fig. 3). On media containing this fraction, growth of L530 is particularly abundant (Tempé et al. 1982).

The purified extract contains several components which can be characterized by high voltage paper electrophoresis. Among these, one is particularly easy to recognize since it reacts with silver nitrate, and since silver nitrate-positive cations are not very many in our extracts. The presence of this compound is easily demonstrated in crude extracts, without purification (Fig. 3). In the purified extracts, ninhydrin-positive compounds have been detected, which are also specific catabolic substrates for L530 (Fig. 4).

Fig. 2. Cluster of nodules formed on *Medicago sativa* grown under aeroponic conditions and inoculated with *Rhizobium meliloti* L530. Photograph taken 6 months after inoculation (real size: 12 mm across, fresh weight: 370 mg)

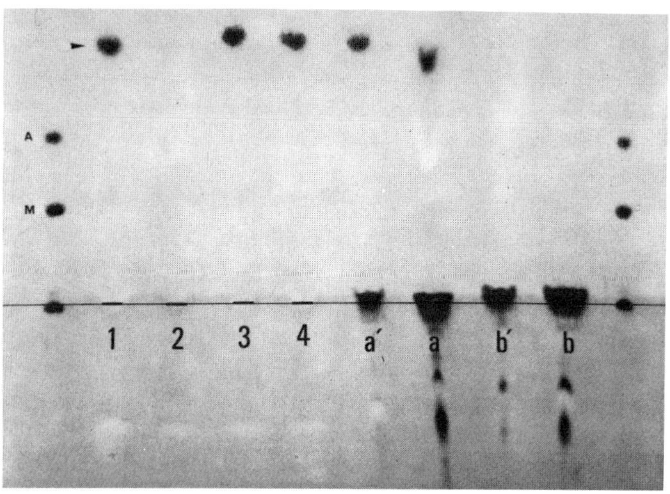

Fig. 3. Nodule-specific silver nitrate-positive compound in *M. sativa* nodule extract. High voltage paper electrophoresis at acid pH, silver nitrate staining. *a* nodule extract; *b* root extract; *a′* analysis of supernate after biological purification of nodule extract; *b′* analysis of supernate after biological purification of root extract. *1–4* organic cations from biologically purified nodule extract have been recovered by ion exchange and incorporated as sole carbon and nitrogen sources into fresh culture medium. The medium was inoculated with various strains and the culture medium was analyzed by high voltage paper electrophoresis, 72 h after inoculation: *1* inoculated with *A. tumefaciens* C58C1; *2* inoculated with *R. meliloti* L530; *3* inoculated with *R. meliloti* L522nod-22; *4* uninoculated. *Extreme left and right lanes:* standard containing mannitol (neutral marker at origin), mannopine *M*, agrocinopine *A*

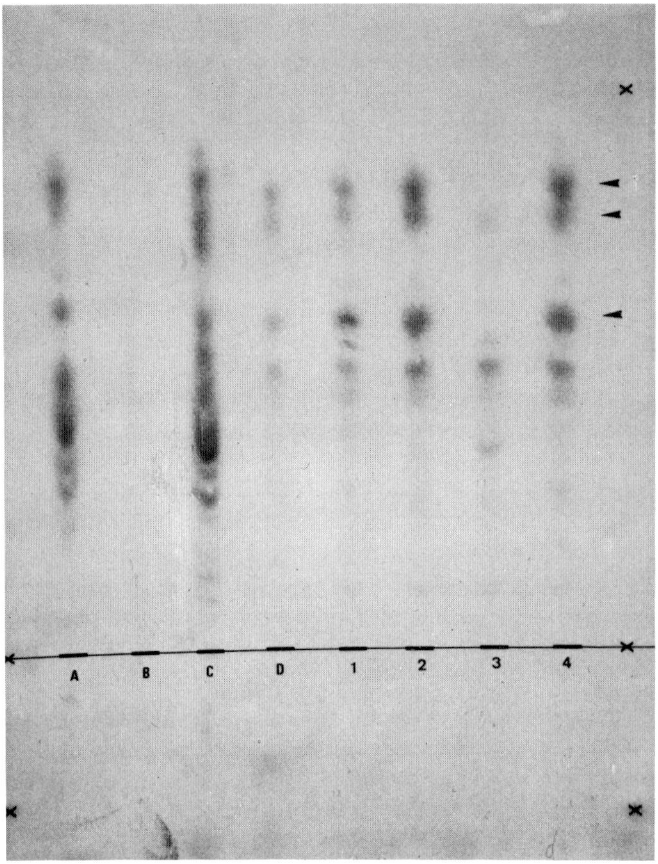

Fig. 4. Nodule-specific ninhydrin-positive compounds in *M. sativa* nodule extract. Same conditions as in Fig. 3, except for staining. *A* root extract; *B* biologically purified root extract; *C* nodule extract; *D* biologically purified nodule extract. *1–4* organic cations from biologically purified nodule extract have been recycled into fresh culture medium as sole carbon and nitrogen sources. The medium was inoculated with various strains and the culture medium was analyzed 72 h after inoculation. *1* inoculated with *A. tumefaciens* C58C1; *2* inoculated with *R. meliloti* L522; *3* inoculated with *R. meliloti* L530; *4* inoculated with *R. meliloti* 41. *Arrows at right* mark the positions of the most conspicuous ninhydrin-positive catabolic substrates for *R. meliloti* L530

Où va la Piste des Opines?

Although we are aware of having left behind us some unexplored parts of the *Agrobacterium* section of the opine trail (there are still some null-type tumours), we feel the urge to step further into the more speculative *Rhizobium* section.

It is very clear to us that, not only, the opine-like compounds that we have found in alfalfa nodules have still to be established as bona fide opines, but also that the role of the opines as described by the opine concept has not yet been demonstrated in nature. Obviously, there is an urgent need for ecological studies in this area.

In the absence of such studies, the best argument we can find for the validity of the opine concept is the important part opines and the opine concept have played in the development of crown gall and hairy root research.

Because opines have been important for the development of crown gall research we find the possibility that opines may be involved in the *Rhizobium*-legume interaction should be thoroughly explored. Not only because this could result in a new extension of the opine concept, but also because of the possible consequences on the development of research on symbiotic nitrogen fixation.

Such an extension, besides providing new tools for research in this particular area, would also stimulate investigations in other systems of interaction between plants and microorganisms. It could also provide an approach to artificial symbiosis.

Acknowledgements. The work reported here has been supported by grants from CNRS, DGRST, INRA, and NIH (USA). We thank Rick L. Pesano for reviewing this manuscript.

References

Biemann K, Lioret C, Asselineau J, Lederer E, Polonski J (1960) Sur la structure chimique de la lysopine, nouvel acide aminé isolé des tissus de crown gall. Bull Soc Chim Fr 17:979–991

Bomhoff GH (1974) Studies on crown gall – a plant tumour. Investigations on protein composition and on the use of guanidine compounds as a marker for transformed cells. Ph D Thesis, Univ Leiden, Netherlands

Bomhoff GH, Klapwijk PM, Kester HC, Schilperoort RA, Hernalsteens JP, Schell J (1976) Octopine and nopaline synthesis and breakdown genetically controlled by a plasmid of *Agrobacterium tumefaciens.* Mol Gen Genet 145:177–181

Braun AC (1947) Thermal studies on the factors responsible for tumour initiation in crown gall. Am J Bot 34:234–240

Braun AC (1959) A demonstration of the recovery of the crown gall tumour cell with the use of complex tumours of single cell origin. Proc Natl Acad Sci USA 45:932–938

Braun AC (1974) The biology of cancer. Addison-Wesley, Reading, Mass

Chilton MD, Farrand SK, Eden F, Bendich AJ, Gordon MP, Nester EW (1975) Is there foreing DNA in crown gall tumour DNA? In: Markham, Davies, Hopwood, Horne (eds) Modification of the information content of plant cells. Elsevier/North Holland Biomedical Press, Amsterdam New York, pp 297–311

Chilton MD, Drummond MH, Merlo DJ, Sciaky D, Montoya AL, Gordon MP, Nester EW (1977) Stable incorporation of plasmid DNA into higher plant cells: the molecular basis of crown gall tumourigenesis. Cell 11:263–271

Chilton MD, Tepfer DA, Petit A, David C, Casse-Delbart F, Tempé J (1982) *Agrobacterium rhizogenes* inserts T-DNA into the genomes of the host-plant root cells. Nature295:432–434

Coxon DT, Davies AMC, Fenwick GR, Self R, Firmin JL, Lipkin D, Janes NF (1980) Agropine, a new amino acid derivative from crown gall tumours. Tetrahedron Lett 21:495–498

Dons JJM (1975) Crown gall – a plant tumour. Investigations on the nuclear DNA content and on the presence of *Agrobacterium tumefaciens* DNA and phage PS8 DNA in crown gall cells. Ph D Thesis, Univ Leyden, Netherlands

Ellis JG, Murphy PJ (1981) Four new opines from crown gall tumours – their detection and properties. Mol Gen Genet 181:36–43

Ellis JG, Kerr A, Petit A, Tempé J (1982) Conjugal transfer of nopaline and agropine Ti-plasmids – The role of agrocinopines. Mol Gen Genet 186:269–274

Engler G, Holsters M, Montagu M Van, Schell J, Hernalsteens JP, Schilperoort RA (1975) Agrocin 84 sensitivity: a plasmid determined property in *Agrobacterium tumefaciens*. Mol Gen Genet 138:345–349

Firmin JL, Fenwick RG (1977) N_2-(1-3-dicarboxypropyl) ornithine in crown gall tumours. Phytochemistry 16:761–762

Firmin JL, Fenwick RG (1978) Agropine – a major new plasmid-determined metabolite in crown gall tumours. Nature 276:842–844

Goldmann A, Tempé J, Morel G (1968) Quelques particularités de diverses souches d'*Agrobacterium tumefaciens*. CR Soc Biol 162:630–631

Goldmann A, Thomas DW, Morel G (1969) Sur la structure de la nopaline, métabolite anormal de certaines tumeurs de crown gall. CR Acad Sci (Paris) Ser D 268:852–854

Goldmann-Ménagé A (1970) Recherches sur le métabolisme azoté des tissus de crown gall cultivés in vitro. Ann Sci Nat Bot Biol Veg 11:223–310

Greve H De, Dhaese P, Seurinck J, Montagu M Van, Schell J (1983) Nucleotide sequence and transcript map of the *Agrobacterium tumefaciens* Ti-plasmid encoded octopine synthase gene. J Mol Appl Genet (submitted)

Guyon P, Chilton MD, Petit A, Tempé J (1980) Agropine in "null type" crown gall tumours: evidence for the generality of the opine concept. Proc Natl Acad Sci USA 77:2693–2697

Hernalsteens JP, Vliet F Van, Beuckeleer M De, Depicker A, Engler G, Lemmers M, Holsters M, Montagu M Van, Schell J (1980) The *Agrobacterium tumefaciens* Ti-plasmid as a host vector system for introducing foreign DNA in plant cells. Nature 287:654–656

Johnson R, Guderian RH, Eden F, Chilton MD, Gordon MP, Nester EW (1974) Detection and quantitation of octopine in normal plant tissue and in crown gall tumours. Proc Natl Acad Sci USA 71:536–539

Kemp JD (1977) A new amino acid derivative present in crown gall tumour tissue. Biochem Biophys Res Commun 74:862–868

Kerr A (1969) Transfer of virulence between isolates of *Agrobacterium*. Nature 223/1175–1176

Kerr A (1971) Acquisition of virulence by non-pathogenic isolates of *Agrobacterium*. Physiol Plant Pathol 1:241–246

Kerr A, Htay K (1974) Biological control of crown gall through bacteriocin production. Physiol Plant Pathol 4:37–44

Kerr A, Roberts WP (1976) *Agrobacterium:* Correlation between and transfer of pathogenicity, octopine and nopaline metabolism and bacteriocin 84 sensitivity. Physiol Plant Pathol 9: 205–211

Klapwijk PM, Hooykaas PJJ, Kester HCM, Schilperoort RA, Roersch A (1976) Isolation and characterisation of *Agrobacterium tumefaciens* mutants affected in the utilization of octopine, octopinic acid and lysopine. J Gen Microbiol 96:155–163

Klein RM (1954) Mechanisms of crown gall induction. Brookhaven Symp Biol No 6: Abnormal and pathological plant growth, pp 97–114

Lejeune B, Jubier MF (1968) Etude de la dégradation de la lysopine par *Agrobacterium tumefaciens*. CR Acad Sci (Paris) Ser D 264:1803–1805

Lemmers M, Beuckeleer M De, Holsters M, Zambrisky P, Depicker A, Hernalsteens JP, Montagu M Van, Schell J (1980) Internal organization, boundaries and integration of Ti-plasmid DNA in nopaline crown gall tumours. J Mol Biol 144:353–376

Ley J De, Bernaerts M, Rassel A, Guilmot A (1966) Approach to an improved taxonomy of the genus *Agrobacterium*. J Gen Microbiol 43:7–17

Lioret C (1956) Sur la mise en évidence d'un acide aminé non identifié particulier aux tissus de crown gall. Bull Soc Fr Physiol Veg 2:76

Lioret C (1957) Les acides aminés libres des tissus de crown gall. Mise en évidence d'un acide aminé particulier à ces tissus. CR Acad Sci (Paris) 244:2171–2174

Lioret C (1966) Recherches sur la signification physiologique de la lysopine dans les tissus de crown gall de scorsonère cultivés in vitro. Physiol Veg 4:89–103

Loening UE, Butcher DN, Schuck W, Satirana ML (1975) Experiments on the homology between DNA from *Agrobacterium tumefaciens* and crown gall tumour cells. In: Markham, Davies, Hopwood, Horne (eds) Modification of the information content of plant cells. Elsevier/North Holland Biomedical Press, Amsterdam New York, pp 297–311

Ménagé A, Morel G (1964) Sur la présence d'octopine dans les tissus de crown gall. CR Acad Sci (Paris) 259:4795–4796

Ménagé A, Morel G (1965) Sur la présence d'un acide aminé nouveau dans le tissu de crown gall. CR Soc Biol Ses Fil 159:561–562

Ménagé A, Morel G (1966) Sur quelques particularités du métabolisme azoté des tissus de tumeurs d'*Opuntia vulgaris*. CR Soc Biol 160:52–54

Ménagé A, Tempé J, Morel G (1966) Analyse des acides aminés par chromatographie. II. Détection automatique des guanidines. Physiol Veg 6:317–322

Metzler DE (1977) Biochemistry, the chemical reactions of living cells. Academic Press, London New York

Milo GE, Srivastava BIS (1969) RNA-DNA hybridization studies with the crown gall bacteria and tobacco tumour tissue. Biochem Biophys Res Commun 34:196–199

Montoya A, Chilton MD, Gordon MP, Sciaky D, Nester EW (1977) Octopine and nopaline metabolism in *Agrobacterium tumefaciens* and crown gall tumour cells: role of plasmid genes. J Bacteriol 129:101–107

Moore L, Warren G, Strobel G (1979) Involvement of a plasmid in the hairy root disease of plants caused by *Agrobacterium rhizogenes*. Plasmid 2:617–626

Morel G (1956) Métabolisme de l'arginine par les tissus de crown gall de topinambour. Bull Soc Fr Physiol Veg 2:76

Morel G (1970) Déviations du métabolisme azoté des tissus de crown gall. In: Les cultures de tissus de plantes. Centre National de la Recherche Scientifique, Paris, pp 463–471

Murai N, Kemp JD (1982) Octopine synthase m-RNA isolated from sunflower crown gall callus is homologous to the Ti-plasmid of *Agrobacterium tumefaciens*. Proc Natl Acad Sci USA 79:86–90

Petit A, Tempé J (1978) Isolation of *Agrobacterium* Ti-plasmid regulatory mutants. Mol Gen Genet 167:147–155

Petit A, Delhaye S, Tempé J, Morel G (1970) Recherches sur les guanidines des tissus de crown gall. Mise en évidence d'une relation biochimique spécifique entre les souches d'*Agrobacterium tumefaciens* et les tumeurs qu'elles induisent. Physiol Veg 8:205–213

Petit A, Dessaux Y, Tempé J (1978a) The biological significance of opines. I. A study of opine catabolism by *Agrobacterium tumefaciens*. In: Ridé M (ed) Proc 4th Int Conf Plant Pathog Bact. INRA, Angers, pp 143–152

Petit A, Tempé J, Kerr A, Holsters M, Montagu M Van, Schell J (1978b) Substrate induction of conjugative activity of *Agrobacterium tumefaciens* Ti-plasmids. Nature 271:570–571

Petit A, David C, Dahl GA, Ellis JG, Guyon P, Casse-Delbart F, Tempé J (1983) Further extension of the opine concept: Plasmids in *Agrobacterium rhizogenes* cooperate for opine degradation. Mol Gen Genet 190:204–214

Quétier F, Huguet T, Guillé E (1969) Induction of crown gall: partial homology between tumour-cell DNA, bacterial DNA and the G+C rich DNA of stressed normal cells. Biochem Biophys Res Commun 34:128–133

Roberts WP, Tate ME, Kerr A (1977) Agrocin 84 is a 6-N-phosphoroamidate of an adenine nucleotide analogue. Nature 265:379–381

Schell J (1978) The use of the Ti-plasmid as a vector for the introduction of foreign DNA into plants. In: Ridé M (ed) Proc 4th Int Conf Plant Pathog Bact. INRA-Angers, pp 115–126

Schell J, Montagu M Van, Beuckeleer M De, Block M De, Depicker A, Wilde M De, Engler G, Genetello C, Hernalsteens JP, Holsters M, Seurinck J, Silva B, Vliet F Van, Villaroel R (1979) Interaction and DNA transfer between *Agrobacterium tumefaciens*, the Ti-plasmid and the plant host. Proc R Soc London Ser B 204:251–266

Schilperoort RA, Veldstra H, Warnaar SO, Mulder G, Cohen SA (1967) Formation of complexes between DNA isolated from tobacco crown gall tumours and RNA complementary to *Agrobacterium tumefaciens*. Biochim Biophys Acta 145:523–525

Schilperoort RA, Meijs WH, Pippel GMW, Veldstra H (1969) *Agrobacterium tumefaciens* cross-reacting antigens in sterile crown gall tumours. FEBS Lett 3:173–176

Schilperoort RA, Sittert NJ Van, Schell J (1973) The presence of both phage PS8 and *Agrobacterium tumefaciens* A6 DNA base sequences in A6-induced sterile crown gall tissue cultured in vitro. Eur J Biochem 33:1–7

Schroeder J, Schroeder G, Huisman H, Schilperoort RA, Schell J (1981) The m-RNA for lysopine dehydrogenase in plant tumour cells is complementary to a Ti-plasmid fragment. FEBS Lett 129:166–238

Sciaky D, Montoya AL, Chilton MD (1978) Fingerprints of *Agrobacterium* Ti-plasmids. Plasmid 1:238–253

Seitz EW, Hochster RM (1964) Lysopine in normal and in crown gall tumour tissue of tomato and tobacco. Can J Bot 42:999–1004

Spano L, Pomponi M, Costantino P, Slogteren GMS Van, Tempé J (1982) Identification of T-DNA in the root-inducing plasmid of the agropine type *Agrobacterium rhizogenes* 1855. Plant Mol Biol 1:291–300

Tate ME, Ellis JG, Kerr A, Tempé J, Murray KE, Shaw KJ (1982) Agropine: a revised structure. Carbohydr Res 104:105–120

Tempé J, Ménagé A, Morel G (1966) Analyse des acides aminés par chromatographie. III. Adaptation de l'autoanalyzer à une double coloration permettant la révélation simultnée des acides aminés et des guanidines. In: Automation in analytical chemistry, vol 2. Mediad, Ardsley, pp 421–425

Tempé J, Estrade C, Petit A (1978) The biological significance of opines. II. The conjugative activity of the Ti-plasmids of *Agrobacterium tumefaciens*. In: Ridé M (ed) Proc 4th Int Conf Plant Pathog Bact. INRA-Angers, pp 153–160

Tempé J, Guyon P, Tepfer D, Petit A (1979) The role of opines in the ecology of the Ti-plasmids of *Agrobacterium*. In: Timmis KN, Pühler A (eds) Plasmids of medical, environmental, and commercial importance. Elsevier/North Holland Biomedical Press, Amsterdam New York, pp 353–363

Tempé J, Petit A, Bannerot H (1982) Présence de composés semblables à des opines dans des nodosités de luzerne. CR Acad Sci (Paris) Ser III 295:413–416

Tepfer DA, Tempé J (1981) Production d'agropine par des racines formées sous l'action d'*Agrobacterium rhizogenes*. CR Acad Sci (Paris) Ser III 292:153–156

Thomashow MF, Nutter R, Montoya AL, Gordon MP, Nester EW (1980) Integration and organization of Ti plasmid sequences in crown gall tumours. Cell 19:729–739

Wendt-Gallitelli MF, Dobrigkeit I (1973) Investigations implying the invalidity of octopine as a marker for transformation by *Agrobacterium tumefaciens*. Z Naturforsch C Biosci 28C: 768–771

White FF, Nester EW (1980) Hairy root: Plasmid encodes virulence traits in *Agrobacterium rhizogenes*. J Bacteriol 141:1134–1141

White FF, Ghidossi G, Gordon MP, Nester EW (1982) Tumour induction by *Agrobacterium rhizogenes* involves the transfer of plasmid DNA to the plant genome. Proc Natl Acad Sci USA 79:3193–3197

Willmitzer L, Simons G, Schell J (1982a) The TL-DNA in octopine crown gall tumours codes for seven well defined polyadenylated transcripts. EMBO J 1:139–146

Willmitzer L, Sanchez-Serrano J, Buschfeld E, Schell J (1982b) DNA from *Agrobacterium rhizogenes* is transferred to and expressed in axenic hairy root plant tissues. Mol Gen Genet 186: 16–22

Zaenen I, Larebeke N Van, Teuchy H, Montagu M Van, Schell J (1974) Supercoiled circular DNA in crown gall inducing *Agrobacterium* strains. J Mol Biol 86:109–127

B. The *Rhizobium*-Plant Interaction

Studies on *Rhizobium meliloti* Plasmids and on Their Role in the Control of Nodule Formation and Nitrogen Fixation: The pSym Megaplasmids and the Other Large Plasmids

T. HUGUET [1], C. ROSENBERG [1], F. CASSE-DELBART [2], P. DE LAJUDIE [1], L. JOUANIN [2], J. BATUT [1], P. BOISTARD [1], J.-S. JULLIOT [3], and J. DÉNARIÉ [1]

Introduction

At the moment the *Rhizobium*-legume symbiosis is the most efficient agricultural system for the biological fixation of atmospheric nitrogen. Lucerne is, with soyabean, the crop which provides the largest amount of fixed nitrogen throughout the world (Hanson 1972). Strains of the symbiotic partner of lucerne, *Rhizobium meliloti,* are fast-growing rhizobia and constitute, on the basis of symbiotic properties, numerical taxonomy, nucleic acid hybridization and genetic studies, a homogenous cluster clearly different from the *R. leguminosarum* grouping of strains of *R. leguminosarum, R. phaseoli* and *R. trifolii* (Gibbins and Gregory 1972, Kondorosi et al. 1980).

Large plasmids of molecular weight greater than 90 million have been described in all the strains of fast-growing rhizobia investigated so far (for a review see Dénarié et al. 1981). In the *R. leguminosarum* grouping, plasmids of various sizes in the range of M.W. 120–550 million have been shown to carry genes controlling such symbiotic properties as host-specificity and nodule formation (Beynon et al. 1980, Hooykaas et al. 1981) and nitrogen fixation (Nuti et al. 1979, Prakash et al. 1981, Krol et al. 1982). No size class could be established among these plasmids, and no correlation was found between the presence of symbiotic genes and the size of the plasmids (Krol et al. 1982).

In this paper we review the results from our laboratory which show that in *R. meliloti* two size classes of large plasmids exist: (1) megaplasmids (M.W. > 450 million) which are present in all the strains and carry *nif* and *nod* genes and (2) plasmids in the M.W. range of 90 million to about 250 million which do not carry nitrogenase genes, share limited DNA homology and seem not to be essential for effective nodulation of *Medicago sativa.*

1 Laboratoire de Biologie Moléculaire des Relations Plantes-Microorganismes, C.N.R.S.-I.N.R.A., 31320 Castanet-Tolosan, France

2 Present Adress: Laboratoire de Biologie Cellulaire, I.N.R.A., Route de Saint-Cyr, 78000 Versailles, France

3 This paper is dedicated to the memory of our friend Jean-Simon Julliot who died in an accident on February 21st,1982

Molecular Genetics of the Bacteria-Plant Interaction
(ed. by A. Pühler)

Indigenous Plasmids of *R. meliloti:* Two Size Classes

To ensure maximum genetic diversity among the *R. meliloti* strains examined, thirty isolates were selected from various geographical origins (Table 1) and from the three nodulation groupings proposed by Brockwell and Hely (1966) according to their nodulation properties on various *Medicago, Melilotus* and *Trigonella* species.

Table 1. List of *R. meliloti* strains [a] and their plasmid content

Geographical origin	Code number
South Africa	Rf22 (3) [b]
North America	12 (2); V7 (2); S14 (2) Balsac (3); 102F51 [c] (2) 102F28 (4); S26 [c] (1) S33 [c] (2); 54032 (2) I1 (3); 3DoA20a (2)
South America	U45 [c] (2); B294 (2) U54 (2)
Oceania	1322 (2); RCR2011 [c] (= SU47) (1) CC169 (2); CC2003 (2) CC2013 (1); CC2090 (2) CC2093 (2)
Europe	L5-30 (2); 41 (2) 311 (3); A145 (1) Sa10 (2); Lb1 (2) Ls2a (3); Ve8 (3)

[a] The strains have been described in Casse et al. (1979) and Rosenberg et al. (1981)

[b] The number of ccc DNA bands observed after agarose gel electrophoresis according to Rosenberg et al. (1981) is given in parentheses

[c] Used in commercial inoculants

Using an alkaline denaturation procedure for DNA extraction it was possible to detect the presence of large plasmids in the M.W. size range from 90 million to around 250 million in all the strains except four: RCR2011, CC2013, S26 and A145 (Casse et al. 1979, Jouanin et al. 1981, Rosenberg et al. 1981). The number of such pRme plasmids varied from 1 to 3 according to the strain (Table 1). It is worth noting that only two strains carried a plasmid of molecular weight lower than 85 million, and no very small plasmids (M.W. $<$ 10 million) suitable as cloning vehicles could be detected.

Slight modifications by Banfalvi et al. (1981) and Rosenberg et al. (1981, 1982) of the analytical 'in gel' lysis technique of Eckhardt (1978) allowed the detection, in addition to the pRme plasmids already described, of plasmids of very high molecular weight greater than 450 million. One such plasmid was present in every strain (Rosenberg et al. 1981, 1982). We suggested calling them *megaplasmids* to distinguish them from the other pRme plasmids.

The Megaplasmids

Localization of Nitrogenase Structural Genes

A direct physical approach to the location of *Rhizobium nif* genes was made possible by the finding that the DNA sequences of the nitrogenase structural genes were highly conserved among various nitrogen-fixing bacteria (Nuti et al. 1979, Ruvkun and Ausubel 1980). Recombinant plasmids carrying *nif* genes isolated from *Klebsiella pneumoniae* could be used as probes to show that at least the nitrogenase structural genes *nif* D and H are located on plasmids in strains of the *R. leguminosarum* cluster (Nuti et al. 1979, Krol et al. 1982, Prakash et al. 1981).

A recombinant plasmid pRmR2 has been constructed which contains at least part of *R. meliloti* 2011 *nif* D and H genes (Ruvkun and Ausubel 1980, 1981). We used this hybrid plasmid as a probe to investigate the physical location of *nif* genes in eighteen different *R. meliloti* strains: A145, S26, L530, 41, V7, U45, 102F51, 12, 1322, Ls2a, Balsac, Ve8, S14, 102F28, Rf22, RCR2011, CC2013 and CC2020. Lysates obtained by the 'lysis in gel' procedure (Rosenberg et al. 1982), were submitted to electrophoresis, blotted and hybridized with a pRmR2 probe (Fig. 1). The ccc DNA band corresponding to the megaplasmid of each of the eighteen strains hybridized strongly with the *nif* DH probe.

On the contrary no hybridization could be seen with the ccc DNA bands of the pRme plasmids of smaller M.W. (90–250 million), except in one strain V7 in which the pRmV7*a* of M.W. around 200 million hybridized with pRmR2 (Rosenberg et al. 1981). On the other hand this plasmid has been shown not to hybridize with *K. pneumoniae nif* DH (Prakash et al. 1981); it is thus likely that the hybridization observed at the level of the pRmeV7*a* was due to a DNA sequence which is present in addition to the *nif* DH genes in pRmR2.

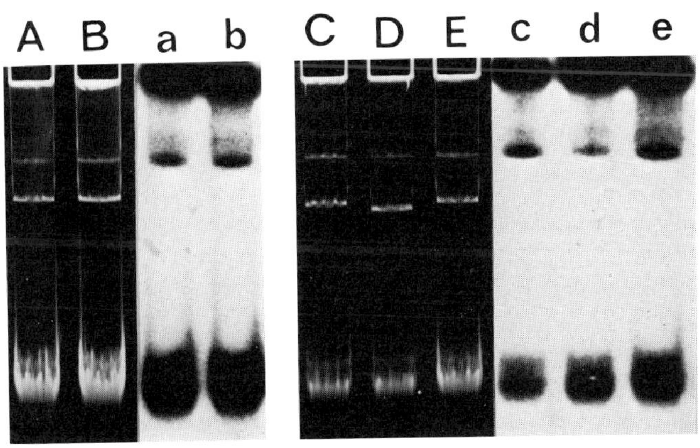

Fig. 1. Agarose gel electrophoresis of lysates from *R. meliloti* strains, *A* = 1322; *B* = 12; *C* = 102F51; *D* = L5-30; *E* = U45. *Small letters* indicate the autoradiograms of the corresponding Southern blots after hybridization with pRmR2

Therefore *nif* genes D and H are located, in all the strains investigated, on the megaplasmid and not on smaller plasmids.

Localization of *nod* Genes

Independent spontaneous Nod⁻ derivatives of L5-30 have been fortuitously isolated after storage in the cold. They were derivatives of L5-30 since they had the same pattern of sensitivity to a set of specific bacteriophages as the wild-type. All still possessed two plasmids: the megaplasmid and the pRmeL5-30*a* (90 million). No deletion of the smaller plasmid could be detected by restriction enzyme analysis. On the contrary, in all the Nod⁻ mutants studied, hybridization with the *K. pneumoniae nif* DH was no longer observed (Rosenberg et al. 1981). These *nif nod* deletions were not large enough to be detected by a difference in electrophoretic mobility of the covalently closed circular forms of the megaplasmid. The same results were obtained with Nod⁻ derivatives of strains Lb1 and Sa10.

Similarly Banfalvi et al. (1981) have observed a simultaneous loss of *nod* and *nif* genes in derivatives of strain 41 which carried a deletion of the megaplasmid. In this book Ausubel et al., Dusha et al., and Long demonstrate that *nod* genes are also located on the megaplasmid in strain RCR2011.

Therefore in the five strains investigated so far (L5-30, 41, Sa10, Lb1 and RCR2011) *nod* genes are located on the megaplasmid.

The fact that spontaneous deletions of the megaplasmid, selected on the Nod⁻ phenotype, result in a loss of the nitrogenase structural genes indicate that these *nif* genes are located only on the megaplasmid and not on the chromosome.

The pRme Plasmids

In the *R. leguminosarum* cluster, some plasmids of molecular weight in the range 90–250 million have been shown to share a considerable degree of homology (Prakash et al. 1981, Prakash and Schilperoort 1982), and to control symbiotic properties see Dénarié et al. 1981). We wanted to know if it was also the case in *R. meliloti* for the pRme plasmids of this size class (= pRme plasmids).

DNA Homology Studies

The study of sequence conservation of pRme plasmids was done with strains of various geographical origin harboring only one pRme plasmid: L5-30 (Europe; M.W. of pRme: 91 million), 41 (Europe, 150 million), 102F51 (North America, 100 million), 12 (North America, 107 million), V7 (North America, 197 million), 54 (South America, 153 million) and 1322 (New Zealand, 121 million). Covalently closed circular DNA was isolated by an alkaline denaturation procedure (Jouanin et al. 1981). DNAs were treated with various restriction endonucleases. The restriction patterns were markedly dissimilar. No band common to all the plasmids studied could be detected whatever the endonuclease used, either giving a great number of fragments such as *Eco*RI, *Hin*dIII or *Bam*HI or a low number such as *Sma*I, *Kpn*I or *Xba*I.

Southern DNA/DNA hybridization experiments were performed for more precise estimation of sequence homologies. Purified pRme molecules, radioactively labelled by nick-translation, were hybridized against Southern blots of restriction fragments of the plasmids from other strains. In all experiments hybridization could be detected suggesting that a certain level of sequence homology exists among all the pRme plasmids studied.

The number of hybridizing bands and the intensity of hybridization were different according to the probe and the plasmid blotted (Fig. 2). There were always only a limited number of restriction fragments which hybridized demonstrating that plasmids share only a limited extent of homology. Experiments on thermal denaturation revealed a high stability of the hybrids suggesting a low sequence divergence within the short hybridizing sequences. No obvious correlation could be found between the extent of pRme homology and the geographical origin of the strains.

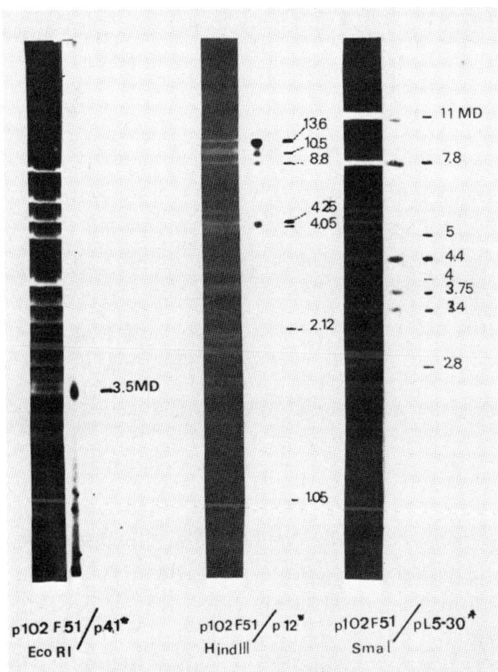

Fig. 2. Restriction endonuclease digests were fractionated on agarose gels, transferred to nitrocellulose sheets and hybridized against ^{32}P-labelled plasmid DNA as probe (Jouanin et al. 1981). *Left track* restriction pattern of plasmid to be blotted; *right track* autoradiogram after hybridization of the plasmid probe with the digested DNA of left track

In order to study the organization of homologous sequences between pRme, we constructed the physical map of one of these plasmids, pRme41 (150 million), isolated from the well studied strain 41 (Banfalvi et al. 1981). This map was constructed by the technique of Depicker et al. (1980): fragments resulting from incomplete digestion of pRme41 by *Hin*dIII were cloned in pBR322 (De Lajudie 1981). Overlapping fragments were used to build the map. Figure 3 summarizes the results concerning the location of pRme41 fragments homologous with pRme from strains L5-30,

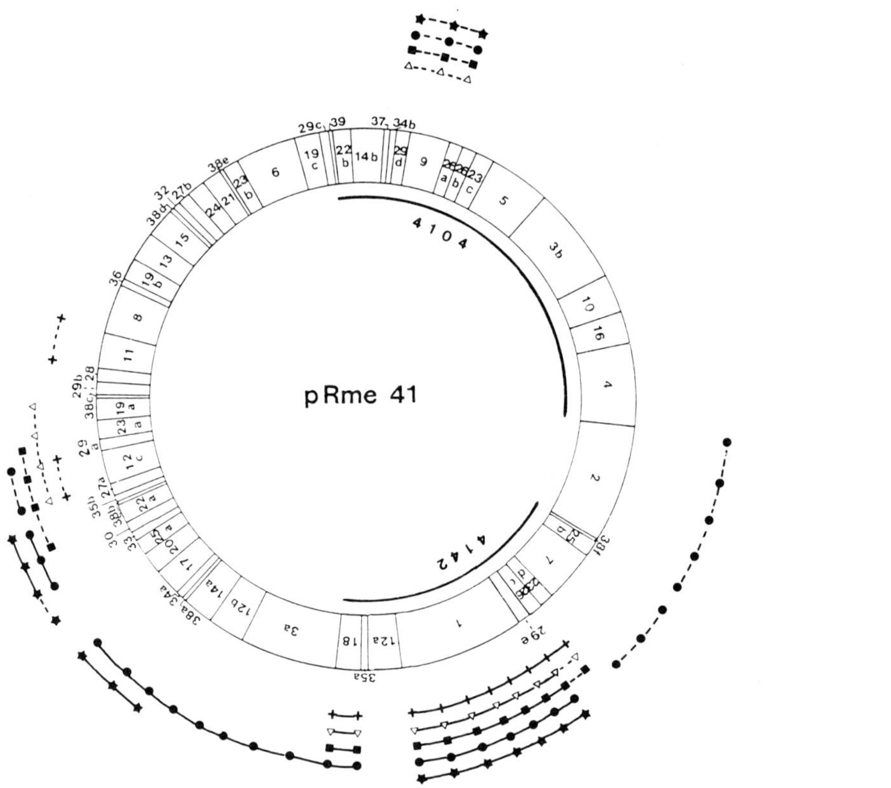

Fig. 3. *Hin*dIII map of pRme41 plasmid DNA and the regions of homology with *Rhizobium* pRme and *Agrobacterium* Ti plasmid DNA. Regions of weak homology are represented by *interrupted lines* (– – –); +++ pRmeL5-30; △ △ △ pRme102F51; ■ ■ ■ pRme1322; ● ● ● pTiC58; * * * pTiB6S3; *lines* and *numbers* inside the physical map refer to the pGMI4104 and pGMI4142 episome used in the genetical experiment

102F51 and 1322, Ti plasmids from *Agrobacterium tumefaciens* C58 (nopaline type) and B6S3 (octopine type). pRme probes hybridized mainly with fragments clustered within three regions of the map, the strongest hybridization being observed within the region cloned in the pGMI4142 episome. Surprisingly the extent of homology of pRme41 DNA was much greater with Ti plasmids than with the other pRme. Both C58 and B6S3 Ti plasmids hybridized strongly with the 4142 region, as did the other pRme plasmids, but also with regions extending further (clockwise).

The use of cloned fragments of pRme41 as probes allowed us to show that the regions of pRme41 from fragment *Hin*dIII 12b to 17 and from 33 to 35b respectively are homologous to the conserved regions C and B of Ti plasmids, regions which are involved in replication and transfer functions (De Greve et al. 1981, Engler et al. 1981). The '4142' region of pRme41 strongly hybridized with regions of Ti plasmids which as far as we know have not yet been genetically characterized. No homology could be detected between pRme41 and T-DNA. The pRme plasmids were not very different from the Nif-plasmids of the *R. leguminosarum* group with respect to their homology with Ti plasmids (Prakash and Schilperoort 1982). In other words the pRme plasmids of *R. meliloti* do not seem to share more homology among themselves than they do with other plasmids of *Rhizobiaceae.*

Genetic Studies

Genetic studies were performed on the plasmid pRme41. The strategy used was to clone the pRme41 plasmid into the broad host-range plasmid RP4 and to use some of the resulting RP4-prime episomes to insert specifically transposons into pRme41 and to mobilize it.

Partial *Hin*dIII digests of pRme41 were cloned into the Km gene (*Hin*dIII site) of RP4. Some of the RP4-primes are illustrated in Fig. 3. The pGMI4104 episome was introduced into an *E. coli* strain carrying transposon Tn5 (Km^R) in the chromosome. pGMI4104::Tn5 hybrids were selected by transfer into another *E. coli* strain, and introduced into a strain 41 recipient by mating. Recombination between the homologous region of pRme41 and the pGMI4104::Tn5 led to the insertion of Tn5 into pRme41 (Fig. 4). Elimination of the RP4-prime in *Rhizobium* was achieved by selecting clones resistant to two phages (PRRI and GU5) specific for RP4-carrying bacteria. Strain 41 (pRme41::Tn5) was fully effective.

Tagging of pRme41 by Tn5 allowed the study of its transfer. It was self-transmissible to another strain of *R. meliloti,* strain L5-30, at a low frequency ($\cong 10^{-7}$). The two plasmids pRme L5-30 and pRme41::Tn5 were compatible. Transfer of pRme41::Tn5 into *A. tumefaciens* C58 could not be detected. pGMI4142, a RP4-

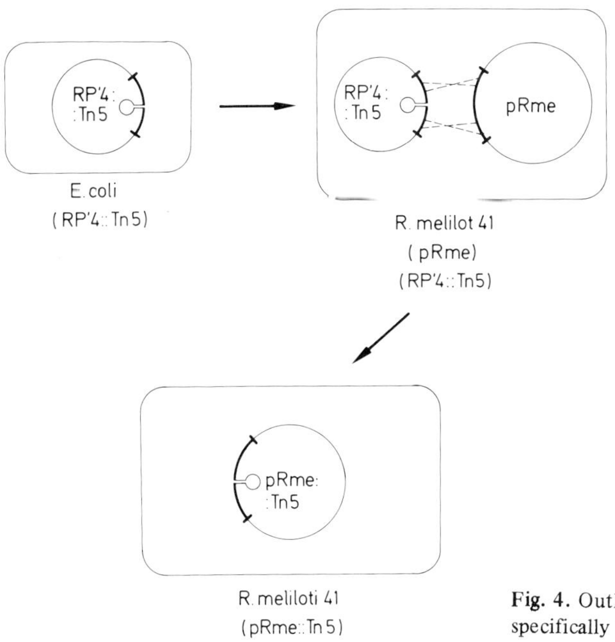

Fig. 4. Outline of strategy used to insert specifically Tn5 into pRme41 plasmid

prime having no common pRme41 *Hin*dIII fragment with pGMI4104 (see Fig. 3), was used to mobilize pRme41: :Tn5 by DNA homology (Julliot and Boistard 1979). A strain 41 (pRme41: :Tn5) (pGMI4142) could transfer pRme41: :Tn5 into *R. trifolii* LPR5001 and *A. tumefaciens* C58 at a frequency of around 5×10^{-7} (Fig. 5). The *R. trifolii* and *A. tumefaciens* pRme41: :Tn5$^+$ transconjugants were unable to form nodules on *M. sativa* (Rosenberg et al., in preparation). pRme41: :Tn5 was compatible with pTiC58 but not with the cryptic large plasmid pAtC58 (Rosenberg et al., in preparation). Strain 41 (pRme: :Tn5) behaves, as does the wild-type 41, like an opine-utilizing strain of *A. tumefaciens:* mutants able to use octopine as a sole carbon and nitrogen source can be isolated. When pRme41: :Tn5 was introduced into C58C1, a nopaline *A. tumefaciens* strain cured of its Ti plasmid, the ability to catabolize opines was not restored.

Attempts to cure strain 41 of the pRme41: :Tn5 plasmid by various treatments (heat, phosphorus starvation, lauryl sulphate, acridine orange, etc.) were unsuccessful: no Km^S derivative could be detected. Insertion of bacteriophage Mu into RP4 is known to prevent replication of the resulting RP4: :Mu in *Rhizobium* (Boucher et al. 1977) and *Agrobacterium* (Van Vliet et al. 1978). Surprisingly, insertion of Mu into the RP4-prime pGMI4142 (carrying a fragment of pRme41) did not significantly decrease its transfer frequency from *E. coli* to *R. meliloti* 41. Nevertheless, the presence of pGMI4142: :Mu led to some instability of pRme41: :Tn5. A Km^S derivative, cured at the same time of pRme41: :Tn5 and pGMI4142: :Mu, could be isolated; it was called 41-Cal (Fig. 5). The fact that the strain was actually cured of the pRme41 was checked by preparing a radioactive probe of purified pRme41 and hybridizing with the *Hin*dIII digested total DNA of strains 41 and 41-Cal; no hybridizing band could be detected with 41-Cal. Strain 41-Cal gave rise to octopine-utilizing clones as

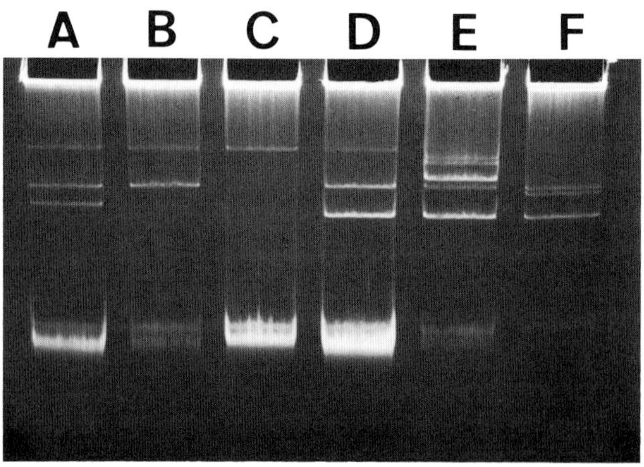

Fig. 5. Agarose gel electrophoresis of lysates from strains *A* = *R. meliloti* L530 (pRme41: :Tn5); *B* = *R. meliloti* 41 (pRme41: :Tn5); *C* = *R. meliloti* 41-Cal; *D* = *R. meliloti* 41 (pRme41: :Tn5) (pGMI4142); *E* = *R. trifolii* LPR5001 (pRme41: :Tn5) (pGMI4142); *F* = *A. tumefaciens* C58 (pRme41: :Tn5) (pGMI4142)

effectively as the wild-type showing that opine catabolism is not controlled by the pRme41 plasmid. The cured strain 41-Cal nodulated ten varieties of *Medicago sativa* and fixed nitrogen as effectively as the wild-type 41 (Rosenberg et al., in preparation). These results indicate that pRme41 does not carry genes essential for nodulation and nitrogen fixation on *M. sativa.*

Conclusions

Studies on the size of plasmids and their role in the control of symbiotic properties for strains of the *R. leguminosarum* cluster present a very diverse picture. No essential plasmid size class has clearly emerged. *nif* DH were shown to be located on plasmids of molecular weight from 130 to 550 million in *R. leguminosarum* (Nuti et al. 1979, Prakash et al. 1981, Krol et al. 1982). Plasmids in the molecular weight range 100–250 million were found to carry genes controlling nodulation and host-specificity in *R. leguminosarum, R. phaseoli* (Beynon et al. 1980) and *R. trifolii* (Hooykaas et al. 1981).

On the contrary in *R. meliloti* a general picture has emerged. There are clearly two size classes of large plasmids.

1. Megaplasmids for which we propose the term pSym (for control of symbiotic properties): (a) their molecular weight is greater than 450 million; (b) they are present in all effective strains; (c) they carry nitrogenase structural genes and, in all the cases studied so far, *nod* genes.

2. Plasmids which we propose to call pRme (for nomenclature see Casse et al. 1979): (a) their molecular weight varies from 90 million to about 250 million; (b) their number varies with the strain and they may be absent; (c) they do not carry nitrogenase structural genes.

An important question to answer is: what functions are carried on these large cryptic pRme plasmids (each carrying more than 100 genes)? Possibly these include: (1) functions related to the infectious process on legumes, for instance host-specificity for hosts other than the ones tested, or non-obligatory functions such as hydrogen-uptake, etc.; (2) life in the rhizosphere and (3) life in the soil.

Genes essential for the symbiotic relationship with *M. sativa,* such as *nod, fix* and *nif* are located on huge plasmids probably carrying more than 400 genes. The fact that *nif* genes have not been observed in effective strains of *R. meliloti* on a plasmid of smaller size indicates that no large segment of the pSym can be deleted without altering bacterial viability or symbiotic effectiveness and suggest that *Rhizobium* genes required for an effective symbiosis with *M. sativa* are numerous and clustered on a single replicon independent of the chromosome. The genetic study of these huge molecules, that is the inventory of genes controlling the sequential stages of the symbiotic process, the study of the organization of these genes on the megaplasmid, of the sequential expression of these genes during the formation of nodules and the establishment of nitrogen-fixation, require other genetic tools. In the following chapter we describe a strategy for the genetic study of the *R. meliloti* pSym megaplasmid.

Acknowledgements. This work was supported by the "Biological Nitrogen Fixation" grant from Société Nationale Elf Aquitaine and Entreprise Minière et Chimique and by an INSERM grant (ATP Recombinaisons Génétiques no. 79-104). Part of the work was performed at the Laboratoire de Biologie Moléculaire Végétale (Laboratoire associé CNRS no. 40) of Université Paris-Sud, Orsay. The authors would like to thank Dr. Bill Broughton for reviewing the manuscript.

References

Banfalvi Z, Sakanyan V, Koncz C, Kiss A, Dusha I, Kondorosi A (1981) Location of nodulation and nitrogen fixation genes on a high molecular weight plasmid of *R. meliloti.* Mol Gen Genet 184:318–325

Beynon IL, Beringer JE, Johnston AWB (1980) Plasmids and host range in *Rhizobium leguminosarum* and *Rhizobium phaseoli.* J Gen Microbiol 120.421–429

Boucher C, Bergeron B, Barate de Bertalmio M, Dénarié J (1977) Introduction of bacteriophage Mu into *Pseudomonas solanacearum* and *Rhizobium meliloti* using the R factor RP4. J Gen Microbiol 98:253–263

Brockwell J, Hely FW (1966) Symbiotic characteristics of *Rhizobium meliloti:* an appraisal of the systematic treatment of nodulation and nitrogen fixation interactions between hosts and rhizobia of diverse origins. Aust J Agric Res 17:885–899

Casse F, Boucher C, Julliot JS, Michel M, Dénarié J (1979) Identification and characterization of large plasmids in *Rhizobium meliloti* using agarose gel electrophoresis. J Gen Microbiol 113: 229–242

Dénarié J, Boistard P, Casse-Delbart F, Atherly AG, Berry JO, Russell P (1981) Indigenous plasmids of *Rhizobium.* Int Rev Cytol Suppl 13:225–245

Depicker A, Wilde M De, Devos G, Montagu M Van, Schell J (1980) Molecular cloning of overlapping segments of the nopaline Ti plasmid pTiC58 as a means to restriction endonuclease mapping. Plasmid 3:193–211

Eckhardt T 1978) A rapid method for the identification of plasmid desoxyribonucleic acid in bacteria. Plasmid 1:584–588

Engler G, Depicker A, Maenhaut R, Villaroel R, Montagu M Van, Schell J (1981) Physical mapping of DNA base sequence homologies between an octopine and a nopaline Ti plasmid of *Agrobacterium tumefaciens.* J Mol Biol 152:183–208

Gibbins AM, Gregory KF (1972) Relatedness among *Rhizobium* and *Agrobacterium* species determined by three methods of nucleic acid hybridization. J Bacteriol 111:129–141

Greve H DE, Decraemer H, Seurinck J, Montagu M Van, Schell J (1981) The functional organization of the octopine *Agrobacterium tumefaciens* plasmid pTiB6S3. Plasmid 6:235–248

Hanson CH (1972) Alfalfa science and technology. Am Soc Agron, Madison, USA

Hooykaas PJJ, Brussel AAN Van, Dulk-Ras H Den, Slogteren GMS Van, Schilperoort RA (1981) Sym-plasmid of *Rhizobium trifolii* expressed in different rhizobial species and in *Agrobacterium tumefaciens.* Nature 291:351–353

Jouanin L, Lajudie P De, Bazetoux S, Huguet T (1981) DNA sequences homology in *Rhizobium meliloti* plasmids. Mol Gen Genet 182:189–195

Julliot JS, Boistard P (1979) Use of RP4-prime plasmids constructed in vitro to promote a polarized transfer of the chromosome in *Escherichia coli* and *Rhizobium meliloti.* Mol Gen Genet 173:289–298

Kondorosi A, Vincze E, Johnston AWB, Beringer JE (1980) A comparison of three *Rhizobium* linkage maps. Mol Gen Genet 178:403–408

Krol AJM, Hontelez JGJ, Kammen A Van (1982) Only one of the large plasmids in *Rhizobium leguminosarum* strain PRE is strongly expressed in the endosymbiotic state. J Gen Microbiol 128:1839–1847

Lajudie P De (1981) Constitution d'une banque de gènes et établissement de la carte des sites de restriction *Hin*d III du plasmide pRme41 de *Rhizobium meliloti.* Thèse Troisième Cycle Université Paris-Sud

Long SR, Buikema WJ, Ausubel FM (1982) Cloning of *Rhizobium meliloti* nodulation genes by direct complementation of Nod⁻ mutants. Nature 298:485–488

Nuti MP, Lepidi AA, Prakash RK, Schilperoort RA, Cannon FC (1979) Evidence for nitrogen fixation genes on indigenous *Rhizobium* plasmids. Nature 282:533–535

Prakash RK, Schilperoort RA (1982) Relationship between Nif plasmids of fast-growing *Rhizobium* species and Ti plasmids of *Agrobacterium tumefaciens.* J Bacteriol 149:1129–1134

Prakash RK, Schilperoort RA, Nuti MP (1981) Large plasmids of fast-growing rhizobia: homology studies and location of structural nitrogen fixation *(nif)* genes. J Bacteriol 145:1129–1136

Rosenberg C, Boistard P, Dénarié J, Casse-Delbart F (1981) Genes controlling early and late functions in symbiosis are located on a megaplasmid in *Rhizobium meliloti*. Mol Gen Genet 184:326–333

Rosenberg C, Casse-Delbart F, Dusha F, David M, Boucher C (1982) Megaplasmids in the plant-associated bacteria *Rhizobium meliloti* and *Pseudomonas solanacearum*. J Bacteriol 150: 402–406

Rosenberg C, Huguet T (1983) The pAtC58 plasmid of *Agrobacterium tumefaciens* is not essential for tumor induction (in preparation)

Ruvkun GB, Ausubel FM (1980) Interspecies homology of nitrogenase genes. Proc Natl Acad Sci USA 77:191–195

Ruvkun BGB, Ausubel FM (1981) A general method for site-directed mutagenesis in prokaryotes: construction of mutations in symbiotic nitrogen fixation genes of *Rhizobium meliloti*. Nature 289:85–88

Vliet F Van, Silva B, Montagu M Van, Schell J (1978) Transfer of RP4::Mu plasmids to *Agrobacterium tumefaciens*. Plasmid 1:446–455

Studies on *Rhizobium meliloti* Plasmids and on Their Role in the Control of Nodule Formation and Nitrogen Fixation: A Strategy for Region-Specific Transposon Mutagenesis of pSym Megaplasmid

I. DUSHA [1], M.H. RÉNALIER [2], B. TERZAGHI [3], A.M. GARNERONE [2], P. BOISTARD [2], and J.-S. JULLIOT [4]

Introduction

As documented in the preceding paper, the pSym megaplasmid of *Rhizobium meliloti* carries genes essential for symbiotic nitrogen fixation. Since few symbiotic genes are likely to be expressed in pure culture, transposon mutagenesis is a tool of choice for their study. Transposon mutagenesis of pSym is rendered difficult by the fact that this plasmid is not self-transmissible at a detectable frequency. By contrast transposon mutagenesis of pRL1J1, a transmissible plasmid of *Rhizobium leguminosarum* could be obtained by plasmid mediated mobilization of a Tn*5* transposon originally located on the chromosome of the donor strain: this made possible the isolation of symbiotically defective mutants (Buchanan-Wollaston et al. 1980).

To date two methods have been used for transposon mutagenesis of pSym. The first method involves in vitro cloning of *R. meliloti* DNA fragments known to belong to pSym, transposon mutagenesis of the cloned fragment and replacement of the wild-type allele by the mutated gene. This method has been successfully used for the genetic analysis of the *nif* region (Ruvkun et al. 1982), but would be difficult to apply to the genetic analysis of the whole pSym. The second method is random mutagenesis of the whole *R. meliloti* genome, using a suicide vector such as pJB4JI. Using this method with a combination of complementation studies and physical mapping it was possible to identify, in the vicinity of the *nif* region, a region of pSym controlling an early event in the symbiotic process (Long et al. 1982). However, for a majority of symbiotic genes it is unlikely that the complementing clones could be easily screened. Therefore we thought it would be desirable to take a more direct approach towards the genetic analysis of pSym.

Here we present a method for the directed mutagenesis of large regions of pSym. Furthermore, our method, which involves the sequential construction of RP4-primes containing large fragments from defined regions of pSym, facilitates the physical mapping of pSym and could be useful for genetic manipulation of pSym in an *Escherichia coli* environment.

1 Present address: Institute of Genetics, Biological Research Center, Hungarian Academy of Sciences, 6701 Szeged, Hungary

2 Laboratoire de Biologie Moléculaire des Relations Plantes-Microorganismes, C.N.R.S.-I.N.R.A., B.P. 12; 31320 Castanet Tolosan, France

3 Present address: DSIR Plant Physiology Division, Private Bag, Palmerston North, New Zealand

4 This paper is dedicated to the affectionate memory of J.S. Julliot, a brilliant young scientist and wonderful colleague who initiated and inspired these studies and was killed in an avalanche on February 21st, 1982

Molecular Genetics of the Bacteria-Plant Interaction
(ed. by A. Pühler)

Cloning of Large Fragments of pSym into RP4

Construction of a pSym-RP4 Cointegrate

We had previously shown that various RP4-primes constructed in vitro and containing different fragments of *R. meliloti* RCR2011 chromosome were able to promote an oriented transfer of the chromosome (Julliot and Boistard 1979). This suggested the integration of the RP4-prime in the homologous region of the chromosome. In order to obtain integration of RP4 into pSym, we used as a source of homology between the two replicons the 5 kb *Eco*RI fragment (R3) of the *nif* region (Ruvkun et al. 1982) of *R. meliloti* 1021, a derivative of *R. meliloti* RCR2011. The *nif* region of *R. meliloti* RCR2011 is known to belong to pSym (Rosenberg et al. 1981).

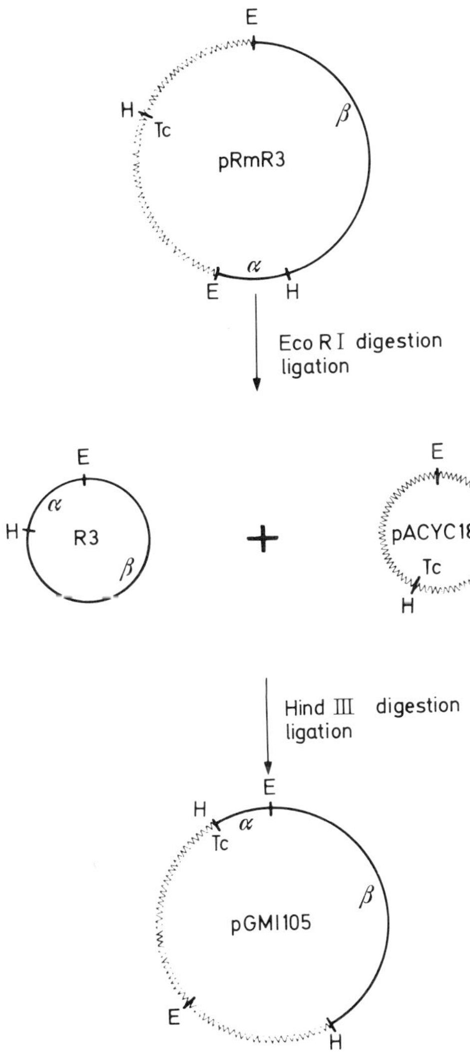

Fig. 1. Permutation of the alpha and beta fragments of R3. pRmR3 (Ruvkun et al. 1982) is a hybrid plasmid containing the 5 kb *Eco*RI fragment (R3) of the *nif* region of *R. meliloti* 1021 cloned into the *Eco*RI site of the $Tc^r Cm^r$ plasmid pACYC184 (Chang and Cohen 1978). Endonuclease restriction sites: *E Eco*RI; *H Hin*dIII

Additive legitimate recombination between two replicons results in the presence of two direct repeats of the homologous fragment at the junctions between the two replicons. The R3 *nif* fragment contains an internal *Hin*dIII site (Ruvkun et al. 1982). Treatment of R3 with *Hin*dIII generates two subfragments of 1 and 4 kb respectively that we will call alpha and beta. In order to create two distinguishable junction fragments in a RP4-prime pSym cointegrate we first changed the order of the alpha and beta fragments of R3 as shown in Fig. 1. Then we inserted the *Hin*dIII permutated fragment thus generated into the *Hin*dIII site of RP4. The structure of this plasmid, pGMI106 is shown in Fig. 2.

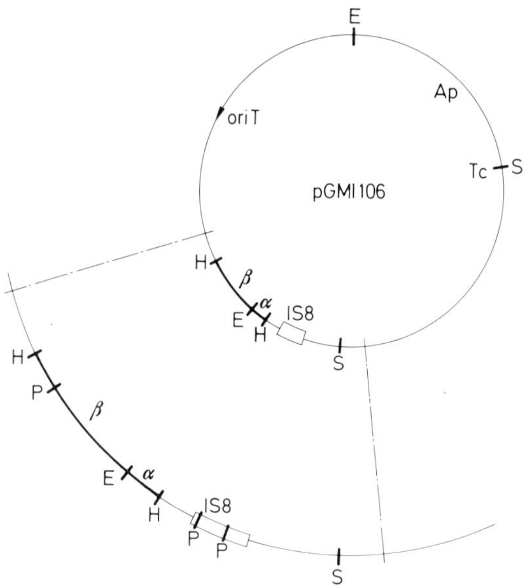

Fig. 2. Structure of pGMI106, an RP4 derivative containing the permutated R3 alpha beta fragments inserted at the *Hin*dIII site of RP4. Endonuclease restriction sites: *E* = *Eco*RI; *S* = *Sal*I; *P* = *Pst*I; *H* = *Hin*dIII

Then pGMI106 was introduced by mating into a Nif$^+$ derivative of *R. meliloti* 1021 containing the *nif*::Tn*5* insertion #20 (Ruvkun et al. 1982). When the Gm^r pPH1J1 plasmid of the same incompatibility group as RP4 (Beringer et al. 1978) was introduced into this strain the Tc^r marker of RP4 was not eliminated. In those Gm^r Tc^r clones no autonomous pGMI106 could be detected by agarose gel electrophoresis suggesting that the maintenance of RP4 Tc^r marker was by integration of pGMI106 into pSym. As a control, the same chase experiment was performed with an isogenic strain containing a Km^s derivative of RP4 carrying no homology with pSym. In this case the Gm^r clones were no longer resistant to tetracycline. The structure of the hypothetical RP4 pSym cointegrate is shown in Fig. 3. Further proof for its formation is given later. When Gm^r Tc^r clones were grown in the presence of tetracycline but without gentamycin pPH1J1 plasmid was gradually lost but some of the clones when analysed by gel electrophoresis procedure did not show the presence of any autonomous RP4-prime plasmid which suggested that excision of the RP4-prime was a slow process.

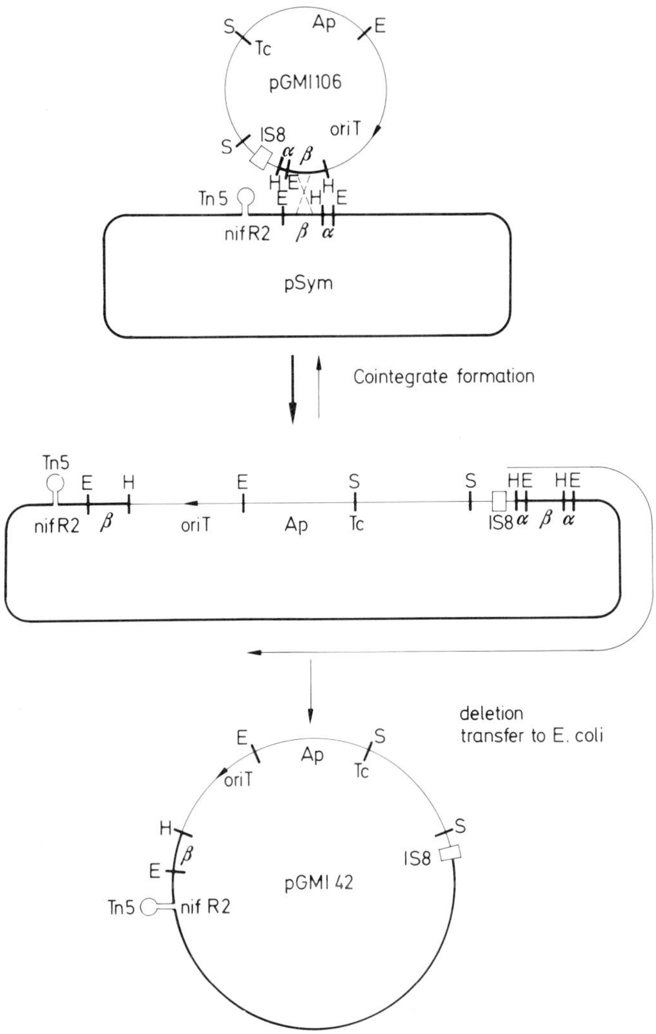

Fig. 3. Proposed model for the cointegration of pGMI106 with pSym and generation of pGMI42. Abbreviations for endonuclease restriction sites are the same as in Fig. 2

Selection of Deleted pSym-RP4 Cointegrates in *E. coli*

One Tc^r clone which did not show the presence of an autonomous pGMI106 was used in a conjugation experiment with a restriction deficient *rec*A strain of *E. coli* as recipient. Kanamycin resistant Tn*5* containing transconjugants were selected. When examined by gel electrophoresis those transconjugants carried various plasmids ranging in size from approximately 150 to 300 kb. One of those hybrid plasmids, pGMI42 which carries a 250 kb insert, was chosen for further characterization.

Physical Characterization of a Deleted RP4-pSym Cointegrate

If pGMI42 was generated by a deletion from a cointegrate having the structure described in Fig. 3, either the deletion was between two pSym sites, leaving intact the RP4 part of the cointegrate, or it started from a site located inside the RP4 part of cointegrate. In order to discriminate between those two possibilities we hybridized a *Sal*I *Hin*dIII digest of pGMI42 with an IS*8* probe. The *Sal*I *Hin*dIII fragment of 4,5 kb which hybridizes with IS*8* in pGMI106 (see Fig. 2) was replaced by a 6 kb fragment, which showed that the *Hin*dIII site of RP4 which was at the junction with the alpha fragment in pGMI106 had been eliminated (Julliot et al., in preparation). Furthermore, hybridization between an *Eco*RI *Hin*dIII digest of pGMI42 and an R3 probe showed that the beta fragment was present unaltered in pGMI42 and that the alpha fragment was missing. Therefore the deletion which generated pGMI42 had its starting point inside RP4. Both those hybridization results were fully compatible with our model of cointegrate formation between pSym and pGMI106.

From these experiments we can conclude that pGMI42 was generated by a deletion in a pGMI106 pSym cointegrate and that the cloned fragment is most probably a genuine pSym fragment. Further hybridization showed that the internal *Pst*I fragment of IS*8* (Depicker et al. 1980), represented in Fig. 2, is present in pGMI42. Therefore the starting point of the deletion is located in a 1 kb segment comprising one end of IS*8* and the *Hin*dIII site of RP4. IS*8* is the most likely causative agent of this deletion.

pGMI42 Contains Nodulation Genes

A pSym region controlling nodulation has been identified and localized 25 kb from the *nif* region. *R. meliloti* 1027 and *R. meliloti* 1126 are two nodulation deficient mutants affected in this pSym region (Long et al. 1982). pGMI42 was able to restore their nodulation ability. This is in full agreement with the proposed structure of pGMI42. pGMI42 was introduced into C58C1, a strain of *Agrobacterium tumefaciens* cured of its Ti plasmid. Root deformations giving rise in some instances to nodule-like structures could be observed when the transconjugants were used to inoculate *Medicago sativa.* This was an indication that pGMI42 carries genes required for the specific recognition of the host plant, for the first steps of the infectious process and for the induction of nodule morphogenesis (Truchet et al., in preparation).

Utilization of pGMI42 for the Introduction of Transposons into pSym

Insertion of Mu into pGMI42

Bacteriophage Mu had been shown to prevent maintenance of Inc-P1 plasmids in *Rhizobium* (Boucher et al. 1977) and this property has been used for the introduction of transposon Tn*5* into *Rhizobium* genome (Beringer et al. 1978). We constructed a pGMI42 derivative in which a Mu prophage was inserted in the *tet* gene of the RP4 component of pGMI42 as outlined in Fig. 4. This new derivative, pGMI104, was used as a suicide vector to introduce Tn7 transposon into the homologous part of pSym as illustrated in Fig. 5. Tn7 was mobilized by pGMI104 from an *E. coli* strain in which it had a chromosomal location, to an *E. coli* C 600 recipient strain. Then individual transconjugants containing pGMI104::Tn7 plasmids obtained independently were

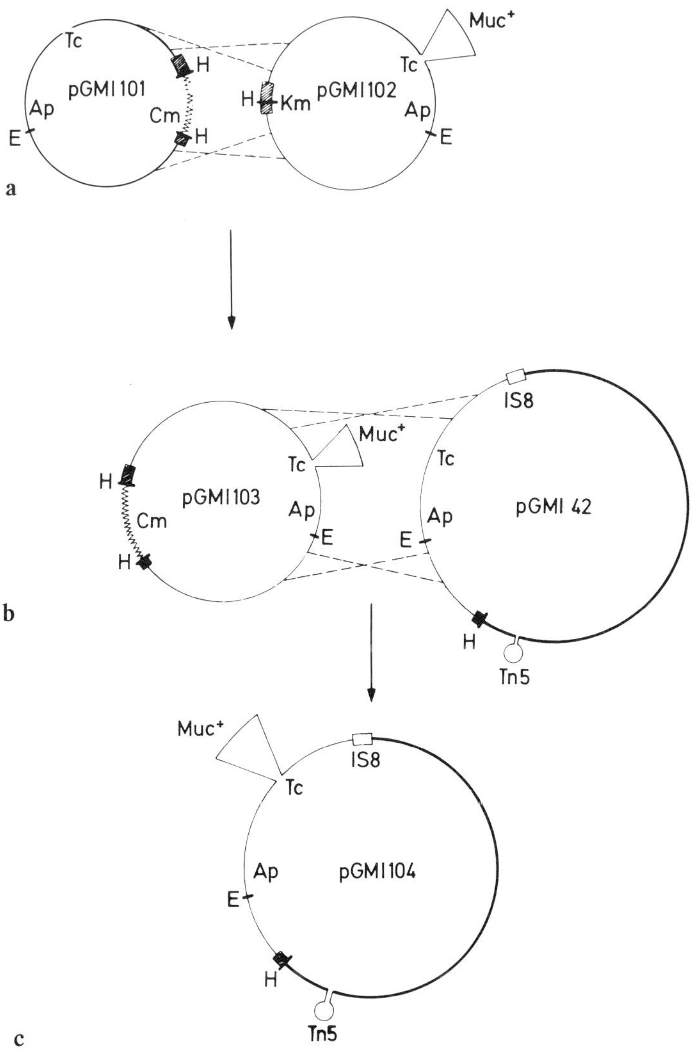

Fig. 4a–c. Construction of the suicide vector pGMI104. pGMI102 is a Tc^S derivative of RP4 obtained by Muc⁺ prophage insertion in the *tet* gene of RP4. pGMI101 is a Km^S Cm^r derivative of RP4 obtained by the insertion of pACYC184 plasmid (Chang and Cohen 1978) into the *Hin*dII site of RP4. **a** pGMI102 is made Km^S Cm^r; **b** and **c** bacteriophage Mu is inserted into the *tet* gene of pGMI42. Endonuclease restriction sites: *E* *Eco*RI; *H* *Hin*dIII. In all the crosses *E. coli* strains lysogen for a Mu *cts* 62 thermoinducible mutant were used: this made possible the selection of clones carrying the Muc⁺ phage inserted in the *tet* gene by growth at 42°C

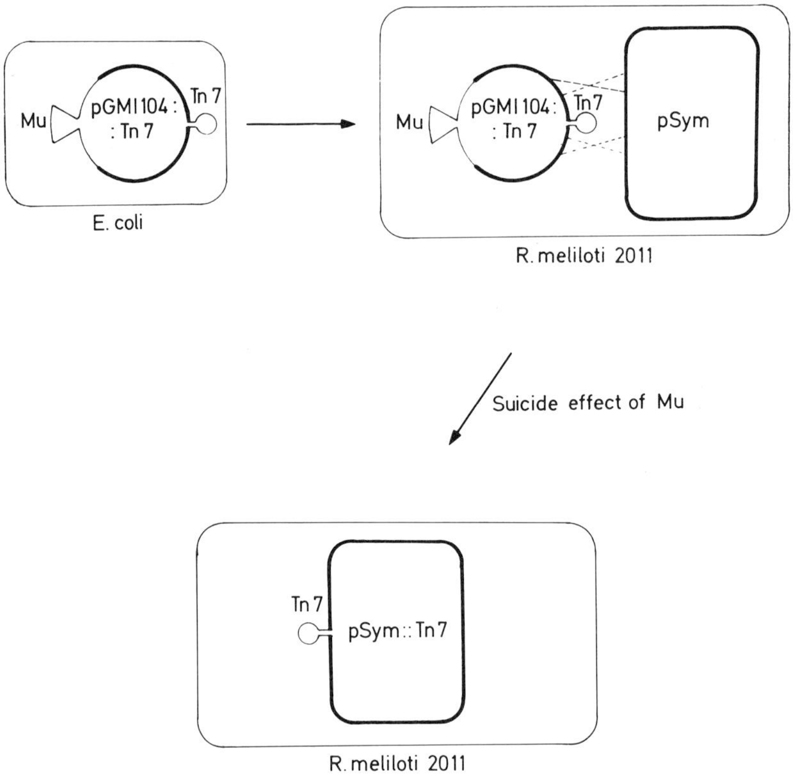

Fig. 5. Use of the suicide vector pGMI104 for the introduction of transposon Tn7 into the region of pSym homologous to the pGMI42 insert

crossed with *Rhizobium meliloti* RCR2011 and acquisition of the transposon was selected. Most pGMI104::Tn7 were able to transfer Tn7 at a high frequency. By contrast pGMI103::Tn7 (see Fig. 4) which did not carry homology with pSym transferred Tn7 at a frequency several orders of magnitude lower. This was a clear indication that Tn7 was acquired by gene replacement and not by random transposition. Most probably the rare *E. coli* (pGMI104::Tn7), about 10% of the clones, which did not transfer Tn7 into the *R. meliloti* recipient had the transposon in the RP4 part of the hybrid plasmid. It is interesting to note that transposon insertion in *R. leguminosarum* using RP4::Mu::Tn7 suicide vector could not be detected (Beringer et al. 1978).

Isolation of Symbiotically Defective Mutants

In a preliminary experiment where 683 clones having acquired the Tn7 transposon were tested for their symbiotic properties on *M. sativa* seedlings, 12 symbiotic mutants were detected: 3 Nod^- and 9 Fix^-.

The 9 Fix^- mutants fell in at least 4 different phenotypes according to the aspect of nodules and of the infected root system: (a) few yellowish nodules without any defined shape. Yellowish abnormal roots; (b) numerous white spherical small nodules; (c) few white nodules of normal shape; (d) few white pink nodules of normal shape. It is worth noting that plants infected by this latter category of mutants reduced acetylene at a low but significant level.

Conclusions

We have presented a strategy for the genetic analysis of the pSym megaplasmid which involved a combination of (1) in vitro and in vivo techniques for cloning of very large fragments of pSym into RP4 and (2) the use of the suicide effect of Mu for efficient regional mutagenesis of pSym.

Fragmentation of the Megaplasmid

One advantage of these methods is that the origin and the orientation of the fragment to be cloned can be chosen. Therefore by starting with the same origin on pSym, *bidirectional cloning* should be possible. Another is that, due to the fact that the deletion which generated the large RP4-prime from an RP4-pSym cointegrate started from inside the RP4 (see Fig. 3), the RP4 genome does not interrupt the cloned fragment. Therefore the part of the fragment distal to the origin of cloning selected can easily be identified and used as a new origin for further *sequential* cloning. Third, although the cloned fragment is about 10 times larger than the size of fragments cloned in cosmids, the resulting plasmid is still manageable by the usual techniques applied to the physical study of plasmids. Physical mapping of the 250 kb fragment of pSym cloned in pGMI42 is nearing completion (Huguet et al., in preparation).

Genetic Studies

Finally our method provides an efficient means for the mutagenesis of pSym and characterization of the mutations. The RP4-prime: :Mu cointegrate so constructed allows efficent introduction of Tn7 into the homolgous part of pSym. Due to the fact that the physical map of the cloned fragment is relatively simple (50 *Hin*dIII fragments in the case of pGMI42) the location of Tn7 in the various RP4-prime: :Mu should be easy to determine. More generally the cloning of large fragments of pSym in RP4 should facilitate genetic manipulation of pSym in the well defined *E. coli* genetic environment.

Acknowledgements. This work was supported by INSERM grant no. 79104 (ATP Recombinaisons génétiques) and by the EMC-ELF-INRA-Institut Pasteur contract: Fixation biologique de l'azote. We thank F.M. Ausubel for the gift of various strains and communication of results prior to publication, A. Pühler for the gift of RP4 Tc: :Mu (= pGMI102) and Cl. Zischek for the gift of pGMI1001. We thank J. Dénarié for constant interest and stimulating discussions and W. Broughton for critical reading of the manuscript.

References

Barth PT, Datta N, Hedges RW, Grinter NJ (1976) Transposition of a deoxyribonucleic acid sequence encoding trimethorpim resistance and streptomycin resistance from R433 to other replicons. J Bacteriol 125:800–810

Beringer JE, Beynon JL, Buchanan-Wollaston AV, Johnston AWB (1978) Transfer of the drug-resistance transposon Tn5 to *Rhizobium*. Nature 276:633–634

Boucher C, Bergeron B, Barate de Bertalmio M, Dénarié J (1977) Introduction of bacteriophage Mu into *Pseudomonas solanecearum* and *Rhizobium meliloti* using the R factor RP4. J Gen Microbiol 98:253–263

Buchanan-Wollaston AV, Beringer JE, Brewin NJ, Hirsch PR, Johnston AWB (1980) Isolation of symbiotically defective mutants in *Rhizobium leguminosarum* by insertion of the transposon Tn5 into a transmissible plasmid. Mol Gen Genet 178:185–190

Chang ACY, Cohen SN (1978) Construction and characterization of amplifiable multicopy DNA cloning vehicles derived from P15A cryptic miniplasmid. J Bacteriol 134:1141–1156

Datta N, Hedges RW, Shaw EJ, Sykes RB, Richmond MH (1971) Properties of an R factor from *Pseudomonas aeruginosa*. J Bacteriol 108:1244–1249

Depicker A, Block M De, Inzé D, Montagu M Van, Schell J (1980) IS-like element IS8 in RP4 plasmid and its involvement in cointegration. Gene 10:329–338

Julliot JS, Boistard P (1979) Use of RP4-prime plasmids constructed in vitro to promote a polarized transfer of the chromosome in *Escherichia coli* and *Rhizobium meliloti*. Mol Gen Genet 173:289–298

Long SR, Buikema WJ, Ausubel FM (1982) Cloning of *Rhizobium meliloti* genes by direct complementation of Nod$^-$ mutants. Nature 298:485–488

Rosenberg C, Boistard P, Dénarié J, Casse-Delbart F (1981) Genes controlling early and late functions in symbiosis are located on a megaplsmid in *Rhizobium meliloti*. Mol Gen Genet 184: 326–333

Ruvkun GB, Sundaresan V, Ausubel FM (1982) Directed transposon Tn5 mutagenesis and complementation analysis of *Rhizobium meliloti* symbiotic nitrogen fixation genes. Cell 29: 551–559

Analysis of Symbiotic Nitrogen Fixation Genes Carried by the *Rhizobium meliloti* Mega-Plasmid

A. KONDOROSI[1,2], E. KONDOROSI[2], Z. BANFALVI[1], W.J. BROUGHTON[2], C.E. PANKHURST[2], G.S. RANDHAWA[1], C.-H. WONG[2], and J. SCHELL[2]

Introduction

The bacterial species *Rhizobium meliloti* reduces molecular nitrogen in symbiotic association with alfalfa *(Medicago sativa).* Development of this N_2-fixing symbiosis is a multistage process, probably coded by a large number of genes. We have been studying those symbiotic genes which reside in the bacterium.

Recently, symbiotically defective (Nod⁻ and Fix⁻) mutants of the wild type *R. meliloti* (AK631) were isolated after random mutagenesis with nitrosoguanidine and with Tn5. Genetic mapping of these mutations revealed that some *fix* genes are located on the bacterial chromosome. The majority of the *fix* genes and *nod* genes, however, were extrachromosomal (Forrai et al. 1982).

We reported earlier (Banfalvi et al. 1981) that *R. meliloti* 41 harbours at least two large indigenous plasmids,. pRme41*a* and pRme41*b*. Several lines of evidence have suggested that pRme41*b* (the mega-plasmid) codes for some *nod* and nitrogen fixation *(nif)* genes (Banfalvi et al. 1981, Rosenberg et al. 1981). For instance, Nod⁻ mutants, isolated after growth of AK631 at elevated temperatures, were shown to have deletions in this plasmid. The presence of the structural genes for enzyme nitrogenase on pRme41*b* was demonstrated by hybridization of a cloned *R. meliloti* DNA fragment carrying part of the *nif* structural genes (Banfalvi et al. 1981) to pRme41*b*. Moreover, it was shown that in those Tn5-induced symbiotic mutants where Tn5 did not map onto the *R. meliloti* linkage map (Kondorosi et al. 1977, 1980, Vincze et al. 1982), a Tn5 DNA probe hybridized specifically to pRme41*b* (Forrai et al. 1982). The *nif* structural genes were analysed in more detail. The nucleotide sequence of *nifH* and part of *nifD* have been determined from which the amino acid sequences of the nitrogenase reductase (M.W. 32,740) were deduced (Török and Kondorosi 1981). A protein of about the same size was detected when the *nifH* gene was expressed from a strong *E. coli* promotor in *E. coli* minicells (Svab and Kondorosi, unpublished).

In this study we present evidence that pRme41*b* carries all genes required for nodule formation and the *nod* genes are located close to the *nif* and other *fix* genes. These conclusions were supported by four types of experiments: (1) mobilization of pRme41*b* to other bacterial species and studies on the symbiotic properties of the

1 Biological Research Centre, Hung. Acad. Sci., P.O.B. 521, 6701 Szeged, Hungary
2 Max-Planck-Institut für Züchtungsforschung, 5000 Köln 30, FRG

Molecular Genetics of the Bacteria-Plant Interaction
(ed. by A. Pühler)

transconjugants, (2) isolation and analysis of R-prime plasmids carrying the *nod-nif* region, (3) cloning and physical mapping of the *nod-nif* DNA segment, (4) studies on the transcription of this region.

Mobilization of the Mega-Plasmid Carrying Nodulation and Nitrogen Fixation Genes into Other Rhizobia and *Agrobacterium*

We observed that pRme41*b* exhibited a very low level of self-transmissibility. To increase its transfer frequency we made it susceptible to mobilization with the P-1 type plasmid pJB3J1 (obtained from J. Beringer) by inserting the mobilization *(mob)* region of RP4 into it.

First the *mob* region together with the kanamycin resistance marker (from a recombinant clone p1011, kindly provided by R. Simon) was inserted in vitro into a fragment of pRme41*b* cloned into pBR322. The recombinant plasmids so formed (pAK11 and pAK12) were then mobilized into *R. meliloti.* Since these recombinant plasmids were unable to replicate in *R. meliloti,* selection for kanamycin resistant derivatives allowed the isolation of pRme41*b*::pAK11 and pRme41*b*::pAK12 cointegrates. It was shown that in the majority of the recombinants pAK11 or pAK12 were integrated into the homologous fragment of pRme41*b*.

The pRme41*b* cointegrates were transferred into *nod-nif* deletion mutants of *R. meliloti* 41 where it was shown that both Nod^+ and Fix^+ phenotypes could be restored. The cointegrates were also transferred into two other *Rhizobium* strains (*Rhizobium* sp. PN4003, Pankhurst 1977, and MPIK3030 = NGR234, Trinick 1980) (Fig. 1) and into *Agrobacterium tumefaciens* (Fig. 2). The *Rhizobium* strains and *A. tumefaciens* carrying pRme41*b* formed nodules of variable size on *Medicago sativa* roots, indicating that at least the early steps of nodulation of *M. sativa* are coded by pRme41*b* and are expressed in these bacteria. These nodules developed only to a certain extent (Fig. 3). Bacteria were found amongst root-hairs and within the intercellular spaces of the outer cells. There was some indication of infection thread initiation. Bacteria were not found within the developing nodule cells, but abundant proplastids were evident. Extensive starch deposition eventually occurred in these cells. Although the transconjugants contained the *nif* region, as demonstrated by hybridization with a cosmid clone carrying a 49 kb segment of pRme41*b* containing the *nif* gene cluster (pEK5022, Kondorosi et al. 1982a, and Fig. 4), the nodules did not fix N_2 (Kondorosi et al. 1982b).

Transfer of pRme41*b* allowed demonstration of other genes on the plasmid. In this way, it was shown that the octopine utilization ability of *R. meliloti* 41 is coded by pRme41*b*.

Fig. 3a–d. Electron micrographs of nodules induced by *A. tumefaciens* carrying pRme41*b*. **a** Electron micrograph of outermost cells of ineffective nodule showing presence of bacteria in intercellular space. Magnification × 6000. **b** Electron micrograph of a cross section of a root-hair cell showing the possible beginning of an infection thread. Magnification × 2400. **c** Electron micrograph of developing cells from ineffective nodule showing numerous proplastids containing starch grains. Magnification × 6000. **d** Electron micrograph of older cells from ineffective nodule showing plastids with large starch grains and the presence of abundant mitochondria. Magnification × 6000 ▶

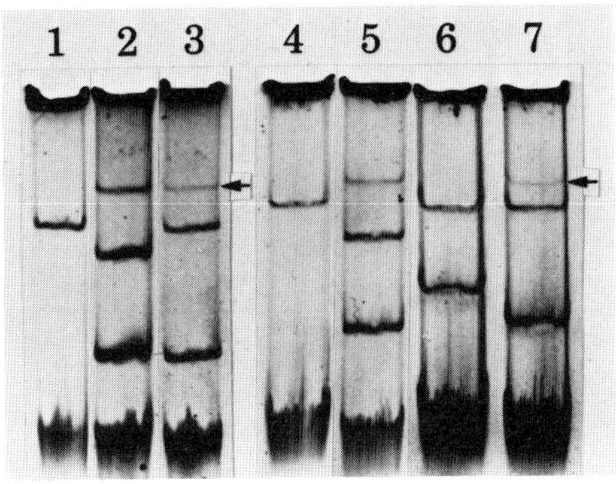

Fig. 1

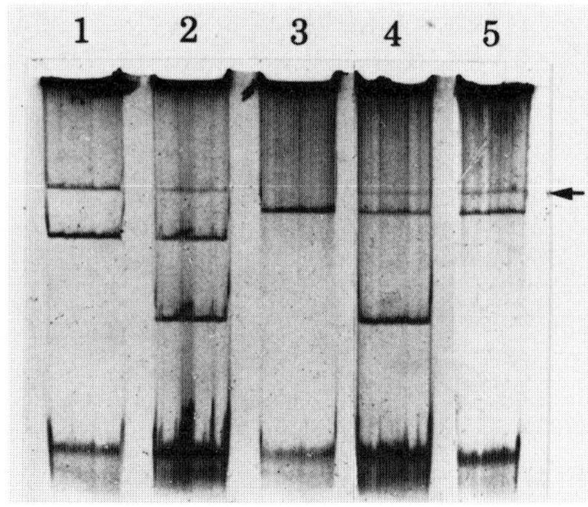

Fig. 2

Fig. 1. Detection of plasmids in *R. meliloti, Rhizobium* sp., and in their transconjugants. *Lane 1* *Rhizobium* sp. strain PN4003; *lanes 2* and *5* AK631 containing pRme41*b*::pAK11 and pJB3JI; *lane 3* PN4003 containing pJB3JI and pRme41*b*::pAK11 *(arrow)*; *lane 4* *Rhizobium* sp. strain MPIK3030; *lane 6* MPIK3030 containing pJB3JI::pAK11; *lane 7* MPIK3030 containing pJB3JI and pRme41*b*::pAK11 *(arrow)*

Fig. 2. Detection of plasmids in *R. meliloti, A. tumefaciens* and their transconjugants. Preparation of crude lysates and gel electrophoresis was according to the modified procedure (Banfalvi et al. 1981) of Eckhardt (1978). *Lane 1* AK631 *(upper band)* pRme41*b*; *lower band* pRme41*a*; *lane 2* AK631 (pRme41*b*::pAK11) containing pJB3JI; *lane 3 A. tumefaciens* GV3101; *lane 4* GV3101 containing pJB3JI and pRme41*b*::pAK11; *lane 5* GV3101 containing only pRme41*b* ::pAK11. *Arrow* indicates the large plasmid band which in *lanes 1* and *2* is pRme41*b* and in *lanes 4* and *5* pRme41*b*::pAK11

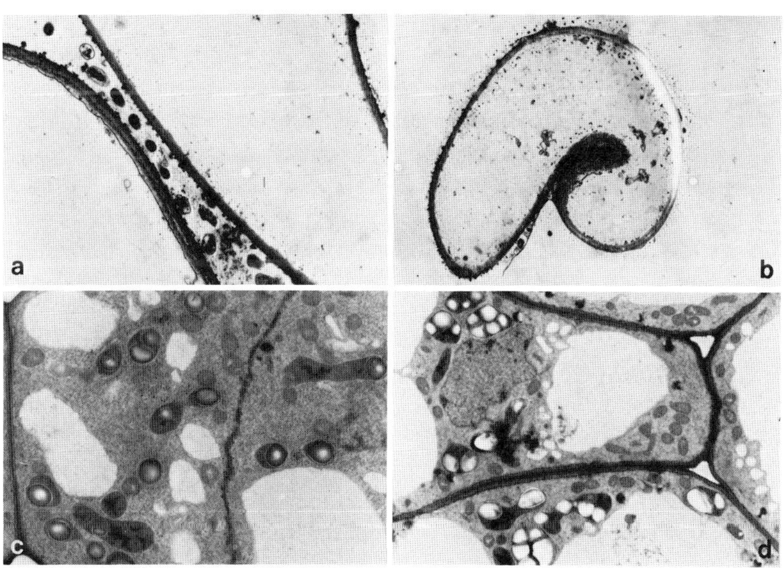

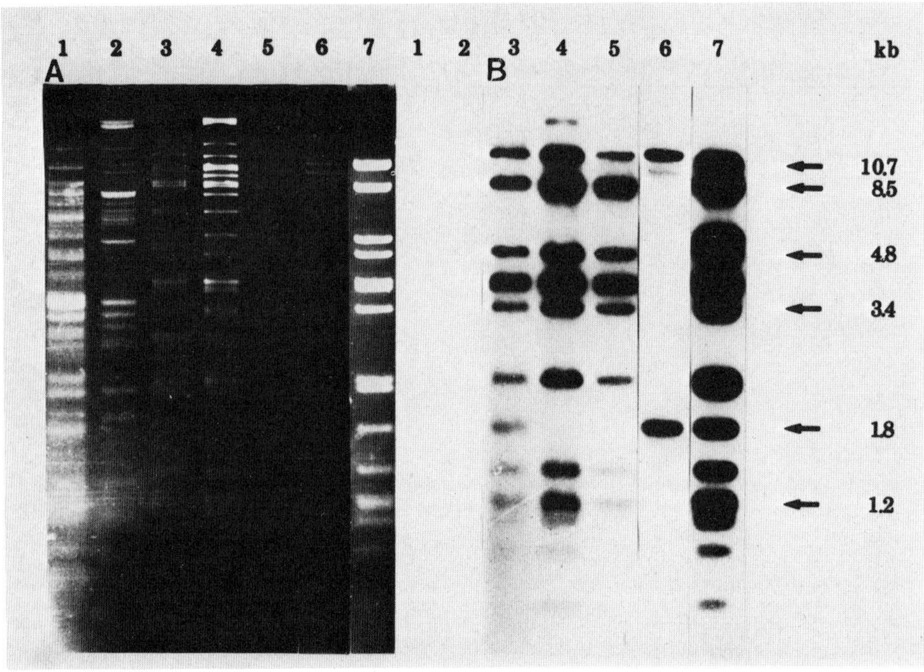

Fig. 4A, B. Hybridization of the *R. meliloti nif* region to purified R-primes. *Eco*RI restriction pattern (**A**) and hybridization pattern (**B**) of different DNA's, using pEK5022 as hybridization probe. *Lane 1* pAK13; *2* pGR4; *3* pGR2; *4* pGR1; *5* pZB1; *6* pGR3; *7* pEK5022 DNA

Construction of R-Prime Plasmids Carrying Symbiotic Genes of *R. meliloti*

Previous studies on symbiotic genes of *R. meliloti* indicated that some of the genes are clustered (Banfalvi et al. 1981, Rosenberg et al. 1981, Corbin et al. 1982, Ruvkun et al. 1982). We thought that transmissible R-prime plasmids carrying sets of symbiotic genes might be valuable for further studies. In vivo-constructed R-primes carrying *R. meliloti* chromosomal regions have been reported (Johnston et al. 1978, Kiss et al. 1980, Kondorosi et al. 1980). In these studies the ability of R68.45 to interact with the chromosome was exploited. Symbiotic genes, however, have not been identified on these R-primes.

In our experiments pJB3JI-primes were derived from strains containing Tn5-induced symbiotic mutations. Since Tn5 is unable to transpose into new sites in *R. meliloti* (Meade et al. 1982, Forrai et al. 1982), transfer of kanamycin resistance into *E. coli* could be followed, providing a simple and efficient way to isolate R-primes.

Such R-primes were generated from both the chromosome and from the megaplasmid at a frequency of 10^{-7}/recipient. The size of the insert was different in various isolates (up to 200–250 kb). It was demonstrated that the regions inserted were identical with those DNA segments in which Tn5 was located in the respective parental strain.

Several R-primes from the mega-plasmid were shown to hybridize with the cloned *nif* region (Fig. 4). When these R-primes were transferred into different Nod$^-$ deletion mutants, a Nod$^+$ phenotype was restored in all cases. The plasmids were also introduced into *A. tumefaciens* and the transconjugants formed small nodules on *M. sativa.* The smallest R-prime of this type was 128 kb. From the physical analysis of the inserted region (see below) we concluded that nodulation and nitrogen fixation genes of the mega-plasmid are clustered on less than a 70 kb DNA segment.

Cloning and Physical Mapping of the *Nod-Nif* Region

A *R. meliloti* 41 gene library, consisting of 860 clones has been constructed by using the cosmid vehicle pJB8 (Ish-Horowitz and Burke 1981). To map pRme41*b*, a recombinant cosmid, carrying the *nif* structural genes (pEK5022) has been isolated by colony hybridization (Kondorosi et al. 1982a) using a 4.8 kb *nif* fragment (Banfalvi et al. 1981) as a probe. Then recombinant cosmids, overlapping pEK5022, were identified by using pEK5022 and the *nod-nif* R-prime pGR1 as hybridization probes.

By determining the cleavage sites of four restriction endonucleases (*Eco*RI, *Kpn*I, *Bst*EII and *Xba*I) on these cosmid DNAs, the physical map of about a 140 kb segment of pRme41*b*, including the *nod-nif* region has been established (Fig. 5).

These cosmid clones were useful to determine the extent of deletions in some Nod$^-$ deletion mutants isolated earlier (Banfalvi et al. 1981). All deletions studied started from the region covered by pEK3089 but their lengths were different. For instance, in two derivatives the deletions ended at the "left" end of pEK5022, in three others at the "right" distal *Eco*RI fragment of pEK3048. In several mutants the endpoint was outside the region mapped. We have reported (Banfalvi et al. 1981) that several Nod$^-$ mutations of *R. meliloti* are suppressed upon the introduction of pJB5JI, a *sym* plasmid of *R. leguminosarum*. In some Nod$^-$ deletion mutants the ability to nodulate alfalfa was restored by the *sym* plasmid but the nodules did not fix nitrogen. In an other deletion derivative, the transconjugants remained Nod$^-$ but the *nif* region was present. These complementation data, together with the deletion maps allowed us to deduce the *nod-fix-nod-nif* gene (set) order. Site-directed mutagenesis studies on two other *R. meliloti* strains (Ruvkun and Ausubel 1981, Ruvkun et al. 1982, Corbin et al. 1982) indicate that there are numerous *fix* genes in the vicinity of the *nif* structural genes. Furthermore, two Nod$^-$ mutations were mapped about 30 kb away from the *nif* structural genes (Long et al. 1982). The comparison of *Eco*RI sites reported for this region with the map of pRme41*b* indicates strong homology between the two strains, although some regions seem to be quiet different. From the data presented above, and from the fact that pGR3 which carries the *nod* region extending to *nif*H (Fig. 3) still codes for the host specificity of nodulation, we suggest that the second *nod* cluster might contain the host specificity genes *(hsp).*

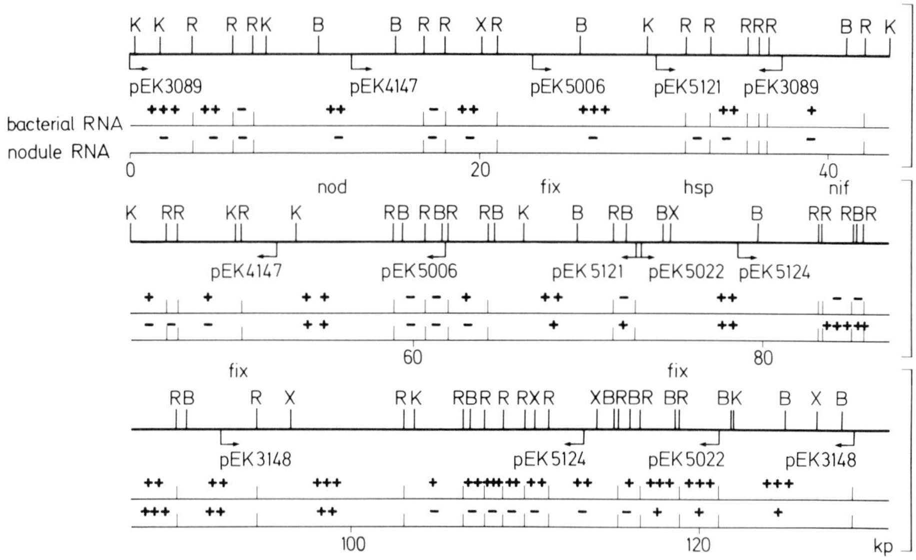

Fig. 5. Physical map of the *nod-nif* region of *R. meliloti* 41 and transcription of the *Eco*RI fragments in vegetatively grown bacteria and in the nodules. The map was constructed by determining the *Eco*RI *(R)*, *Kpn*I *(K)*, *Bst*EII *(B)* and *Xba*I *(X)* cleavage sites. The extent of inserts in the different cosmids is indicated by *arrows*. To determine transcribed regions, RNA from vegetatively grown bacteria and from nitrogen-fixing nodules were isolated, end-labelled with ^{32}P and hybridized to the *Eco*RI-digested recombinant DNAs. + + +, very strong hybridization, –, hybridization was not detected

Expression of Symbiotic Genes

We observed that only a relatively small region of the mega-plasmid is transcribed actively in nitrogen-fixing nodules. In these experiments RNA was isolated from nodules and from vegetatively grown bacteria, labelled with ^{32}P and hybridized to various R-primes or cosmid clones.

Figure 5 shows that when RNA from vegetatively grown bacteria was used, numerous hybridizing fragments were observed, even in the case of the *nif* cosmid clone. The *nif* structural genes, however, did not express, as expected. When RNA from the nodules was hybridized, the *nif* structural genes gave a very strong hybridizing signal (Fig. 6), indicating a high level of transcription of these genes. Similar observations were reported by Corbin et al. (1982). Moreover, about a 20 kb region in the vicinity of the *nif* structural genes hybridized. This also suggests that in this DNA region several *fix* (probably *nif*) genes are clustered.

The recombinant cosmids presented in Fig. 4 were also tested for RNA-DNA hybridization. Figure 5 shows that nearly each DNA fragment hybridized with RNA from vegetative bacteria but only the *nif* and two other regions (probably carrying *fix* genes) hybridized with nodule RNA.

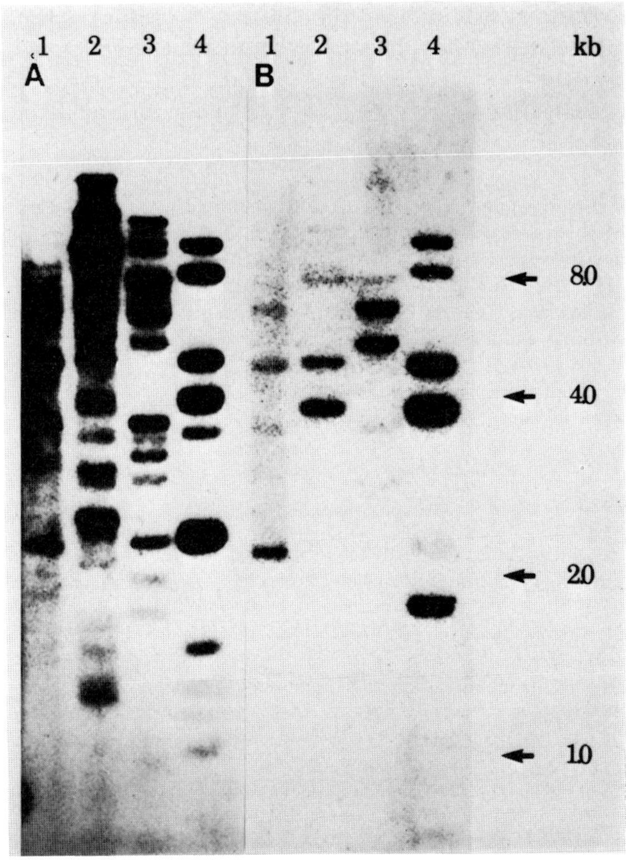

Fig.6A, B. Hybridization of ^{32}P-labelled RNA from vegetatively grown bacteria (**A**) and from nodules (**B**) to *Eco*RI-digested *R. meliloti* DNA clones. *Lanes 1* and *2* R-primes from the mega-plasmid; *3* R-prime from the chromosome; *4* pEK5022 (Fig. 5)

R-primes carrying other mega-plasmid regions showed little hybridization with nodule RNA. In Fig. 6/lane 3 a chromosomal R-prime contains two *Eco*RI fragments which hybridize to nodule RNA. Tn5 mutagenesis indicated that at least one of these fragments carries symbiotic genes. This gene maps between the *pur-1* and *cys-46* chromosomal markers (Forrai et al. 1982). These studies suggest that hybridization of labelled nodule RNA to *R. meliloti* gene library might be useful to identify genes expressed in the nodules.

Conclusions

Genes coding for early functions of nodulation, the *nif* structural genes and some other *fix* (probably *nif*) genes are clustered on about a 70 kb region of the mega-plasmid. Transfer of this region into other *Rhizobium* species or into *Agrobacterium* resulted in Nod^+ Fix^- nodules.

Other *fix* genes were localized at four different chromosomal regions. We assume that some of these genes are different or not present in other rhizobia or in *Agrobacterium*. This implies that some chromosomal *fix* genes would exhibit plant host specificity, which should be considered in attempts to transfer symbiotic nitrogen-fixing ability from one *Rhizobium* species to other bacterial species.

Acknowledgements. The work performed in Szeged was partially supported by O.M.F.B. grant No. 13017/FPI. One of us (G.S.R.) was recipient of a fellowship from the International Training Course of the Hungarian Academy of Sciences. Much of this work was performed when four of us (A.K., E.K., C.E.P. and C.H.W.) were on leave at the Max-Planck-Institut für Züchtungsforschung, Köln, as recipients of fellowships from EMBO, from the Max-Planck-Gesellschaft, from the Alexander-von-Humboldt-Stiftung, and from University Sains Malaysia respectively.

References

Banfalvi Z, Sakanyan V, Koncz C, Kiss A, Dusha I, Kondorosi A (1981) Location of nodulation and nitrogen fixation genes on a high molecular weight plasmid of *Rhizobium meliloti*. Mol Gen Genet 184:318–325

Corbin D, Ditta G, Helinski DR (1982) Clustering of nitrogen fixation *(nif)* genes in *Rhizobium meliloti*. J Bacteriol 149:221–228

Eckhardt T (1978) A rapid method for identification of plasmid deoxyribonucleic acid in bacteria. Plasmid 11:584–588

Forrai T, Vincze E, Banfalvi Z, Kiss GB, Randhawa GS, Kondorosi A (1982) Location of symbiotic mutations in *Rhizobium meliloti*. J Bacteriol 153:635–643

Ish-Horowitz D, Burke JF (1981) Rapid and efficient cosmid cloning. Nucleic Acids Res 9: 2989–2998

Johnston AWB, Setchell SM, Beringer JE (1978) Interspecific crosses between *Rhizobium leguminosarum* and *R. meliloti:* formation of haploid recombinants and of R. primes. J Gen Microbiol 104:209–218

Kiss GB, Dobo K, Dusha I, Breznovits A, Orosz L, Vincze E, Kondorosi A (1980) Isolation and characterization of an R-prime plasmid in *Rhizobium meliloti*. J Bacteriol 141:121–128

Kondorosi A, Kiss BG, Forrai T, Vincze E, Banfalvi Z (1977) Circular linkage map of *Rhizobium meliloti* chromosome. Nature 264:525–527

Kondorosi A, Vincze E, Johnston AWB, Beringer JE (1980) A comparison of three *Rhizobium* linkage maps. Mol Gen Genet 178:403–408

Kondorosi A, Banfalvi Z, Broughton WJ, Forrai T, Kiss GB, Kondorosi E, Pankhurst C, Randhawa G, Svab Z, Vincze E (1982a) Localization and molecular genetic analysis of symbiotic nitrogen fixation genes in *Rhizobium meliloti*. In: Ciferri O, Dure L (eds) Structure and function of plant genomes. (in press)

Kondorosi A, Kondorosi E, Pankhurst CE, Broughton WJ, Banfalvi Z (1982b) Mobilization of a *Rhizobium meliloti* megaplasmid carrying nodulation and nitrogen fixation genes into other rhizobia and *Agrobacterium*. Mol Gen Genet 88:433–439

Long SR, Buikema WJ, Ausubel FM (1982) Cloning of *Rhizobium meliloti* genes by direct complementation of Nod⁻ mutants. Nature 298:485–488

Meade HM, Long SR, Ruvkun GB, Brown SE, Ausubel FM (1982) Physical and genetic characterization of symbiotic and auxotrophic mutants of *Rhizobium meliloti* induced by transposon Tn5 mutagenesis. J Bacteriol 149:114–122

Pankhurst CE (1977) Symbiotic effectiveness of antibiotic-resistant mutants of fast- and slow-growing strains of *Rhizobium*-nodulating *Lotus* species. Can J Microbiol 23:1026–1033

Rosenberg C, Boistard P, Dénarié J, Casse-Delbart FL (1981) Genes controlling early and late functions in symbiosis are located on a mega-plasmid in *Rhizobium meliloti*. Mol Gen Genet 184:326–333

Ruvkun GB, Ausubel FM (1981) A general method for site-directed mutagenesis in prokaryotes. Nature 289:85–88

Ruvkun GB, Sundaresan V, Ausubel FM (1982) Directed transposon Tn5 mutagenesis and complementation analysis of *Rhizobium meliloti* symbiotic nitrogen fixation genes. Cell 29: 551–559

Török I, Kondorosi A (1981) Sequence of the *R. meliloti* nitrogenase reductase/*nifH*/gene. Nucleic Acids Res 9:5711–5723

Trinick MJ (1980) Relationships amongst the fast-growing rhizobia of *Lablab purpureus, Leucaena leucocephala, Mimosa* spp.,*Acacia farnesiana* and *Sesbania grandiflora* and their affinities with other rhizobia groups. J Appl Bacteriol 49:39–53

Vincze E, Koncz C, Kondorosi A (1982) Construction in vitro of R-prime plasmids and their use for transfer of chromosomal genes and plasmids of *Rhizobium meliloti*. Acta Biol Acad Sci Hung 32:195–204

Characterization of *Rhizobium meliloti* Genetic Loci Essential for Symbiotic Nitrogen Fixation

W.W. SZETO [1], J.L. ZIMMERMAN [2], and F.M. AUSUBEL [1]

Rhizobium meliloti fixes nitrogen (N_2) in symbiosis with *Medicago sativa* (alfalfa). The developmental process that culminates in symbiotic N_2 fixation is controlled by specific genes found in both the invading bacteria and the host plant (Vincent 1980). A major area of research in our laboratory has been the identification and characterization of *R. meliloti* symbiotic *(sym)* genes, including those *sym* genes involved directly in nitrogen fixation (*nif* genes).

The manipulation of *Rhizobium* symbiotic genes is difficult because these genes are not expressed by the bacteria in its free-living state. Furthermore, very few classical genetic techniques are available for *Rhizobium.* To overcome these problems, experimental strategies which combine recombinant DNA methods with transposon mutagenesis were devised for the study of *Rhizobium sym* genes (Ruvkun and Ausubel 1980, 1981). Briefly, a cloned *R. meliloti* DNA fragment containing part of the nitrogenase structural gene operon *(nifHDK)* (Ruvkun and Ausubel 1980) was used to probe a cosmid clone bank of total *R. meliloti* DNA to screen for recombinant plasmids that carry overlapping DNA sequences (Buikema et al. 1983). Several recombinant plasmids containing DNA contiguous to and including the nitrogenase gene cluster *(nifHDK)* were obtained. Selected recombinant plasmids were then mutagenized with transposon Tn5 in *E. coli.* Each DNA fragment carrying a Tn5 insertion was re-introduced into *R. meliloti* by homologous recombination (Ruvkun and Ausubel 1981) and the symbiotic phenotypes of the resulting *Rhizobium* strains [i.e., whether they nodulated alfalfa (Nod^+ or Nod^-) and/or whether the induced nodules fixed N_2 (Fix^+ or Fix^-)] were analyzed by inoculation of each strain onto alfalfa plants.

Using this approach, several loci essential for symbiotic nitrogen fixation, in addition to the nitrogenase structural genes, have been identified in the DNA region of approximately 90 kilobases, surrounding and including the nitrogenase gene cluster (Buikema et al. 1983, Ruvkun et al. 1982). The results of studies investigating the functions of some of these loci during symbiosis will be described below. The most interesting finding is the identification of a potential regulatory locus which appears to be required for the expression of the *nifHDK* nitrogenase operon.

A series of Nod^+ Fix^- *R. meliloti* strains carrying selected Tn5 insertions in different regions (generated by the site-directed mutagenesis method described above)

1 Current Address: Department of Molecular Biology, Massachusetts General Hospital, Boston, Massachusetts 02114, USA

2 Current Address: United States Department of Agriculture, Agricultural Research Service, Beltsville, Maryland 20705, USA

Molecular Genetics of the Bacteria-Plant Interaction
(ed. by A. Pühler)

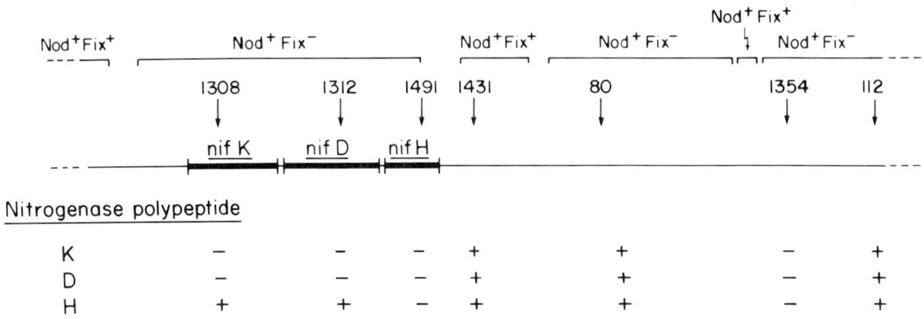

Fig. 1. The symbiotic phenotype of various Tn5 induced *R. meliloti* symbiotic mutants. ↓ indicates the position of Tn5 insertion; *Nod*$^+$ is the ability to induce nodules; *Fix*$^+$ or *Fix*$^-$ indicates the ability or inability of the nodules to reduce acetylene

was used to induce ineffective (Fix$^-$) nodules in alfalfa. The positions of Tn5 insertions in these strains relative to the nitrogenase operon are shown in Fig. 1. Both plant-derived and bacteria-derived (bacteroid) proteins were purified from these Fix$^-$ nodules and compared with those obtained from nodules induced by wild-type *Rhizobium.* In all of our experiments, a *Rhizobium* strain carrying a Tn5 insertion in a DNA region not essential for symbiosis [strain 1431 – (Nod$^+$ Fix$^+$)] was used as a positive control. This was to ensure that any defect that may be associated with the mutants is a true symbiotic lesion and not a phenotype attributable to Tn5 insertions per se. See the legend to Fig. 2 for experimental details.

Analysis of the plant-derived proteins in mutant nodules showed that the Fix$^-$ nodules induced by all of the *Rhizobium* mutants examined thus far synthesized significantly less leghemoglobin [the major plant-specific nodule protein (Dilworth and Appleby 1977)] than nodules induced by wild-type *Rhizobium* (data not shown). Experiments are now in progress to determine whether all of the isoelectric variants of leghemoglobin (Jing et al. 1982) are present in the Fix$^-$ nodules. In addition, the transcriptional properties of leghemoglobin in the ineffective nodules are being examined. It is hoped that such studies will elucidate whether the decrease in the amount of leghemoglobin in the Fix$^-$ nodules is a direct result of a decrease in the rate of transcription of leghemoglobin structural genes.

Analyses of the bacteroid-derived proteins obtained from the various Fix$^-$ nodules demonstrated that *R. meliloti* strains carrying Tn5 insertions in various regions of the nitrogenase structural gene *(nif)* operon lacked some or all of three nitrogenase polypeptides; the particular polypeptide missing being dependent on the position of the Tn5 insertion in the operon (Fig. 1 and 2). Bacteroids produced by a strain carrying a Tn5 insertion in the *nifH* gene (1491) did not synthesize detectable amounts of *nifH, nifD* nor *nifK* polypeptides (Fig. 2). The bacteroids of a strain carrying a Tn5 insertion in *nifD* (1312) contained *nifH* but not *nifD* nor *nifK* polypeptides (Fig. 2). Because Tn5 insertions are polar (Berg et al. 1980), these data support previous results which indicate that the *nifHDK* cluster is transcribed as a single unit and that transcription starts at *nifH* (Ruvkun et al. 1982, Sundaresan and Ausubel 1981). Surprisingly, strains carrying Tn5 insertions in *nifK* (represented by

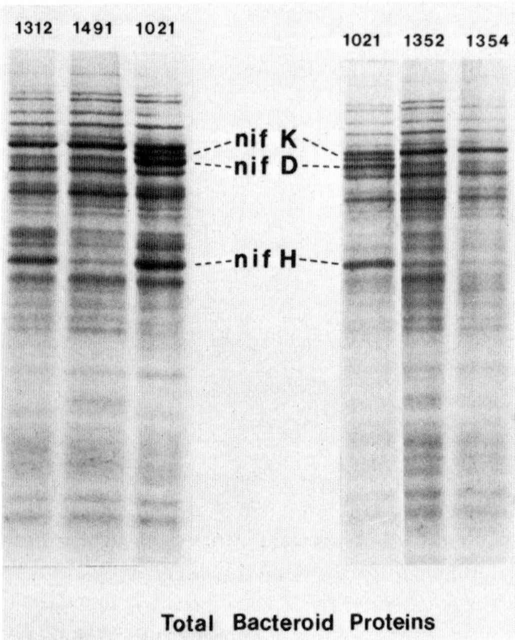

Fig. 2. SDS-polyacrylamide gel profiles of bacteroid proteins purified from nodules induced by different *Rhizobium* strains. Nodules were picked and frozen in vials of liquid nitrogen. The frozen nodules were ground into a fine powder and then homogenized in a buffer containing 50 mM Tris pH 7.5, 20 mM KDl, 10 mM $MgCl_2$, 0.5 M mannitol, 0.1% PVP and 1 mM PMSF (phenylmethyl sulfonyl fluoride). The suspension was filtered through 2 layers of miracloth and bacteroids were collected from the filtrate by spinning at 3000 rpm for 10 min at 4°C (in the Sorvall HB4 rotor). The crude bacteroid pellet thus obtained was further purified by centrifugation at 3000 rpm for 10 min through a 5%–20% sucrose gradient (containing 0.025 M Tris pH 8.0 and 0.025 M beta-mercaptoethanol) [in the HB4 rotor (4°C)]. Fractions were collected from the gradient and assayed for the presence of bacteroids by microscopy. Bacteroid-containing fractions were pooled and centrifuged at 8000 rpm for 10 min to collect the bacteroids. Bacteroids were lysed in a buffer containing 25 mM Tris pH 7.5, 1 mM DTT, 20 mM $Na_2S_2O_4$, 1 mM PMSF (phenylmethyl sulfonyl fluoride) and the solubilized proteins were analyzed on SDS-polyacrylamide gels

1308) differentiated into bacteroids which accumulated wild-type levels of *nifH* but accumulated neither *nifD* nor *nifK* polypeptides (data not shown). [The presence of *nifD* polypeptides in 1308 is expected because *nifD* is transcribed before *nifK;* therefore, Tn5 insertion in *nifK* should not have a polar effect on *nifD.*] Although we have no definitive explanation for this result, there are several possible explanations. *nifD* and *nifK* code for subunits of the iron-molybdenum (FeMo) cofactor of the nitrogenase complex (Roberts et al. 1978), and each of these polypeptides may be unstable in the absence of the other. It is also possible that each subunit polypeptide (*nifD* or *nifK*) by itself cannot be processed [the FeMo protein is processed before its assembly into the nitrogenase complex (P. McLean, personal communication)] and the unprocessed polypeptides may be unstable. Similar results have also been obtained with *Klebsiella pneumoniae nifK* insertion mutants.

Fix$^-$ mutants carrying Tn5 insertions in a DNA region located at about 1000 to 3000 base pairs to the right of the *nifH* promoter (represented by strain 80, Fig. 1) produced nodules containing bacteroids which synthesized all three of the nitrogenase polypeptides (data not shown). Mutant 112 (Fig. 1) which carries a Tn5 insertion about 8000 base pairs to the right of the *nifHDK* operon also had the same symbiotic phenotype (data not shown). Experiments to determine whether the nitrogenase synthesized by these mutants is enzymatically active are currently in progress.

In contrast to the results obtained with Tn5 mutants 80 and 112, a DNA region approximately 6000 base pairs to the right of the *nifH* promoter (marked by the Tn5 insertions in strains 1354 and 1352) is essential for the production of nitrogenase polypeptides. Neither *nifH, nifD* nor *nifK* polypeptides were detected in the bacteroid proteins purified from nodules induced by 1354 and 1352 (Fig. 2). Physical mapping studies and genetic complementation analyses (Ruvkun et al. 1982, Fig. 1) indicated that this symbiotic locus is quite distinct from the *nifHDK* operon and that the absence of *nifKDH* polypeptides in the bacteroids of 1354 and 1352 is not the result of a secondary (not Tn5 produced) mutation in the *nifH* promoter.

It is possible that the DNA region marked by the Tn5 insertions carried in 1352 and 1354 ("regulatory" locus) codes for an activator of the nitrogenase *(nif)* operon. Recent data from our laboratory have demonstrated that the *R. meliloti nifH* promoter can be activated in vitro by either the *Klebsiella pneumoniae nifA* or *glnG* proteins (Sundaresan et al. 1983a,b). [*K. pneumoniae* (in contrast to *R. meliloti*) fixes nitrogen in the free-living state. The *K. pneumoniae nifA* gene codes for the activator of all of the *K. pneumoniae nif* operons (MacNeil et al. 1978, Dixon et al. 1980, Buchanan-Wollaston et al. 1981). The *glnG* product is the regulator of the major nitrogen utilization pathways of enteric bacteria (Magasanik 1982).] In these in vitro activation experiments, the *R. meliloti nifH* promoter was fused to the *E. coli lacZ* gene. It was found that in *E. coli* strains which produce either the *nifA* or *glnG* proteins constitutively, the *R. meliloti nifH* promoter, as monitored by the production of beta-glactosidase, was activated. Therefore, to test if the *R. meliloti* "regulatory" locus codes for a *nif* activator analogous to *K. pneumoniae nifA* or *glnG,* the following experiments are now in progress to determine whether the *R. meliloti* locus possesses *glnG* or *nifA* properties: (1) plasmids which produce the *K. pneumoniae nifA* and *E. coli glnG* products constitutively will be independently introduced into 1354 to test if any of the symbiotic defects associated with 1354 are corrected in the "complemented" strain; (2) the growth characteristics of 1354 and 1352 on media with different nitrogen sources (a *glnG* controlled function) are being examined; and (3) DNA from the *R. meliloti* "regulatory" locus will be hybridized to purified *K. pneumoniae glnG* and *nifA* DNA fragments to check for sequence homology.

References

Berg DF, Weiss A, Crossland L (1980) Polarity of Tn5 insertion mutations in *Escherichia coli*. J Bacteriol 142:439–446

Buchanan-Wollaston V, Cannon MC, Beynon JL, Cannon FC (1981) Role of the *nifA* gene product in the regulation of *nif* expression in *Klebsiella pneumoniae*. Nature 294:776–778

Buikema WB, Long SR, Brown SE, Bos RC van den, Earl CD, Ausubel FM (1983) Cosmid cloning of a large region of symbiotic genes from the megaplasmid of *Rhizobium meliloti*. J Mol Appl Genet (in press)

Dilworth MJ, Appleby CA (1977) Leghemoglobin and *Rhizobium* hemoproteins. In: Burns RC (ed) A treatise on dinitrogen fixation. John Wiley, New York, p 691

Dixon R, Merrick M, Filser M, Kennedy C, Postgate J (1980) Transcriptional organization of the *Klebsiella pneumoniae nif* gene cluster. In: Newton WE, Orme-Johnson WH (eds) Nitrogen fixation, vol I. University Park Press, Baltimore, p 71

Jing Y, Paau AS, Brill WJ (1982) Leghemoglobins from alfalfa (*Medicago sativa* L. vernal) root nodules. I. Purification and in vitro synthesis of five leghemoglobin compounds. Plant Sci Lett 25:119–132

MacNeil T, MacNeil D, Roberts GP, Supiano MA, Brill WJ (1978) Fine structure mapping and complementation analysis of *nif* (nitrogen fixation) genes in *Klebsiella pneumoniae*. J Bacteriol 136:253–266

Magasanik B (1982) Nitrogen assimilation in bacteria. Annu Rev Genet 16:135–168

Roberts GP, MacNeil T, MacNeil G, Brill WJ (1978) Regulation and characterization of protein products coded by the *nif* (nitrogen fixation) genes in *Klebsiella pneumoniae*. J Bacteriol 136:267–279

Ruvkun GB, Ausubel FM (1980) Interspecies homology of nitrogenase genes. Proc Natl Acad Sci USA 77:191–195

Ruvkun GB, Ausubel FM (1981) A general method for site-directed mutagenesis in prokaryotes. Nature 289:85–89

Ruvkun GB, Sundaresan V, Ausubel FM (1982) Directed transposon mutagenesis and complementation analysis of *Rhizobium meliloti* symbiotic nitrogen fixation genes. Cell 29:551–559

Sundaresan V, Ausubel FM (1981) Nucleotide sequence of the gene coding for the nitrogenase iron protein from *Klebsiella pneumoniae*. J Biol Chem 256:2808–2812

Sundaresan V, Jones JDG, Ow DW, Ausubel FM (1983a) *Klebsiella pneumoniae nifA* product activates the *Rhizobium meliloti* nitrogenase promoter. Nature 301:728–731

Sundaresan V, Ow DW, Ausubel FM (1983b) Activation of *Klebsiella pneumoniae* and *Rhizobium meliloti* nitrogenase promoters by *gln (ntr)* regulatory proteins. Proc Natl Acad Sci USA 80:4030–4034

Vincent JM (1980) Factors controlling the legume *Rhizobium* symbiosis. In: Newton WE, Orme-Johnson WH (eds) Nitrogen fixation, vol II. University Park Press, Baltimore, p 103

Mapping and Regulation of the Structural Genes *nif*K, *nif*D, and *nif*H of *Rhizobium meliloti*

A. PÜHLER, W. KLIPP, and G. WEBER[1]

Introduction

Klebsiella pneumoniae (known for free-living nitrogen fixation) and *Rhizobium meliloti* (capable of fixing nitrogen in symbiosis with *Medicago sativa*) are two model systems used by several groups to study the genetics of biological nitrogen fixation. The gene regions for nitrogen fixation (*nif*-regions) of *K. pneumoniae* (Cannon et al. 1979, Pühler et al. 1979) and of *R. meliloti* (Ruvkun et al. 1980, Banfalvi et al. 1981, Ruvkun et al. 1982) were cloned in *E. coli* and have been analyzed in detail. In this paper we describe the mapping of the *R. meliloti nif*H, *nif*D, and *nif*K coding regions. Of further interest is the transcriptional regulation of *K. pneumoniae* and *R. meliloti nif* genes. We show here that the *R. meliloti* as well as the *K. pneumoniae nif*H promotor is activated by the *Klebsiella nif*A gene product.

Results and Discussion

An Experimental System to Map Coding Regions on Fragments Cloned into *E. coli* Plasmid Vectors

In order to analyze *R. meliloti* nitrogen fixation *(nif)* genes in *E. coli* we employed the system outlined in Fig. 1. This system allows the transcription of coding regions by using constitutive plasmid promotors and the identification of gene products by the *E. coli* minicell technique. For this technique *E. coli* minicells carrying the hybrid plasmid in question are isolated and incubated in the presence of radioactively labelled aminoacids. The synthesized polypeptides are separated on SDS polyacrylamide gels and visualized by autoradiography. By this procedure the molecular weight of the gene products encoded by a cloned fragment can be determined. Using Tn5 induced mutations and analyzing them in the minicell system it is even possible to map their coding regions. The rationale of this procedure is outlined in Fig. 1c: The insertion of the transposon Tn5 into the cloned fragment terminates the transcription at the insertion point. Thus, no coding region downstrem from the Tn5 insertion is transcribed. If the Tn5 transposon is inserted directly into a coding region only a truncated polypeptide is produced. The molecular weight of such a truncated polypeptide and the

1 Lehrstuhl für Genetik, Universität Bielefeld, Postfach 8640, 4800 Bielefeld 1, FRG

Molecular Genetics of the Bacteria-Plant Interaction
(ed. by A. Pühler)

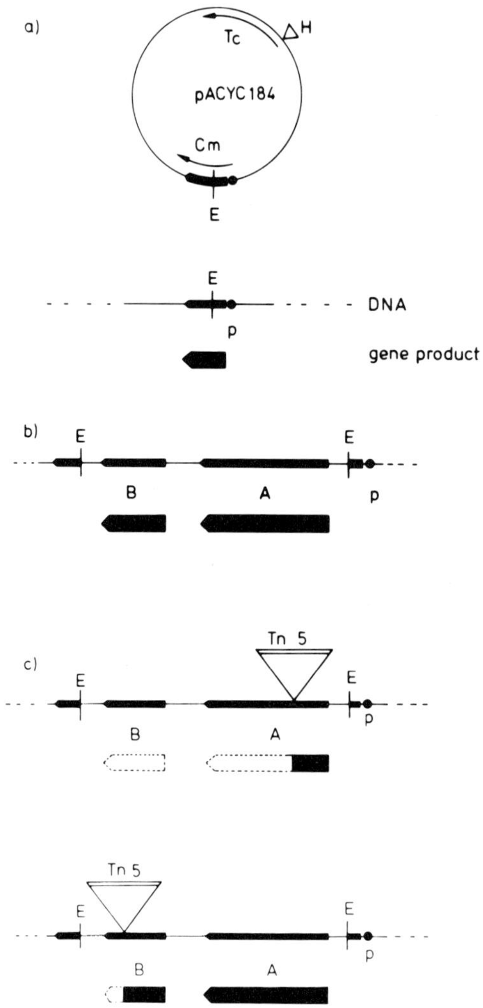

Fig. 1a–c. Determination of coding regions located on fragments cloned into the *Eco*RI site of plasmid pACYC184. **a** The vector plasmid pACYC184 carrying genes for tetracycline resistance *(Tc)* and chloramphenicol resistance *(Cm)* is shown. *H* and *E* indicate restriction sites for *Hin*dIII and *Eco*RI respectively. If plasmid pACYC184 is analyzed in the minicell system (see text) only one strong protein band of 22,000 molecular weight can be identified. This protein represents chloramphenicol acetyltransferase (CAT), the gene product of the Cm gene. **b** Insertion of a DNA fragment into the *Eco*RI site of the Cm gene results in the fusion of the constitutive Cm promotor to the cloned fragment. If the cloned fragment is free of transcriptional termination signals the hypothetical genes *A* and *B* are under the control of the Cm promotor. **c** The mapping of coding regions located on the cloned fragment can be achieved by analyzing different Tn5 insertion mutants in the minicell. For details see text

exact location of the Tn5 insertion finally allow the mapping of the corresponding coding region.

This system has already been employed previously to map coding regions on cloned fragments. For example 15 out of 17 *K. pneumoniae nif* conding regions could be mapped (Pühler and Klipp 1981). Similarly, an *Eco*RI fragment of the T-DNA region of the octopine Ti-plasmid pTiAch5 was analyzed and part of the coding region for the lysopine dehydrogenase was mapped (Schröder et al. 1981). Further examples are presented in this volume: *R. japonicum nif* DNA (Hennecke et al., this vol.) as well as T-DNA regions isolated from octopine and nopaline Ti-plasmids (Schröder, this vol.) were analyzed by the system described in this section.

Expression of *R. meliloti nif* Genes in *E. coli* Minicells

For the expression of *R. meliloti nif* genes in *E. coli* we first worked with plasmid pRmR2 (Ruvkun et al. 1980), consisting of the vector plasmid pACYC184 and an *R. meliloti* DNA insert hybridizing to *Klebsiella nif*H and *nif*D DNA. The *Eco*RI insert of pRmR2 can be integrated in the vector plasmid in two orientations resulting in plasmid pRmR2.1 (orientation 1) and pRmR2.2 (orientation 2). Analysis of gene products produced by plasmid pRmR2.1 (Fig. 2) in the *E. coli* minicell system revealed that the DNA insert is responsible for a double band on a SDS polyacrylamide gel of 30,000 and 31,000 molecular weight (Weber and Pühler 1982). Another plasmid, designated pID1, containing a *Pst*I fragment of *R. meliloti* DNA hybridizing to *Klebsiella nif*H and *nif*D DNA was constructed by Banfalvi et al. (1981). The pID1.1 version of this plasmid (Fig. 2) was induced in our investigations. For further experiments we used a cosmid clone (pRmR29D5), containing an *R. meliloti* DNA fragment hybridizing to *Klebsiella nif*K, *nif*D, and *nif*H DNA (Ruvkun et al. 1982). In Fig. 2 part of the restriction map of this *R. meliloti* fragment hybridizing to *Klebsiella nif* DNA is shown. We subcloned these DNA into various *E. coli* plasmid vectors. The hybrid plasmids that we obtained and used for studies of expression in *E. coli* minicells are described in Fig. 2. Their properties, in particular the plasmid promotors which are linked to the *R. meliloti* inserts, are listed. All these plasmids

Plasmid designation	Vector plasmid	Cloning site	Used promotor
pRmR2.1	pACYC184	*Eco*RI	P_{Cm}
pRmW58.1	pBR325-derivative	*Hind*III	P_{Cm}
pID1.1	pBR322	*Pst*I	P_{Ap}
pRmW61.1	pACYC184-derivative	*Sma*I	P_{Cm}
pRmW53.1	pACYC184	*Eco*RI	P_{Cm}

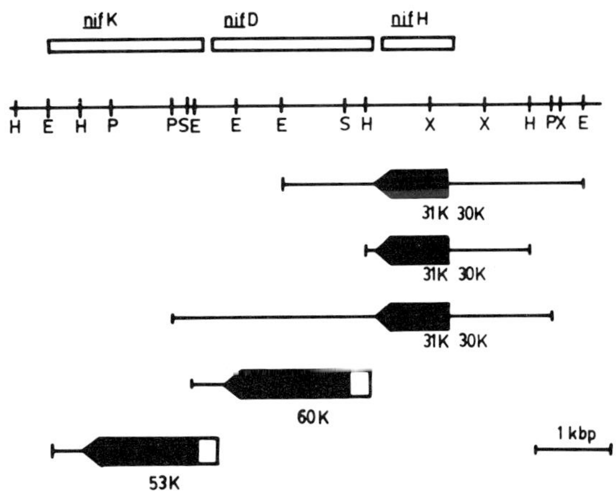

Fig. 2. Mapping of the *R. meliloti nif*H, *nif*D, and *nif*K coding region. The restriction map of the *R. meliloti* DNA region hybridizing to *Klebsiella nif*K, *nif*D, and *nif*H DNA *(open boxes)* is indicated as published by Ruvkun et al. (1982). Subfragments of this region were cloned in different *E. coli* plasmid vectors. The resulting plasmids pRmW58.1, pRmW61.1, and pRmW53.1 together with pRmR2.1 and pID1.1 are listed. The cloned *R. meliloti* fragments are all fused to special plasmid promotors, either to P_{Cm} or P_{Ap}. In the case of pRmW58.1 a derivative of pBR325 was used where the Tc gene is under the control of the Cm promotor. For pRmW61.1 an additional DNA-fragment containing a *Sma*I site was cloned into the *Eco*RI site of pACYC184. Coding regions located on these *R. meliloti* subfragments and expressed in *E. coli* minicells are shown by *black thick arrows.* The molecular weights of the corresponding gene products are indicated. *E Eco*RI; *H Hind*III; *P Pst*I; *S Sma*I; *X Xho*I; P_{AP} ampicillin promotor; P_{Cm} chloramphenicol promotor

were analyzed in the minicell system. The results can be summarized as follows: The inserts of the plasmids pRmW58.1 and pID1.1 produce the same two protein bands on SDS gels as plasmid pRmR2. For plasmid pRmW61.1 as well as for plasmid pRmW53.1 only a very strong protein band was found on polyacrylamide gels in each case with molecular weights of 60,000 and 53,000 respectively.

It sould be mentioned that the same experiments were repeated with plasmids where the *R. meliloti* DNA was inserted in opposite orientation (orientation 2). For all these plasmids no insert specific gene products could be identified. Only the double band of 30,000 and 31,000 molecular weight was weakly expressed by the plasmids pRmR2.2, pRmW58.2, and pID1.2. This finding indicates that a promotor located on the *R. meliloti* insert is weakly recognized by the *E. coli* polymerase.

The *R. meliloti* fragments analyzed so far all hybridize to *Klebsiella nif* DNA coding for the structural genes of the nitrogenase (Fig. 2). Therefore, one can assume that the polypeptides specified by these fragments represent the gene products of the *R. meliloti nif*H, *nif*D, and *nif*K genes. In order to confirm this assumption the coding regions of the polypeptides in question were mapped on the different *R.meliloti* fragments.

Mapping of the *R. meliloti nif*H, *nif*D, and *nif*K Coding Region

As outlined in Fig. 1, the coding regions of the polypeptides identfied in *E. coli* minicells were mapped using Tn5 insertions. In Fig. 3 the mapping of the coding region of the 53,000 molecular weight polypeptide on plasmid pRmW53.1 is shown. Ten different Tn5 insertions were located on the *R. meliloti* DNA insert and the

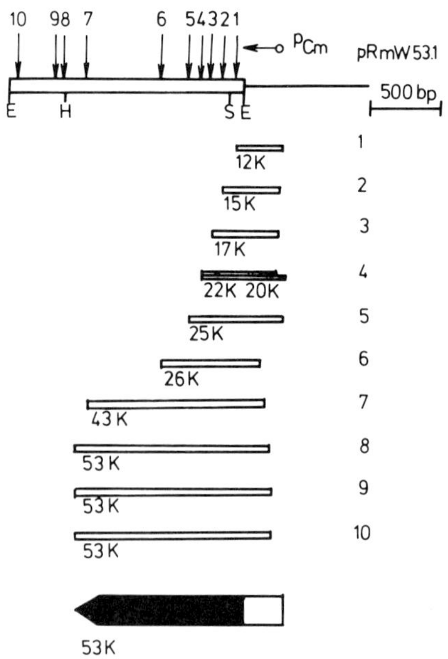

Fig. 3. Mapping of the *R. meliloti nif*K coding region on plasmid pRmW53.1. Part of the linearized restriction map of plasmid pRmW53.1 is shown. Ten different Tn5 insertions were mapped *(arrowed)*. Beneath the restriction map the coding region expressed in minicells of *E. coli* from the Cm promotor p_{Cm} are drawn to scale. The molecular weights of the gene products are indicated. for abbreviations see Fig. 2

molecular weights of the corresponding truncated polypeptides were determined by the minicell technique. All data are presented in Fig. 3. From this figure it is evident that the 53,000 polypeptide is a fusion protein whose coding region starts on the plasmid vector pACYC184 and covers most of the *R. meliloti* DNA insert.

We assume that the fusion protein contains part of the chloramphenicaol acetyltransferase (CAT) specified by plasmid pACYC184 (Fig. 1) and part of the *R. meliloti nif*K gene product. The latter assumption is based on hybridization studies with *Klebsiella nif*K DNA (Fig. 2) (Ruvkun et al. 1982). This assumption of a fusion protein is also supported by the strong expression of the 53,000 polypeptide.

A different situation was found for plasmid pRmR2.1. Here a coding region was mapped which is located completely on the insert (Weber and Pühler 1982). In Fig. 2 the exact location of this coding region which specified 2 protein bands of 30,000 and 31,000 molecular weights is indicated. There are two reasons why we conclude that this coding region is identical with the *R. meliloti nif*H gene: one is its hybridization with *Klebsiella nif*H DNA (Fig. 2) (Ruvkun and Ausubel 1981); the other comes from sequencing this region (Török and Kondorosi 1981) which shows an open reading frame coding for a protein highly related to the *Klebsiella nif*H gene product.

It is necessary to discuss the 60,000 protein specified by plasmid pRmW61.1 in more detail. Although we did not perform mapping experiments, we conclude from the strong expression of this protein in minicells of *E. coli* that the 60,000 polypeptide is again a fusion protein containing part of the CAT-protein and part of the *R. meliloti nif*D gene product. The identification of the *R. meliloti nif*D gene is again based on hybridization data with *Klebsiella nif*D DNA (Ruvkun et al. 1982). In Fig. 2 the presumed location of the coding region of the 60,000 fusion polypeptide is indicated.

Finally, the following result is worthy of comment. Plasmid pID1.1 contains an insert which spans the *R. meliloti nif*H and *nif*D genes (Fig. 2). This plasmid was analyzed by the minicell technique and it was found that in contrast to the *nif*H gene products the *nif*D gene product was not synthesized. Since it is known that the *R. meliloti nif*H, *nif*D, and *nif*K genes are located in a single operon (Ruvkun et al. 1982) one possible conclusion is that the *nif*D gene is not translated in *E coli.* One explanation could be a weak or regulated ribosomal binding site in front of the *nif*D gene.

The Regulation of the *K. pneumoniae nif*YKDH Promoter by the *nif*A and *nif*L Gene Product in *E. coli*

The next important question after mapping the *nif*H, *nif*D, and *nif*K coding regions of *R. meliloti* is the transcriptional control of these genes. For this purpose we constructed special plasmids, designated pWK130 and pWK131, which were first used to investigate the transcription of *Klebsiella nif* operons. The restriction maps of the plasmids are presented in Fig. 4. Plasmid pWK130 is derived from the plasmid vector pACYC177 (Chang and Cohen 1978). The ampicillin resistance gene (Ap) was inactivated by inserting a *Pst*I fragment carrying chloramphenicol resistance (Cm) which originated from the P1Cmts prophage (Rosner 1972). The *Hin*dIII site of the

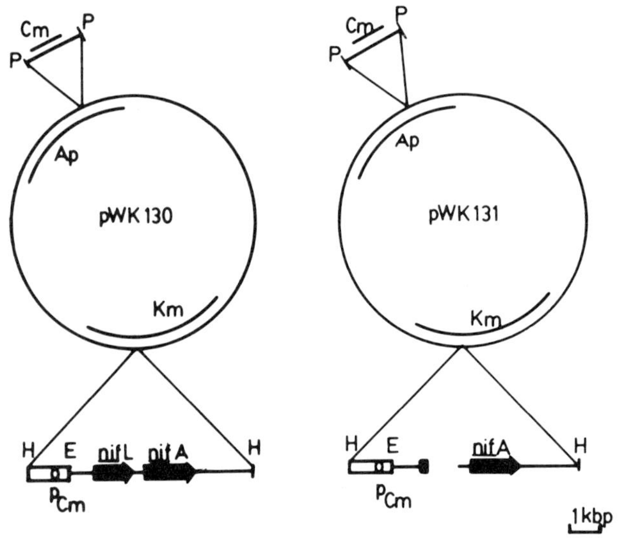

Fig. 4. Restriction maps of plasmid pWK130 and plasmid pWK131 expressing constitutively the *Klebsiella nif* genes *nif*L and *nif*A. pWK130 and pWK131 contain the *E. coli* vector plasmid pACYC177. The Ap gene was inactivated by inserting a *Pst*I fragment carrrying the Cm gene of phage P1Cmts. The *Hin*dIII site of the Km gene was used in the case of pWK130 to clone a *Hin*dIII fragment coding for *nif*L and *nif*A. These genes are under the control of the constitutive Cm promotor of plasmid pACYC184. pWK131 differs from pWK130 by a deletion spanning most of the *nif*L coding region. For abbreviations see Fig. 2

kanamycin resitance (Km) gene was used for the cloning of a *Hin*dIII fragment which contains the chloramphenicol promotor (p_{Cm}) linked to the *Klebsiella* genes *nif*L and *nif*A. In contrast to pWK130 plasmid pWK131 contains a deletion which removes most of the *nif*L gene.

The plasmids pWK130 and pWK131 are important since they express the regulatory proteins necessary for the transcription of the *Klebsiella nif* region (Kennedy et al. 1981). The *nif*A gene product (*nif*Agp) is believed to act as a transcriptional activator of *Klebsiella nif* operons whereas the *nif*L gene product (*nif*Lgp) acts as antiactivator or transcriptional repressor. This can be shown with the help of plasmid pWK132, which was constructed as follows. We cloned a *Hin*dIII-*Eco*RI fragment of the *Klebsiella nif* region containing part of the *nif*YKDH operon into pBR322. By this cloning step the *nif*YKDH promotor was linked to the promotorless Tc gene (Fig. 5). Activation of the *nif*YKDH promotor should therefore give rise to tetracycline resistance. The activation of the nifYKDH promotor by the *Klebsiella nif*Agp was demonstrated by constructing an *E. coli* strain containing pWK131 and pWK132. Both plasmids are compatible since they contain the different replicons pACYC177 and pBR322 respectively. As a control experiment an *E. coli* strain containing pWK130 and pWK132 was used. In this case the presence of the *nif*Lgp under appropriate oxygen conditions should inhibit the *nif*A dependent activation. The results are summarized in Fig. 5. The *nif*Agp activates the *nif*YKDH promotor.

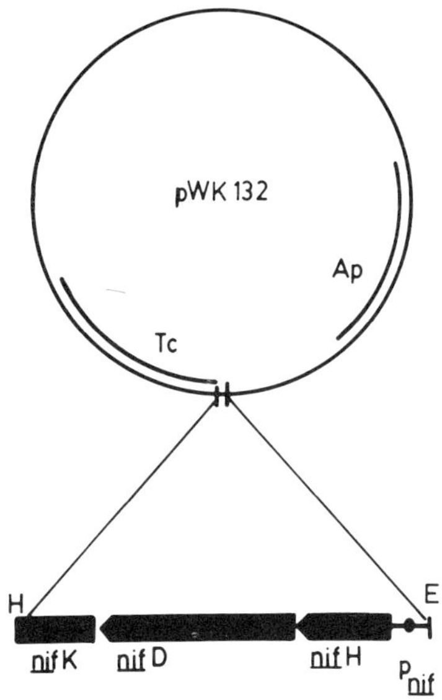

Fig. 5. Analysis of the regulation of the *K. pneumoniae nif*YKDH promotor in *E. coli.* Plasmid pWK132 contains the *E. coli* plasmid vector pBR322. The *Eco*RI/*Hind*III fragment carrying part of the Tc gene promotor is replaced by an *Eco*RI/*Hind*III fragment of the *Klebsiella nif* region containing part of the operon *nif*YKDH. By this manipulation the Tc coding region is linked to the *Klebsiella nif*YKDH promotor. Since plasmid pWK132 is compatible with pWK130 and pWK131 the influence of the *Klebsiella nif*Agp and *nif*Lgp on the regulation of the *nif*YKDH promotor could be studied. For abbreviations see Fig. 2

E. coli HB101 containing the plasmids	temperature 30°C	44°C
pWK 131: nif A pWK 132: Tc	Tc^r	Tc^s
pWK 130: nif A nif L pWK 132: Tc	Tc^s	Tc^s

If in addition to the *nif*Agp the *nif*Lgp is also present no activation of the *nif* promotor can be detected. A further control experiment used the temperature sensitivity of the *nif*Agp (Zhu and Brill 1981). The activation of the *nif* promotor only occurred at 30°C but not at 45°C. These experiments clearly demonstrate the role of *nif*Agp and of *nif*Lgp in the regulation of the *Klebsiella nif*YKDH promotor. In the next section we will show that the *Klebsiella nif*Agp and *nif*Lgp also regulate the *R. meliloti nif*H promotor.

The Regulation of the *R. meliloti nif*H promoter by the *K. pneumoniae nif*A and *nif*L Gene Product in *E. coli*

In order to study the regulation of the *R. meliloti nif*H promotor a plasmid similar to pWK132 was constructed. The restriction map of this plasmid, designated pGW5, is shown in Fig. 6. We started again with pBR322 and cloned a small *Hind*III-*Eco*RI

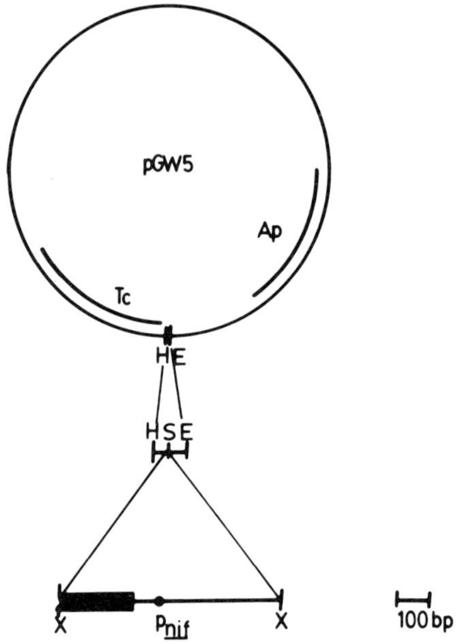

Fig. 6. Analysis of the regulation of the *R. meliloti nif*H promotor by the *K. pneumoniae nif*A and *nif*L gene products in *E. coli*. Plasmid pGW5 contains the *E. coli* vector plasmid pBR322. The Tc gene is inactivated by insertion of a small *Hin*dIII/*Eco*RI fragment of plasmid pUC8. This insertion destroys the promotor of the Tc gene. An *Xho*I fragment carrying the *R. meliloti nif*H promotor and part of the *nif*H coding region was cloned into a *Sal*I site *(S)* present on the *Hin*dIII/*Eco*RI fragment. By this manipulation the Tc-coding region is controlled by the *R. meliloti nif*H promotor. Plasmid pGW5 is compatible with pWK130 and pWK131. Thus, the influence of the *Klebsiella nif*Agp and *nif*Lgp on the regulation of the *R. meliloti nif*H promotor could be studied. For abbreviations see Fig. 2

E.coli HB101 containing the plasmids	temperature 30°C	44°C
pWK 131 : nif A pGW 5 : Tc	Tc^r	Tc^s
pWK 130 : nifA nifL pGW5 : Tc	Tc^s	Tc^s

fragment originating from plasmid pUC8 (Messing, unpublished) containing a *Sal*I site. The resulting plasmid is tetracycline sensitive. Into this *Sal*I site we cloned an *Xho*I fragment (Fig. 2) containing the *R. meliloti nif*H promotor. Thus the *nif*H promotor is again linked to the promotorless Tc gene. Plasmid pGW5 was now tested in combination with pWK130 or pWK 131 for the expression of tetracycline resistance. Again the same results as in the case of plasmid pWK132 were obtained (Fig. 6): *Klebsiella nif*Agp can activate the *R. meliloti nif*H promotor at 30°C. At 45°C or in the presence of *Klebsiella nif*Lgp this activation is inhibited. It should be noted that in this experiment cellular components from 3 different bacterial species are working together: The *R. meliloti nif*H promotor activated by the *K. pneumoniae nif*A gene product is recognized by the *E. coli* RNA polymerase.

Conclusions

The experiments reported so far demonstrate how coding regions of *R. meliloti nif* genes were mapped on cloned fragments in *E. coli.* In addition, by fusing the *R. meliloti nif*H promotor to a promotorless Tc gene, the regulation of the promotor was studied. This system can now be used in further experiments to clone a *R. meliloti* "*nif*A-like" gene. Furthermore, the question of whether *R. meliloti* posseses a "*nif*L-like" gene can also be investigated. Finally, it is evident from this work that genetic engineering techniques developed in *E. coli* are very successful for the analysis of the *R. meliloti* nitrogen fixation region. Similar techniques should also prove useful in the analysis of *R. meliloti* nodulation genes.

Acknowledgements. We wish to express our sincere gratitude to colleagues who helped us in carrying out these experiments: F. Ausubel who provided pRmR2 and pRmR29D5 and A. Kondorosi who supplied plasmid pID1. We also acknowledge the help of V. Krishnapillai and M. O'Connell for discussing and critically reading the manuscript. In addition, we would like to thank Ms. B. Gronemeier, one of our students, for isolating Tn5 mutations, Ms. U. Beigel for technical assistance and Ms. S. Malmivaara for patiently typing the manuscript. These investigations were supported financially by Deutsche Forschungsgemeinschaft (Pu 28/9) and Bundesministerium für Forschung und Technologie (PTB 8301).

References

Banfalvi z, Sakanyan V, Koncz C, Kiss A, Dusha I, Kondorosi A (1981) Location of nodulation and nitrogen fixation genes on a high molecular weight plasmid of *R. meliloti.* Mol Gen Genet 184:318–325

Cannon FC, Reidel GE, Ausubel FM (1979) Overlapping sequences of *Klebsiella pneumoniae nif* DNA cloned and characterised. Mol Gen Genet 174:59–66

Chang ACY, Cohen SN (1978) Construction and characterization of amplifiable multicopy DNA cloning vehicles derived from the p15A cryptic miniplasmid. Bacteriology 134:1141–1156

Kennedy C, Cannon F, Cannon M, Dixon R, Hill S, Jensen J, Kumar S, McLean P, Merrick M, Robson R, Postgate J (1981) Recent advances in the genetics and regulation of nitrogen fixation. In: Gibson AH, Newton WE (eds) Current perspectives in nitrogen fixation. Aust Acad Sci, Canberra, pp 146–156

Pühler A, Klipp W (1981) Fine structure analysis of the gene region for N_2-fixation *(nif)* of *Klebsiella pneumoniae.* In: Bothe H, Trebst A (eds) Biology of inorganic nitrogen and sulfur. Springer, Berlin Heidelberg New York, pp 275–286

Pühler A, Burkardt HJ, Klipp W (1979) Cloning of the entire region for nitrogen fixation from *Klebsiella pneumoniae* on a multicopy plasmid vehicle in *E. coli.* Mol Gen Genet 176: 17–24

Rosner JL (1972) Formation, induction and curing of bacteriophage P1 lysogens. Virology 48: 679–689

Ruvkun GB, Ausubel FM (1980) Interspecies homology of nitrogenase genes. Proc Natl Acad Sci USA 77:191–195

Ruvkun GB, Ausubel FM (1981) A general method for site-directed mutagenesis in prokaryotes. Nature 289:85–89

Ruvkun GB, Sundaresan V, Ausubel FM (1982) Directed transposon Tn5 mutagenesis and complementation analysis of *Rhizobium meliloti* symbiotic nitrogen fixation genes. Cell 29: 551–559

Schröder J, Hillebrand A, Klipp W, Pühler A (1981) Expression of plant tumor-specific proteins in minicells of *Escherichia coli:* a fusion protein of lysopine dehydrogenase with chloramphenicol acetyltransferase. Nucleic Acids Res 9:5187–5194

Török I, Kondorosi A (1981) Sequence of the *R. meliloti* nitrogenase reductase (*nif*H) gene. Nucleic Acids Res 9:5711–5723

Weber G, Pühler A (1982) Expression of *Rhizobium meliloti* genes for nitrogen fixation in *Escherichia coli* minicells: Mapping of the subunit of the nitrogenase reductase. Mol Plant Biol 1:305–320

Zhu J, Brill WJ (1981) Temperature sensitivity of the regulation of nitrogenase synthesis by *Klebsiella pneumoniae.* J Bacteriol 145:1116–1118

Early Nodulation Genes of *Rhizobium meliloti*

S.R. LONG, T. JACOBS, D. BEEBE, and T. EGELHOFF [1]

Introduction

We are interested in the early stages of *Rhizobium*-legume interaction, such as recognition, attachment, and invasion by bacteria of the host. The early events of nodule formation have been studied by biochemistry and microscopy but are not well understood at a molecular level. We hope to use mutants and cloned genes to understand these early stages better.

To date, most studies of early nodulation steps have focussed on the chemistry of bacterial and host surfaces, and its possible relationship to nodulation; see Bauer (1981), Dazzo and Hubbell (1982), for a recent review. By extension, most of these studies are concerned with the intriguing question of host range: why do individual *Rhizobium* strains form nodules on some but not all host plants? There is evidence that in some systems, surface properties of the bacteria, such as polysaccharide composition, may account for specific interaction with plant proteins such as lectins. However, unambiguous biochemical correlations are not found in all *Rhizobium*-host systems. Partly this may be due to the fact that modification of bacterial polysaccharides occurs during culture growth, and in some systems can occur when bacteria are treated with exudate from roots.

From this and other evidence, it has been proposed that recognition occurs in several steps during which bacterium and host signal and respond to each other (Bauer 1981, Dazzo and Hubbell 1982). The nature of the signals and responses is unknown and could involve gene expression. One important question which can be answered with cloned nodulation genes is whether these genes are expressed in free-living bacteria, or only in the presence of the host plant. Such regulation may explain the differing behavior of various of *Rhizobium* strains.

Initiation of the *Rhizobium*-plant symbiosis thus appears to be complex and transient. It is difficult, in such systems, to establish cause and effect relationships by biochemical analysis alone. In such systems, genetic analysis can make it possible to dissect individual steps from the process as a whole, and to establish the chain of events.

1 Department of Biological Sciences, Stanford University, Stanford, CA 94305, USA

Molecular Genetics of the Bacteria-Plant Interaction
(ed. by A. Pühler)

Genetic Studies of Nodule Formation

Research on nodulation genes has taken advantage of the existing variety of *Rhizobium* species with varying host ranges, as well as generating mutants. It is important in these studies to be able to move genes from one strain to another to study their effect in different backgrounds and combinations. Additionally, it is desirable to be able to clone the genes of interest, which refines the study of gene properties, and makes it possible to follow regulation at the molecular level.

One way in which nodulation genes have been identified and their activity assayed is that in several types of *Rhizobium,* they are located on transmissible plasmids (Johnston et al. 1978, Brewin et al. 1980, Beynon et al. 1980, Zurkowski and Lorkiewicz 1979). Transfer of such plasmids can restore function to certain nodulation mutants and can alter the host range of recipient *Rhizobium* cells. Such plasmids have primarily been studied in the related fast-growing strains *R. leguminosarum, R. trifolii,* and *R. phaseoli.* Similar results are now being found with some broad host-range *Rhizobium* strains (Pankhurst et al., Ronson and Scott, Rolfe et al., this vol.). Some indigenous *Rhizobium* plasmids can confer upon *Agrobacterium* the ability to nodulate a legume host (Hooykaas et al. 1981).

In *R. meliloti,* the situation has been different because these bacteria have symbiotic genes, including nitrogen fixation and nodulation genes, on a very large "megaplasmid" which is not self-transmissible in conditions which work for other plasmids (Rosenberg et al. 1981, Banfalvi et al. 1981). Therefore, the study of these genes, including the demonstration of their plasmid location and their study in different backgrounds, has required other methods than those used other fast-growing strains. For example, Banfalvi et al. (1981) and Rosenberg et al. (1981) have generated nodulation mutants by methods which often lead to plasmid curing or deletion, such as heat treatment. They have shown that these mutants are indeed deletion derivatives, which have lost part of the megapalsmid.

Another approach to the study of early nodulation genes is to generate mutations at individual loci in the *Rhizobium* genome, screening for mutants which are defective in nodulation ability. This permits identification of individual steps in the early stages of nodule formation, and analysis of the events (Vincent 1980). To generate and analyze mutants of the symbiosis, many genetic studies have used transposon mutagenesis (Beringer et al. 1978). Transposon insertion generates null mutations which can be mapped using the transposon's drug-resistance function.

Meade et al. (1982) and Long et al. (1981) described using this technique to identify symbiotic mutants of *R. meliloti* strain 1021. To map the Tn5-induced symbiotic mutants, a combination of physical and genetic tracking were used. Loci which were on the chromosome were mapped by frequency of co-transfer with known chromosomal markers (Meade et al. 1982). Mutants which appeared not to be on the chromosome were found to be located on the megaplasmid by carrying out Eckhardt gel analysis of the mutant strain DNA, which permits separation of the megaplasmid band position on a Southern filter transfer of the gel (Buikema et al., in press). Under these conditions, no chromosomal DNA is present in the megaplasmid band in the gel, as shown by the lack of hybridization of chromosomal DNA sequences to that position on the filter.

By this method of tracking Tn5 insertions, the *nif* genes and also genes for root hair curling (Hac) were located on the megaplasmid. Tn5 mutants have been used to demonstrate plasmid location in other systems as well, by transmissibility studies in strains of *R. leguminosarum* (Johnston et al. 1978, Buchanan-Wollaston et al. 1981) and in *R. trifolii* and *R. parasponia* (Rolfe et al., this vol.).

Nodulation Mutants 1027 and 1126

We have focused our attention on two mutants, *R. meliloti* 1027 and 1126. These two mutants are both Tn5-containing derivatives of parental strain 1021 (Long et al. 1981, Meade et al. 1982). Both strains are Nod$^-$, that is, they form no nodules when inoculated onto host alfalfa plants. Microscopic examination shows that these strains do not induce host root-hair curling, or other marked root-hair deformation of lysis (Hirsch et al. 1982).

Preliminary studies of the cell surface of mutants 1027 and 1126 show them to be indistinguishable from the wild-type parent 1021. They produce the same amount of extracellular polysaccharide (J. Darvill and P. Albersheim, pers. commun.). The lipopolysaccharide has not been characterized. We have found that both have identical phage sensitivities to the parent strain (M. Martin and S. Long, unpubl.). Also, E. Johanson and E. Signer (pers. commun.) have found that monoclonal antibodies raised against wild type *R. meliloti* surface components bind normally to both mutants. Rolfe et al. (1980) have reported that Nod$^-$ mutants of *R. trifolii* divide into two groups, according to whether or not they can be "helped" into nodules by co-infection with Fix$^-$ or wild-type *Rhizobium* cells. We have found that mutants 1027 and 1126 fall into the category which can be "helped" into nodules.

Cloning Nodulation Genes

To better understand the mutations in these strains, and thereby to characterize the steps they affect, we are studying the mutant regions at the DNA level. The cloning of this region is described by Long et al. (1982). Briefly, the method is as follows: a clone bank of wild-type DNA fragments was constructed in a broad host-range cosmid vector, pLAFR1 (Friedman et al. 1982) which is a *cos*-containing derivative of pRK290 (Ditta et al. 1980). This clone bank was conjugated into each of the Nod$^-$ mutants.

Among the Nod$^-$ cells containing recombinant plasmids, a small subpopulation will contain a plasmid bearing the wild type genes for nodulation. If this wild-type DNA segment complements the mutant genes in the cell, then it will act as if it were Nod$^+$. The plant, presented with a mixture of Nod$^-$ mutants and "Nod$^+$" complemented mutants, should select the "Nod$^+$" subpopulation, forming nodules which contain those cells. The bacteria included in the nodules can be isolated and the plasmids analyzed. By this means a recombinant clone, pRmSL26, was isolated.

It confers on both mutants 1126 and 1027 the ability to curl root hairs and form nodules.

To assay the effect of the *R. meliloti* genes cloned in pRmSL26, we have moved it into two other *Rhizobium* strains, *R. trifolii* Rt37 and *R. leguminosarum* 897. The conjugants were purified to single colonies and inoculated onto alfalfa seedlings. No nodules were observed on these plants. There are several possible explanations of this: either the genes cloned in pRmSL26 do not suffice for nodule formation on alfalfa, or they are not expressed, or the gene product of the *R. meliloti nod* region does not dominate phenotypically in the wild-type of the other strains (see below). The latter two explanations seem more likely, since Hirsch and Wilson (this vol.) have found that pRmSL26 can confer upon *Agrobacterium tumefaciens* the ability to nodulate alfalfa at low frequency.

We have mapped several restriction sites in pRmSL26, as shown in Fig. 1. Using digested insert DNA from this plasmid, we have begun to study expression of the nodulation gene(s) in *R. meliloti* cells. Expression was measured by hybridization of RNA from free-living cells or from bacteroids with cloned DNA fragments. RNA extracted from cells was lightly hydrolyzed and labelled with ^{32}P, and used as probe with Southern transfer filters bearing DNA sequences from the nodulation region, from the *nif* region (Ruvkun and Ausubel 1982) and from tryptophan metabolism loci (Friedman et al. 1982).

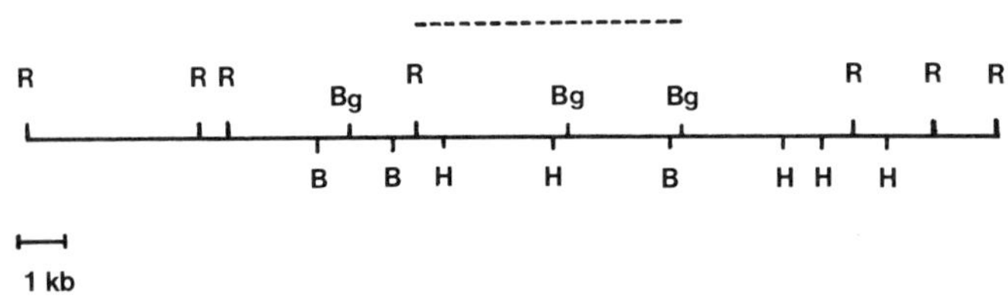

Fig. 1. Restriction map of the *R. meliloti* DNA inserted in recombinant plasmid pRmSL26. These sequences are located on the *R. meliloti* megaplasmid; the *nif* sequences are found approximately 18 kilobase pairs to the right of the rightmost border shown here. Tn5 insertions in the largest *Eco*R1 fragment, which is about 8.7 kb in length, produce *nod* mutations. *Dotted line* shows region transcribed in bacteroids. The position of the left-most two fragments may possibly be reversed. *R Eco*R1; *B Bam*H1; *H Hind*III; *Bg Bgl*II. *Eco*R1 fragment sizes *(from left)* = 3.5 kb, 0.6 kb, 3.8 kb, 8.7 kb, 1.7 kb, 1.25 kb

We found that RNA from bacteroids hybridized strongly to the *nif* sequences, as predicted, and that it hybridized also to an 8.7 kb *Eco*R1 fragment contained in pRmSL26. Double digests of pRmSL26 insert DNA, followed by hybridization with labelled bacteroid RNA, indicate that only part of the 8.7 kb fragment is expressed; this region is indicated by the dotted line in Fig. 1. When this experiment was tried with RNA from free-living cells, we found that it hybridized to the tryptophan metabolism clone, as expected; however there was no detectable hybridization to any inserted sequences in pRmSL26. At this level of detection, then, it appears that the *R. meliloti* nodulation genes used for root-hair curling of alfalfa may not be expressed, or are expressed very weakly, in free-living cells. To further characterize these regions, we have generated Tn5 insertions in the various fragments, and will describe the phenotype of these insertional mutations by the method of Ruvkun and Ausubel (1981). As an adjunct study, we are using bacteroid RNA to map the extent of transcription in clone pRmSL26.

Generality of the Cloning Technique

Using complementation host-plant selection to obtain *nod* clones ought to be broadly applicable for certain types of genes. It should permit cloning genes identified by point mutants or small deletions. This may be of use in working with mutant strains isolated after chemical mutagenesis or heat treatment.

For the technique to work, several conditions must be met. These set constraints on which types of genes can be cloned by this method. First, it is essential that the complemented mutant cells ("wild type") be able to outcompete the other cells. The most obvious consequence of this is that only mutants defective in early steps, where the plant's selectivity probably occurs, are likely to be outcompeted: a mutant whose only defect occurs at a late stage will appear indistinguishable from wild type at early stages. Therefore the complementation technique can probably be used only to clone genes for early steps in the symbiosis.

It is possible that even among mutants defective in early steps, wild-type cells would have no advantage over mutants and would therefore not be able to selectively nodulate the host. For example, some mutants could bind the host at a hypothetical "attachment site", saturating these sites, but then would not complete the next step of the symbiotic interaction. No wild-type cells could get to sites, and the plant would remain unnodulated. Another possibility is that some mutant types might provoke host defense reactions, which would prevent the wild-type cells from successfully entering the plant as symbionts.

There is a further requirement, which is for dominance of the wild-type gene over the mutant gene within the bacterial cell. In the case of transposon mutants, this is probably not a problem, since the Tn-induced mutants are nulls: there is no gene product. But in the case of trying to complement mutants generated by mutagens such as NTG, or others which result in missense mutations, it is possible that a defective product made from the mutant gene could be phenotypically dominant over the normal product arising from the cloned wild-type gene. If the nodulation gene product is one which requires assembly with other components in order to be active, for example, such as assembly into a membrane or a multi-peptide complex, the defective product might decrease the likelihood that the normal one would be active. Finally there is the possibility that some nodulation gene products might be regulatory, and an abnormal gene product might be dominant due to the nature of its action.

Complementation of Other Mutants

We have attempted to use the method described in Long et al. (1982) to obtain cloned *R. meliloti* genes which restore function to three NTG derived mutants of *R. meliloti* strain 102F51. These mutants, isolated and studied by W. Leps, A Paau and W. Brill (Paau et al. 1981) are all Nod$^-$ and are designated WL113, Wl131, and WL188. WL113 and WL131 do not bind to alfalfa agglutinin, nor do they adhere to alfalfa root hairs. WL188 binds agglutinin and root hairs, but does not nodulate; this also describes the behavior of our mutants 1027 and 1126 with respect to agglutinin binding (Paau et al. 1981 and pers. commun.).

In preliminary trials, we found that pRmSL26 did not complement the defect of these strains. Therefore we conjugated into them a clone bank of *R. meliloti* 1021 DNA borne on vector pLAFR1 (Friedman et al. 1982). Exconjugants were selected, mixed together in groups of 200, and each mixture was inoculated onto groups of 5 matched alfalfa seedlings. The results are shown in Table 1.

Table 1. Trials with pRmSL26

Mutant	Mixture	Plants with nodules/total plants
WL113	A	0/5
	B	2/5
	C	5/5
	D	3/5
	E	4/5
WL131	A	0/5
	B	0/5
	C	0/5
	D	0/5
	E	0/5
WL188	A	0/5
	B	0/5
	C	5/5
	D	0/5
	E	4/5

Nodules from plants in group 113D and 188E were sterilized, squashed and the contents were plated on non-selective media. Tetracycline-resistant colonies, containing pLAFR1 plasmids, were purified and re-inoculated onto alfalfa seedlings. These were found to be Nod$^+$. We are currently analyzing the plasmids from these strains. Our next goal will be to insert Tn5 into the cloned *R. meliloti* genes on these plasmids, and re-insert them into the genome of *R. meliloti* by the method of Ruvkun and Ausubel (1981), which will permit them to be mapped.

It is not clear why mutant WL131, which is phenotypically similar to WL113, could not be complemented by the clone bank. Currently, we are testing the plasmid obtained in complementation of WL113 to see if it will confer upon WL131 the ability to form nodules. An alternate strategy will be to use the clone bank of wild-type DNA sequences to complement agglutinin-binding activity alone, using the agglutinin-binding to directly select complemented mutants. Any clones obtained will be tested to see if they also confer nodulation ability on the mutants.

Other Uses for Complementation Strategies

The procedure described here for cloning genes for early nodulation stages by complementation of mutants may be applicable to the study of invasion genes of bacterial plant pathogens. The same constraints – competitiveness of the wild-type with respect to the mutant, and dominance of the normal gene product in the establishment of the bacterial phenotype – would apply to this type of study.

Another possible application might be the direct cloning of *Rhizobium* nodulation host-range genes. For example, a clone bank of one *Rhizobium* strain's DNA could be conjugated into another *Rhizobium* with a different range, and the mixture could be placed on a host plant receptive to only the first strain. Nodule formation could be used to select bacteria containing cloned host-range genes of the first strain.

There are several potential problems with this scheme. First is a finding reported in the plasmid-transfer studies of Johnston et al. (1978) and Beynon et al. (1980). If native plasmids of two host ranges are put into the same cell, the it is often found that the resulting cell no longer nodulates either plant host very well. Thus, the plasmid-containing cells might not be highly competitive.

At the base of this is the question as to whether nodulation genes would be expected to be co-dominant. If nodulation genes code for enzymes which act on the cell surface of *Rhizobium,* modifying it, then in distantly related *Rhizobium* strains the appropriate substrate might not be on the surface to be modified by the products of each other's nodulation genes. (It is also possible that some nodulation gene products are not enzymes per se, but are more analogous to immunologically active proteins which bind but do not modify their target molecules. Whether there could be co-dominance of such products as these might depend on assembly into surface arrays.)

These considerations make it seem likely that cloning host-range genes directly would be difficult. One possible solution might be to use nodulation mutants of the second strain as recipients, to give the incoming cloned genes a better chance to be expressed phenotypically. Experiments to complement mutations across "species" boundaries may also give us clues as to which steps in the nodulation process, if any, are performed in a common way by all *Rhizobium* strains. This will, in turn, provide us with an idea of what it is that allows the legumes as a group to interact with these bacteria.

Conclusions

Genetic studies will be helpful in analyzing early events in nodule formation, which appear to be multi-step. Surface properties of *Rhizobium* and host do not always show specific binding, based on current methods of study; it is possible that transient changes in gene activity could influence early interactions such as binding. In *Rhizobium meliloti,* whose large symbiotic "megaplasmid" is not easily transmissible, it will be useful to have recombinant plasmids which carry the nodulation genes. These can be used to complement mutants and study genetic effects on host range, and will make it possible to study the regulation of early nodulation events. We have been focussing on two mutants of *Rhizobium meliloti* which do not curl plant root hairs or form nodules. The genes which complement function in these mutants have been isolated as recombinant plasmids in a broad host-range cosmid vector. We have found that part of the sequences cloned in this plasmid are expressed in bacteroids, and are in the process of mapping the symbiotic genes in the cloned region.

Acknowledgements. We are grateful to A. Paau, W. Leps and W. Brill for sending us strains WL113, WL131 and WL188. We would also like to thank Ann Hirsch, Kate Wilson, Jonathan Jones, Bill Buikema and Fred Ausubel for sharing results with us prior to this symposium.

References

Banfalvi Z, Sakanyan V, Koncz C, Kiss A, Dusha L, Kondorosi A (1981) Location of nodulation and nitrogen fixation genes on a high molecular weight plasmid of *R. meliloti.* Mol Gen Genet 184:318–325

Bauer WD (1981) Infection of legumes by Rhizobia. Annu Rev Plant Physiol 32:407–449

Beringer JE, Beynon JL, Buchanan-Wollaston AV, Johnston AWB (1978) Transfer of the drug-resistance transposon Tn5 to *Rhizobium.* Nature 276:633–634

Beringer JE, Brewin NJ, Johnston AWB (1980) The genetic analysis of *Rhizobium* in relation to symbiotic nitrogen fixation. Heredity 45:161–186

Beynon JL, Beringer JE, Johnston AWB (1980) Plasmids and host range in *Rhizobium leguminosarum* and *Rhizobium phaseoli.* J Gen Microbiol 120:421–429

Brewin NJ, Beringer JE, Johnston AWB (1980a) Plasmid-mediated transfer of host-range specificity between two strains of *Rhizobium leguminosarum.* J Gen Microbiol 120:413–420

Brewin NJ, Beringer JE, Buchanan-Wollaston AV, Hirsch PR (1980b) Transfer of symbiotic genes with bacteriocinogenic plasmids in *Rhizobium leguminosarum.* J Gen Microbiol 116:261–270

Buchanan-Wollaston AV, Beringer JE, Brewin NJ, Hirsch PR, Johnston AWB (1981) Isolation of symbiotically defective mutants in *Rhizobium leguminosarum* by insertion of the transposon Tn5 into a transmissible plasmid. Mol Gen Genet 178:185–190

Buikema WJ, Long SR, Brown SE, Bos R van den, Earl C, Ausubel FM (1983) Cosmid cloning of a large region of symbiotic genes from the megaplsmid of *R. meliloti.* J Mol Appl Genet (in press)

Dazzo FB, Hubbell DH (1982) Control of root hair infection. In: Broughton W (ed) Nitrogen fixation, vol II: *Rhizobium.* University Press, Oxford, pp 275–309

Ditta G, Stanfield S, Corbin D, Helinski D (1980) A broad host-range DNA cloning system for Gram-negative bacteria: construction of a gene bank of *Rhizobium meliloti.* PNAS 77: 7347–7351

Friedman AM, Long SR, Brown SE, Buikema WJ, Ausubel FM (1982) Construction of a broad host range cosmid cloning vector and its use in the genetic analysis of *Rhizobium* mutants. Gene 18:289–296

Hirsch A, Long SR, Bang M, Haskins N, Ausubel FM (1982) Structural studies of alfalfa roots infected with nodulation mutants of *Rhizobium meliloti*. J Bacteriol 151:411–419

Hooykaas PJJ, Brussel AAN van, Dulk-Ras H den, Slogetern GMS van, Schilperoort RA (1981) Sym plasmid of *Rhizobium trifolii* expressed in different rhizobial species and *Agrobacterium tumefaciens*. Nature 291:351–353

Johnston AWB, Beynon JL, Buchanan-Wollaston AV, Setchell SM, Hirsch PR, Beringer JE (1978) High frequency transfer of nodulating ability between strains and species of *Rhizobium*. Nature 276:1635–1636

Long SR, Meade HM, Brown SE, Ausubel FM (1982) Transposon-induced symbiotic mutants of *Rhizobium meliloti*. In: Panopoulos N (ed) Genetic engineering in the plant sciences. Praeger Press, New York, pp 129–143

Meade HM, Long SR, Ruvkun GB, Brown SE, Ausubel FM (1982) Isolation of symbiotic and auxotrophic mutants of *Rhizobium meliloti* using transposon Tn5 mutagenesis. J Bacteriol 149:114–122

Paau AS, Leps WT, Brill WJ (1981) Agglutinin from alfalfa necessary for binding and nodulation by *Rhizobium meliloti*. Science 213:1513–1514

Rolfe BG, Gresshoff PM (1980) *Rhizobium trifolii* mutant interactions during the establishment of nodulation in white clover. Aust J Biol Sci 33:491–504

Rolfe BG, Gresshoff PM, Shine J (1980) Rapid screening for symbiotic mutants of *Rhizobium* and white clover. Plant Sci Lett 19:277–284

Rolfe BG, Djordjevic M, Scott KF, Hughes JE, Jones J, Badenoch J, Gresshoff PM, Cen Y, Dudman WF, Zurkowski W, Shie J (1981) Analysis of the nodule forming ability of fast-growing *Rhizobium* strains. In: Gibson AH, Newton WE (eds) Current perspectives in nitrogen fixation. Elsevier-North Holland Biomedical Press, Amsterdam New York, pp 142–145

Rosenberg C, Boistard P, Denarie J, Casse-Delbart F (1981) Genes controlling early and late functions in symbiosis are located on a megaplasmid in *Rhizobium meliloti*. Mol Gen Genet 184:326–333

Ruvkun GB, Ausubel FM (1981) A general method for site-directed mutagenesis in prokaryotes. Nature 289:85–88

Vincent JM (1980) Factors controlling the legume *Rhizobium* symbiosis. In: Newton WE and Orme-Johnson WH (eds) Nitrogen fixation, vol II. University Park Press, Baltimore, pp 103–129

Zurkowski W, Lorkiewicz Z (1979) Plasmid-mediated control of nodulation in *Rhizobium trifolii*. Arch Microbiol 123:195–201

Symbiotic Nitrogen Fixation Genes of *Rhizobium meliloti*

G. DITTA [1], D. CORBIN [1], S. LEONG [1], L. BARRAN [2], and D.R. HELINSKI [1]

Introduction

Gram-negative bacteria of the genus *Rhizobium* are unique in that they have the ability to infect and form nitrogen fixing nodules on the roots of leguminous plants. This symbiotic association is of fundamental importance since it represents one of the main avenues for entry of atmospheric nitrogen into the biosphere. Agriculturally, there is great interest in *Rhizobium* symbiosis not only because legumes are major sources of dietary protein, but because a detailed understanding of this symbiosis could lead to the development of new types of plant-bacteria interactions. A major task which confronts us is to try to understand the complex molecular mechanisms which are essential for nodule function. Identifying symbiotically essential plant and bacterial genes is a first step toward this goal and the advent of modern recombinant DNA techniques has tremendously accelerated the targeting of such genes. Our laboratory has contributed to this process through the development of a broad host range plasmid cloning system for rhizobia and through the construction of a gene bank of the alfalfa symbiont *Rhizobium meliloti* (Ditta et al. 1980). In this report we present a summary of recent work involving the use of this bank to investigate symbiotically essential *Rhizobium* genes.

nif Genes of *Rhizobium meliloti*

Identification of Nitrogenase

The *nif* gene cluster of *Klebsiella pneumoniae* has been extensively studied as a model system for nitrogen fixation, and much is known about the organization and regulation of expression of these genes (for recent review see Roberts and Brill 1981). Progress is now being made toward a similar understanding of the molecular events underlying symbiotic nitrogen fixation between rhizobia and legumes. Work in several laboratories has focused on investigating cloned DNAs representing the nitrogenase structural genes of *Rhizobium* since these genes are essential to any biological

1 University of California, San Diego La Jolla, CA 92093, USA
2 Current Address: Chemical and Biological Research Institute, Agriculture Canada, Ottawa, Ontario, Canada K1A0C6

Molecular Genetics of the Bacteria-Plant Interaction
(ed. by A. Pühler)

nitrogen fixation process. The nitrogenase genes are highly conserved at the DNA sequence level among all nitrogen fixing organisms (Ruvkun and Ausubel 1980), and cloned DNA from *Klebsiella* can be used as a hybridization probe to identify *Rhizobium* nitrogenase sequences. We have used such an approach to identify the nitrogenase genes in a previously constructed *R. meliloti* gene bank carried by the broad host-range cloning vector pRK290 (Ditta et al. 1980). An analysis of DNA fragments surrounding and including nitrogenase indicated that there is a clustering of symbiotically specific nitrogen fixation genes in this region (Corbin et al. 1982). We report here the full extent of this cluster. A stretch of approximately 14 kb of DNA containing predominantly symbiotically essential genes including nitrogenase is bounded on either side by at least 10 kb of symbiotically unessential DNA. Two centrally located promoters control divergent transcription of this region.

Transposon Mutagenesis

A useful technique has been described for creating transposon insertion mutations in selected regions of genomic DNA of Gram-positive bacteria (Ruvkun and Ausubel 1981). In this procedure, termed marker exchange, transposon insertions are obtained within a selected cloned DNA fragment located on a broad host-range plasmid vector in *E. coli.* The position of the insertion can be determined by restriction enzyme ananysis. The cloned DNA is introduced into the parental organism and selection is made for recombinants which replace the wild-type genomic allele with the plasmid-borne transposon-inserted allele. The latter step involves introducing a second incompatible plasmid to effect elimination of the original plasmid. The final result is a genomic transposon insertion mutation which can be scored for a particular phenotype.

Figure 1 shows a marker exchange analysis of the nitrogenase region of *R. meliloti* 102F34 where a total of 69 Tn5 insertions spanning 34 kb of *Rhizobium* DNA are represented. Tn5 is a 5.6 kb transposon conferring resistance to kanamycin and

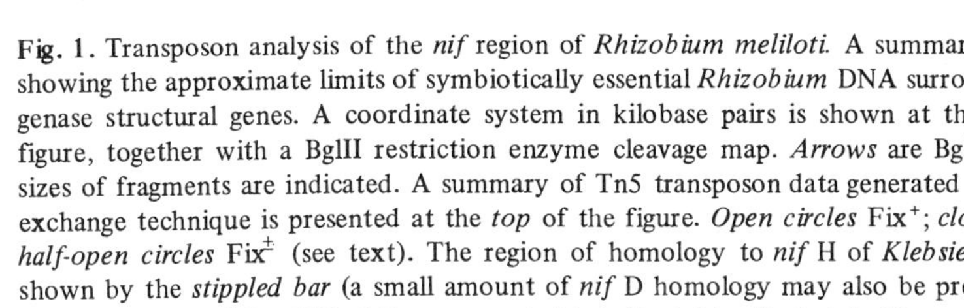

Fig. 1. Transposon analysis of the *nif* region of *Rhizobium meliloti.* A summary of genetic data showing the approximate limits of symbiotically essential *Rhizobium* DNA surrounding the nitrogenase structural genes. A coordinate system in kilobase pairs is shown at the *bottom* of the figure, together with a BglII restriction enzyme cleavage map. *Arrows* are BglII cleavage sites; sizes of fragments are indicated. A summary of Tn5 transposon data generated using the marker exchange technique is presented at the *top* of the figure. *Open circles* Fix^+; *closed circles* Fix^-; *half-open circles* $Fix^\pm$ (see text). The region of homology to *nif* H of *Klebsiella pneumonia* is shown by the *stippled bar* (a small amount of *nif* D homology may also be present); homology to *nif* D (and possibly *nif* K of *K. pneumoniae*) is indicated by the *hatched bar*

neomycin. The placement of insertion sites within the *Rhizobium* genome was verified by digesting total DNA from each mutant with an appropriate restriction enzyme (either BglII, HindIII or BamHI), followed by: (1) fractionation of the DNA fragments by agarose gel electrophoresis, (2) transfer of the DNA to nitrocellulose by the technique of Southern (1975) and (3) probing with radioactively labeled *nif* DNA lacking Tn5. In each case we saw the disappearance of a particular restriction enzyme fragment containing the wild-type allele and the appearance of two new fragments containing portions of Tn5. The symbiotic phenotype of each insertion mutant was assayed by inoculating alfalfa seedlings with the marker-exchanged strain and monitoring plant growth and acetylene reduction after 4–6 weeks. Three classes of insertions were found: (1) Fix^+, those that had no discernible effect on either plant growth or acetylene reduction; i.e., they were as effective as wild-type *R. meliloti;* (2) Fix^-, those that were totally ineffective on alfalfa, showing plant growth and acetylene reduction indistinguishable from uninoculated controls; and (3) $Fix^{\pm}$, a set of three insertions at coordinate 23.4 which were very weakly positive (i.e., $< 5\%$ of Fix^+) for plant growth and acetylene reduction. These three insertions map very close to one another ($<$ 100 bp) but are distinguishable. All symbiotically defective mutants formed small white nodules (i.e., were Nod^+), suggesting that genes involved in the early stages of invasion and nodule formation are not located in this region.

The two most distant Fix^- mutations in Fig. 1 are 14.3 kb apart. This region clearly represents a unique cluster of *nif* genes in *R. meliloti* 102F34, as it is bounded on both sides by at least 10 kb of symbiotically unessential DNA. An additional 10 kb of DNA to the right of that depicted in Fig. 1 has also been examined and found to contain sequences that are not needed for symbiosis.

A number of Fix^+ insertions were obtained which are bounded on either side by Fix^- insertions. If Tn5 mutations have strong polar effects on the expression of downstream genes in *Rhizobium,* then these Fix^+ Tn5's can be taken to delimit operon boundaries. In *E. coli,* Tn5 insertions are strongly polar (Berg 1977), although weak transcription out of the transposon was observed approximately one third of the time for insertions into *lac* (Berg et al. 1980). We have carefully examined the effect of Tn5 insertions in our strain of *R. meliloti* and have found that this transposon is surprisingly quite capable of effecting high levels of transcription of adjacent DNA sequences. This analysis was done using a cloned 0.8 kb HindIII fragment carrying the structural gene and presumptive ribosome binding site for the enzyme chloramphenicol acetyltransferase (CAT) (Close and Rodriguez, manuscript submitted). This enzyme confers resistance to the antibiotic chloramphenicol(CM). Since the promoter region for CAT is missing on the fragment, expression of CM resistance is dependent on transcription from an outside promoter. When the CAT fragment was inserted into the HindIII site at coordinate 13.5 in the nitrogenase structural gene region, cells expressed a level of CM resistance which was barely above background (as measured without the chloramphenicol fragment). This is consistent with the fact that the *nif* genes of *R. meliloti* are not transcribed under vegetative growth conditions (Corbin et al. 1982). The introduction of Tn5 at various positions upstream from the CAT gene, however, resulted in the expression of significantly higher levels of CM resistance. Similar data were obtained with the well-characterized δ-aminolevulinic acid synthetase (δ-ALAS) gene (Leong et al.

1982) which indicated that Tn5 insertions between the δ-ALAS promoter and the CAT gene could yield high levels of CAT expression and were not necessarily polar. At this time, it is not clear whether the orientation of Tn5 has an effect on outward transcription since not all insertions in the above study elicited the same level of CM resistance. In general, there was a decline in the expression of CAT the further upstream was the insertion. While these studies have yet to be completed, it is clear that no definitive conclusions can be drawn about operon structure in the *nif* region from transposon insertion analysis alone.

RNA Studies

Direct analysis of the RNA population in a developmental system can reveal much about the organization and expression of important genes. In *R. meliloti* (Corbin et al. 1982) and *R. leguminosarum* (Krol et al. 1980), nodule-specific transcription of specific DNA sequences, including nitrogenase, has been reported. Using Northern blot analysis we have identified the nitrogenase transcript in *R. meliloti* (Fig. 2, lanes a, b) and it appears to be appreciable larger than that of *R. leguminosarum* (Krol et al. 1982). Our experiments were done by isolating total nodule RNA from 4–6 week old nodules, removing trace amounts of DNA, running the RNA in a denaturing formaldehyde gel, transferring the RNA to a nitrocellulose filter and probing the filter with nick-translated DNA containing the nitrogenase genes. For the latter purpose we used the 3.6 kb BglII fragment at coordinates 11.1 to 14.7 since it contains all of the homology to the cloned nitrogenase gene region of *K. pneumoniae* (Corbin et al. 1982). The unavailability of appropriate RNA markers precludes a precise assign-

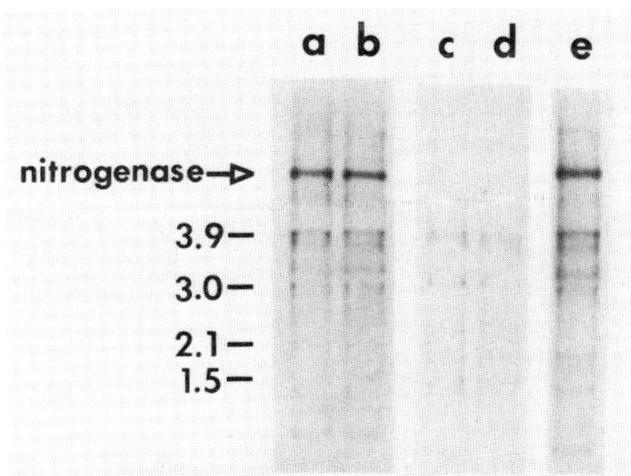

Fig. 2. Northern analysis of *R. meliloti* nodule RNA. *Lanes a,b* total nodule RNA from 3- and 6-week-old nodules, respectively, probed with the 3.6 kb BglII fragment (Fig. 1) carrying the nitrogenase structural genes. *Lanes c,d* total RNA as above probed with the 8.0 kb BglII fragment (Fig. 1). *Lane e* control lane identical to lane a, but without treating the RNA to remove trace DNA contamination. *Numbers to the left of lane a* denote the position of bacterial and plant ribosomal RNAs visible in the ethidium bromide stained gel. The sizes of these RNAs (in kb) are assumed to be approximately the same as for *E. coli* and higher plants

ment of size, but based on the position of ribosomal RNA present in the sample, the *R. meliloti* nitrogenase transcript is estimated to be approximately 5.5–6.9 kb.

Since RNA homologous to other portions of the *nif* cluster could not be unambiguously detected by Northern analysis (Fig. 2, lanes c, d), we employed a different strategy to identify additional *nif*-specific transcripts. Individual double-stranded DNA fragments were radioactively labeled at either their 5′ or 3′ ends, denatured, and hybridized with total nodule RNA from 4–6 week old plants. The conditions of the hybridization reaction (temperature, formamide concentration) were chosen such that RNA-DNA duplex formation was favoured over DNA-DNA formation. Unhybridized DNA was eliminated after hybridization by treatment with SI nuclease and the "protected" DNA was analyzed by alkaline agarose gel electrophoresis followed by autoradiography. This procedure, originally described by Berk and Sharp (1977), is a relatively simple yet sensitive way to detect and map RNA transcripts to specific DNA fragments. If a transcript originates within a particular 5′ end-labeled DNA fragment and extends past one end, then a labeled hybrid of reduced size relative to the unhybridized fragment will be observed. The size of the hybrid allows precise positioning of the 5′ end of the RNA relative to one of the 5′ ends of the DNA fragment. The same technique will map the 3′ end of RNAs on 3′ end-labeled fragments. The direction of transcription can be deduced by asymmetrically cleaving the DNA fragment with a restriction enzyme prior to hybridization.

Such an analysis has been applied to different DNA fragments from the *R. meliloti nif* cluster and a summary of the results is shown in Fig. 3. Two divergent transcripts (RNA1 and RNA2) with 5′ ends 1.9 kb apart have so far been detected. The DNA between these transcripts is either not transcribed in the nodule, or the steady state level of RNA from this region is exceedingly low relative to RNA1 and RNA2. This agrees with the results of transposon analysis (Fig. 1) indicating that no symbiotically essential genes are contained here.

RNA1 covers the region coding for the nitrogenase structural genes. The presence of a large transcript in this region agrees with the results from Northern analysis shown in Fig. 2 and with the genetic data of Ruvkun et al. (1982). The latter authors showed that genomic transposon insertions to the left of coordinate 14.4 could be complemented by cloned DNA encompassing this region only if DNA sequences to the right were also present in *cis*. DNA from the latter region would include the presumptive promoter for RNA1. Ruvkun et al. (1982) also report the isolation of two closely spaced transposon insertions near coordinate 13.5 which had weakly positive symbiotic phenotypes. It was suggested that these might represent intergenic insertions coupled with incomplete polarity of Tn5. We have obtained a single Tn5 insertion at coordinate 13.5 which has a totally normal symbiotic phenotype. This is consistent with the fact that Tn5 mutations in our *Rhizobium* strain can be nonpolar. All three of these insertions probably define an intergenic region between *nif*H and *nif*D. DNA sequence analysis indicates that the structural gene for *nif*H terminates 200 bp to the right of a HindIII site a coordiante 13.5 (Török and Kondorosi 1981).

The size of RNA1 (> 5.5 kb) suggests that it may code for symbiotically essential gene product(s) other than the nitrogenase polypeptides. The latter would require only a 4.3 kb coding sequence in *K. pneumoniae* corresponding closely to the reported size of nitrogenase transcript from *R. leguminosarum* (Krol et al. 1982).

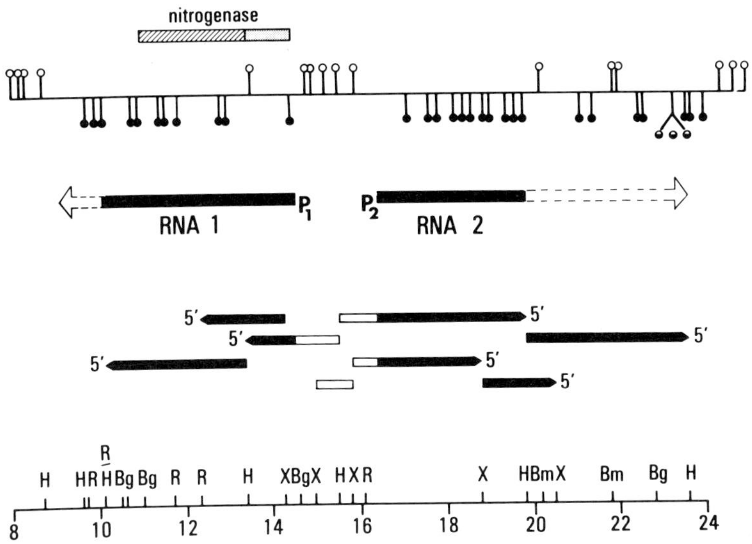

Fig. 3. S1 nuclease protection assay of the *R. meliloti* region. A summary of the DNA fragments and labeling scheme used to define RNA1 and RNA2. *Upper portion* of the drawing is an enlargement of part of the transposon data in Fig. 1; the symbols are the same. The coordinate system at the *bottom* is also the same as in Fig. 1. A more complete restriction enzyme map is shown here: *Bg* BglII; *R* EcoRI; *H* HindIII; *Bm* BamHI; *X* XhoI. *Bars* with 5′ ends represent DNA fragments labeled at their 5′ termini which were used in an S1 protection assay with total nodule RNA; the *solid regions* were protected by hybridization with RNA, the *open regions* unprotected; the *pointed ends* indicate that polarity of the transcript was determined relative to the restriction map. *Solid bars* labeled RNA1 and RNA2 represent minimum single transcripts based on these data. *Dashed arrows* show the direction of transcription and indicate possible transcript extensions. P_1 and P_2 are presumptive promoter regions

In *K. pneumoniae,* the nitrogenase genes, K, D and H are transcribed as part of an operon coding for another gene of 24,000 d molecular weight, *nif*Y. There is obviously room for a similar protein on the nitrogenase trascript of *R. meliloti.*

RNA2 seems to be less abundant than RNA1 in our experiments. Minimally, it covers 3.4 kb of DNA, from coordinate 16.4 to coordinate 19.8 in Fig. 3. We have observed transcription from RNA2 to extend an additional 3.8 kb beyond this, to at least coordinate 23.6, since DNA fragments at coordinates 19.8 to 23.6 and 18.8 to 20.5 (Fig. 3) can be fully protected in an S1 protection assay. It is not clear at the moment, however, how much of this transcription is essential for gene expression and to what extent it represents read-through due to leaky termination. Since Northern analysis failed to reveal a unique transcript for this region (Fig. 2) the data obtained to date must be interpreted with caution. Other studies using radioactively labeled total nodule RNA as probe indicate that a significant level of transcription does not extend beyond the BglII cleavage site at coordinate 25.4 (Fig. 1) which is consistent with the transposon insertion data (data not shown). Symbiotically positive Tn5 insertions were obtained at coordinates 20.2, 21.9, and 22.0, indicating that these are within or near intergenic regions. No function has yet been assigned to any of the gene products encoded by DNA to the right of coordinate 16.4.

It is well to consider the implications and limitation of these data. A number of similarities exist between the *R. meliloti nif* cluster and the *nif* cluster of *K. pneumoniae.* By definition, both are essential for nitrogen fixation. Both are expressed only under specific conditions and both contain the nitrogenase structural genes. Aside from nitrogenase, however, it is not known whether the remaining gene products of the *R. meliloti nif* cluster are indeed concerned with biochemical aspects of nitrogen fixation during symbiosis or whether they have some more unique role in nodule function. The maximal single strand coding capacity of the *R. meliloti nif* cluster examined here is only approximately 13.8 kb (15.7 kb as defined for the entire region by transposon insertion minus 1.9 kb between RNA1 and RNA2). This is considerable less than the nearly 24 kb required for the 17 tightly packed *nif* genes of *Klebsiella* (Cannon et al. 1979, Riedel et al. 1979). Either fewer genes are required for nitrogen fixation in *R. meliloti,* or additional symbiotically specific genes have yet to be discovered. Another possibility is that the informational density of this *nif* cluster is greater than supposed here. There is ample precedent for overlaping genes in other systems (Barrell et al. 1976), even for overlapping convergent transcripts (Rak et al. 1982). The latter could be present at lower abundance than RNA1 and RNA2, making detection and assignment difficult.

In the absence of an RNA processing event, the 5′ ends of RNAs 1 and 2 can be assumed to define *Rhizobium nif* promoter regions (P_1 and P_2, Fig. 3). Control mechanisms operating at these promoters are of fundamental importance for symbiosis. While it is clear that RNA1 and RNA2 represent important transcription of the *nif* cluster in mature bacteroids, it should be kept in mid that transcription from other less active promoters could also play a role, particularly early in nodule establishment. The steps leading to initial activation of the *nif* region in *R. meliloti* will have to be carefully examined either by developing the means to active *nif* transcription in vitro, or by the more difficult process of directly examining the earliest stages of nodule formation.

δ-Aminolevulinic Acid Synthetase

Heme biosynthesis in *Rhizobium* is intimately involved in symbiotic nitrogen fixation. On the one hand, heme is an essential component of bacterial cytochromes and plays a direct role in bacteroid respiration. On the other, bacterial heme is also a component of leghemoglobin, the nodule-specific hemoprotein responsible for regulating oxygen tension during symbiosis. It has been estimated that a 10–20-fold increase in heme synthesis by the bacteria must occur to keep pace with the demands of leghemoglobin production (Dilworth and Appleby 1979). Our laboratory is investigating the regulatory pathways controlling heme biosynthesis during symbiosis and we have cloned the structural gene for δ-aminolevulinic acid synthetase (δ-ALAS), the first enzyme in the heme biosynthesis pathway (Leong et al. 1982). This enzyme catalyses the condensation of glycine and succinyl coenzyme A to form δ-aminolevulinic acid (δ-ALA), and has been implicated as a major control point for heme biosynthesis in a number of organisms. In *R. japonicum,* δ-ALAS undergoes a substantial increase in activity both during symbiosis (Nadler and Avissar 1977) and during vegetative growth under reduced oxygen tension (Avissar and Nadler 1978).

The first step in cloning δ-ALAS involved isolating and characterizing a mutant defective in the enzyme. Two such mutants were obtained by mutagenizing *R. meliloti* 102F34 with N-methyl-N′-nitro-N-nitrosuguanidine and screening survivors for δ-ALA auxotrophy during vegetative growth. Both were also symbiotically defective in the absence of exogenously supplied δ-ALA. Subsequently, the mutants were complemented with cloned DNA fragments contained in the *R. meliloti* gene bank and a single hybrid plasmid was identified which contained the wild type δ-ALAS allele. This plasmid, pRK290.790, is shown in Fig. 4. Localization of the δ-ALAS gene on the plasmid was accomplished by extensive Tn5 mutagenesis. Approximately 90 δ-ALAS-negative mutants were identified and used to map the limits of the gene to a 1.5 kb stretch of DNA contained within the largest BglII fragment. Precise 5′ and 3′ ends for the δ-ALAS transcript have been determined and these suggest that δ-ALAS is coded for by a single gene operon of approximately 1.5 kb which is transcribed in the direction shown.

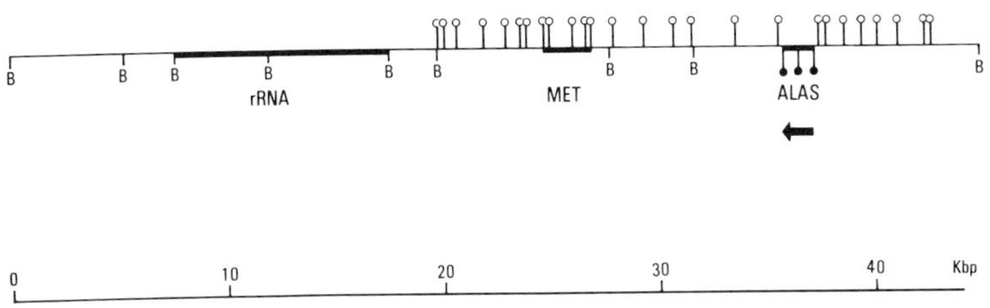

Fig. 4. Analysis of the δ-ALAS genetic region in *R. meliloti.* Sequences surrounding δ-ALAS were analyzed for symbiotic essentiality by marker exchange with Tn5. *Open circles* Fix$^+$ insertions, *solid circles* Fix$^-$. *Arrow* direction and limits of transcription of δ-ALAS. *MET* a region essential for methionine prototrophy; *rRNA* approximate region of homology to ribosomal RNA. A coordinate system is provided at the *bottom* in kb

It has been demonstrated that the cloned δ-ALAS gene is in fact the structural gene for δ-ALAS and not a regulatory controlling biosynthesis or enzyme activity (Leong et al. 1982). As such, it is only the second example of a cloned symbiotically essential gene which has been explicitly defined biochemically (the other being nitrogenase). An immediate question which comes to mind is whether there might be additional symbiotically essential genes located near δ-ALAS, one possibility being other genes involved in heme biosynthesis. To address this point, the technique of marker exchange was used to mutagenize the region surrounding δ-ALAS with Tn5. The symbiotic phenotype of various Tn5 mutants created in this way is shown in Fig. 4. Except for insertions directly into the δ-ALAS gene, all marker-exchanged strains were totally effective. We did find that methionine auxotrophs were created by insertions covering a 2.7 kb range, from coordinate 24.7 to 27.4. RNA Northern blots have revealed that pRK290.790 carries homology to both 23S and 16S ribosomal RNAs of *Rhizobium.* A comparison of gene bank clones which strongly hybridize with radioactively labeled total RNA suggests that this homology is probably contained somewhere within the two BglII fragments spanning coordinates 7.6 to 17.5.

Work is continuing in our laboratory to detail the regulation of expression of δ-ALAS during both vegetative growth and symbiosis. As an example of a normal "housekeeping" gene of *Rhizobium* which is called upon to play a key role during symbiotic development, studies of δ-ALAS will provide an interesting counterpoint to the rapidly unfolding story of the *Rhizobium nif* region.

Summary

The nitrogenase genes are part of a large cluster of symbiotically essential nitrogen fixation *(nif)* genes in *Rhizobium meliloti.* This *nif* region spans 14–16 kb of *Rhizobium* DNA with divergent transcription occuring from two centrally located promoters. Previous work has shown that transcription of the *nif* region is specific for the symbiotic state. Another symbiotically essential gene is that encoding δ-aminolevulinic acid (δ-ALAS), the first gene in the heme biosynthetic pathway. In contrast to the *nif* region, however, δ-ALAS is also required for normal vegetative growth of *Rhizobium.* δ-ALAS is organized as a single gene operon and is closely linked to ribosomal RNA genes and a methionine biosynthetic gene. There is no clustering of symbiotically essential function near δ-ALAS. An understanding of the organization and expression of symbiotically important genes is essential to the manipulation of the *Rhizobium*-legume symbiosis. This ability will lead to innovative technology for the enhancement of agricultural productivity.

Acknowledgements. We gratefully acknowledge the excellent technical assistance of Marty Yanofsky and Patty Tooker. This work was supported by grant NSF PCM 82-04730 from the National Science Foundation. Sally Leong is a recipient of an NIH Postdoctoral Fellowship.

References

Avissar YJ, Nadler KD (1978) Stimulation of tetrapyrole formation in *Rhizobium japanicum* by restricted aeration. J Bacteriol 135:782–790

Barrell BG, Air GM, Hutchinson CA (1976) Overlapping genes in bacteriophage X174. Nature 264:33–41

Berg DE (1977) Insertion and excision of the transposable kanamycin resistance determinant Tn5. In: Bukhari AL, Shapiro JA, Adhya SL (eds) DNA insertion elements, plasmids, and episomes. Cold Spring Harbor Laboratory, Cold Spring Harbor, New York, p 205

Berg DE, Weiss A, Crossland L (1980) Polarity of Tn5 mutations in *Escherichia coli.* J Bacteriol 142:439–446

Berk AJ, Sharp PA (1977) Sizing and mapping of early adenovirus mRNAs by gel electrophoresis of S1 endonuclease-digested hybrids. Cell 12:721–732

Cannon FC, Riedel GE, Ausubel FM (1979) Overlapping sequences of *Klebsiella pneumoniae nif* DNA cloned and characterized. Mol Gen Genet 174:59–66

Close T, Rodriguez R (submitted)

Corbin D, Ditta G, Helinski DR (1982) Clustering of nitrogen fixation *(nif)* genes in *Rhizobium meliloti.* J Bacteriol 149:221–228

Dilworth MJ, Appleby CA (1979) Leghemoglobin and *Rhizobium* hemoproteins. In: Hardy RWF, Bottomley F, Burns RC (eds) A treatise on dinitrogen fixation, sect. I and II. John Wiley, New York, p 691

Ditta G, Stanfield S, Corbin D, Helinski DR (1980) Broad host range DNA cloning system for Gram-negative bacteria: Construction of a gene bank of *Rhizobium meliloti*. Proc Natl Acad Sci USA 77:7347–7351

Krol AJM, Hontelez JFJ, Bos RC Van den, Kammeu A Van (1980) Expression of large plasmids in the endosymtiotic form of *Rhizobium leguminosarum*. Nucleic Acids Res 8:4337–4347

Krol AJM, Hontelez JGJ, Roozendaal B, Kammen A Van (1982) On the operon structure of the nitrogenase genes of *Rhizobium leguminosarum* and *Azotobacter vinelandii*. Nucleic Acids Res 14:4147–4157

Leong SA, Ditta GS, Helinski DR (1982) Heme biosynthesis in *Rhizobium*. Identification of a cloned gene coding for δ-aminolevulinic acid synthetase from *Rhizobium meliloti*. J Biol Chem 257:8724–8730

Nadler KD, Avissar YJ (1977) Heme biosynthesis in soybean root nodules. I. On the role of bacteroid δ-aminolevulinic acid synthetase and δ-aminolevulinic acid dehydrase in the synthesis of the heme of leghemoglobin. Plant Physiol 60:433–436

Rak B, Lusky M, Hable M (1982) Expression of two proteins from overlapping and oppositely oriented genes on transposable DNA insertion element IS5. Nature 297:124–128

Riedel GE, Ausubel FM, Cannon FC (1979) Physical map of chromosomal nitrogen fixation *(nif)* genes of *Klebsiella phneumoniae*. Proc Natl Acad Sci USA 76:2866–2870

Roberts GP, Brill WJ (1981) Genetics and regulation of nitrogen fixation. Annu Rev Microbiol 35:207–235

Ruvkun GB, Ausubel FM (1980) Interspecies homology of nitrogenase genes. Proc Natl Acad Sci USA 77:191–195

Ruvkun GB, Ausubel FM (1981) A general method for site-directed mutagenesis in prokaryotes. Nature 289:85–88

Ruvkun GB, Sundaresan V, Ausubel FM (1982) Directed transposon Tn5 mutagenesis and complementation analysis of *Rhizobium meliloti* symbiotic nitrogen fixation genes. Cell 29: 551–559

Southern EM (1975) Detection of specific sequences among DNA fragments separated by gel electrophoresis. J Mol Biol 98:503–517

Török I, Kondorosi A (1981) Nucleotide sequence of the *R. meliloti* nitrogenase reductase (*nif*H) gene. Nucleic Acids Res 9:5711–5723

Vector Plasmids for in-Vivo and in-Vitro Manipulations of Gram-Negative Bacteria

R. SIMON, U. PRIEFER, and A. PÜHLER [1]

Introduction

In the last few years there has been an increasing interest in the molecular genetics of Gram-negative bacteria other than *E. coli,* especially of soil bacteria that exhibit important biological properties, e.g., symbiotic nitrogen fixation or induction of plant disease. Until recently, modern techniques such as transposon mutagenesis and molecular gene cloning have been restricted to *E. coli* and closely related organisms. However, they are now being adopted for use in a variety of other Gram-negative bacteria: special plasmids have been constructed which allow the introduction of transposons into different bacterial genera (Beringer et al. 1978, Van Vliet et al. 1978) and cloning vectors which are maintainable in a wide range of Gram-negative bacteria are also available (Bagdasarian et al. 1979, 1981, 1982, Ditta et al. 1980).

In this paper we describe the construction and use of some new vector plasmids that might facilitate in-vivo manipulation by transposon mutagenesis and also allow mobilisation of cryptic non-selftransmissible plasmids. We also describe new cloning vectors which combine the advantages of the usual *E. coli* vectors with the broad host range mobilisation and replication functions of the IncQ/P plasmid RSF1010.

Results and Discussion

The Mobilising System

Construction of Mobilisable Vector Plasmids

The usual *E. coli* vector plasmids such as pBR325 (Bolivar et al. 1977) or pACYC184 (Chang and Cohen 1978) are neither selftransmissible nor efficiently mobilisable.

In order to provide these very versatile vector plasmids with the ability to be mobilised we cloned small restriction fragments of the broad host range plasmid RP4 into linearised vector DNA. The principle of this shot gun cloning experiment is outlined in Fig. 1. The resulting mobilisable hybrid plasmids carry a cloned DNA fragment believed to contain the *ori*T region that enables the vector plasmids to be efficiently transferred in the presence of RP4. Since we believe this DNA region to be

1 Lehrstuhl für Genetik, Fakultät für Biologie, Universität Bielefeld, 4800 Bielefeld, FRG

Molecular Genetics of the Bacteria-Plant Interaction
(ed. by A. Pühler)

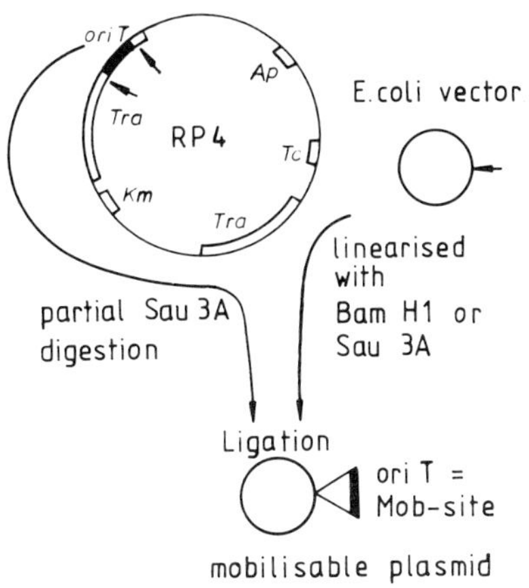

Fig. 1. Construction of mobilisable vector plasmids. The RP4 DNA was partly digested with *Sau*3A and the desired *E. coli* vector was linearised with *Bam*H1 or by very mild treatment with *Sau*3A. After ligation and transformation the resulting hybrid plasmids were screened for the ability to be mobilised. *Ori* T refers to the origin of transfer replication which has been mapped within the transfer *(Tra)* region (Guiney and Helinski 1979)

recognized in trans by the transfer functions of conjugative IncP plasmids, we call this DNA sequence the Mob-site. In mating experiments between *E. coli* strains the frequency with which RP4 mobilises the Mob-site containing vectors is nearly 100%. However, in most cases the mobilising plasmid RP4, which may be undesirable in some experiments, we constructed special *E. coli* donor strains which carry RP4 derivatives integrated in their chromosomes.

The Mobilising Strains

If two different derivatives of plasmid RP4 are forced to co-exist in a recA deficient *E. coli* strain the cells grow very poorly under coselective pressure. In rare cases normally growing colonies can be isolated, some of which contain one of the two incompatible RP4 derivatives integrated in the chromosome (R. Simon, unpubl. observation). We took advantage of these incompatibility reactions to enforce the integration of RP4 plasmids into the chromosomes of *E. coli* strains. Table 1 shows the properties of the resulting strains. The transfer genes of the integrated RP4 derivatives are still expressed and can be used in conjugation experiments to mobilise vector

Table 1. The mobilising strains

Strain number	Integrated plasmid	Resistance markers		Other important properties
		RP4	Chrom.	
S 68-7	RP4-2 (Tc: :Mu)	Km	Sm	Thi, Pro, His, su$^-$
SM10	RP4-2 (Tc: :Mu)	Km	–	Thi, Thr, Leu, su$_{III}$
S 17-1	RP4-2 (Tc: :Mu) (Km: :Tn7)	Tp-Sm	–	Pro, res$^-$ mod$^+$

The *E. coli* donor strains listed in the table are recA deficient. S68-7 originates from CSH52, SM10 from C600, and S17-1 from *E. coli* 294. The chromosomally integrated RP4 derivatives are derived from RP4-2. This plasmid has a deletion in Tn*1* causing Ap sensitivity. In addition Tc resistance was inactivated by phage Mu insertion (Tc: :Mu) and in S17-1 the Km resistance was eliminated by Tn7 insertion (Km: :Tn7) plasmids carrying the RP4 specific Mob-site. We call this type of donor a mobilising-strain. The use of mobilising strains has the important advantage that the mobilising plasmid (e.g., RP4) is not transferred to the recipients, whereas the transfer frequency of the mobilisable vector plasmid is as high as with wild type RP4.

Class I Vector Plasmids

We define class I vectors as plasmids that carry the RP4 specific Mob-site and that replicate only in *E. coli.* The basic replicons are pACYC184, pACYC177, and pBR325. The Mob-site was cloned into them as has been shown schematically in Fig. 1. Figure 2 shows the genetic and physical maps of the class I vectors constructed so

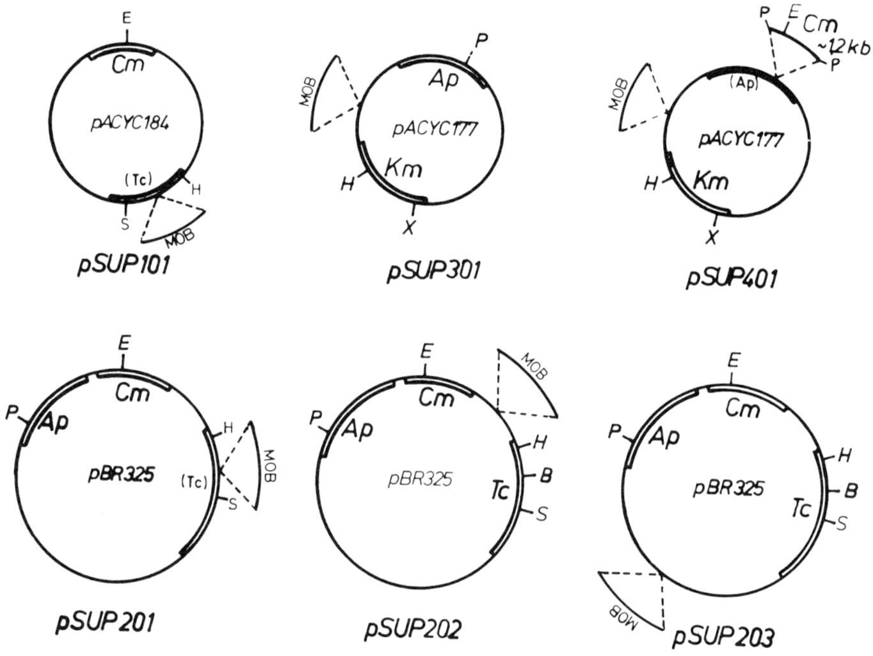

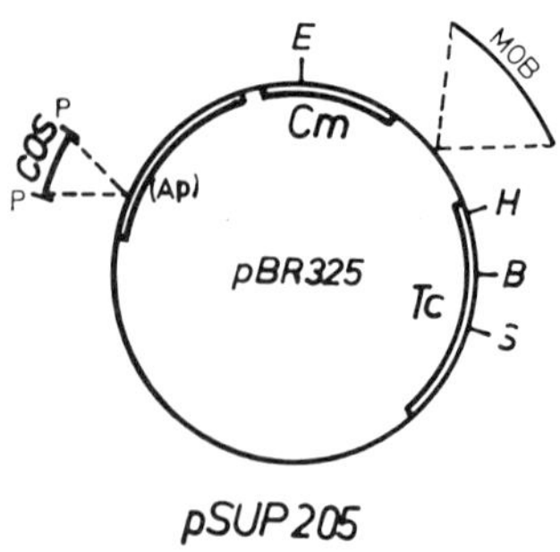

Fig. 2. Class I vector plasmids. The plasmid name within the *circle* refers to the basic replicon. The Mob-sites are located on *Sau*3A generated fragments of RP4. In pSUP101, 301, 401, and 201 the Mob fragments were cloned into the unique *Bam*H1 sites of the vector plasmids. In pSUP202, 203, and 205 one of the numerous *Sau*3A sites was used to clone the Mob-site. The sizes of the Mob-site containing DNA fragments are in the range of 1.9–2.6 kb. The *Pst*I fragment containing the λ cohesive ends *(Cos)* was taken from the plasmid MUA-10 (Meyerowitz et al. 1980). The *Pst*I fragment coding for Cm resistance in pSUP401 originates from the phage P1cmts (W. Klipp, unpubl.). *Ap* Ampicillin resistance; *Km* Kanamycin resistance; *Tc* Tetracycline resistance; *Cm* Chloramphenicol resistance; *E Eco*RI; *B Bam*HI; *H Hin*dIII; *P Pst*I; *S Sal*I; *X Xho*I

far. Because class I vectors are based on *E. coli* specific narrow host range replicons they cannot replicate in bacteria which are not closely related to *E. coli.* However, since the broad host range transfer functions of RP4 are used to mobilise these vectors it should be possible to introduce them into any Gram-negative bacterium that can act as recipient for conjugal transfer of RP4. Therefore class I vector plasmids are most useful for transposon mutagenesis of many Gram-negative bacteria, for example the Rhizobiaceae or other soil bacteria. The usefulness of class I vectors for transposon mutagenesis of non-*E. coli* strains is demonstrated by the following experiment. Plasmid pSUP2021 is a mobilisable pBR325 derivative carrying the Nm resistance transposon Tn*5* (see Fig. 3a). It was mobilised from the mobilising strain S17-1 into *R. meliloti* 2011 (Sm^r). The mating was done using filter membranes on rich medium for about 5 h at 32°C. The cells were then suspended in sterile water and plated onto selective medium containing 500 mg/l Sm and 100 mg/l Nm. Nm^r transconjugants were obtained at a frequency of 10^{-5} per recipient. About 1%–2% of these Tn*5* containing clones were auxotrophic mutants demonstrating the potential of the system for the isolation of transposon induced mutants. Class I vector plasmids can also be used for site-specific transposon mutagenesis as shown in the following model experiment. A DNA fragment from *R. meliloti* containing the *nif*H gene (Ruvkun and Ausubel 1980) was cloned into the mobilisable vector pSUP2121 (Fig. 3b) was then mobilised from strain S17-1 into *R. meliloti* 2011, the strain from which the *nif*H gene originated. Selection for Nm resistance (i.e., for the introduction of Tn*5* into the genome of the recipient) yielded transconjugants at a frequency of about 10^{-3}. This is about 100 times higher than the frequency obtained with a random Tn*5* transposition experiment using pSUP2021 (see above). This increase indicates that in the second experiment homologous recombination has occurred between the wild type *nif* DNA region of the recipient and the Tn*5* mutated *nif* sequence present on pSUP2121.

Several hundred Nm^r transconjugants were tested for the presence of the vector encoded resistance markers (Ap, Tc). Most of the clones were Ap^r and Tc^r which can be explained by a single recombination event leading to the integration of the whole vector plasmid. DNA sequence exchange between the plasmid borne *nif*H::Tn*5* and the wild-type *nif*H of the host requires a double cross over and is indicated by ampicillin and tetracycline sensitivity of the Nm^r transconjugants (loss of the vector plasmid). This process, which is called homogenotisation, occurred in about 2% of the

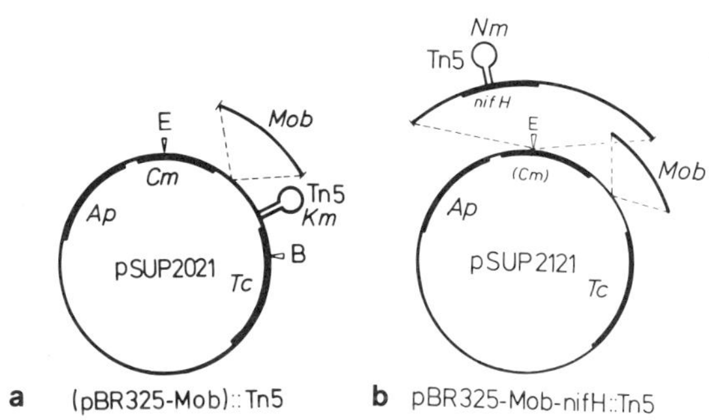

Fig. 3a, b. Class I vector plasmids for Tn*5* mutagenesis. **a** pSUP2021 is pSUP202 (see Fig. 2), but carrying the Nm resistance transposon Tn*5*. **b** pSUP2121 is also derived from pSUP202. The DNA fragment carrying the *nif*H gene of *R. meliloti* was cloned into the unique *Eco*RI site. The *nif*H gene was mutagenized by Tn*5* insertion

Tn5 containing transconjugants. Ten Nm^r Tc^S Ap^S clones were tested in plant inoculation experiments. All of them showed the expected Nif^- phenotype.

Class II Vector Plasmid

We define class II vectors as plasmids which are mobilisable by IncP plasmids and which are able to replicate in a wide range of Gram-negative bacteria. We combined the useful properties of *E. coli* vector plasmids (i.e., the antibiotic resistance markers and the cloning sites) with the mobilisation and replication properties of the broad host range plasmid RSF1010 (Guerry et al. 1974). The basic *E. coli* vectors again were pACYC184, pACYC177, and pBR325, whereas the replication and mobilisation functions were taken from pKT210, a chloramphenicol resistant derivative of RSF1010 (Bagdasarian et al. 1979). The cloning strategy was as follows: *E. coli* vector DNA was treated with very low concentrations of the four base specific restriction enzyme *Sau*3A so that, on average, only one cut per molecule was introduced. This linearised vector DNA was ligated to pKT210 DNA partially digested with *Sau*3A and the ligation mix was used to transform the mobilising strain S17-1. The transformed bacteria subsequently were crossed en masse with an *R. meliloti* recipient selecting for one of the vector encoded resistance markers. This selection resulted in the isolation of transconjugants which had received a vector derivative with the broad host range replication and mobilisation properties of RSF1010. Figure 4 shows the genetic

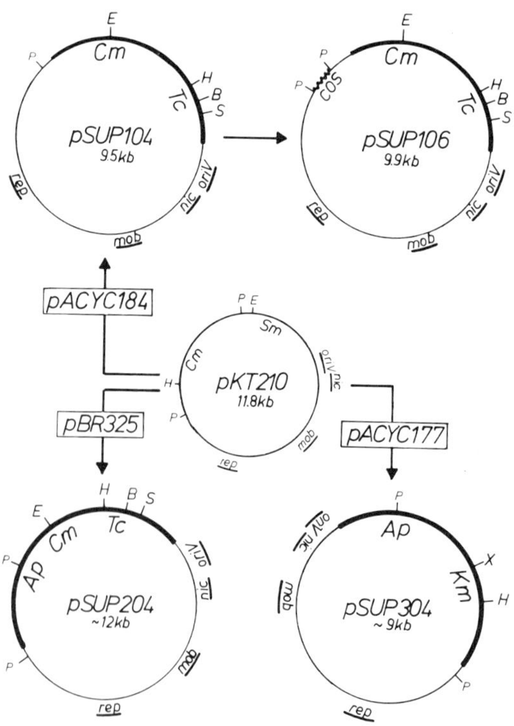

Fig. 4. Class II vector plasmids. The vector plasmids pSUP204 and pSUP-304 have been constructed by cloning the replication and mobilisation functions of pKT210 into pACYC184, pBR325, and pACYC177 respectively *(thick lines)*. Plasmid pSUP106 is derived from pSUP104 by inserting the λ-Cos site from plasmid MUA-10 (Meyerowitz et al. 1980) into the single *Pst*I site. *Cm, Tc, Ap, Km, Sm* genes for chloramphenicol, tetracycline, ampicillin, kanamycin, and streptomycin, respectively. *E, H, B, S, P, X* restriction sites for *Eco*RI, *Hin*dIII, *Bam*HI, *Sal*I, *Pst*I, and *Xho*I, respectively; *ori V* origin of vegetative replication; *rep* transacting replication function(s); *nic* origin of transfer replication (*ori*T); *mob* transacting mobilisation function(s)

and physical maps of the constructed class II vectors. They are mobilised at high frequency from an *E. coli* donor and are maintained in a variety of Gram-negative bacteria such as *R. meliloti, R. leguminosarum, R. trifolii,* and *Agrobacterium tumefaciens.* Other hosts have not been tested as yet but we expect our class II vectors to have retained the broad host range of RSF1010.

Construction and Use of Tn5-Mob

Genetic analysis of many plasmids is difficult if not impossible because of their lack of selectable markers and, additionally, in many cases they are transfer deficient. This is especially so for some of the symbiotic plasmids of Rhizobia, whose analysis is of great importance. One previously established way of genetic transfer is via cointegrate formation by means of P-type plasmids, e.g., R68.45 (Scott and Ronson 1982). We have established a system which not only confers selectable markers to these plasmids but in addition the ability to be mobilised. The idea was to introduce the Mob-site of RP4 into the transposon Tn*5*, so that both, the Nm resistance and the ability to be mobilised, could be translocated by Tn*5* transposition. This was verified by cloning the RP4 specific Mob-site into the unique *Bam*HI site of Tn*5* (in principle the same strategy as shown in Fig. 1). We call this transposon Tn*5*-Mob. The Nm resistance of Tn*5*-Mob is still retained and so also are the transposition properties of Tn*5*. Thus the Mob insertion has not affected those properties.

The Tn*5*-Mob loaded plasmid pSUP5011 is shown in Fig. 5. It belongs to the class I vector family and can therefore be used for transposon mutagenesis. In order to demonstrate the application of Tn*5*-Mob we show here the mobilisation of a megaplasmid from *R. meliloti.* The *R. meliloti* strain 2011 was mutagenised with Tn*5*-Mob. We isolated 20 Nm^r transconjugants with Tn*5*-Mob at unknown positions and introduced plasmid RP4-4 (a Nm sensitive derivative of RP4) by conjugation. These clones were then mated with *A. tumefaciens* and Nm^r transconjugants were selected. Seven clones showed a significant Nm resistance transfer (10^{-6}–10^{-7} per donor). We concluded from these data that in these clones Tn*5*-Mob had transposed to the megaplasmid which subsequently was mobilised by the coexisting RP4-4 plasmid. This assumption was confirmed by the analysis of the plasmids present in the Nm^r *A. tumefaciens* transconjugants. They all showed an additional plasmid corresponding to the megaplasmid of the *R. meliloti* 2011 donor (Fig. 6). The megaplasmid of *R. meliloti* is stably maintained in the *A. tumefaciens* clones and these strains have simultaneously acquired the ability to nodulate alfalfa plants although these nodules are ineffective.

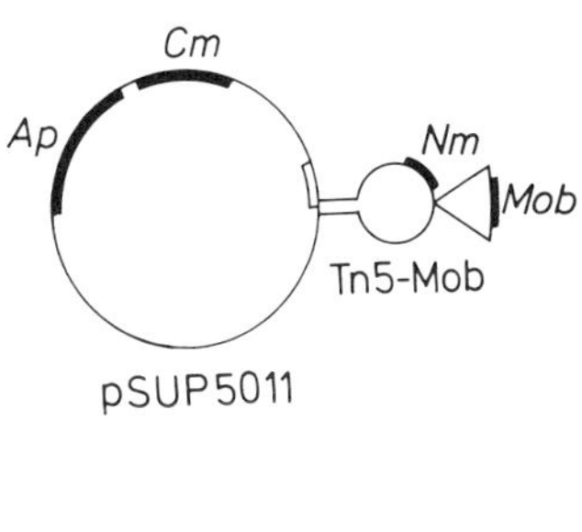

Fig. 5. The class I vector carrying Tn*5*-Mob. The basic replicon of pSUP5011 is pBR325. A deletion was introduced which removed part of the Tc resistance gene including the *Bam*H1 site. Tn*5* was inserted at the position shown in the figure and the Mob-site was cloned into the unique *Bam*H1 site of Tn*5*

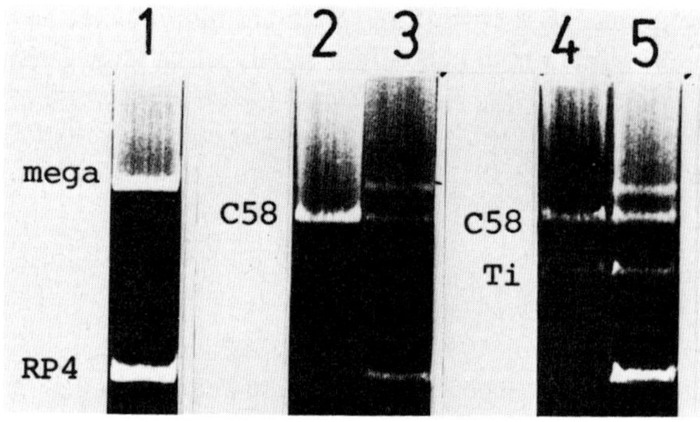

Fig. 6. Mobilisation of the *R. meliloti* megaplasmid into *Agrobacterium tumefaciens*. The strains were lysed using a modified Eckhardt lysis procedure (Eckhardt 1978). The DNA was run in a 0.8% agarose gel in tris-borate buffer for 3 h at 120 V, stained with ethidium bromide and photographed under UV light. *1* Donor *R. mel* 2011. It carries RP4-4 and Tn*5*-Mob inserted in the megaplasmid; *2* Recipient *A. tum* LBA 290, a Ti-plasmid free *A. tum* strain; *3* Nm resistant transconjugant from the mating between the *R. mel* donor and *A. tum* LBA 290; *4* Recipient *A. tum* LBA 275 which carries the Ti-plasmid and the cryptic plasmid C58; *5* Nm-resistant transconjugant from the mating between the *R. mel* donor and *A. tum* LBA 275

By mobilising the megaplasmid of *R. meliloti* to *A. tumefaciens* by means of Tn*5*-Mob we could show genetically that at least the genes responsible for host specificity and recognition and for the early steps of nodulation are located on the megaplasmid and are expressed in *A. tumefaciens*.

Concluding Remarks

In this paper we describe the potential of some new vector plasmids which were developed as tools for the in-vivo genetic manipulation of Gram-negative bacteria unrelated to *E. coli*. The initial idea was to provide usual *E. coli* vectors with a DNA sequence (the Mob-site) which facilitates their mobilisation by RP4 into many Gram-negative genera. On this basis we constructed two classes of vector plasmids. Class I vectors replicate only in *E. coli* whereas class II plasmids are transferred and maintained in many other bacteria. Class I vectors are useful vehicles for the introduction of transposons into many hosts. Here we have discussed random Tn*5* mutagenesis of *R. meliloti*. But the methodology is also applicable to other transposons as well as to other bacteria. To demonstrate the use of class I vectors in site-specific transposon mutagenesis experiments we showed the isolation of *nif*H::Tn*5* mutants of *R. meliloti*. Similar gene replacement experiments should be possible with many other cloned DNA fragments from a wide range of Gram-negative bacteria. The main advantages of class I vectors compared to the pRK290 system (Ruvkun and Ausubel 1981) are: they are multicopy plasmids, they have more cloning sites, and the homogenotisation experiments are quicker (they need only one bacterial mating). Class II vectors can also be easily mobilised from *E. coli* to many other bacteria. They have broad host range replication properties. This fact allows genetic analysis (e.g., complementation studies) of cloned DNA fragments of different origin in the original or in many other host organisms. Finally, the Tn5-Mob transposon described in this paper should be of great value if one wants to mobilise non-selftransmissible plasmids. We demonstrated the labelling of the *R. meliloti* megaplasmid with Tn5-Mob and the subsequent mobilisation into other strains. Again, the Tn5-Mob system is not restricted to *R. meliloti* plasmids. Tn5-Mob might also facilitate genetic analysis of many other replicons (including the chromosome), whenever conjugal transfer is the method of choice.

Acknowledgements. We wish to thank Mrs. K. Aufderheide for technical assistance, Mrs. A. Schulte for photographic aid and Mrs. S. Malmivaara for typing the manuscript. We would like to mention the help of Mrs. M. Labes, one of our students, in constructing strain SM10. We gratefully acknowledge the help of Mr. V. Krisnapillai in preparing the manuscript. This work was supported by Deutsche Forschungsgemeinschaft grant No. Pu 28/12-2.

References

Bagdasarian M, Bagdasarian MM, Coleman S, Timmis KN (1979) New vector plasmids for cloning in Pseudomonas. In: Timmis KN, Pühler A (eds) Plasmids of medical, environmental and commercial importance. Elsevier/North Holland Biomedical Press, Amsterdam New York, pp 411–422

Bagdasarian M, Lurz R, Rückert B, Franklin FCH, Bagdasarian MM, Frey J, Timmis KN (1981) Specific-purpose plasmid cloning vectors. II. Broad host range, high copy number, RSF1010-derived vectors, and host-vector system for gene cloning in Pseudomonas. Gene 16:237–247

Bagdasarian M, Lurz R, Nordheim A, Frey J, Timmis KN (1982) Molecular and functional analysis of the broad host range Sm-Su plasmid RSF1010, and construction of derivative vectors for gene cloning in Gram-negative bacteria. In: Mitsuhashi S (ed) Microbial drug resistance. Proc 3rd Int Symp, Tokyo

Beringer JE, Beynon JL, Buchanan-Wollaston AV, Johnston AWB (1978) Transfer of the drug resistance transposon Tn5 to Rhizobium. Nature 276:633–634

Bolivar F, Rodriguez RL, Greene PJ, Betlach MC, Heyneker HL, Boyer HW, Crosa J, Falkow S (1977) Construction and characterization of new cloning vehicles. II. A multi-purpose cloning system. Gene 2:95–113

Chang ACY, Cohen SN (1978) Construction and characterization of amplifiable multicopy DNA cloning vehicles derived from the P15A cryptic miniplasmid. J Bacteriol 134:1141–1156

Ditta G, Stanfield S, Corbin D, Helinski DR (1980) Broad host range DNA cloning system for Gram-negative bacteria: Construction of a gene bank of *Rhizobium meliloti*. Proc Natl Acad Sci USA 77:7347

Eckhardt T (1978) A rapid method for the identification of plasmid deoxyribonucleic acid in bacteria. Plasmid 1:584–588

Guerry P, Embden J Van, Falkow S (1974) Molecular nature of two non-conjugative plasmids carrying drug resistance genes. J Bacteriol 117:619–630

Guiney DG, Helinski DR (1979) The DNA-protein relaxation complex of the plasmid RK2: Location of the site-specific nick in the region of the proposed origin of transfer. Mol Gen Genet 176:183–189

Meyerowitz EM, Guild GM, Prestidge LS, Hogness DS (1980) A new cosmid vector and its use. Gene 11:271–282

Ruvkun GB, Ausubel FM (1980) Interspecies homology of nitrogenase genes. Proc Natl Acad Sci USA 77:191–195

Ruvkun GB, Ausubel FM (1981) A general method for site-directed mutagenesis in prokaryotes. Nature 289:85

Scott DB, Ronson CW (1982) Identification and mobilization by cointegrate formation of a nodulation plasmid in *Rhizobium trifolii.* J Bacteriol 151:36–43

Vliet F Van, Silva B, Montagu M Van, Schell J (1978) Transfer of RP4::Mu plasmids to *Agrobacterium tumefaciens.* Plasmid 1:446–455

Genetic Analysis of pRL1JI, a Symbiotic Plasmid of *Rhizobium leguminosarum*

J.A. DOWNIE, G. HOMBRECHER, Q.-S. MA, C.D. KNIGHT, and A.W.B. JOHNSTON [1]

Introduction

The fact that many of the genes concerned with nodulation and nitrogen fixation are located on large indigenous *Rhizobium* plasmids is well established at least for the fast-growing species such as *R. leguminosarum* and *R. phaseoli* (e.g., Hombrecher et al. 1981), *R. trifolii* (Hooykaas et al. 1981) and *R. meliloti* (Banfalvi et al. 1981, Rosenberg et al. 1981). In this paper we describe some of the properties of the *R. leguminosarum* plasmid pRL1JI, which were initially detected following a survey of *R. leguminosarum* field isolates for their ability to produce bacteriocin. One isolate (strain 248) synthesised a polypeptide bacteriocin whose production could be transferred at high frequency (1%) by conjugation to strains of *R. leguminosarum, R. trifolii* and *R. phaseoli* (Hirsch 1979). Examination of the bacteriocinogenic transconjugants on agarose gels revealed the presence of an extra plasmid band whose size was estimated to be 220 kb (Hirsch et al. 1980). This bacteriocinogenic plasmid was termed pRL1JI.

When pRL1JI was transferred to non-nodulating mutants (Nod$^-$) and to some non-fixing mutants (Fix$^-$) of *R. leguminosarum,* the transconjugants could nodulate peas and fix nitrogen (Johnston et al. 1978, Brewin et al. 1980). Furthermore when a derivative of pRL1JI containing the transposon Tn5 was transferred to strains of *R. trifolii* or *R. phaseoli,* the transconjugants nodulated peas (Johnston et al. 1978, Beynon et al. 1980). It was found also that pSA30 (the plasmid carrying the *Klebsiella pneumoniae* structural nitrogenase genes; Cannon et al. 1979) hybridised to pRL1JI indicating that it carried *nif* genes (Hombrecher et al. 1981).

Buchanan-Wollaston et al. (1980) isolated some symbiotically defective mutants by inserting Tn5 into pRL1JI. In the present study, we describe the isolation of further Fix$^-$ and Nod$^-$ mutations in pRL1JI and describe some of their properties.

Results

Plasmid pRL1JI is highly transmissable and it is possible to select for Tn5 transpositions from a chromosomal location into pRL1JI by selecting for the transfer of kanamycin resistance to an *R. leguminosarum* recipient that contains neither Tn5

1 John Innes Institute, Colney Lane, Norwich, NR4 7UH, UK

Molecular Genetics of the Bacteria-Plant Interaction
(ed. by A. Pühler)

nor pRL1JI. By such means, 200 independent insertions of Tn5 into pRL1JI were obtained (Ma et al. 1982). Three of these insertions resulted in Nod$^-$ mutants and nine generated mutants that were Fix$^-$.

The plasmids of these mutant strains were examined on agarose gels. It was shown by DNA hybridisation, using radioactively labelled probe (plasmid pKan2; Scott et al. 1982) which contained Tn5 DNA, that Tn5 was present in pRL1JI. By co-transduction, it was confirmed that Tn5 was responsible for the symbiotic defects.

Characterisation of the Fix$^-$ Mutants

The Fix$^-$ mutant strains could be distinguished into two groups on the basis of the morphology of the nodules. The strains with the *fix-78* and *fix-17* alleles induced small, pale pink nodules indicating that they contained less leghaemoglobin than normal. When thin sections of such nodules were examined in the electron microscope by the method of Beringer et al. (1977) it was found that a proportion of the plant cells did contain bacteroids of normal appearance and that such plant cells differed from cells containing wild type *R. leguminosarum* only in that they contained large starch grains (Fig. 1a). However, these non-fixing nodules contained many empty plant cells whereas in normal nodules, all the cells in the nitrogen-fixing zone contained many bacteroids.

The Fix$^-$ strains containing the alleles *fix-3, -15, -16, -37, -52, -67,* or *-137* induced green or yellow nodules. A minority of the plant cells contained undifferentiated bacteria with no peribacteroid membrane. These plant cells as well as the major class of uninfected cells had very little organised structure. It was striking that the infection threads of some nodules contained many more bacteria than did those in wild type nodules (Fig. 1b).

Genomic DNA was isolated from each of the mutant strains and was digested with *Eco*RI or *Bam*HI. The digested DNAs were ligated to small multicopy *Escherichia coli* plasmids and selection made for kanamycin resistant transformants. Since *Eco*RI does not cut Tn5 and *Bam*HI cuts it once, but outside the kanamycin resistance gene (Jorgensen et al. 1979), it follows that the "*Eco*RI" clones would contain *Rhizobium* DNA on both sides of Tn5 and the "*Bam*HI" clones would contain *Rhizobium* DNA on one side of the transposon. Thus, a series of clones was constructed and these were mapped using *Eco*RI, *Bam*HI and *Hin*dIII. A comparison of the maps of the different clones revealed strong similarities among some of them. For example, it appeared that the clones of the *fix-15, -16, -37, -67* and *-137* alleles all contained similar restriction fragments. By cross-hybridisation, they were confirmed to contain the same region of pRL1JI DNA. Similarly, the *fix-137* and *fix-52* alleles were found to be in the same *Eco*RI fragment. The three other *fix* alleles *(fix-3, -17* and *-78)* were closely linked on DNA which did not overlap with that described above (see Fig. 2). Using probes derived from the clones described above, several cosmid clones carrying normal DNA were identified. One of these cosmids hybridised to both regions and mapping of this clone indicated that the two regions were separated by about 30 kb of DNA.

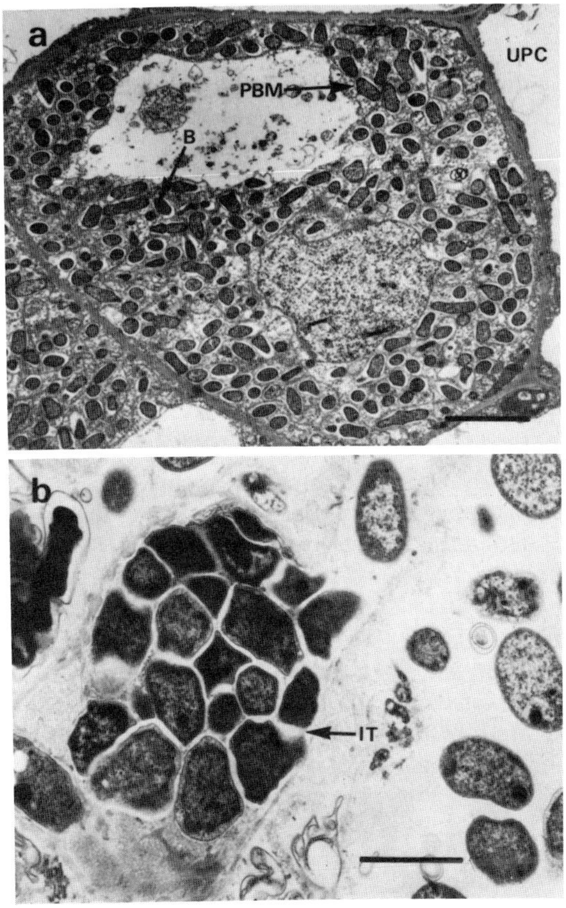

Fig. 1a, b. Electron micrographs of nodules induced by two mutants carrying **a** *fix-78;* **b** *fix-3* alleles. *B* bacteroid; *PBM* peribacteroid membrane; *IT* infection thread; *UPC* uninfected plant cells. *Bars* in **a** and **b** represent 5 microns and 1 micron respectively

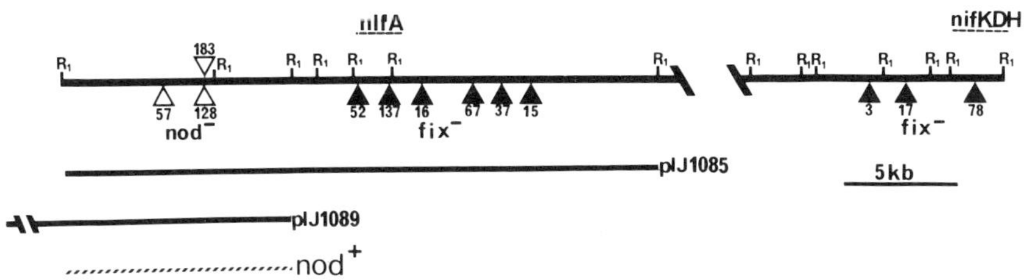

Fig. 2. Region of plasmid pRL1JI concerned with nodulation and nitrogen fixation. The *fix* alleles are represented by *closed triangles* and the *nod* alleles are represented by *open triangles*. The restriction sites were obtained by the mapping of two series of cosmids, constructed using either pLAFRI (Long et al. 1981) or pSAE1 (Grosveld et al. 1981) as vector. The regions of homology to the *K. pneumoniae* genes *nifKDH* and *nifA* were determined using the plasmids pSA30 (Cannon et al. 1979) or pMC71a (Buchanan-Wollaston et al. 1981) as hybridisation probes. The relative orientation of the two regions was not established

Nod⁻ Mutants

The three alleles *(nod-57, nod-128* and *nod-183)* were all in the same 6.6 kb *Eco*RI fragment. DNA from this fragment was used as a hybridisation probe to an *R. leguminosarum* gene bank cloned in the wide host-range cosmid pLAFR1 (Long et al. 1982). Two cosmid clones designated pIJ1085 and pIJ1089 were found to hybridise. Mapping of these revealed that they had only two *Eco*RI fragments in common; the 6.6 kb fragment in which the *nod* alleles were located, and a 3.3 kb *Eco*RI fragment.

It was clear from the restriction map of pIJ1085 that it contained a region of DNA in which the *fix* alleles *-52, -137, -67, -16, -37* and *-15* had been located. Thus the linkage of the *nod* and these *fix* alleles was determined (see Fig. 2).

Transfer of Nodulation Ability and Host Range Using Cloned DNA

When pIJ1085 or pIJ1089 was transferred into the mutant strains carrying the *nod-57, -128* or *-183* alleles, the transconjugants induced nitrogen-fixing nodules on peas. These two plasmids were introduced into *R. leguminosarum* strain B151 (Brewin et al. 1982), a strain that has been cured of its symbiotic plasmid; again nodules were formed but they failed to fix nitrogen since neither pIJ1085 nor pIJ1089 carries the nitrogenase genes.

Strain 8401 is a derivative of the *R. phaseoli* strain 8002 which has lost pRP2JI, the symbiotic plasmid of this strain (Lamb et al. 1982). When pIJ1085 or pIJ1089 was introduced into *R. phaseoli* strain 8401, the transconjugants nodulated peas. This indicates that host range and pea nodulation ability are determined by a maximum of 10 kb of DNA.

Regions of Homology with *Klebsiella nif* Genes

Using the DNA homology between the *nifKDH* genes of *K. pneumoniae* and *R. leguminosarum* (Nuti et al. 1979, Ruvkun and Ausubel 1980), it was possible to identify that region of the cloned *R. leguminosarum* DNA which is homologous to the probe pSA30. Thus, the 2.5 kb *Eco*RI fragment in which the *fix-78* allele is located was identified as containing DNA homologous to the *nifKDH* region of *K. pneumoniae.*

Similarly, it has been possible to identify a region of DNA homology using the plasmid pMC71a (Buchanan-Wollaston et al. 1981) which contains a fragment of DNA representing the entire *nifA* gene and parts of the *nifB* and *nifL* genes. As illustrated in Fig. 2, there is homology between this probe and the 1.8 kb fragment of DNA in which the *fix-52* and *fix-137* alleles are located.

Discussion

From an analysis of the location of Tn5-induced symbiotically-defective mutants in pRL1JI, it is clear that *nod* and *fix* genes are closeley linked on this plasmid confirming earlier genetic evidence (Hirsch et al. 1980). Linkage between *nod* and *fix* genes has also been noted in *R. meliloti* (Banfalvi et al. 1981, Rosenberg et al. 1981, Long et al. 1982).

It has been reported (Nuti et al. 1979, Ruvkun and Ausubel 1980) that only the *nifD* and *nifH* genes of *K. pneumoniae* hybridise to *Rhizobium* DNA. However, we have observed hybridisation between plasmid pMC71a (which carries the entire *nifA* gene) and pRL1JI in a region containing *fix* alleles. In *K. pneumoniae* the *nifA* gene product activates *nif* expression (see Buchanan-Wollaston et al. 1981) and the hybridization between this gene and pRL1JI DNA suggests that there is a similar gene in *R. leguminosarum.* Recently, the *nifA* region cloned in a broad range vector has been shown to activate *nif* expression in *Azotobacter* species (C. Kennedy and R. Robson, submitted for publication).

It was of interest to note that pIJ1085 and pIJ1089, which overlap by only 10 kb of DNA, could both confer the ability to nodulate peas when transferred to strains of *R. leguminosarum* cured of their symbiotic plasmids or indeed to a derivative of *R. phaseoli.* Thus nodulation ability and host range appears to be determined by the relatively small region of pRL1JI DNA common to these two clones.

Acknowledgements. We thank S. Long for plasmid pLAFR1, M. and F. Cannon for plasmid pMC71a and C. Kennedy and R. Robson for communicating unpulished results.

References

Banfalvi Z, Sakanyan V, Koncz C, Kiss A, Dusha I, Kondorosi A (1981) Location of nodulation and nitrogen fixation genes on a high molecular weight plasmid of *Rhizobium meliloti.* Mol Gen Genet 184:318–325

Beringer JE, Johnston AWB, Wells B (1977) The isolation of conditional ineffective mutants of *Rhizobium leguminosarum.* J Gen Microbiol 98:339–343

Beynon JL, Beringer JE, Johnston AWB (1980) Plasmids and host range in *Rhizobium leguminosarum* and *Rhizobium phaseoli.* J Gen Microbiol 120:421–429

Brewin NJ, Beringer JE, Buchanan-Wollaston AV, Johnston AWB, Hirsch PR (1980) Transfer of symbiotic genes with bacteriocinogenic plasmids in *Rhizobium leguminosarum.* J Gen Microbiol 116:261–270

Brewin NJ, Wood EA, Johnston AWB, Dibb NJ, Hombrecher G (1982) Recombinant nodulation plasmids in *Rhizobium leguminosarum.* J Gen Microbiol 128:1817–1827

Buchanan-Wollaston AV, Beringer JE, Brewin NJ, Hirsch PR, Johnston AWB (1980) Isolation of symbiotically defective mutants in *Rhizobium leguminosarum* by insertion of the transposon Tn5 into a transmissible plasmid. Mol Gen Genet 178:185–190

Buchanan-Wollaston V, Cannon MC, Beynon JL, Cannon FC (1981) Role of the *nifA* gene product in the regulation of *nif* expression in *Klebsiella pneumoniae.* Nature 294:776–778

Cannon FC, Reidel GE, Ausubel FM (1979) Overlapping sequences of *Klebsiella pneumoniae nif* DNA cloned and characterised. Mol Gen Genet 174:59–66

Grosveld FG, Dahl H-HM, Boer E de, Flavell RA (1981) Isolation of β-globin-related genes from a human cosmid library. Gene 13:227–237

Hirsch PR (1979) Plasmid-determined bacteriocin production in *Rhizobium leguminosarum.* J Gen Microbiol 112:135–142

Hirsch PR, Montagu M Van, Johnston AWB, Brewin NJ, Schell J (1980) Physical indentification of bacteriocinogenic nodulation and other plasmids in strains of *Rhizobium leguminosarum.* J Gen Microbiol 120:403–412

Hombrecher G, Brewin NJ, Johnston AWB (1981) Linkage of genes for nitrogenase and nodulation ability on plasmids in *Rhizobium leguminosarum* and *R. phaseoli*. Mol Gen Genet 182: 133–136

Hooykaas PJJ, Brussel AAN Van, Dulk-Ras H Den, Slogteren GMS Van, Schilperoort RA (1981) Sym plasmid of *Rhizobium trifolii* expressed in different rhizobial species and *Agrobacterium tumefaciens*. Nature 291:351–354

Johnston AWB, Beynon JL, Buchanan-Wollaston AV, Setchell SM, Hirsch PR, Beringer JE (1978) High frequency transfer of nodulation ability between strains and species of *Rhizobium*. Nature 276:634–636

Jorgensen RA, Rothstein SJ, Reznikoff WS (1979) A restriction enzyme map of Tn5 and location of a region encoding neomycin resistance. Mol Gen Genet 177:65–72

Lamb J, Hombrecher G, Johnston AWB (1982) Transmissable nodulation plasmid in a strain of *Rhizobium phaseoli*. Mol Gen Genet 186:449–452

Long SR, Buikema WJ, Ausubel FM (1982) Cloning of *Rhizobium meliloti* nodulation genes by direct complementation of nod$^-$ mutants. Nature 198:485–488

Ma Q-S, Johnston AWB, Hombrecher G, Downie JA (1982) Molecular genetics of mutants of *Rhizobium leguminosarum* which fail to fix nitrogen. Mol Gen Genet 187:166–171

Nuti MP, Lepidi AA, Prakash RK, Schilperoort RA, Cannon FC (1979) Evidence for nitrogen fixation *(nif)* genes on indigenous *Rhizobium* plasmids. Nature 282:533–535

Rosenberg C, Boistard P, Dénarié J, Casse-Delbart F (1981) Genes controlling early and late functions in symbiosis are located on a megaplasmid in *Rhizobium meliloti*. Mol Gen Genet 184:326–333

Ruvkun GB, Ausubel FM (1980) Interspecies homology of nitrogenase genes. Proc Natl Acad Sci USA 77:191–195

Scott KF, Hughes JE, Gresshoff PM, Beringer JE, Rolfe B, Shine J (1982) Molecular cloning of *Rhizobium trifolii* genes involved in symbiotic nitrogen fixation. J Mol Appl Genet 1:315–326

Natural Variation in *Rhizobium* Plasmids

N.J. BREWIN, N.J. DIBB, and G. HOMBRECHER [1]

Introduction: The Plasmid-Borne Cluster of Symbiotic Genes

A substantial weight of evidence has shown that, in the fast-growing rhizobia, the structural genes for nitrogenase are on large plasmids. On the same plasmids are also found genetic determinants essential for the induction of legume root nodules *(nod)* and the development of nitrogen-fixing bacteroids within those nodules *(fix)*. All these symbiotic determinants *(sym)* are apparently clustered in one region, representing some 50 kb, or less than a quarter of the smallest known 'symbiotic' plasmid (see Downie et al., this vol.). In the case of *R. leguminosarum* strains, similar (if not identical) symbiotic determinants are to be found on plasmids from at least three different incompatibility groups, suggesting that under natural conditions the symbiotic cluster may be exchanged between plasmids by in vivo genetic recombination. Indeed, the transfer of symbiotic determinants between plasmids of different incompatibility groups has been demonstrated in the laboratory (Brewin et al. 1980a, 1982).

The main thrust of research in *Rhizobium* is currently directed towards the mutagenesis and mapping of genes essential for symbiotic nitrogen fixation. However, instead of examining in detail the symbiotic determinants specified by one particular 'model' symbiotic plasmid, this paper will emphasise the surprising range of natural genetic variability that has been encountered in a superficial survey of different field isolates of *R. leguminosarum.* Symbiotic *(sym)* plasmids vary with respect to size, transmissibility and incompatability group; in addition, allelic variation exists in genes essential for symbiosis; moreover, other genes that might contribute to the overall efficiency of symbiosis are present on some *sym* plasmids, but absent from others. In particular, two *sym* plasmids from *R. leguminosarum* will be compared: pRL1JI, a relatively small but highly transmissible plasmid that carries determinants for bacteriocin production, and pRL6JI, a larger non-transmissible *sym* plasmid that carries determinants for hydrogen recycling. The significance of this genetic variability will be discussed in relation to our long-term efforts towards the genetic improvement of the *Rhizobium*-legume symbiosis.

1 John Innes Institute, Colney Lane, Norwich, NR4 7UH, UK

Molecular Genetics of the Bacteria-Plant Interaction
(ed. by A. Pühler)

Identification of a Symbiotic Plasmid Bearing Hydrogenase Determinants

In some strains of *Rhizobium,* hydrogen, evolved from bacteroids as a by-product of nitrogen fixation, is recycled through an uptake hydrogenase *(hup)* system that regenerates ATP (Emerich et al. 1979). *R. leguminosarum* strain 128C53 is a Hup$^+$ field isolate, and we sought to investigate whether the *hup* determinants were genetically linked to plasmid-borne *nod* and *fix* determinants. Examination of lysates of strain 128C53 on agarose gels revealed the presence of two plasmid bands (Brewin et al. 1980c). The *sym* plasmid was identified because DNA from the smaller of these plasmids (pRL6JI) hybridised to a radioactive probe containing cloned *nif* DNA (Hombrecher et al. 1981). pRL6JI was then eliminated from strain 128C53 in a two-step procedure: firstly, it was replaced after conjugation by an incompatible plasmid, a derivative of pRL3JI that carried a determinant for kanamycin resistance; in the second stage, a spontaneous kanamycin-sensitive derivative was obtained, which was shown to have lost the plasmid that had been introduced in stage 1. The resultant 'cured' strain, B151, lacked *nod* and *fix* determinants and is a convenient recipient strain in which to test the phenotypes of a range of transmissible symbiotic plasmids. Of particular significance was the observation that, when a *sym* plasmid was introduced into strain B151 from a Hup$^-$ *R. leguminosarum* field isolate (248), the transconjugant was Nod$^+$ Fix$^+$ but Hup$^-$, indicating that the *hup* determinants had probably been lost from strain 128C53 when plasmid pRL6JI was eliminated. The properties of the plasmids used in these genetic crosses are summarised in Table 1.

Table 1. *Rhizobium leguminosarum* plasmids from a single incompatibility group

Plasmid	Strain of origin	Approx. size (Md)	Markers [a]				
			Cin/Tn5	Tra	Nod	Fix	Hup
pRL6JI	128C53	190	–	–	+	+	+
pRL1JI	248	130	+	+	+	+	–
pRL3JI	306	130	+	+	–	–	–
pRL4JI	309	160	+	+	–	–	–
pIJ1008	recombinant [b]	195	+	+	+	+	+

[a] Cin = non-dialysable bacteriocin; Tn5 = transposon-coded kanamycin resistance (derivatives of each of the plasmids exist as Cin::Tn5); Nod = pea root nodule induction; Fix = fixes N_2 within nodules (includes structural genes for nitrogenase); Hup = hydrogen uptake activity

[b] pIJ1008 is an in vivo recombinant formed between pRL6JI and pRL4JI Cin::Tn5

Mobilization of *sym* and *hup* Determinants from 128C53

Regrettably, pRL6JI was not self-transmissible from strain 128C53. It was therefore necessary to mobilise the *sym* determinants by recombination with a transmissible plasmid bearing kanamycin-resistance as a marker that permitted selection for plasmid transfer. Introduction of the transmissible plasmid, a derivative of pRL4JI, was always associated with immediate recombination with pRL6JI, because they belonged to the same incompatibility group: the first product of recombination was interpreted to be a cointegrate of these two plasmids (although such a species could not be visualised as a discrete band on agarose gels). After subsequent selection for transfer of *kan* to a new recipient strain, recombinant plasmids were obtained that had acquired *nod,* and in addition *fix* and *hup* determinants. Interestingly, these recombinant plasmids were of uniform size (195 Md) and one of them, pIJ1008, has been studied further (Table 1).

Coordinate Expression of *hup* and *fix* Determinants

Not only are the *hup* determinants of 128C53 genetically linked to *nod* and *fix* determinants, but the expression of hydrogenase activity within the pea root nodule is closely coupled to the expression of *fix* determinants. Some of the data is summarised in Table 2. The experiment involved strain 116, a mutant of Hup^- strain 1062 that formed bacteroids but failed to fix nitrogen (Nadler 1981). Introduction of pIJ1008 into strain 116 failed to suppress the Fix^- phenotype and the transconjugant also showed less than 1% of the hydrogenase activity normally associated with pIJ1008 in a Fix^+ genetic background. Even preincubation of excised nodulated roots for 6 or 24 h with 15% H_2 was not sufficient to induce hydrogenase activity. Similarly in six other Fix^- genetic backgrounds, the hydrogenase determinants from pIJ1008 were only expressed at very low levels, and comparable results have been obtained with a Fix^- mutant of *R. japonicum* (Ruiz-Arguëso et al. 1981).

Table 2. Coordinate expression of determinants for hydrogenase uptake and nitrogen fixation within pea root nodules

Strain [a]	Observed phenotype [b]	
	Fix nmol C_2H_4/plant/h	Hup $\times 10^3$ cpm H-T./50 mg nodules/h
1062	395 ± 88	0.03 ± 0.01
1062 pIJ1008	630 ± 144	62.7 ± 16.3
116	⩽ 1.0	0.03 ± 0.02
116 pIJ1008	⩽ 1.0	0.04 ± 0.02

[a] pIJ1008 = nod^+ fix^+ hup^+; 1062 = *R. leguminosarum* wild type (derivative of strain 300); 116 = 1062 fix^- (Nadler 1981)
[b] See Brewin et al. (1982) for conditions for acetylene reduction and tritium incorporation assays

Natural Variation in Nitrogenase Proteins

Examination of bacteroid proteins by SDS polyacrylamide gel electrophoresis (Fig. 1) showed that the presence of *hup* was not the only detectable difference between pIJ1008 and other *sym* plasmids such as pRL1JI and pRL10JI. Comparison of

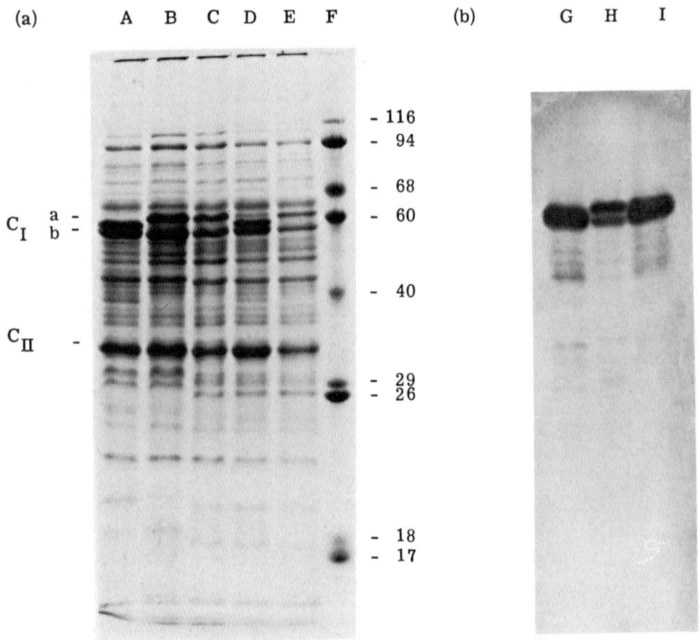

Fig. 1a. Natural variation in plasmid-determined bacteroid proteins. SDS polyacrylamide gels of proteins from bacteroids containing different *sym* plasmids, stained with Coomassie blue. *A* B151 pIJ1008; *B* B151 pRL1JI; *C* 1062 (*sym* plasmid = pRL10JI). The positions of nitrogenase components Ia, Ib and II are indicated. Tracks *D* and *E* are from merodiploids containing pRL10JI as well as an introduced *sym* plasmid: *D* 1062 pIJ1008; *E* 1062 pIJ1009 (i.e., pIJ1008 Δ fix). Track *F* contains molecular weight markers as indicated. **b** Specific staining of nitrogenase component 1 proteins from bacteroids of *G* B151 pIJ1008; *H* 1062; *I* 1062 pIJ1008. Proteins from an SDS-PAGE were transferred to nitrocellulose treated with goat anti-rabbit antiserum conjuggated to peroxidase and stained with 25 μg/ml chloronaphthol and 0.01% hydrogen peroxide according to the method of Towbin et al. (1979)

bacteroids obtained from derivatives of the cured strain B151 containing either pIJ1008 or pRL1JI revealed a number of different polypeptide bands that were plasmid coded. The most conspicuous difference was that pIJ1008 specified a nitrogenase component Ia of slightly higher electrophoretic mobility. Proof that the relevant bands corresponded to nitrogenase was obtained by blotting the proteins to nitrocellulose (Bowen et al. 1980), and then using antiserum to component I (generously provided by Dr. T. Bisseling) followed by a second antibody staining reaction (Towbin et al. 1979).

Because pRL10JI and pIJ1008 are compatible *sym* plasmids, it was possible to create strains that were merodiploid for these nitrogenase determinants. Fig. 1 shows that, in the merodiploid, the nitrogenase form with higher mobility was predominant, although a minor amount of pRL10JI-coded component Ia was still present, as revealed by Western blotting. In a merodiploid containing pRL10JI and pIJ1009 (i.e., pIJ1008 Δfix) the CIa isomer of higher mobility was absent and only the component characteristic of pRL10JI could be seen. It is not yet known whether the differences in mobility seen for nitrogenase component Ia represent differences in primary sequence of the structural genes or differences in post-translational modification of those polypeptides.

Improved Symbiotic Performance by *sym* Plasmid Transfer

So far, we have demonstrated that natural genetic variability exists among *R. leguminosarum sym* plasmids, both with respect to essential symbiotic genes (e.g., nitrogenase) and inessential symbiotic genes (e.g., hydrogenase). Because the *sym* plasmids are sometimes transmissible by conjugation between strains, and because the cluster of *sym* genes is sometimes transmissible by in vivo recombination between plasmids, there are opportunities for *Rhizobium* strain improvement as a result of the reassortment of natural genetic variation using natural methods of bacterial sexual reproduction. In collaboration with D.A. Phillips and T.M. DeJong (U.C. Davis, California), we have undertaken such a study in "*Rhizobium* breeding" by *sym* plasmid transfer. As with any other breeding programme, it is to be expected that progeny inferior to both parents will emerge as well as those that are superior (DeJong et al. 1981, 1982). Some of the more positive results are presented in Table 3.

Table 3. *Rhizobium* strain improvement by plasmid transfer

Strain of *R. leguminosarum*	Total plant dry weight ((g/plant ± SE) [a]	Total plant N (mg/plant ± SE)
300	3.98 ± 0.10	153 ± 4.0
300 pIJ1008	4.58 ± 0.26	198 ± 9.1
3622	2.00 ± 0.11	60 ± 3.3
3622 pIJ1008	3.75 ± 0.11	127 ± 2.6
128C53	4.05 ± 0.14	138 ± 4.8

[a] Mean of 7 replications after 35 d growth without combined N

The *sym* plasmid pIJ1008 was introduced into two field isolates of *R. leguminosarum* that each carried a compatible *sym* plasmid. Symbiotic performance was measured as total plant growth dependent on symbiotic N_2 fixation for peas growing in Leonard-jar assemblies under controlled environmental conditions (DeJong et al. 1982). The performance of the constructed strains was compared to that of the original field isolates, and also to that of 128C53, from which the *sym* and *hup* determinants of pIJ1008 were derived. After the growth experiment, bacteria recovered from nodules were shown to contain the appropriate inoculant strain. In the case of strain 300, the introduction of pIJ1008 resulted in a strain giving 15% more plant dry weight and 30% more N_2 fixation than 300 itself (Table 3). This strain was in this case superior to either of the parents used in the genetic cross (DeJong et al. 1982).

Transfer of *hup* Determinants to *R. phaseoli*

Because no known field isolate of *R. phaseoli* has been shown to possess a hydrogen recycling *(hup)* system, we introduced pIJ1008 into a series of *R. phaseoli* strains, and tested the progeny for hydrogenase activity within *Phaseolus* nodules. A typical result is presented in Table 4. When introduced into *R. phaseoli*, the *hup* determinants of pIJ1008 did express within *Phaseolus* nodules, although the level of activity was not high. Most significantly however, nitrogenase activity was severely reduced as a result of the introduction of pIJ1008. A possible reason for the deleterious effect of pIJ1008 in *R. phaseoli* may be that, in addition to *hup,* the plasmid also introduced *nod* and *fix* determinants that are adapted to function within pea root nodules: conceivably therefore, the *nod* and *fix* determinants of pIJ1008 interacted unfavourably with the corresponding determinants on the compatible *sym* plasmid of *R. phaseoli.*

Table 4. Hydrogenase and nitrogenase activity of *Phaseolus* nodules following the introduction of pIJ1008 into *R. phaseoli* 8002 (scored 28 days after inoculation). Values are the means (± SE) for six plants grown on 0.5% agar containing Fahraeus medium

Introduced plasmid	Observed phenotype	
	Hup [a]	Fix [b]
None	0.04 ± 0.01	464 ± 154
pIJ1008	10.7 ± 5.1	71 ± 71

[a] $\times 10^3$ cpm H-T/50 mg nodules/h
[b] nmol C_2H_4/plant/h

Effects of Plasmid Content on the Competitiveness of *Rhizobium*

As another approach towards *Rhizobium* breeding, we have examined the effects of *sym* plasmid content on competitiveness for nodulation in mixed inocula (Table 5). Using as a recipient the cured *R. leguminosarum* strain B151, a series of derivatives was constructed that contained six different *sym* plasmids derived from different field isolates of *R. leguminosarum.* B151 is itself streptomycin-resistant, and each of the derivatives was mixed in equal proportion with strain 726, a rifampicin-resistant derivative of strain 300. Each of these mixtures was used to inoculate ten pea seedlings. After 21 days, 15 nodules from each plant were surface sterilised and stabbed with a sterile toothpick to determine whether the contents were streptomycin-resistant or rifampicin-resistant. (The small minority of mixed nodules was disregarded.) Table 5 shows that significant differences in competitiveness for nodulation could be ascribed to the particular *sym* plasmid introduced into B151, despite the fact that the relative proportions of Rif-r and St-r bacteria on the surface of the roots remained unchanged. It is perhaps interesting to note that all the introduced plasmids resulted

Table 5. Effects of plasmid content on the competitiveness of *Rhizobium* for nodulation in mixed inocula

B151 derivative (Str-r)	Strain of origin of nodulation plasmid	Number of nodules containing viable bacteria	% Nodules [a] with Str-r clones
B224	257	73	32 [a]
B166	TOM	81	47 [a]
B200	300	82	74 [b]
B155	248	66	77 [b]
B225	279	61	77 [b]
B164	128C53	107	97 [c]

[a] Groups a, b, c, are significantly different from each other ($p \leqslant 0.05$)

in derivatives that were less competitive than B151 pIJ1008 or 128C53 *str,* the original field isolate from which the cured strain, B151, was derived. Perhaps it is easier to make an inferior strain by plasmid transfer rather than a superior one: this might be the case if a particular *sym* plasmid was co-adapted to the genetic background of the strain that harboured it.

Significance of Natural Variation Among *sym* Plasmids

The natural variation that exists among *Rhizobium sym* plasmids can either be ignored or exploited. Those who choose to ignore it would argue that it is more important to study the major features that *Rhizobium* strains have in common, rather than minor variations in detail among different strains: for such people, the major research goal is to define what is genetically unique about the root nodule symbiosis but common to all rhizobia and legumes.

On the other hand, natural genetic variation can be exploited as a research tool to study structure-function relationships for allelic variants of essential symbiotic genes. For example, what is the significance of the two kinds of nitrogenase identified in this study, and similarly what is the genetic basis for the ability of *R. leguminosarum* TOM to nodulate the 'Afghan' type pea cultivar (Brewin et al. 1980b; Young et al. 1982) and the genetic basis for host-range specificity between *R. leguminosarum, R. trifolii* and *R. phaseoli* species (Johnston et al. 1978)? Another source of natural variation is for genes present on some *sym* plasmids, but absent from others, for example, determinants for bacteriocin production or hydrogen uptake. Such genes may contribute quantitatively to symbiosis by increasing competitiveness or the overall efficiency of symbiotic nitrogen fixation, and therefore merit consideration in any complete description of the interaction between *Rhizobium* and its host.

At a cruder level, the natural genetic variation among plasmid-borne *sym* genes could be exploited as the basis for a *Rhizobium* breeding programme. For such purposes the cluster of symbiotic genes might be treated like a classical linkage group. The approach would be to optimise the combinations of *sym* genes within the *sym* cluster, the *sym* gene cluster with the natural plasmid vector, the *sym* plasmid with the host strain of *Rhizobium*, and the *Rhizobium* strain with the host-legume strain. It should be stressed that, as with plant breeding, such *Rhizobium* breeding could be done empirically, without necessarily defining individual *sym* genes at the molecular level.

Acknowledgements. We thank Miss E.A. Wood for excellent technical assitance and A.W.B. Johnston and J.A. Downie for their comments on the manuscript.

References

Brewin NJ, Beringer JE, Buchanan-Wollaston AV, Johnston AWB, Hirsch PR (1980a) Transfer of symbiotic genes with bacteriocinogenic plasmids in *Rhizobium leguminosarum.* J Gen Microbiol 116:261–270

Brewin NJ, Beringer JE, Johnston AWB (1980b) Plasmid mediated transfer of host-range specificity between two strains of *Rhizobium leguminosarum.* J Gen Microbiol 120:413–420

Brewin NJ, DeJong TM, Phillips DA, Johnston AWB (1980c) Co-transfer of determinants for hydrogenase activity and nodulation ability in *Rhizobium leguminosarum.* Nature 288: 77–79

Brewin NJ, Wood EA, Johnston AWB, Dibb NJ, Hombrecher G (1982) Recombinant nodulation plasmids in *R. leguminosarum.* J Gen Microbiol 128:1817–1827

Bowen B, Steinberg J, Laemmli UK, Weintraub H (1980) The detection of DNA binding proteins by protein blotting. Nucleic Acids Res 8:1–20

DeJong TM, Brewin NJ, Phillips DA (1981) Effect of plasmid content in *Rhizobium leguminosarum* on symbiosis and host plant functions. J Gen Microbiol 124:1–7

DeJong TM, Brewin NJ, Johnston AWB, Phillips DA (1982) *Rhizobium* strain improvement by plasmid transfer. J Gen Microbiol 128:1829–1838

Emerich DW, Ruiz-Arguëso T, Ching TM, Evans H (1979) Hydrogen-dependent nitrogenase activity and ATP formation in *Rhizobium japonicum* bacteroids. J Bacteriol 137:153–160

Hombrecher G, Brewin NJ, Johnston AWB (1981) Linkage of genes for nitrogenase and nodulation ability on plasmids in *Rhizobium leguninosarum* and *R. phaseoli.* Mol Gen Genet 182: 133–136

Johnston AWB, Beynon JL, Buchanan-Wollaston AV, Setchell SM, Hirsch PR, Beringer JE (1978) High frequency transfer of nodulating ability between strains and species of *Rhizobium.* Nature 276:635–636

Nadler KD (1981) a mutant strain of *Rhizobium leguminosarum* with an abnormality in heme synthesis. In: Gibson AH, Newton WE (eds) Current perspectives in nitrogen fixation 414. Aust Acad Sci

Ruiz-Arguëso, T, Cabrera E, Bertalmio MB (1981) Induction of hydrogenase in non-nitrogen-fixing soybean nodules produced by a Hup^{+} strain of *Rhizobium japonicum.* In: Gibson AH, Newton WE (eds) Current perspectives in nitrogen fixation. Aust Acad Sci, p 377

Towbin H, Staehelin T, Gordon J (1979) Electrophoretic transfer of proteins from polyacrylamide gels to nitrocellulose sheets: procedure and some applications. Proc Natl Acad Sci USA 76:4350–4354

Young JPW, Johnston AWB, Brewin NJ (1982) A search for peas showing strain specificity for symbiotic *Rhizobium leguminosarum.* Heredity 48:197–201

Analysis of Nodule-Specific Plant and Bacteroid Proteins in Pea Plants Inoculated by Transposon Mutagenized *Rhizobium leguminosarum*

R.C. VAN DEN BOS, TH.M.P. SCHETGENS, T. BISSELING, J.G.J. HONTELEZ, and A. VAN KAMMEN[1]

Introduction

The process of nodulation and nitrogen fixation in the *Rhizobium* legume symbiosis involves a sequence of interactions between the two symbiotic partners in which each influences gene expression in the other. An important step forward in the study of these processes was made by the possibility to mutagenize the *Rhizobium* genome by the introduction of transposon Tn5 (Ruvkun and Ausubel 1981); this method allows the identification and site-specific mutagenesis of genes involved in nodulation (nod) and nitrogen fixation (fix). A similar technique to study which genes in the host are specifically active during symbiosis does not exist as yet. However, if on the *Rhizobium* level genes essential for the symbiosis are affected by Tn5 insertions, one may expect that the expression of nodulin genes (nodule specific host proteins, Legocki and Verma 1980) beyond a certain point will also be altered. In *Rhizobium* the genes involved in infection and nitrogen fixation are located on a large indigenous plasmid (reviewed by Kondorosi and Johnston 1981, Dénarié et al. 1981). Several genes on this plasmid are only actively transcribed in the endosymbiotic state in *Rhizobium leguminosarum* (Krol et al. 1981); therefore we have intitially chosen this region of the structural *nif*-genes for Tn5-mutagenesis.

In this report we describe experiments to study *nif*-protein synthesis in Tn5-mutagenized *Rhizobium;* furthermore the influence of Tn5-mutagenesis in *Rhizobium* upon synthesis of nodule-specific proteins in the plant cell is studied.

Materials and Methods

Cloning of the *R. leguminosarum nif*-Region. Isolation of total DNA from *R. leguminosarum,* partial EcoR1-digestion, size fractionation and construction of a clonebank in pLAFR1 were as described by Friedman et al. (1982). Colony hybridization of the bank with pRmR2 (Ruvkun et al. 1982a) as a probe was done according to Grunstein and Wallis (1979).

DNA Biochemistry. All manipulations were as described (Ruvkun et al. 1982b).

1 Agricultural University, Department of Molecular Biology, De Dreijen 11, Wageningen, The Netherlands

Molecular Genetics of the Bacteria-Plant Interaction
(ed. by A. Pühler)

Tn5 Mutagenesis of Plasmid Clones. Either of two procedures was used. Tn5 mutagenesis of a costramid clone using λb221 *rex*::Tn5cI8570$_{am8}$ P$_{am29}$, triparental mating into *R. leguminosarum* and marker-exchange by conjugation with Str^S *E. coli* strain 2124/pPH1JI were performed.

EcoR1-fragments from the original costramid clone were subcloned in the EcoR1-site of pSUP201, a conjugative vector before effecting Tn5-transposition. Details of this method for Tn5-mutagenesis and conjugation to *Rhizobium* will be described by Simon et al. (in preparation).

Phenotype of Tn5-Mutagenized *R. leguminosarum*. *Pisum sativum* (Rondo) seeds were surface sterilized and inoculated in sterile vermiculite with *R. leguminosarum* wild type and Tn5-mutagenized strains. For this, one fresh colony was used suspended in 5 ml sterile water. After 15 days plants were observed for nodulation. Acetylene reduction by the whole plant was measured at various times after infection (Bisseling et al. 1978). For large-scale isolation of *Rhizobium* or plant proteins pea plants were grown in gravel.

Fractionation of Nodules. Proteins from plant cytoplasm, peribacteroid space and bacteroids were obtained by pea nodule fractionation as will be described (Bisseling et al., in preparation).

Protein Analysis. SDS-polyacrylamide gel electrophoresis was done as described by Zabel et al. (1982). After electrophoresis proteins were transferred to nitrocellulose filters (Bowen et al. 1980), incubated with a specific antiserum and subsequently with ^{125}I-labeled protein A (Zabel et al. 1982). Radioactive bands on the filter were detected by autoradiography.

Specific Antisera. The production of antisera specific against nodulins, and *Rhizobium* secretory proteins will be described elsewhere (Bisseling et al., in preparation). Antisera specific against nitrogenase components I and II have been described (Bisseling et al. 1980).

Results

Construction of a *R. leguminosarum nif* Clone

The newly described costramid pLAFR1 (Friedman et al. 1982) was chosen as a cloning vehicle for the construction of a *R. leguminosarum* clonebank; in this vector fragments of about 20 kb can be cloned. pLAFR1 is a cosmid derivative of the low copy-number broad host-range cloning vector pRK290 (Ditta et al. 1980); it is 21.6 kb long, confers tetracycline resistance, contains a unique EcoR1 site and can be mobilized into *Rhizobium*. Total DNA from *R. leguminosarum* PRE was partially restricted with EcoR1, fragments of 15–30 kb were isolated by sucrose gradient centrifugation and ligated to EcoR1-digested pLAFR1. The ligated DNA was packaged in vitro in λ heads and used to infect *E. coli* HB101; transformants were selected on LB-plates containing 10 mg/l tetracycline. This resulted in 800 Tc^r-colonies which we expected to represent a 99% complete bank of the *R. leguminosarum* genome (Friedman et al. 1982).

The colonies were transferred to nitrocellulose filters and colony-hybridization using the nick-translated (Rigby et al. 1977) total insert cut from pRmR2 (Ruvkun et al. 1982) with EcoR1 and Bgl1I as a probe resulted in one positive colony. Plasmid DNA from this colony was isolated (Holmes and Quigley 1981) digested with several restriction enzymes and a physical map of pRB10 was made. Hybridization of a Southern blot (Southern 1975) of pRB10 cut with EcoR1 using pRmR2 as probe revealed that the leftmost 6.5 kb EcoR1 fragment of pRB10 is homologous to the *R. meliloti nif*-genes. Hybridization of Southern blots with an XhoI-fragment or an EcoR1-Hind III-fragment from pRmR2, corresponding to *R. meliloti nif*-genes H and D respectively as probes gave the results shown in Fig. 1. Homology to *nif*D is only found in the leftmost EcoR1-Hind III-fragment, while homology to *nif* H is restricted to the Hind III-XhoI-fragment indicated.

The 6.5 kb EcoR1-fragment of pRB10 homologous to pRmR2 corresponds in size to the EcoR1 fragment of *R. leguminosarum* plasmid DNA (pR11; Krol et al. 1981) carrying the structural *nif*-genes.

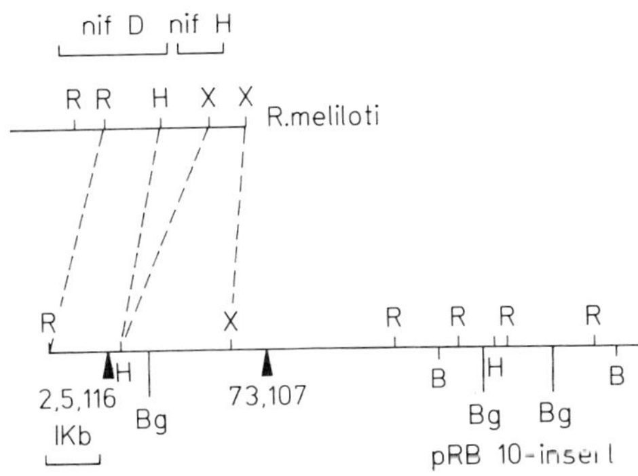

Fig. 1. Restriction map of PRE-*nif*-region cloned in pRB10. *Numbers* and *arrows* indicate Tn5-insertion positions. Above the pRB10 map part of the *R. meliloti nif*-region cloned in pRmR2 is shown. Fragments from pRmR2 as indicated were nick-translated and used to probe Southern blots of restricted pRB10-DNA. Regions of homology are indicated by *dotted lines*. *R Eco*RI; *H Hind* III; *X Xho*I; *B Bam*HI; *Bg Bgl*II

Tn5-Mutagenesis and Conjugation of PRB10; Marker Exchange in *R. leguminosarum*

Transposon Tn5 was introduced into pRB10 in various positions of the insert as described under Methods. From a total of 150 $Tc^R Km^R$ colonies resulting from transformation of *E. coli* HB101 with plasmid DNA isolated from λ::Tn5-infected HB101/pRB10 eight were identified as Tn5-insertions in the left 6.5 kb EcoR1-fragment by restriction analysis; these strains were chosen for further analysis. A mutagenized plasmid was conjugated into *Rhizobium* and the Tn5-containing fragment was exchanged against a wild-type fragment by homologous recombination (Ruvkun and Ausubel 1981). To determine the exact position of Tn5 in the *Rhizobium* genome total DNA of wild-type or mutant strains was digested with various restriction enzymes or combinations of enzymes and electrophoresed in an agarose gel. A Southern blot of the gel was hybridized with nick-translated pRB10 or with restriction fragments derived from pRB10. The hybridization pattern in the Tn5-mutagenized strains changes in a characteristic way from wild-type and the hybridization pattern allows the determination of the site of Tn5 insertion (Ruvkun and Ausubel 1981). The map positions of the Tn5 insertions is shown in Fig. 1. For strains 3, 6 and 26 the exact position in the left 6.5 kb EcoR1-fragment of pRB10 has yet to be determined.

Phenotype of *nif*-Tn5 Mutagenized Strain PRE

The different Tn5-mutagenized strains were inoculated on *P. sativum* in vermiculite. All strains were found to induce nodules (nod^+) and by microscopic observation were not morphologically different from the characteristic wild-type bacteroid form. Acetylene reduction of the different *Rhizobium*-host combinations after various times was measured; strains 2 and 5 were fix^- and the other strains were all fix^+, although the onset of acetylene reduction was delayed by about two days. Strain 116 (fix^+) behaved anomalously in that after three times of subculturing under Str-Km-selection the phenotype changed to fix^-. The resulting mutant is designed 116A; determination of the Tn5-insertion position in strains 116 and 116A by hybridizing Southern blots with ^{32}P-labeled pRB10 showed no difference between the two strains. The rate of acetylene reduction of the fix^+-strains did not differ appreciably from that of the parent strain. The presence of nitrogenase proteins CI and CII in bacteroids of the various mutant strains was assayed by blotting SDS-polyacrylamide gels of bacteroid proteins from 21 day nodules onto nitrocellulose filters. The filters were subsequently incubated with antisera specific against Ci or CII and finally with ^{125}I-protein A (Fig. 2). Strains 2, 5 and 116A at this stage contain small amounts of CII and traces of CI. After a few days of further development CI disappears from the

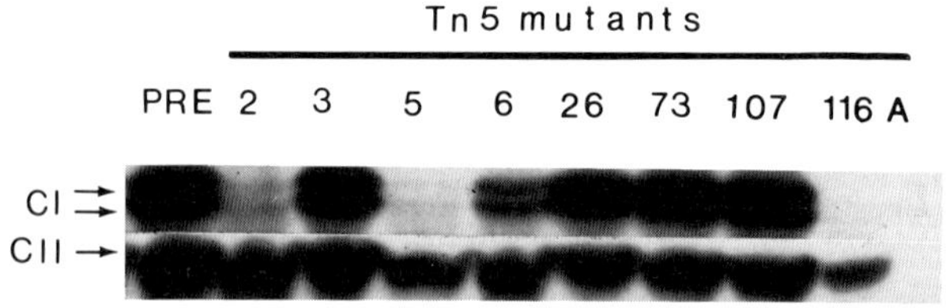

Fig. 2. Bacteroid proteins analysed for the presence of nitrogenase. *P. sativum* was nodulated by the strains indicated; at 21 days after inoculation soluble bacteroid proteins were separated by SDS-polyacrylamide gel electrophoresis. Per blot 10 μg protein was loaded. Proteins were blotted onto nitrocellulose and the filter was probed with antisera specific against CI and CII; antigen-antibody complexes were then visualized by reaction with ^{125}I-protein A and autoradiography

bacteroids and CII diminishes (not shown). The other Tn5-mutagenized strains (3, 26, 73, 107, 116) all appear to be comparable to wild-type in their nodulation and nitrogen fixation characteristics; nitrogenase proteins are synthesized although their appearance and concomittant activity is slightly delayed.

Leghemoglobin in Pea Nodules Induced by Tn5-PRE-Mutants

Cytoplasmic proteins were isolated from pea nodules after infection with the different Tn5-PRE-mutants. The proteins were separated by SDS-polyacrylamide gel electrophoresis and blotted onto nitrocellulose. The blot was then incubated with antiserum specific against Lb and subsequently with ^{125}I-protein A. An autoradiogram of the parts of this blot containing Lb (Fig. 3) shows the amounts of Lb induced by the different PRE-strains at various times after seeding and inoculation. At 16 days the wild type shows an amount of Lb which does not increase appreciably during development; strains 3, 26, 107 induce Lb concentrations comparable to the wild type strain after a short delay. In the fix$^-$-strains 2 and 5 the amount of Lb initially is similar to wild type and then declines.

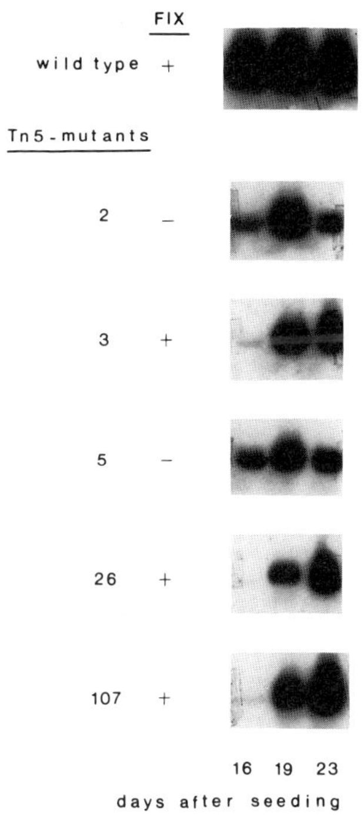

Fig. 3. Lb concentration in mutant *Rhizobium*-induced nodules. Soluble proteins were isolated from *P. sativum* nodules induced by the *Rhizobium* strains indicated at various intervals after seeding. 10 μg of protein per blot was probed with antiserum specific against Lb as in Fig. 2. Part of the autoradiogram, corresponding to Lb, is shown

Nodulins Induced by Tn5-Mutagenized PRE

With specific antisera about 30 nodulins can be detected in the cytoplasm of pea nodule cells (Bisseling et al., in preparation). As a step toward elucidating the function of these nodulins we analysed changes in nodulin proteins induced by Tn5-mutagenesis of the infecting *Rhizobium*. Nodulins were therefore isolated from plants at 21 days after infection with the different Tn5-PRE-strains and analysed by protein blotting and incubation with antinodulin serum and ^{125}I-protein A. The autoradiogram (Fig. 4) shows a number of nodulins in each case, of which Lb is the most prominent. In the fix$^-$-strains 2 and 5 the amount of Lb is reduced. The same holds for nodulins 15 and 17 corresponding to proteins of approximately 43×10^3 and 40×10^3 molecular weight. This might mean that these nodulins play a role in nitrogen fixation. The other nodulins which are not repressed in the mutants may be involved in the earlier stages of nodule development.

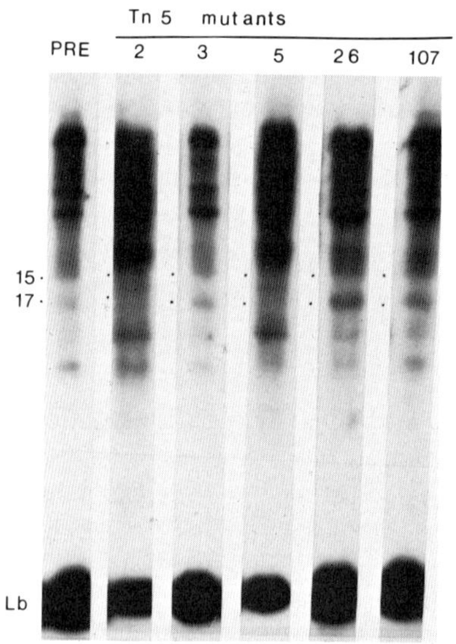

Fig. 4. Nodulins in mutant *Rhizobium*-induced nodules. Soluble proteins were isolated from *P. sativum* nodules induced by the *Rhizobium* strains indicated at 21 days after seeding; 10 μg protein per blot was analysed as in Figs. 2 and 3 probing the filter with antiserum specific against pea nodulins. *Numbers* on the left correspond to those in Bisseling et al. (in preparation)

Rhizobium Secretory Proteins

Rhizobium leguminosarum PRE in culture secretes a number of proteins which react with an antiserum specific against nodulins. Therefore these proteins are probably also secreted by bacteroids in the host cytoplasm. To prove this we raised an antiserum specific against the secretory proteins of free-living strain PRE and found that a number of proteins in the host cytoplasm and in the peribacteroid space react with this antiserum (Bisseling et al., in preparation). We investigated if the Tn5-mutagenized *Rhizobium* strains described are changed with regard to the secretion of these proteins inside the plant cell. Soluble nodule cytoplasmic proteins were separated by SDS-gel electrophoresis and after protein blotting the filter was incubated with anti-secretory protein serum. The autoradiogram after ^{125}I-protein A incubation (Fig. 5) shows that in strain PRE a very strong band of 38 kd is visible; besides that a number of other proteins are seen to react. In mutant 5, the protein at 55 kd is weak and the proteins below 38 kd are missing, indicating that protein secretion into the host cytoplasm is influenced by mutagenesis of the *nif*-region.

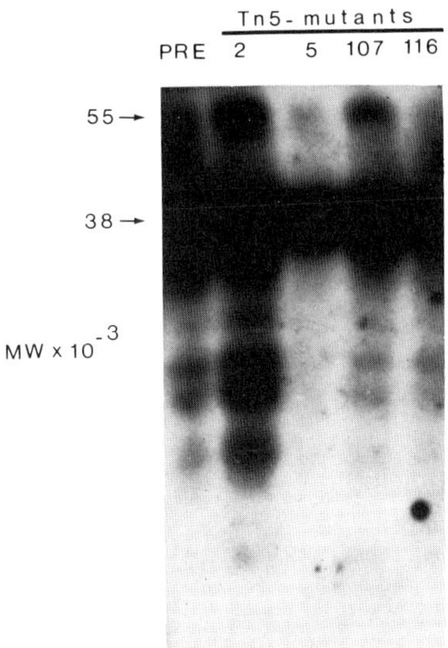

Fig. 5. *Rhizobium* secretory proteins in mutant *Rhizobium* induced nodules. Soluble proteins were isolated from *P. sativum* nodules induced by the *Rhizobium* strains indicated and analysed as in Figs. 2–4; 10 μg protein per blot. The filter was probed with antiserum specific against *Rhizobium leguminosarum* PRE secretory proteins

Discussion

Transposon mutagenesis is a very useful manner to introduce site-specific alterations in a region of the *Rhizobium* genome. We have begun our investigations concerning the influence of mutations in *Rhizobium* on genome expression in *Rhizobium* and plant by mutagenizing a recombinant plasmid pRB10. The conclusion that the *Rhizobium* DNA cloned in pRB10 contains the *nif*-region is based on the observations that the left-hand part of pRB10 is homologous to the *R. meliloti nif*-region (pRmR2, Ruvkun and Ausubel 1981) and that Tn5-insertions in this region cause a fix$^-$phenotype. The hybridisation data (Fig. 1) allow an approximate localization of the *nif* D and *nif* H-genes; probably the *nif* K-gene will not or only partially be located on pRB10.

Although these insertions were randomly chosen from a population of 150 Tn5-insertions in a 6.5 kb EcoR1-fragment of pRB10, we have not been able to detect differences in the insertion positions of mutations 2, 5 and 116. Possibly this site represents a "hot spot" for Tn5-insertions in *Rhizobium.* From the phenotype of mutations 2 and 5 (fix$^-$) we conclude that the *nif*-region was hit; the fact that a protein is synthesized of mobility identical to that of CII (Fig. 2), while small amounts of CI can be detected suggests that the insertions are in a region between genes D and H. These insertions are therefore probably analogous to those in the *R. meliloti* genome between D and H reported by Ruvkun et al. (1982a); the weak acetylene reduction observed in this case is explained by incomplete polarity on *Rm nif* D of these Tn5 insertions. We found no acetylene reduction by mutants 2 and 5 between the period of 14 to 35 days after inoculations, although both CI and CII are present albeit in unequal amounts. On the other hand in insertion strain number 116 both nitrogenase components are present in the same amounts as in the wild-type and the rate of acetylene reduction is normal. This brings us to propose that under normal conditions the *nif*-operon in *R. leguminosarum* PRE is transcribed into one single m RNA of about 1.2×10^6 M.W. (Krol et al. 1982) but that for genes D (and probably K) a second promoter is present which is active in mutant 116 and at a very low rate in 2 and 5. This model would be in accordance with our earlier observations that the synthesis of CI and CII in strain PRE is not coordinated in a strict way; the ratio of CI/CII synthesis varies with the stage of bacteroid development (van den Bos et al. 1978) and during nodule development the presence of CI proceeds that of CII (Bisseling et al. 1980). We have no explanation yet for the different phenotypes of 116 and 116A. Except for strains 2, 5 and 116A all others appear to have a fix$^+$-phenotype, although nitrogen fixing activity develops a few days later than in the parent strain. This may be related to the changes in nodulins we observe. As for the amounts of Lb we conclude that strains 2 and 5 initially induce (a normal rate of) Lb synthesis, which subsides when no active nitrogen fixation develops, in agreement with our finding that Lb synthesis preceeds nitrogenase synthesis (Bisseling et al. 1980). The other Tn5 insertion strains develop a Lb-concentration in the normal way. As for the other nodulins we observe changes in proteins 15 and 17, which are virtually absent in the fix$^-$-mutants; at the same time the intensity of other nodulin bands is also changed in these strains as well as in the fix$^+$-strains. We assume that in the interaction between *Rhizobium* and host proteins secreted by the endosymbiont may be involved. With an antiserum specifically directed against *Rhizobium* secretion proteins we indeed observed a strong decrease in concentration of a 55×10^3 M.W. protein in the host cytoplasm, while a number of proteins smaller than 38×10^3 M.W. are not detected after infection of the pea with Tn5 mutant number 5. At present we cannot assign any function however to these nodulins or *Rhizobium* secretory proteins.

Acknowledgements. We thank Drs. G.B. Ruvkun, F. de Bruijn and F.M. Ausubel for support in the cloning of pRB10. This research was financially supported by the Netherlands Organisation for the Advancement of Pure Research (Z.W.O.) and by the Netherlands organisation for Biological Research (B.I.O.N.). We thank Drs. R. Simon and A. Pühler for making available details on their method of gene mobilisation before publication. Rob Krutzen is thanked for technical assistance in the nodulin analysis, P. Madern for making illustrations and M.J. van Neerven for typing the manuscript.

References

Bisseling T, Bos RC van den, Kammen A van (1978) The effect of ammonium nitrate on the synthesis of nitrogenase and the concentration of leghaemoglobin in pea root nodules induced by *Rhizobium leguminosarum*. Biochim Biophys Acta 539:1–11

Bisseling T, Moen AA, Bos RC van den, Kammen A van (1980) The sequence of appearance of leghemoglobin and nitrogenase components I and II in root nodules of *Pisum sativum*. J Gen Microbiol 118:377–381

Bos, RC van den, Bisseling T, Kammen A van (1978) Analysis of DNA content, nitrogenase activity and in vitro protein synthesis of *Rhizobium leguminosarum* bacteroids in sucrose gradients. J Gen Microbiol 19:131–139

Bowen B, Steinberg J, Laemmli UK, Weintraub H (1980) The detection of DNA-binding proteins by protein-blotting. Nucleic Acids Res 8:1–20

Dénarié J, Boistard P, Casse-Delbart F, Atherly AG, Berry JO, Russel P (1981) Indigenous plasmids of *Rhizobium.* Int Rev Cytol Suppl 13:225–246

Ditta G, Stanfield S, Corbin D, Helinski DR (1980) Broad host range DNA cloning sytem for Gram-negative bacteria: Construction of a gene bank of *Rhizobium meliloti.* Proc Natl Acad Sci USA 77:7347–7351

Friedman AM, Long SR, Brown SE, Buikema WJ, Ausubel FM (1982) Construction of a broad host range cloning vector and its use in the genetic analysis of *Rhizobium* mutants. Gene 18: 289–296

Grunstein M, Wallis J (1979) Colony hybridization. Meth Enzymol 68:379–389

Holmes DS, Quigley D (1981) A rapid boiling method for the preparation of bacterial plasmids. An Biochem 114:193–197

Kondorosi A, Johnston AWB (1981) The genetics of *Rhizobium.* Int Rev Cytol Suppl 13:191–223

Krol AJM, Hontelez JGJ, Bos RC van den, Kammen A van (1981) Expression of large plasmids in the endosymbiotic form of *Rhizobium leguminosarum.* Nucleic Acids Res 8:4337–4347

Krol AJM, Hontelez JGJ, Roozendaal B, Kammen A van (1982) On the operon structure of the nitrogenase genes of *R. leguminosarum* and *Azotobacter vinelandii.* Nucleic Acids Res 10: 4147–4157

Legocki RP, Verma DPS (1980) Identification of "nodule-specific" host proteins (nodulins) involved in the development of *Rhizobium*-legume symbiosis. Cell 20:153–163

Rigby PWJ, Dieckmann M, Rhodes C, Berg P (1977) Labeling deoxyribonucleic acid to high specific activity in vitro by nick-translation with DNA polymerase I. J Mol Biol 113:237–251

Ruvkun GB, Ausubel FM (1981) A general method for site-directed mutagenesis in prokaryotes. Nature 289:85–88

Ruvkun GB, Sundaresan V, Ausubel FM (1982a) Site directed transposon Tn5 mutagenesis and complementation analysis of the *Rhizobium meliloti* symbiotic nitrogen fixation *(nif)* genes. Cell 29:551–559

Ruvkun GB, Long SR, Meade HM, Bos RC van den, Ausubel FM (1982b) ISRml: A *Rhizobium meliloti* insertion sequence which preferentially transposes into nitrogen fixation *(nif)* genes. J Mol Appl Genet (in press)

Southern EM (1975) Detection of specific sequences among DNA fragments separated by gel electrophoresis. J Mol Biol 98:503–517

Zabel P, Moerman M, Straaten F van, Goldbach R, Kammen A van (1982) Antibodies against the genome-linked protein VPg of cowpea mosaic virus recognize a 60,000-dalton precursor polypeptide. J Virol 41:1083–1088

Rhizobia Use Free-Living N_2 Fixation to Promote Growth by a Mechanism of Crossfeeding

R.A. LUDWIG[1]

Introduction

Rhizobia are unique among characterized N_2-fixing prokaryotes in that they have not been shown to be diazotrophs, i.e. organisms that utilize N_2 as sole N-source. Rather N_2 fixation primarily functions during the symbiotic state of these organisms as root nodule "bacteroids". During symbiotic N_2 fixation *Rhizobium* bacteroids exist as differentiated organelles that reduce N_2 to NH_3 and fail to assimilate such NH_3 into amino acids. They export NH_3 into the host plant cell cytosol where assimilation occurs via plant-coded enzymes (Brown and Dilworth 1975).

Certain slow-growing *Rhizobium* strains can also express N_2 fixation in microaerobic culture (Pagan et al. 1975, Kurz and LaRue 1975, McComb et al. 1975, Tjepkema and Evans 1975, Keister 1975). Since rhizobia are obligate aerobes, the large ATP requirement for N_2 fixation must be obtained by oxidative phosphorylation wherein electron transport utilizes O_2 as terminal electron acceptor. Microaerobic conditions are required since intracellular O_2 readily redoxes with electron transport proteins supplying nitrogenase during N_2 fixation resulting in free-radicals and subsequent lethal reactions.

Why then are such *Rhizobium* strains not microaerophilic diazotrophs? In this article I shall discuss experiments that investigate the microaerobic physiology of *Rhizobium* sp. strains under supposed free-living, N_2 fixing conditions. These detailed experiments demonstrate conclusively that such *Rhizobium* strains do not simultaneously grow and fix N_2. Moreover evidence is presented which suggests that free-living N_2 fixation by *Rhizobium* sp. ensues in a cell-state metabolically distinct from that of the vegetatively growing bacterium. Herein the ammonium produced is exported, and can be used to crossfeed other rhizobia, N-starved, which remain in in a vegetatively growing state. I suggest that sibling cross-feeding in rhizosphere colony populations is the selective advantage resulting from *Rhizobium* free-living N_2 fixation.

Comparative N-Assimilation of *Rhizobium* sp. Strains

Recently a mutant strain of *Rhizobium* sp. 32H1 (RC3200) has been described that fixes N_2 continuously in culture when supplied with excess carbon (as succinate and

1 Department of Biology, University of California, Santa Cruz, CA 95064, USA

Molecular Genetics of the Bacteria-Plant Interaction
(ed. by A. Pühler)

and L-arabinose) and nitrogen (as glutamate) sources under conditions of limiting-O_2 (Ludwig 1980a). This strain, RC3205,was obtained as a revertant of a glutamine (gln) auxotroph and possessed a constitutively unadenylated glutamine synthetase (GlnS I) but remains GlnS II-defective, like its parent strain RC3201 (Ludwig 1980b). Wild-type RC3200 under these culture conditions assimilates no ammonium produced by N_2 fixation into amino acids, rather it exports ammonium (Ludwig 1980a).

Since under certain culture conditions of carbon (succinate, L-arabinose) and nitrogen (glutamate) excess and low O_2 RC3205 can continuously biosynthesize the nitrogenase enzyme complex and can assimilate ammonium, neither of which can be accomplished by wild-type RC3200, what would occur when glutamate was exhausted? Would RC3205 continue to grow with N_2 as sole N-source? Experiments presented here clearly demonstrate that the answer is no. The more general question asks: can rhizobia indeed grow and fix N_2 at the same time, or do these activities represent mutually exclusive metabolic states of the organism? A definitive answer to this more general paradigm remains elusive, but the experiments presented here build a circumstantial case that such metabolic states are indeed mutually exclusive.

Microaerobic Growth Experiments

Rhizobium strains were cultured in constant temperature water baths at 28°C with continuous sparging by indicated gas mixtures. Gas mixtures were first passed through a high pressure nitrocellulose filter (0.45 micron) and then 20% sterile H_2SO_4 to strip the gas of contaminating NH_3. Dissolved O_2 concentrations were monitored by O_2 electrode (Clark type) corrected for ionic strength of the medium and barometric pressure. Viable cell count determinations of high reproducibility were obtained with automatic pipettes maintained at 30°C. Multiple, independent trials showed reproducibility in viable cell counts within 5% of the mean of three independent trials. Triton X-100 at 0.1% was included in growth media to suppress cell aggregation which adversely affected viable cell count determinations but was otherwise without effect when compared with control cultures. Total protein concentrations were determined in ultrasonically disrupted cell extracts by a modified Coomassie blue spectrophotometric procedure.

RC3200 and RC3205 Microaerobic Cultures in Limiting-N Induce Loss of Colony-Forming Ability

Aerobic cultures of RC3200 grown in minimal, defined medium with succinate and L-arabinose as excess carbon sources and L-glutamate as limiting nitrogen sources (MAS+Glt) enter stationary phase upon glutamate exhaustion. Thereafter for periods up to 2 weeks, cell viabilities, measured as colony-forming abilities, remain constant, fluctuating only within experimental error in measurement (data not presented).

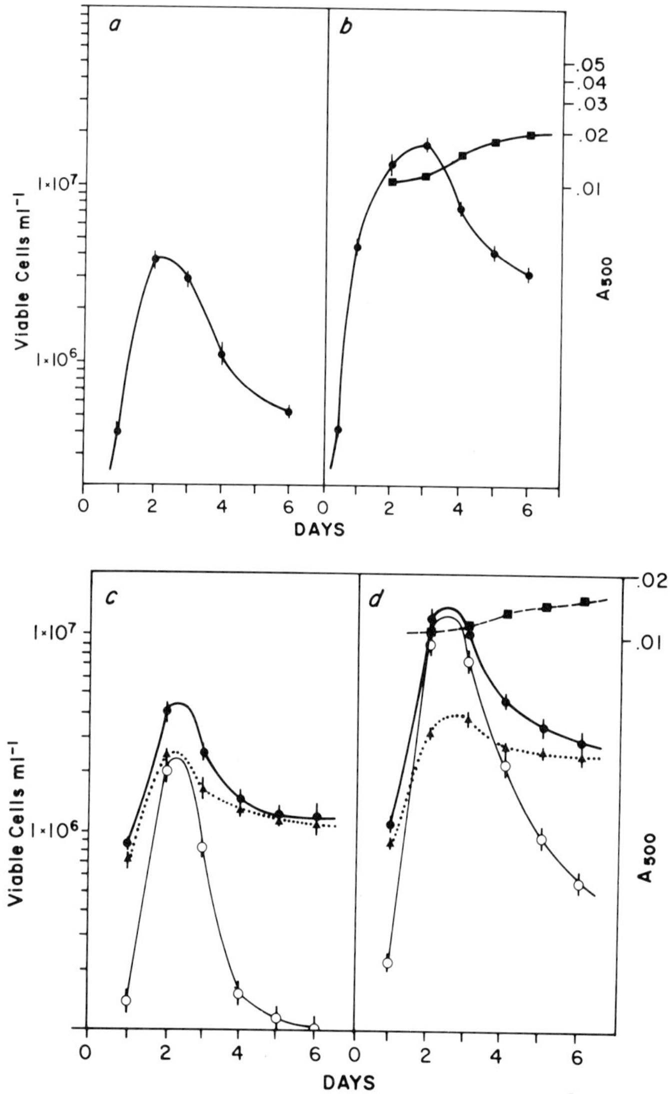

Fig. 1a–d. Viable cell counts of RC3200 and RC3205 cultures in N-limited microaerobiosis. RC3200 was cultured in MAS + **a** 6.6 μM Glt, or **b** 66 μM Glt as N-source at 28°C under continuous sparging with: 0.2% O_2, 1.0% CO_2, and 98.8% N_2. Aliquots were removed at times indicated, diluted, and plated intriplicate for viable cells on GYPC solid medium. When culture turbidity exceeded A_{500} nm = 0.01, this was monitored (■ ■). RC3205 was cultured in MAS + c 6.6 μM Glt, or **d** 66 μM Glt. Viable cell counts were subdivided into those that yielded colonies at normal times (▲ ▲), delayed times (Dcf phenotype) (○ ○), and total viable cells (● ●). Minimal growth medium MAS contained: 0.2% K succinate pH 6.3, 0.4% L-arabinose, 7.5 mM K_X PO_4 pH 6.3, 0.2 mM $MgSO_4$, 1 μM $FeCl_3$, 1 μM Na_2MoO_4, plus indicated amounts of L-glutamate (Glt) added as N-source. GYPC solid medium contains: 0.4% D-glucose, 0.2% yeast extract, 7.5 mM K_X PO_4 pH 6.3, and 0.2% casamino acids

Microaerobic cultures of RC3200 however exhibited a dramatic loss of colony-forming ability under these conditions (Fig. 1). Microaerobiosis was attained by continuous sparging of the gas mixture: 0.2% O_2, 1.0% CO_2, and 98.8% N_2. Irrespective of initial limiting-N (glutamate) levels supplied to RC3200, these cultures exhibited an exponential loss of colony-forming ability upon exhaustion of glutamate (Fig. 1a, b). The horizontal asymptotes of exponential growth curves reached by such cultures were indeed dependent upon amounts of limiting-N (glutamate) supplied and were identical whether growth was aerobic or microaerobic. Therefore exhaustion of limiting-N (glutamate) triggered subsequent loss of colony-forming ability under microaerobiosis. Cell titers obtained from these cultures were essentially identical whether rich (GYPC) or chemically-defined minimal media were employed to measure viability.

Under continuous sparging in these N-limited conditions, dissolved O_2 concentrations remained constant at approximately 3 μM. Culture turbidity was monitored by light scattering at 500 nm where possible. When glutamate exhaustion occurred turbidity nevertheless continued to increase, albeit slowly, with time (Fig. 1). Total cell protein remained constant during time of loss of colony-forming ability. Thus cell lysis during this time could be ruled out. Increased culture turbidity probably resulted from continued extracellular polysaccharide synthesis promoted by excess L-arabinose in the medium.

RC3200 N-limited cultures were assessed for their abilities to resume growth with readdition of glutamate during the period of loss of colony-forming ability (data not presented). After ever-increasing lag times proportional to time after glutamate exhaustion, culture growth resumed. Lag times were such that it was not possible to determine if only residual viable cells in starved cultures resumed growing, or whether all (viable + presumed non-growing) cells did so. Since loss of colony-forming ability under N-limitation was not observed in aerobic cultures, it was hypothesized that RC3200 under microaerobic N-limitation was entering a metabolically distinct cell-state unlike that of vegetative growth under these conditions. Under these conditions as many as 90% of total cells could be considered to have entered this distinct cell-state (Fig. 1).

RC3205 N-limited microaerobic cultures also exhibited loss of colony-forming ability. However the behavior of RC3205 under these conditions was more complex. Viable rhizobia could be divided into two classes according to their colony-forming abilities: those that produced colonies at normal times (5 days at 30°C) and those that required two additional days to do so [delayed colony-forming phenotype (Dcf), Fig. 1c, d]. These assignments were unambiguous. Dcf rhizobia first appeared at the time of glutamate exhaustion, and at their zenith comprised 50% of total cells, but were always transient (Fig. 1c, d). Thereafter Dcf rhizobia rapidly decreased in numbers. When total (Dcf + normally viable) RC3205 cells were plotted versus time, a loss of colony-forming ability in the population was observed that was similar to that observed for RC3200 (Fig. 1a–d). These results demonstrated that strain RC3205 behaved similarly to RC3200 under N-limited microaerobiosis with the additional observation of a transient population of Dcf cells. One interpretation of such results is that Dcf cells are in limbo between distinct, metastable cell states.

RC3200 and RC3205 Cultures in Limiting-O_2 Fix N_2, Fail to Exhaust Utilizable-N, and do not Induce Measurable Loss of Colony-Forming Ability

When similar microaerobic cultures were established and glutamate was increased to 0.7 mM, cultures became O_2 limited (Fig. 2). (Since the genus *Rhizobium* is comprised of obligate aerobes, RC3200 must employ O_2 as terminal electron acceptor under these conditions in order to retain redox balance.) Under these conditions dissolved O_2 concentrations dropped from 3 μM, observed under N-limitation, to below 1 μM. Growth continued unabated however. Stationary phase of such cultures was reached and exhibited a viable cell density of 3×10^8 cells ml^{-1}, which corresponded to predicted cell density at glutamate exhaustion. Under these conditions negligible loss of colony-forming ability occurred during O_2 limitation, despite extending observations up to two weeks. Here cell densities (above 10^8 ml^{-1}) were sufficient to allow measurement of nitrogenase activities as direct ammonium export, and such results have been reported (Ludwig 1980a). To test whether N_2 fixation contributed to cell growth under these conditions, N_2 was replaced with argon as predominant sparge gas in such O_2-limited cultures. Under argon, both RC3200 and RC3205 exhibited marginally slower growth (Fig. 2). Moreover stationary phases for both RC3200 and RC3205 under N_2 contained viable cell densities twice those observed under argon (Fig. 2). Therefore fixed-N contributed to cell mass under O_2 limitation.

These results suggest that N_2 fixation did not occur in N-limited culture (Fig. 1) at 3 μM O_2 in amounts sufficient to alleviate N-limitation. (Low cell densities in

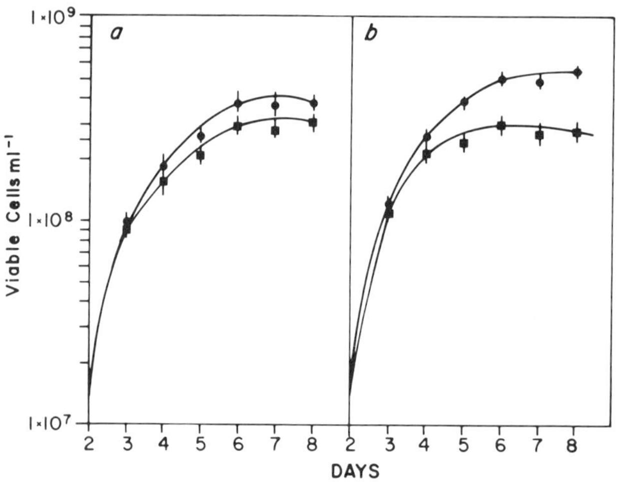

Fig. 2a, b. Viabilities of RC3200 and RC3205 cultures under limiting O_2. **a** RC3200, and **b** RC3205 were cultured in MAS + 0.7 mM Glt under conditions described in Fig. 1 except for C-source enrichment: L-arabinose (1.0%) and succinate (0.4%). At time = 0, cultures were split and sparged under: 0.2% O_2, 1.0% CO_2, and balance N_2 (● ●) or argon (■ ■) at 28°C. Viable cell counts versus time were determined as above. Nitrogenase activities as direct ammonium export have been described under these conditions previously (Ludwig 1980a)

N-limited cultures prevented detection of any N_2 fixation activities.) Secondly, since viable cell counts continued to increase in O_2-limited cultures, it was not possible to ascertain what amounts, if any, of rhizobia might be switching to the putative non-growing cell-state.

Neither RC3200 nor RC3205 Grow Directly on N_2 as Sole N-Source

As demonstrated, when RC3205 was microaerobically cultured at 3 μM O_2 with either 6.6 μM or 66 μM glutamate as N-source (limiting-N), growth ceased after glutamate was exhausted (Fig. 1c, d). As discussed, N-limited microaerobiosis did not allow either strain to switch to N_2 as N-source on which to continue to grow. Presumably this condition does not facilitate N_2 fixation in substantial amounts. If an as yet hypothetical O_2-mediated repression of *Rhizobium nif* genes exists that is similar to that observed in *Klebsiella pneumoniae* (Hill 1976), 3 μM O_2 might still constitute an O_2-repressed condition in *Rhizobium.* Since additional experiments comparing viable cell yields in N-limited microaerobiosis (3 μM O_2) under N_2 versus argon indicated no differences (data not presented) it was concluded that if it occurred at all, N_2 fixation does not contribute to cell yields under this condition. The sum total of experiments indicate that N_2 fixation requires the O_2-limited condition (below 1 μM).

To definitely ascertain if either RC3200 or RC3205 might grow on N_2 as sole N-source, it was therefore necessary to establish N-limited, O_2-limited culture conditions. This was accomplished by sparging with: 20 ppm O_2, 1.0% CO_2, and 99% N_2 or argon. Under this culture condition the O_2 concentration (undetermined) was nevertheless still sufficient to allow slow growth while other nutrients persisted! In this condition at onset of glutamate exhaustion cultures were O_2-limited as well (below 1 μM O_2). RC3200 cultures showed no significant differences in viable cell counts nor cell yields under N_2 as compared to argon (Fig. 3). On the other hand, loss of colony-forming ability was slight and slow to manifest itself (Fig. 3). Low cell densities again prohibited direct measurement of nitrogenase activites under these conditions. As compared to RC3200, RC3205 grew to viable cell densities under N_2 twice those under argon. Total cell protein was also commensurately greater. This suggested that for RC3205 under this condition, fixed-N contributed to cell mass. Nevertheless RC3205 cultures reached stationary phases at cell densities of approximately 6×10^7 cells ml^{-1} and very slow losses of colony-forming ability ensued thereafter (Fig. 3). Thus no continuous N_2-dependent growth could be demonstrated for either strain, but neither did loss of colony-forming ability predominate in such N-limited, O_2-limited cultures.

The behavior of *Rhizobium* sp. cultures under microaerobiosis was clearly complex vis-à-vis N_2 fixation, whose metabolic role under these conditions remained unclear. However a few simplifying assumptions could be made that facilitated interpretations of these results:

(1) Free-living N_2 fixation occurs under O_2-limitation and contributes to cell yield, yet the cultures as a whole do not grow in N-limitation, even in the presence of N_2.

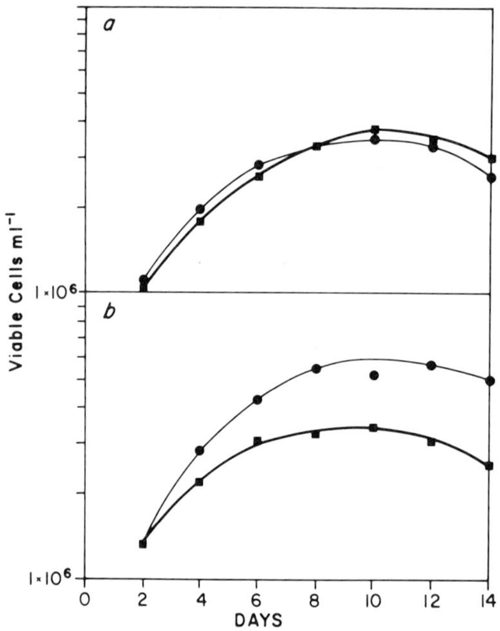

Fig. 3a, b. Growth of RC3200 and RC-3205 in N-limited, O_2 limited culture. **a** RC3200, and **b** RC3205 were cultured as described in Fig. 1 in MAS + 6.6 μM Glt and cultures were continuously sparged under: 20 ppm O_2, 1.0% CO_2, and either 99% N_2 (● ●) or argon (■ ■). Viable cell counts were determined versus time as in Fig. 1

(2) At 3 μM O_2, where it can be inferred that N_2 fixation is absent or at best suboptimal, N-limitation results in dramatic loss of colony-forming ability among these rhizobia. Either an N-excess condition or N_2 fixation obscures this phenomenon because amongst the total cell population, some cells continue to grow.

Given these simplifying assumptions, one may conclude that such cultures contain mixtures of rhizobia in distinct cell-states, that switching between cell-states occurs with some predictability, and that N_2 fixation occurs in a cell-state distinct from that of vegetative growth. Since N_2 fixation supplies fixed-N as exported ammonium under these conditions, this hypothesis necessitates that N_2 fixation promotes growth of vegetative rhizobia by a crossfeeding mechanism.

Free-Living *Rhizobium* N_2 Fixation Can Promote Growth by Crossfeeding

To test this hypothesis, N-limited microaerobic cultures were established with identified Nif⁻ mutant strains of RC3200. One such Nif⁻ strain, RC3312, exhibited behavior distinct from wild type RC3200 in N-limitation at both 3 μM O_2 and under O_2 limitation (below 1 μM O_2) (Fig. 4a). RC3312 exhibited no loss of colony-forming ability under 3 μM O_2 conditions (Fig. 4a) in contrast to RC3200. Neither did O_2-limited culture under N_2 yield enhanced growth when compared with argon, behavior consistent with the Nif⁻ phenotype of this strain, RC3312 (data not presented).

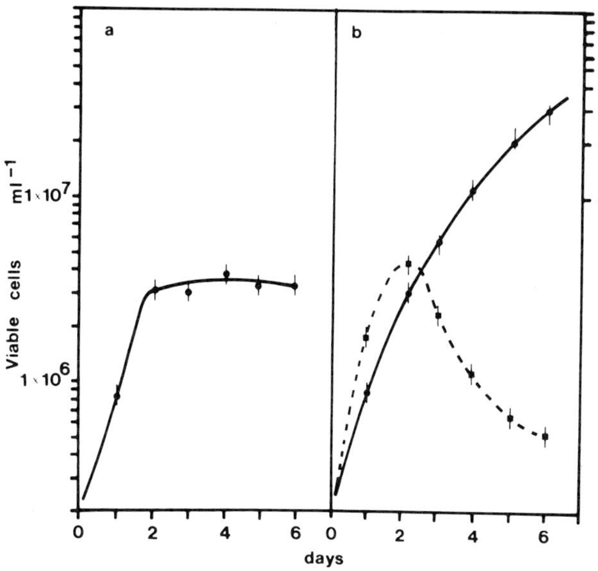

Fig. 4a, b. Nif-dependent crossfeeding between *Rhizobium* strains. **a** RC3312 Ery^R, nif⁻was cultured in MAS + 6.6 μM Glt as described in Fig. 1. Viable cell counts were determined versus time as described in Fig. 1. **b** Mixed RC3312 (● ●) and RC3505 Cam^R, Nif⁺ (■ ■) N-limited, O_2-limited culture was conducted as described in Fig. 3, under N_2 sparging. Viable cell counts versus time were determined as described in Fig. 1, by double-plating on GYPC + Ery (200 μg ml^{-1}) and GYPC + Cam (200 μg ml^{-1})

RC3312 was co-cultured with RC3505, a chloramphenicol-resistant (Cam^R), Nif⁺ derivative of RC3200, under N-limited, O_2-limited conditions. Whereas RC3505 exhibited loss of colony-forming ability after glutamate exhaustion, RC3312 continued to grow (Fig. 4b)! It was inferred that RC2505 Nif⁺ was cross-feeding RC3312 (pleiotrophic) Nif⁻ with ammonium under these conditions.

Since microaerobic culture of RC3312 failed to elicit loss of colony-forming ability, the Nif⁻ mutant phenotype in RC3312 is presumably pleiotrophic (other isolated Nif⁻ strains do exhibit loss of viability) and includes an impaired ability to switch cell-state, according to the presented hypothesis. Alleviation of N-depletion in the mixed culture by virtue of RC3505-mediated N_2 fixation resulted in facilitated selective growth of RC3312, again presumably because RC3312 exhibited no propensity to switch cell-state under these conditions. Mutant allele(s) in RC3312 remain uncharacterized except that Nif⁺ revertants may be selected in planta. RC-3505 on the other hand, did exhibit loss of colony-forming ability in the mixed culture, suggesting that the culture as a whole remained N-limited – the ammonium produced by RC3505 N_2 fixation having been withdrawn by RC3312.

Therefore what might be concluded about N-assimilation during ammonium crossfeeding? Ammonium produced by N_2 fixation is available in theory to contribute to *Rhizobium* growth. Whether it does so has been arguable since previous experiments demonstrated that ammonium can only be assimilated by GlnS I and II in these rhizobia, and that when such microaerobic cultures were analyzed, GlnS I proved predominantly adenylylated (and thus inactive) and GlnS II was undetectable (Ludwig 1980a, b). However such enzyme assays average whole populations of cells. Since experiments reported here suggest that it is possible to observe rhizobia in distinct cell states constituting heterogeneous populations, previous results must be reinterpreted. Ammonium crossfeeding requires that non-growing cells actively fixing N_2 contain adenylylated GlnS I, and that growing cells assimilating ammonium contain at least partially unadenylylated GlnS I.

Table 1. Bacterial strains

Strain	Phenotype	GlnS I	GlnS II	Nif
RC3200	wt	+	+	+
RC3201	Gln^-	ad	–	–
RC3205	Gln^+ rev of RC3201	un	–	++
RC3312	Ery^R, Nif^-	+	+	–
RC3505	Cam^R	+	+	+

ad = adenylylated; un = constitutively unandelylylated; Nif phenotype refers to both in culture and in planta

Discussion

Specific loss of colony-forming ability of *Rhizobium* sp. strains RC3200 and RC3205 in N-limited microaerobic culture has been observed. Because such loss of colony-forming ability occurs in microaerobic culture where these strains can otherwise grow perfectly well in N-excess conditions, and because RC3205 exhibits an additional Dcf state, it follows that such a non-growing state is deliberate and programmed by microaerobic N-limitation. Characterization of distinct cell-states, factors involved in transitions, and whether such interconversions are reversible remain to be addressed. Obviously one would like to be able to fractionate heterogeneous populations into their components, but this has not yet proven possible.

Is there a causal relationship between a non-growing cell-state and active N_2 fixation? It seems reasonable to infer that such a cell-state would be a prerequisite to N_2 fixation. Herein catabolic energy could be efficaciously directed to provide N_2 fixation with high reducing-potential electrons and ATP without competition from growth processes.

Is it reasonable to conclude that active N_2 fixation occurs at dissolved O_2 concentrations below 1 μM? Slow-growing *Rhizobium* strains have been found to possess multiple cytochrome oxidases with extremely high affinities for O_2 ($K_m = 10^{-8}$ M) that would allow efficient respiratory-linked oxidative phosphorylation to occur at these extremely low O_2 levels (Bergersen and Turner 1980).

Why does a strain such as RC3205, which can continuously biosynthesize nitrogenase and assimilate ammonium under N_2 fixing conditions, not grow directly on

N_2 as N-source? RC3205 indeed enters a stationary phase (when considered as a homogeneous culture) in O_2-limited conditions (Fig. 3) and displays loss of colony-forming ability in N-limited microaerobiosis. One must conclude that there exist additional control mechanisms that predicate such changes under these conditions, apart from an open cycle of N-regulation. In any case the assimilation of ammonium is not sufficient to engender growth of RC3205 under these conditions. Although this strain contains a constitutively unadenylylated GlnS I, non-growth under these conditions would result in accumulations of nucleotides and amino acids that would mediate allosteric inhibition of GlnS I and eventually arrest ammonium assimilation (Ginsburg and Stadtman 1974).

If indeed a distinct, non-growing cell-state proves a prerequisite for free-living *Rhizobium* N_2 fixation, it would explain the inability of liquid suspension cultures to simultaneously grow and fix N_2. Therein all cells experience the same milieu. This immediately allows one to speculate on the natural ecology of these processes in the rhizosphere. *Rhizobium* cells could assume specialized functional states within a colonial organization that would be determined by nutrient- and O_2-accessibility. Superficial cells of a defined colonial organization would experience greatest nutrient- and O_2-accessibility and their growth would be correspondingly promoted. Some interior stratum of cells under N- and O_2-limitation would be optimally suited for N_2 fixation, could then switch to the required cell-state, and could export produced ammonium to superficial cells to allow their continued growth. This would confer on *Rhizobium* strains able to fix N_2 by such a mechanism an appropriate selective advantage provided such exported ammonium was utilized to preferentially crossfeed their own siblings. Such a theory then must presuppose that N_2-fixing rhizobia would tend to be surrounded by their own siblings, which seems reasonable, but requires that we learn a great deal about homogeneous versus heterogeneous organizations in soil microorganism communities, their population dynamics, and their functional organizations.

Acknowledgements. This work was supported by Grants to the author from the Competitive Research Grants Office of the U.S. Department of Agriculture, the U.S. National Science Foundation, and the Camille and Henry Dreyfus Foundation. The author thanks Gert deVries for illuminating discussions.

References

Appleby CA (1969) Electron transport systems of *Rhizobium japonicum*, I. Biochim Biophys Acta 172:71–87

Bergersen FJ, Turner GL (1980) Properties of terminal oxidase systems of bacteroids from root nodules of soybeans and cowpea and of N_2-fixing bacteria grown in continuous culture. J Gen Microbiol 118:235–252

Brown CM, Dilworth MJ (1975) Ammonium assimilation by *Rhizobium* cultures and bacteroids. J Gen Microbiol 86:38–48

Ching TM, Henke S, Newcomb W (1977) Isolation of bacteria, transforming bacteria, and bacteroids from soybean nodules. Plant Physiol 60:771–780

Ginsburg A, Stadtman ER (1974) Regulation of glutamien synthetase in *E. coli.* In: Prusiner S, Stadtman ER (eds) Enzymes of glutamine metabolism. Academic Press, London New York, pp 9–43

Hill S (1976) Influence of atmospheric O_2 concentration on acetylene reduction and efficiency of N_2 fixation in *Klebsiella pneumoniae*. J Gen Microbiol 93:335–345

Keister DL (1975) Acetylene reduction by pure cultures of rhizobia. J Bacteriol 123:1265–1268

Kurz WGW, LaRue T (1975) Nitrogen fixation activity in *Rhizobium* in absence of plant host. Nature 256:407–408

Ludwig RA (1980a) Regulation of *Rhizobium* nitrogen fixation by the unandelylylated glutamine synthetase I system. Proc Natl Acad Sci USA 77:5817–5821

Ludwig RA (1980b) Physiological roles of glutamien synthetases I and II in ammonium assimilation in *Rhizobium* sp. 32H1. J Bacteriol 141:1209–1216

McComb JA, Elliott JJ, Dilworth MJ (1975) Acetylene reduction by *Rhizobium* in pure culture. Nature 256:409–410

Pagan J, Child JJ, Scowcroft WR, Gibson AH (1975) Nitrogen fixation by *Rhizobium* cultured in a defined medium. Nature 256:406–407

Tjepkema J, Evans HJ (1975) Nitrogen fixation by *Rhizobium* in a defined liquid medium. Biochem Biophys Res Commun 65:625–628

The Nitrogenase Genes of *Rhizobium japonicum*

H. HENNECKE, M. FUHRMANN, M. HAHN, K. KALUZA, B. REGENSBURGER, and R. WIRTH[1]

Introduction

Rhizobium japonicum elicits the formation of nitrogen-fixing nodules in soybeans *(Glycine max)*. As with all other rhizobia, the molecular analysis of its nitrogen fixation *(nif)* genes is hampered by the fact that it is difficult to obtain *nif*⁻ mutants by classical genetic techniques. A further disadvantage of using *Rhizobium japonicum* for genetic analyses is its slow growth. This is partly compensated by the ability of some strains to derepress nitrogenase activity under free-living culture conditions, because this unique trait will facilitate biochemical investigations once *nif*⁻ mutants are obtained.

The best approach towards the elucidation of *R. japonicum nif* genes has been the use of recombinant DNA techniques. Based on the *Klebsiella/Rhizobium* nitrogenase DNA *(nifHDK)* homology (Ruvkun and Ausubel 1980) we were able to detect nitrogenase genes in a clone bank of *R. japonicum* strain 110 (Hennecke and Mielenz 1981, Hennecke 1981). The recombinant plasmid obtained (pRJ676) carries a molecular weight 5.3×10^6 *Hin*dIII insert within the vector pBR322. Interspecies hybridization analysis has revealed that strong homology exists only with the *Klebsiella* gene *nifD*. Little, if any, homology was found with *nifH* and apparently none with *nifK*. This paper presents our recent progress on the structural analysis of these genes. In particular, we were interested (1) to learn more about the coding properties of the cloned region, (2) to support previous interspecies hybridization data by comparing nucleotide sequences, and (3) in finding ways to exploit the cloned genes for site-directed mutagenesis in *R. japonicum*.

Expression of Cloned Nitrogenase Genes in Minicells of *Escherichia coli* Strain DS410

The purpose of this study was to analyze the gene products encoded by the cloned *R. japonicum* DNA of pRJ676. We hoped to overcome problems of regulation and promoter recognition by cloning a large number of subfragments of pRJ676 into

1 Lehrstuhl für Mikrobiologie der Universität München, Maria-Ward-Str. 1a, 8000 München 19, FRG

Molecular Genetics of the Bacteria-Plant Interaction
(ed. by A. Pühler)

Vector	Recombinant Plasmids and Direction of Transcription from the Fused Promoter ← Plasmid Designation	← Gene Product(s) (kd)	→ Plasmid Designation	→ Gene Product(s) (kd)
pACYC177	pRJ4019	25	pRJ4018	17
pACYC177	pRJ4025	n.p.	pRJ4022	n.p.
pBR322	pRJ4040	58,55 (18,16)	pRJ4041	58,55 (weak)
pACYC177	pRJ4039	n.p.	pRJ4038	n.p.
pACYC177	pRJ4034	55,47	pRJ4035	n.p.
pACYC177	pRJ4042	47		
pACYC177	pRJ4043	n.t.		
pACYC177	pRJ4045	58,55		
pBR322	pRJ4017	55 (18,16)	pRJ4007	55(weak) (18,16)
pACYC177	pRJ4021	55	pRJ4024	n.p.
pACYC184	pRJ4006	29	pRJ4011	n.p.
pACYC184	pRJ4012	27	pRJ4013	n.p.

Physical Map ←

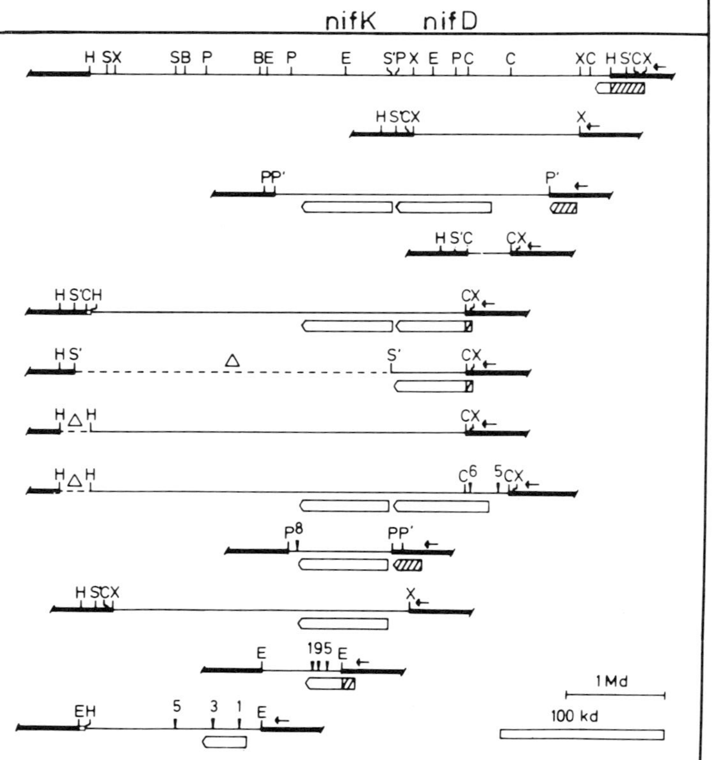

suitable vectors (pBR322, pACYC184, pACYC177) in such a way that they were transcribed from the strong promoters of antibiotic resistance genes. Where appropriate each fragment was cloned in the two possible orientations with respect to the outside promoter. Recombinant plasmid-encoded polypeptides were then synthesized in *E. coli* minicells and identified on polyacrylamide gels. In addition, Tn*5* insertion mutations were used to determine the direction of transcription, the transcriptional organization, and the exact termination sites of genes.

Figure 1 shows the results of this analysis, and Fig. 2 shows some examples. The general direction of transcription proceeds from right to left (Fig. 1). There is a strong transcription termination site far on the right. No readthrough was observed beyond this point, leaving a gap of molecular weight 10^6 length before the next gene products appeared. Further downstream we located a transcriptional unit consisting of two cistrons: *nifD* and *nifK;* these two genes respectively code for the 58 $\times$ 10^3 M.W. β subunit and the 55 $\times$ 10^3 M.W. α subunit of component 1 of the *R. japonicum* nitrogenase complex (Fuhrmann and Hennecke 1982). On Fig. 2 both proteins are seen in lane d (pRJ4045); the α subunit is also seen in lanes a (pRJ4017) and b (pRJ4034), and truncated derivatives of the β subunit are seen in lanes b (pRJ4034) and c (pRJ4042). A transposon (Tn*5*) insertion near the beginning of the *nifD* gene prevents synthesis of both subunits (lane e: pRJ4045::Tn*5*-6). Left of the *nifDK* operon there is evidence for another transcriptional unit the promoter distal cistron of which codes for a 27 kd protein of unknown function.

◀ **Fig. 1.** A collection of cloned *R. japonicum* DNA fragments derived from pRJ676, and their gene products synthesized in *E. coli* minicells. The *second* and *fourth columns* show the names of the new recombinant plasmids where most fragments are cloned in two orientations with respect to the promoter of the antibiotic resistance gene. Full length (intact) gene products of the respective vectors as well as polypeptides expressed by Tn*5*-mutated plasmids are not included in the list. Truncated vector-encoded proteins, and strong background minicell proteins are in parentheses. The physical map shows only the plasmids named in the second column. *R. japonicum* DNA fragments *(thin lines)* are flanked by vector DNA, which is not drawn to completion. Deletions ($\triangle$) are *dashed.* In pRJ4034 and pRJ4012, the cloned fragment contains on its far left-hand end a tiny piece of pBR322 DNA originating from pRJ676. Expression starts from promoters located on the right of the cloned fragments (indicated by small leftward-oriented *arrows*). Tn*5* insertion sites are marked by *numbered arrowheads.* Gene products encoded by *R. japonicum* DNA are drawn as *open bars,* whereas *hatched regions* denote peptides specified by fractions of the antibiotic resistance genes. *n.p.* no significant protein detectable; *n.t.* not tested; *Md* megadaltons; *kd* kilo M.W. Restriction sites are: *Eco*RI (E), *Pst*I (P), *Sma*I (S'), *Bam*HI (B), *Hin*dIII (H), *Cla*I (C), *Sal*I (S), *Xho*I (X), *Pvu*I (P'). (Modified from Fuhrmann and Hennecke 1982)

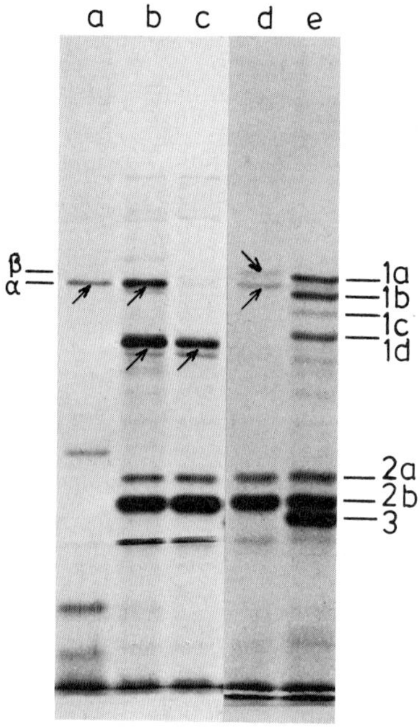

Fig. 2. SDS gel electrophoretic analysis (autoradiogram) of (^{35}S)-labelled extracts from minicells: plasmid-directed synthesis of *R. japonicum* nitrogenase proteins. Polypeptides are shown encoded by pRJ-4017 *(a)*, pRJ4034 *(b)*, pRJ4042 *(c)*, pRJ4045 *(d)*, pRJ4045::Tn*5*-6 *(e)*. α and β denote the positions of the constituent subunits of component 1. Proteins designated on the right margin are: Tn*5*-encoded transposase proteins *(1a* to *1d)* and aminoglycoside 3′-phosphotransferase II *(3)*, and vector-encoded precursor *(2a)* and mature *(2b)* β-lactamase

DNA Sequence Analysis

Based on DNA/DNA hybridization data the *R. japonicum nifD* gene is strongly homologous to its *K. pneumoniae* counterpart (Hennecke 1981). We have started to determine the DNA sequence of this gene in order to find an explanation for this homology on the molecular level. We used the dideoxy/chain termination methods of Sanger et al. (1977) employing phages M13mp7, -mp8, and -mp9, and an Apple II computer for data processing (Messing 1982). Fig. 3 shows a stretch of 57 bases from within the *Eco*RI-*Cla*I fragment of the *R. japonicum nifD* gene. It shows strong homology to a corresponding segment of *K. pneumoniae nifD*, and can be translated into a corresponding cysteine-containing peptide of 19 amino acids. Seventeen of these are identical with those from *K. pneumoniae* (Scott et al. 1981), and 14 are identical with those of a tryptic peptide isolated from component 1 of *Azotobacter vinelandii* (Lundell and Howard 1981). The sequence Glu-Cys-Pro-Val-Gly-Leu-Ile is unique to 4Fe:4S proteins (Lundell and Howard 1981).

The nucleotide sequence of the region upstrem of *nifDK* was also of interest to us. In *K. pneumoniae* the *nifH* gene coding for component 2 of nitrogenase is located there. Interspecies DNA/DNA hybridization within this area, however, has been found to be very weak, and may not be enough proof for an analogous location of *nifH* in *R. japonicum* (Hennecke 1981). Moreover, we have not obtained expression of the *R. japonicum* component 2 protein in minicell experiments (Fig. 1). A DNA segment of 1082 basepairs on the right of *nifD* was sequenced in order to obtain evidence for or against the presence of *nifH*. The sequenced DNA was translated in

NIF D - HOMOLOGY

```
K.p.a  GGG ATC ACC ATT CAG TCG GAA TGC CCG GTG GGG CTG ATC GGT GAT GAT ATC AGC GCG
       GLY ILE THR ILE GLN SER GLU CYS PRO VAL GLY LEU ILE GLY ASP ASP ILE SER ALA

R.j.   GGC TTT ACG ATA CAA TCA GAG TGC CCG GTA GGC TTG ATC GGT GAC GAT ATC GAG GCG
       GLY PHE THR ILE GLN SER GLU CYS PRO VAL GLY LEU ILE GLY ASP ASP ILE GLU ALA

A.v.b  GLY ILE SER VAL GLN SER GLU CYS PRO ILE GLY LEU ILE GLY ASP ASP ILE GLU SER
```

a) Scott et al., 1981
b) Lundell and Howard, 1981

Fig. 3. Nucleotide sequence of a strongly conserved region from the *R. japonicum (R.j.) nifD* gene. Homologous sequences from *K. pneumoniae (K.p.)* and *Azotobacter vinelandii (A.v.)* are shown above and below

12 reading frames (6 for each strand). The comparison of these sequences with those of *Klebsiella nifH* (Scott et al. 1981, Sundaresan and Ausubel 1981) failed to reveal any homology between them, both on the nucleotide as well as on the amino acid sequence level (not shown). Thus, it is highly unlikely that a *nifH*-analogous gene is encoded in this area.

In a further experiment some of the single-stranded M13 clones of this region were used to determine the coding strand by "dot" hybridization (Messing 1982). The (^{32}P)-labelled probe was mRNA produced in vitro in a homologous transcription system with linear DNA fragments of this area and purified RNA polymerase from *R. japonicum.* The DNA strand thus identified is transcribed in the same direction as *nifDK* and may code either for two 15×10^3 M.W. polypeptides or, in case of a frameshift sequence error, for one 30×10^3 M.W. polypeptide (data not shown).

Site Directed Mutagenesis

Ruvkun and Ausubel (1981) have outlined a method for the localized mutagenesis in Gram-negative prokaryotes. The principle is to mutagenize cloned fragments in *E. coli,* transfer them back into the bacterium of origin, and select for the replacement of wild-type DNA against the homologous mutated DNA. We have tried to adopt this technique for *R. japonicum* by using a vector system which is efficiently mobilized into *Rhizobium* but is not stably maintained there (Simon, this vol.). When Tn*5*-mutagenized fragments are transferred defined mutants may be found among the kanamycin/neomycin-resistant exconjugants.

Fig. 4 shows the Tn*5* mutagenized regions of cloned *R. japonicum* DNA which were transferred to a Nod$^+$ Fix$^+$ spectinomycin-resistant derivative of *R. japonicum* strain 110. Neomycin-resistant *R. japonicum* strains were obtained at various frequencies. Total DNA of such strains was then digested with suitable restriction

vector	plasmid designation	physical map of the cloned R. japonicum - fragments	transfer-rate of neo-resistance	recombination event observed as: cointe-gration	marker exchange	designation of the neo-resistant transconjugants
pBR322	pRJ 676	H E E EC C C H (nif K, nif D)	———	———	———	———
p 201	pRJ4029::Tn5-6	E ▼ H	$3 \cdot 10^{-6}$	2	2	strain A1 to A4
p 201	pRJ6201::Tn5-301	E ▼E	$5 \cdot 10^{-7}$	10	1	strain B1 to B11
p 201	pRJ6201::Tn5-401	E ▼ E	$1 \cdot 10^{-7}$	0	1	strain C1
p 201	pRJ6000::Tn5-5	E ▼ E	$2 \cdot 10^{-7}$	2	2	strain D1 to D4
p 201	pRJ4029::Tn5-5	E ▼ H	$2 \cdot 10^{-7}$	6	0	strain E1 to E6
p201	pRJ6201::Tn5-302	E ▼ E	$5 \cdot 10^{-8}$	8	0	strain F1 to F8
p 201	pRJ6000::Tn5-1	E ▼E	$2 \cdot 10^{-6}$	18	0	strain G1 to G18
p 201	pRJ6201::Tn5-201	E▼ E	$1 \cdot 10^{-6}$	1	0	strain H1

Fig. 4. Summary of data on the localized mutagenesis of cloned *R. japonicum nif* DNA. *Eco*RI or *Eco*RI-*Hin*dIII fragments of pRJ676 were cloned into the mobilizable vector p201. The names of the new plasmids are shown in the *second column;* each one contains a Tn*5* insertion *(arrowheads)* at a known location as depicted on the physical map. All plasmids were transferred from *E. coli* Sm10 donor strain (Sinmon, this vol.) to *R. japonicum* strain 110 spc^R by conjugation. Neomycin-resistant *R. japonicum* exconjugants were obtained at a frequency shown in the *fourth column.* DNA of such strains was scrutinized for the presence or absence of p201 material by hybridization, and by further analysis as described in the text: this helped to distinguish strains with co-integrates from those where a real marker exchange event had occurred *(fifth* and *sixth columns)*

enzymes, separated on agarose gels and analyzed by Southern blot hybridization (Southern 1975). The following nick-translated DNAs (Maniatis et al. 1975) were used as radioactive hybridization probes: vector p201; the *Hpa*I fragment of Tn*5; Eco*RI or *Eco*RI-*Hin*dIII fragments from pRJ676. As a result of a homologous marker exchange event the DNA of a "true" *R. japonicum* Tn*5* mutant should meet the following requirements: (1) the Tn*5*-containing restriction fragment should migrate into the same position as the corresponding fragment of the recombinant clone; (2) unmutated wild-type fragments should not be present; (3) no remnants of the transfer vector should be present. As can be seen from the results in Fig. 4 only a minority of mutants with Tn*5* inserts at four different locations fulfil these requirements. In all other strains the predominant recombination event must have been the cointegration of homologous mutated DNA together with vector DNA into the genome. In such strains one copy of the wild-type DNA fragment is still present (as evidenced by Southern blot hybridization; not shown). The frequent formation of cointegrates seems to be an intrinsic property of the vector system used.

Tn*5*-containing *Eco*RI fragments from DNA of two of the "clean" replacement mutants, namely from strains A3 and D3 (Fig. 4), were recloned into pACYC184. Tn*5* was found to be located in the same sites as known from the original recombinant clones (pRJ4029::Tn*5*-6 and pRJ6000::Tn*5*-5, respectively). The phenotypes of all true replacement mutants with respect to nodulation and nitrogen fixation is now under investigation.

Concluding Remarks

The results presented in this report have shown that the structural genes for the β and the α subunits of component 1 of *R. japonicum* nitrogenase are organized in one transcriptional unit, *nifDK.* In contrast to *K. pneumoniae* or *R. meliloti* (Ausubel et al., this vol.) a *nifH* analogous gene is not part of the same operon; rather it must be located elsewhere on the genome. Although it is quite surprising that the nitrogenase genes are rearranged relative to each other this may not be an uncommon feature: recent work on *Anabena variabilis* has shown that *nifK* is separated from *nifHD* (R. Haselkorn, pers. commun.).

It is also interesting to note that, in contrast to the gene/protein relationships found in *Klebsiella* (Kennedy et al. 1981), the *R. japonicum nifD* gene codes for the larger β subunit and *nifK* for the smaller α subunit of component 1. DNA sequence analysis, however, has clearly verified *nifD* homology. Thus, evolution has permitted a certain variation in the length of the individual subunits without affecting conserved structures.

The availability of cloned and physically characterized *R. japonicum nif* DNA in *E. coli* has made it possible to mutate this region specifically. The experiments presented here to obtain *R. japonicum* mutants by site-directed mutagenesis have turned out to be even more laborious than with fast-growing *Rhizobium* strains (Ruvkun and Ausubel 1981). Nevertheless, there is the great potential to construct defined *nif*$^-$ mutants and to assign physiological functions to genes in the vicinity of *nifDK.*

Acknowledgements. We thank R. Simon for donating bacterial strains prior to publication, J. Messing for the computer program, and A. Böck for support. Grants for various parts of this work were provided by the Deutsche Forschungsgemeinschaft (He1036/3 and SFB145), and by the Bundesministerium für Forschung und Technologie (PTB8427).

References

Fuhrmann M, Hennecke H (1982) Coding properties of cloned nitrogenase structural genes from *Rhizobium japonicum.* Mol Gen Genet (in press)

Hennecke H (1981) Recombinant plasmids carrying nitrogen fixation genes from *Rhizobium japonicum.* Nature 291:354–355

Hennecke H, Mielenz JR (1981) Molecular cloning of *Rhizobium japonicum* DNA in *E. coli* and identification of nitrogen fixation *(nif)* genes. In: Lyons et al (eds) Genetic engineering of symbiotic nitrogen fixation and conservation of fixed nitrogen. Plenum Press, New York London, p 41

Kennedy C, Cannon F, Cannon M, Dixon R, Hill S, Jensen J, Kumar S, McLean P, Merrick M, Robson R, Postgate J (1981) Recent advances in the genetics and regulation of nitrogen fixation. In: Gibson AH, Newton WE (eds) Current perspectives in nitrogen fixation. Australian Acad Sci, Canberra, p 146

Lundell DJ, Howard JB (1981) Isolation and sequences of the cysteinyl tryptic peptides from the MoFe-protein of *Azotobacter vinelandii* nitrogenase. J Biol Chem 256:6385–6391

Maniatis T, Jeffrey A. Kleid DG (1975) Nucleotide sequence of the rightward operator of phage λ. Proc Natl Acad Sci USA 72:1184–1188

Messing J (1982) An integrative strategy of DNA sequencing and experiments beyond. In: Hollaender A, Setlow J (eds) Genetic engineering, principles and methods, vol IV. Plenum Press, New York London (in press)

Ruvkun GB, Ausubel FM (1980) Interspecies homology of nitrogenase genes. Proc Natl Acad Sci USA 77:191–195

Sanger F, Nicklen S, Coulson AR (1977) DNA sequencing with chain terminating inhibitors. Proc Natl Acad Sci USA 74:5463–5467

Scott KF, Rolfe BG, Shine J (1981) Biological nitrogen fixation: primary sequence of the *Klebsiella pneumoniae nifH* and *nifD* genes. J Mol Appl Gen 1:71–81

Southern EM (1975) Detection of specific sequences among fragments separated by gel electrophoresis. J Mol Biol 98:503–517

Sundaresan V, Ausubel FM (1981) Nucleotide sequence of the gene coding for the nitrogenase iron protein from *Klebsiella pneumoniae.* J Biol Chem 256:2808–2812

The Structure and Organization of the Soybean Leghemoglobin Genes

K.A. MARCKER, K. BOJSEN, E.Ø. JENSEN, K. PALUDAN, and O. WIBORG [1]

Introduction

Leghemoglobins (Lbs) are myoglobin-like proteins synthesized exclusively in the root nodules that develop due to the symbiotic association of *Rhizobium* with leguminous plants. Evidence suggests that their function is to ensure an adequate oxygen supply for bacteroid respiration in nodule tissue (Appleby 1974). Soybean nodules contain four major species of Lbs called Lba, Lbc_1, Lbc_2 and Lbc_3, respectively (Fuchsman and Appleby 1979). In addition several minor Lb-components are present in the nodule, but some of these components most likely represent post-translational modification products of some of the major components. The differences in the amino acid sequences among the various Lb-components are small corresponding to 6–8 amino acid substitutions only (Sievers et al. 1978).

The Lb-Gene Family

The Lbs are encoded in the plant genome by a small family of genes (Sullivan et al. 1981, Marcker et al. 1981). The activation of these genes occurs only after the infection of the plant by *Rhizobium*. Southern blotting analysis of EcoR1 digested soybean DNA revealed the presence of seven DNA fragments hybridizing to an Lb cDNA probe. The sizes of these fragments are 1.4, 4.2, 5.5, 6.0, 7.5, 10 and 12 kb, respectively. All seven DNA fragments have now been isolated from various soybean DNA libraries. For all seven fragments a restriction endonuclease map has been constructed. In addition DNA sequence analysis has been completed for the Lb-genes located on five fragments. Sequence analysis of the 4.2 and 1.4 kb fragments revealed that both fragments contained incomplete genes, the 4.2 kb fragment carrying an 5′ end and the 1.4 kb fragment carrying the corresponding 3′ end. The 5.5 kb fragment contains a truncated gene consisting of a 3′ end only. The 6.0 kb fragment contains a complete gene. The 7.5 and 12 kb fragment are linked and contain two complete genes, one on each fragment and an Lb-like gene with its 5′-end on the 7.5 kb fragment and the corresponding 3′ end on the 12 kb fragment. The 10 kb fragment most likely

1 Department of Molecular Biology and Plant Physiology, University of Aarhus, 8000 Aarhus C, Denmark

Molecular Genetics of the Bacteria-Plant Interaction
(ed. by A. Pühler)

contains a complete Lb-gene. Brisson and Verma (1982) have also isolated a 4.2 kb fragment which contains a truncated Lb-gene consisting of a 3′ end only. Thus, the Lb-gene family in soybeans consists of five complete genes, one Lb-like gene and at least two truncated genes.

DNA Sequence Analysis of the Lb-Genes

DNA sequence analysis of the Lb-genes located on the 1.4, 4.2, 6.0, 7.5 and 12 kb fragment have been completed (Hyldig-Nielsen et al. 1982, Wiborg et al. 1982). Comparison with known Lb-amino acid sequences indicates that the Lb-genes on the 6.0, 7.5 and 12 kb fragments correspond to Lbc_2, Lbc_1 and Lbc_3, respectively. The gene located on the combined 4.2, 1.4 kb fragment corresponds to Lba. The gene extending across the 7.2–12 kb junction corresponds to an Lb-like gene which quite possibly is an Lb-pseudo gene.

The general DNA sequence organization of the Lba, Lbc_1, Lbc_2 and Lbc_3 genes is shown in Fig. 1. The coding sequences are in all cases interrupted by intervening sequences at three identical positions. The locations of the introns are at codon 32 (IVS-1), between codons 68–69 (IVS-2) and between codons 103–104 (IVS-3). The positions of IVS-1 and IVS-3 correspond precisely to the positions of the two intervening sequenes found in all globin genes sequenced so far (Jensen et al. 1981).

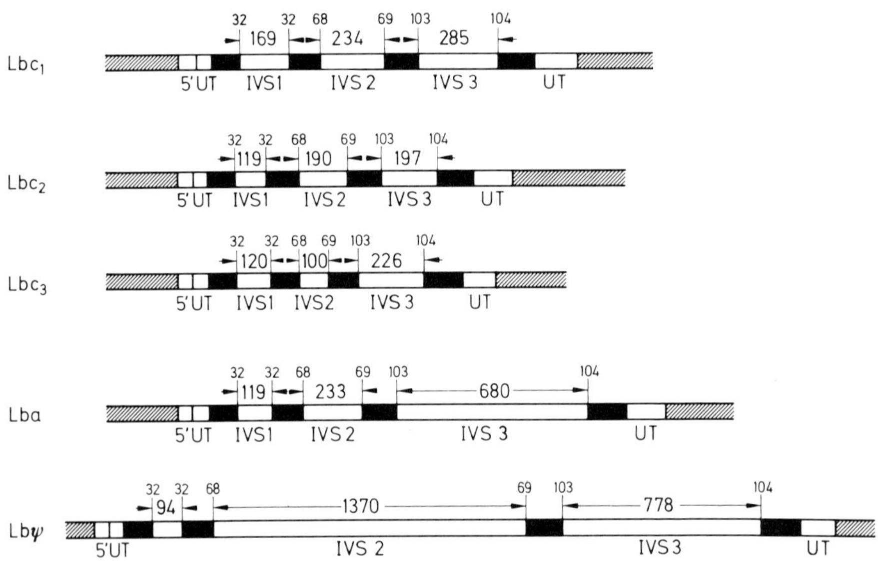

Fig. 1. The general DNA sequence organization of five Lb-genes. The introns are denoted IVS-1, IVS-2 and IVS-3, respectively. The exons are shown as *solid boxes*. *UT* denotes untranslated regions. The length of the introns are indicated *above the bars* and their location with respect to codon numbers, *below the bars*. The DNA sequence of Lbc_3 has also been determined by Brisson and Verma (1982)

This supports the notion that all globin genes including the Lb-genes are derived from a common ancestral gene. Gō (1981) recently suggested that the central exon in globin genes might be the result of a fusion between two exons with a division somewhere between amino acid residues 66–71. The location of IVS-2 between codons 68–69 in the Lb-genes is in excellent agreement with Gō's proposal. The evolutionary implications of the structure of the Lb-genes have been discussed by us recently (Hyldig-Nielsen et al. 1982).

Breathnach et al. (1978) have noted the tendency for intervening sequences to begin with the dinucleotides GT and end with the dinucleotide AG. For the four Lb-genes all sequences can be aligned such that intervening sequences start with GT and terminate with AG. The length of the intervening sequences are quite variable. Thus a considerable size variation is noted for both the IVS-2 and the IVS-3 sequences.

All four Lb-genes contain putative regulatory sequences identical to or very similar to the corresponding sequences found in other eukaryotic genes. Thus a cap addition site TGCATAAC (Fig. 2) is located about 46–53 nucleotides upstream from the ATG initiation codon. The canonical sequence TATAA is found about 30 nucleotides upstream from the cap addition site. About 39–56 nucleotides upstream from the TATAA sequence the pentanucleotide CCAAG occurs. In most eukaryotic genes a homologous sequence occurs at or close to this position. In the 3′ non-coding region proximal to the putative polyA addition site the hexanucleotide GATAAA occurs in all four genes. This sequence is probably analogous to the ubiquitous AATAAA hexanucleotide found in a similar position in other eukaryotic genes where it probably functions as polyA addition signal (Proudfoot and Brownlee 1976). Thus the putative regulators signals in the Lb-genes are very similar to the corresponding signals in other eukaryotic genes. This strongly suggests that the mechanisms for gene transcription in plants are very similar to those found in other eukaryotes.

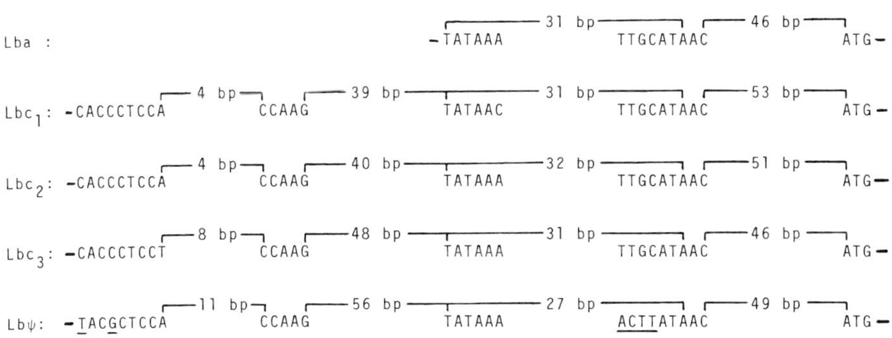

Fig. 2. Comparison of putative regulatory sequences on the 5′ flanking region of five Lb-genes. The nucleotide sequences correspond to regulatory sequences cited in the text. Distances are given in base pairs (bp). The *underlined* nucleotides in Lb ψ indicate mutations

The Structure of the Unusual Lb-Gene

The general DNA sequence organization of the Lb-like gene extending across the 7.5, 12 kb junction is shown in Fig. 1. The locations of the intervening sequences are identical to those found in the other Lb-genes. However, the sizes of IVS-3 and in particular of IVS-2 are greatly increased when compared to the corresponding sequences in the other Lb-genes. Because of these increases, the size of this gene is around 3 kb which is about twice the size of a normal Lb-gene. The sequences around splicing junctions in this gene conform with the Breathnach-Chambon Rule (Breathnach et al. 1978). The intervening sequences do not show any homology with the corresponding sequences in the other Lb-genes except close to the splicing junctions. The amino acid sequence derived from the coding sequence (Fig. 3) although showing 85% homology with Lbc_1 does not correspond to any known soybean Lb. There are 29 amino acid replacements which is about 20–25 more than found among the protein sequences of the other soybean Lbs. In addition there are seven extra amino acids present at the 3′ end of the gene. The third exon of the gene is only about 79% homologous with the corresponding exon in the other Lb-genes. This exon was cloned into M13 and the resulting clone was subsequently used to screen a soybean nodule cDNA library. However no positive signals were detected suggesting that soybean nodules do not contain any corresponding mRNA or alternatively that the mRNA is present in very minor amounts only. These findings suggest that this particular gene is non-functional either by not being transcribed or by having a transcript which turns over rapidly or cannot be transported. The most plausible explanation is that the gene is not transcribed and it is therefore likely that mutations have occurred in sequences important for transcription. Fig. 2 shows the sequences of the 5′ franking regions of the five Lb-genes. In the Lb-like gene there are several mutations in the

```
Gly-Ala-Phe-Thr-Glu-Lys-Gln-Glu-Ala-Leu-Val-Asn-Ser-Ser-Phe-
Glu-Ala-Phe-Lys-Ala-Asn-Leu-Pro-His-His-Ser-Val-Val-Phe-Phe-
Asn-Ser-Ile-Leu-Glu-Lys-Ala-Pro-Ala-Ala-Lys-Asn-Met-Phe-Ser-
Phe-Leu-Gly-Asp-Ala-Val-Asp-Pro-Lys-Asn-Pro-Lys-Leu-Ala-Gly-
His-Ala-Glu-Lys-Leu-Phe-Gly-Leu-Val-Arg-Asp-Ser-Ala-Val-Glu-
Leu-Gln-Thr-Lys-Gly-Leu-Val-Val-Ala-Asp-Ala-Thr-Leu-Gly-Pro-
Ile-His-Thr-Gln-Lys-Gly-Val-Thr-Asp-Leu-Gln-Phe-Ala-Val-Val-
Lys-Glu-Ala-Leu-Leu-Lys-Thr-Ile-Lys-Glu-Ala-Val-Gly-Asp-Lys-
Trp-Ser-Glu-Glu-Leu-Ser-Asn-Ala-Trp-Glu-Val-Ala-Tyr-Asp-Glu-
Ile-Ala-Ala-Ala-Ile-Lys-Lys-Ala-Met-Ala-Ile-Gly-Ser-Leu-Val
```

Fig. 3. The amino acid sequence derived from the Lb-like gene. The *arrows* indicate the location of the introns. Underlining with *heavy lines* indicates amino acid replacements when compared to the coding sequence of Lbc_1. *Dotted line* indicates the seven amino acid extension of the 3′ end. Brisson and Verma (1982) have also determined the DNA sequence corresponding to exons 3 and 4 of the Lb-like gene

cap addition site area. This region is important for the transcription of the β-globin genes as shown recently by Talkington and Leder (1982). In addition the Lb-like gene is mutated in the region immediately preceeding the CCAAG sequence. Recently Grosveld et al. (1982) have shown that this region is also very important for transcription of the globin genes. Thus, the Lb-like gene is mutated in two regions which seem to be essential for transcription. We therefore suggest that the Lb-like gene is not functional and consequently is a pseudo gene.

The Arrangement of the Lb-Genes Upon the Soybean Chromosome

The entire Lb-gene region has been investigated by Southern blotting analysis of restriction endonuclease digests of soybean DNA using specific probes for the various Lb-genes. In addition overlapping clones have been obtained from several soybean DNA libraries. The results so far obtained show that all the Lb-genes are linked and that the Lb-genes are contained within a region of approximately 90 kb as shown in Fig. 4. The overlapping clones proving the locations of the gene on the 10 kb fragment and the truncated gene on the 5.5 kb fragment have not yet been obtained. However, Southern blotting analysis of soybean DNA suggests that the indicated positions for these genes are correct. Four genes are very closely linked in the 5′ Lba-Lbc_1-ψLb-Lbc_3-3′. The distances between these genes are about 2.5–3 kb. However, the distance from the Lba gene to the Lb-gene on the 10 kb fragment is about 25 kb while the distance between Lbc_2 and the truncated gene on the 5.5 kb fragment is about 15 kb. The distance between Lbc_2 and Lbc_3 is 25 kb.

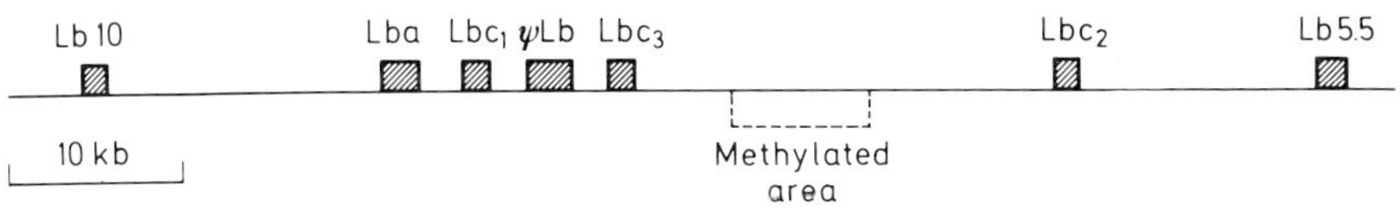

Fig. 4. Arrangement of the Lb-genes upon the chromosone. *Hatched boxes* above the line represent the respective Lb-genes. Lb 5.5 and 10 represent the Lb-genes located on the 5.5 and 10 kb EcoR1 fragments, respectively. The *open box* below the line represents the region exhibiting methylation variance

Unfortunately, sequence analysis has not yet been completed on the Lb-gene present on the 10 kb fragment, but available evidence suggests that this gene is functional. The observed Lb-gene organization on the soybean chromosome may reflect a possible sequential gene activation during nodule development (Fuchsman and Appleby 1979, Verma et al. 1979). Finally interspersed between the Lb-genes several repetitive sequences have been located. However, at present nothing is known concerning the nature of these sequences.

Lb-Gene Activation and Methylation

Most active eukyryotic genes are hypomethylated when compared to the corresponding inactive genes (Razin and Riggs 1980). This also appears to be the case for the Lb-genes. Thus, Southern blotting analysis of nodule and seedling soybean DNA digested with various restriction endonucleases revealed that the Lb-gene region in nodule DNA is hypomethylated when compared to the corresponding region in seedling DNA. The type of sequences displaying methylation variation include CCGG, CXG and CXXG. It should be pointed out that no difference is observed in the Lb-gene region when using HpaII or MspI digested DNA from the same tissue. However, a HpaII or MspI digest of seedling DNA exhibit a clear difference in the Lb-gene region when compared to similar digests of nodule DNA. Thus, in the relevant CCGG sequences in seedling DNA both Cs are methylated, while in the corresponding sequences in nodule DNA both Cs are demethylated. All the sites exhibiting methylation variance have been assigned to a region of 8 kb located 3′ to the Lbc_3 gene (Fig. 4). The study of the methylation variance of the Lb-gene region during the activation of the genes is not yet completed. It is therefore quite conceivable that other areas of the Lb-gene region display variation in the degree of methylation depending upon the state of activation of the genes. However, it is clear from the results so far obtained that active Lb-genes are hypomethylated when compared to the corresponding inactive genes. Thus also in this respect the Lb-genes are analogous to other eukaryotic genes.

Note Added in Proof. Further analysis has shown that the 10 kb EcoR1 fragment contains an Lb pseudo gene and the Lbc_2 gene. During phage propagation the Lb pseudo gene is eliminated resulting in a 6 kb EcoR1 fragment which contains the Lbc_2 gene only. Subsequent investigation has revealed that the six Lb genes are organised in two independent clusters; one containing four genes in the order 5′ Lba-Lbc_1-ψLb_2-Lbc_3-3′ and another cluster containing two genes in the order 5′-ψLb_1-Lbc_2-3′.

Acknowledgements. We would like to thank Drs. R. Goldberg and R. Fischer, UCLA, for making their random EcoR1 and ALu/HaeIII soybean DNA libraries available to us.

References

Appleby CA (1974) In: Queispel A (ed) Leghemoglobin in the biology of nitrogen fixation. North-Holland Publ Comp, Amsterdam Oxford, pp 521–554

Bojsen K, Abildsten D, Jensen EØ, Paludan K, Marcker KA (••••) The chromosomal arrangement of six soybean leghemoglobin genes. EMBO J 2:1165–1168

Breathnach R, Benoist C, O'Hare K, Gannon F, Chambon P (1978) Ovalbumin gene: evidence for a leader sequence in mRNA and DNA sequences at the exon-intron boundaries. Proc Natl Acad Sci USA 75:4853–4857

Brisson N, Verma DPS (1982) Soybean leghemoglobin gene family: normal, pseudo and truncated genes. Proc Natl Acad Sci USA 79:4055–4059

Fuchsman WH, Appleby CA (1979) Separation and determination of the relative concentration of the homogeneous components of soybean leghemoglobin by isoelectric focusing. Biochem Biophys Acta 579:314–324

Gō M (1981) Correlation of DNA exonic regions with protein structural units in haemoglobin. Nature 291:90–92

Grosveld GC, Rosenthal A, Flavell RA (1982) Sequence requirements for the transcription of the rabbit β-globin gene in vivo: the -80 region. Nucleic Acid Res 10:4951–4971

Hyldig-Nielsen JJ, Jensen EØ, Paludan K, Wiborg O, Garrett R, Jørgensen P, Marcker KA (1982) The primary structures of two leghemoglobin genes from soybean. Nucleic Acids Res 10: 689–701

Jensen EØ, Paludan K, Hyldig-Nielsen JJ, Jørgensen P, Marcker KA (1981) The structure of a chromosomal leghemoglobin gene from soybean. Nature 291:677–679

Marcker KA, Gausing K, Jochimsen B, Jørgensen P, Paludan K, Truelsen E (1981) The cloning and organization of soybean leghemoglobin genes. In: Panopoulos NJ (ed) Genetic engineering in the plant sciences. Praeger Publ

Proudfoot NJ, Brownlee GG (1976) 3′ non-coding region sequences in eukaryotic messenger RNA. Nature 263:211–214

Razin A, Riggs AD (1980) DNA methylation and gene function. Science 210:604–610

Sievers G, Huhtala M-L, Ellfolk N (1978) The primary structure of soybean *(Glycine max.).* Acta Chem Scand B32:380–386

Sullivan D, Brisson N, Goodchild B, Verma DPS, Thomas DY (1981) Molecular cloning and organization of two leghaemoglobin genomic sequences of soybean. Nature 289:516–518

Talkington CA, Leder P (1982) Rescuing the in vitro function of a globin pseudogene promotor. Nature 298:192–195

Verma CPS, Ball S, Guérin C, Wanamaker L (1979) Leghemoglobin biosynthesis in soybean root nodules, characteriaztion of the nascent and released peptides and the relative rates of synthesis of the major leghemoglobins. Biochemistry 18:476–483

Wiborg O, Hyldig-Nielsen JJ, Jensen EØ, Paludan K, Marcker KA (1982) The nucleotide sequences of two leghemoglobin genes from soybean. Nucleic Acids Res 10:3487–3494

The Role of Plant Genes in Soybean-*Rhizobium* Interactions

D.P.S. VERMA, H. BERGMANN, F. FULLER, and E. PREDDIE [1]

Introduction

The ability to reduce atmospheric dinitrogen is restricted to prokaryotes. Some higher plants have evolved such that they are able to associate with nitrogen fixing microbes in order to obtain reduced nitrogen. These associations range from very loose (e.g., colonization of the plant rhizosphere) to the development of an endosymbiotic state (see Vose and Ruschel 1981, Verma 1982, Verma and Long 1983). A successful interaction between a legume host and its invading *Rhizobium* strain results in the development of a symbiotic nitrogen fixing state. The legume-*Rhizobium* association may represent an ultimate endosymbiotic state in which the autonomy of the micro-symbiont is maintained. Development of this symbiosis requires the expression of a number of plant and bacterial genes, which may have to be coordinated. Results obtained with various plant and *Rhizobium* mutants indicate that the interactions between these two organisms occur at different nested levels and show great specificity. Some of these interactions may be common to all legume-*Rhizobium* systems while others may be unique to a given species of variety/strain interaction.

Involvement of Specific Host Genes in Symbiosis

A number of host genes have been implicated in the development of a successful association between a legume plant and its nodulating *Rhizobium* species and some of them have been identified (see Weber 1966, Caldwell and Vest 1977, Nutman 1981) by classical genetic experiments. Mutations in these genes result in Nod^- phenotype. Some of these mutations are conditional (i.e., plants nodulate with some strains but not with others). Three recessive genes, which influence nodule development at different stages, have been identified in clover (Bergersen and Nutman 1957) and in addition there are certain genes in the plant which affect the function of the nodules after their apparently normal development (Holl 1973, see also Nutman 1981). The characterization of these genes, however, is limited (except for leghemoglobin genes) and their precise functions in synbiosis are practically unknown.

1 Department of Biology, McGill University, 1205 Avenue Docteur Penfield, Montreal, Quebec, Canada H3A 1B1

Molecular Genetics of the Bacteria-Plant Interaction
(ed. by A. Pühler)
Ü Springer-Verlag Berlin Heidelberg 1983

Induction of Host Genes During Nodulation

The induction of leghemoglobin takes place several days prior to and is independent of the appearance of nitrogenase activity in nodules. Leghemoglobin mRNAs have been detected as early as three days after infection (Sutton, Mauro and Verma, unpublished observations) and are present in several ineffective nodules formed by mutant strains of *Rhizobium* (Verma et al. 1981b). A number of other host genes are induced along with leghemoglobin genes and follow similar kinetics of induction (see Verma et al. 1981a). In all cases, the level of induction appears to depend upon the specific strain of *Rhizobium*. This is also apparent at the protein level since the pattern of host proteins is altered significantly by *Rhizobium* mutants (Legocki and Verma 1980, Noel et al. 1982).

Identification of Nodule-Specific Host Gene Products

Host genes have an effect on a number of events in this symbiosis which determine the ability to nodulate, the abundance of nodules, the intracellular organization and differentiation of nodules as well as their efficiency in fixing nitrogen (see Verma 1982, Verma and Long 1983). We have identified (Legocki and Verma 1979, 1980, Auger and Verma 1981, Fuller et al. 1983) a number of host gene products that are induced following infection of soybean by *R. japonicum*. These tissue-specific host proteins, nodulins, appear in parallel with leghemoglobins during the development of nodules (Verma et al. 1981a) which suggests that they may play an important role in the process of symbiosis. Some of these proteins have been purified and characterized with respect to their possible functions.

In order to understand the molecular mechanisms underlying this process, we have attempted to isolate nodulin coding genes (Fuller et al. 1983). A cDNA library constructed from nodule-mRNA was screened for the most abundant nodule-specific sequences after eliminating leghemoglobin clones. We obtained several clones representing unique, abundant sequences ranging from 6.5% to 0.6% of mRNA molecules in nodules.

Isolation and Characterization of Nodulins

The effectiveness of the nodule in fixing nitrogen may be influenced both by structural features and by the activity of enzymes which support this process. A definite interdependence between nodule-efficiency and photosynthesis has been observed (Bethlenfalvay et al. 1978), indicating that major enzymatic activities in nodule tissue are directed towards carbon-nitrogen metabolism. Table 1 lists a number of enzymes whose activities have been observed to be increased several fold in nodule tissue. Most of these enzymes are involved in metabolizing reduced nitrogen to amides or ureides; two major compounds used for translocation of nitrogen to the shoot of the plant. In addition, there are several other enzymes such as catalase, carbonic anhydrase, and leghemoglobin reductase, etc. which may control the efficiency of the tissues in symbiotic nitrogen fixation. Some of the nodulins may be these enzymes.

Table 1. Increase in specific enzymes and proteins which are putative candicates for nodulins

Enzyme	Increase in activity/concentration (nod./root)	Plant source	References
Allantoinase	3	Soybean	Tajima and Yamamoto (1975)
Asparagine synthetase	Several hundred-fold	Lupin	Scott et al. (1976)
Aspartate aminotransferase	(Follows nitrogenase)	Soybean	Werner and Mörschel (1978)
Catalase	5–10	Soybean	Francis and Alexander (1972)
Carbonic anhydrase	12.5	Lupin	Christeller et al. (1977)
Glutamine synthetase	300–500	Lupin	Robertson et al. (1975a)
Glutamate synthase	10	Lupin	Robertson et al. (1975b)
Glutamate decarboxylase	–	Trifolium	Freney and Gibson (1975)
Invertase	3–4	Soybean	Streeter (1977)
Leghemoglobin reductase	–	Soybean	Puppo et al. (1980)
Phosphofructokinase	5	Lupin	Mulongoy and Elkan (1977)
PEP-carboxylase	3	Lupin	Christeller et al. (1977)
Uricase	1.7/0	Soybean	Tajima and Yamamoto (1975)
Xanthine oxidase	0.014/0	Soybean	Tajima and Yamamoto (1975)
Membrane envelope proteins	Several-fold	Soybean	Verma et al. (1978)

The existence of nodulins has now been observed in a number of legume-*Rhizobium* associations (e.g., soybean, pea, clover). These proteins may be divided into three types: (1) structural proteins, (2) enzymes responsible for the specific assimilation of reduced nitrogen and (3) proteins which support bacteroid functions. These proteins are in addition to leghemoglobins which may be considered to be nodule-specific proteins of the third kind. Due to the lack of identifiable functional assays for these proteins, attempts to purify them have been limited to the observation of their abundance and tissue specificity.

Nodulin-35

Analysis of the cytoplasmic proteins from effective root nodules of soybean indicated the presence of an abundant 35,000 MW protein which was referred as Nodulin-35 (Legocki and Verma 1979). This protein is present in all ineffective nodules tested (Legocki and Verma 1979), although in varying amounts, suggesting that its induction may be similar to that of leghemoglobin which is also present in ineffective nodules but in much lower concentration (Verma et al. 1981b). Nodulin-35 was found to be the major component of the 0–30% ammonium sulfate pellet from nodule S100 preparations and could be further purified by passage over DEAE-cellulose as it did not bind to this matrix (Fig. 1). Attempts to find an associated enzyme activity led to the detection of uricase (urate: O_2 oxidoreductase, EC 1.7.3.3) activity in this fraction. Jochimsen and Rasmussen (this vol.) have independently concluded that Nodulin-35 is an uricase. Since the pH optima and Km of the uricase activity found to be induced in nodules are very different from those of uricase activities present in other soybean tissues (Tajima and Yamamoto 1975) and since anti-Nodulin-35 serum does not cross-react with proteins from root or leaf (Legocki and Verma

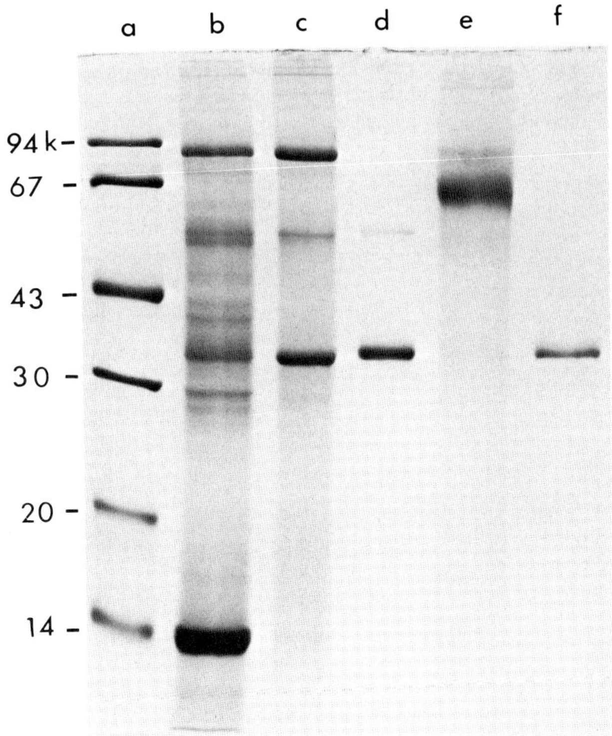

Fig. 1. SDS gel electrophoresis of various fractions during purification of Nodulin-35. Lane *a* molecular weight markers; *b* total cytoplasmic proteins (S-100) from 21-day effective nodules of soybean; *c* 0–30% ammonium sulfate precipitated fraction; *d* material from lane c which did not bind to DEAE-cellulose; *e* material bound to DEAE-cellulose and eluted with 0.5 M NaCl; *f* DEAE unbound material passed through CM-cellulose

1979), the nodule enzyme may be a novel form involved specifically in symbiotic nitrogen fixation. A similar situation exists for glutamine synthetase activities in *Phaseolus* nodules (Cullimore et al. 1982).

Other Nodulins

In vitro translation of nodule polysomes in a wheat germ cell free system followed by immuno-precipitation with a 'nodule-specific' antiserum (Legocki and Verma 1980) led to the detection of a number of other nodule-specific host proteins which appeared to be involved in root nodule symbiosis. These proteins were found to be of relatively low molecular weight (< 20,000 MW). However, a similar experiment carried out using rabbit reticulocyte lysate and purified mRNA detected nodulins ranging from 120,000 to 20,000 MW (see below). It thus appears that earlier results were due to degradation of polysomes during translation or premature termination of nascent peptides in the wheat germ cell free system or proteolysis during immuno-precipitation. In fact, larger MW nodule-spedific peptides were detected by in vivo labelling of the nodule tissue (Legocki, unpubl. data).

Attempts to purify these proteins by classical chromatographic procedures in the absence of a specific assay have met with little success. However, when soluble proteins were analysed on western-blots using antisera either against root or nodule

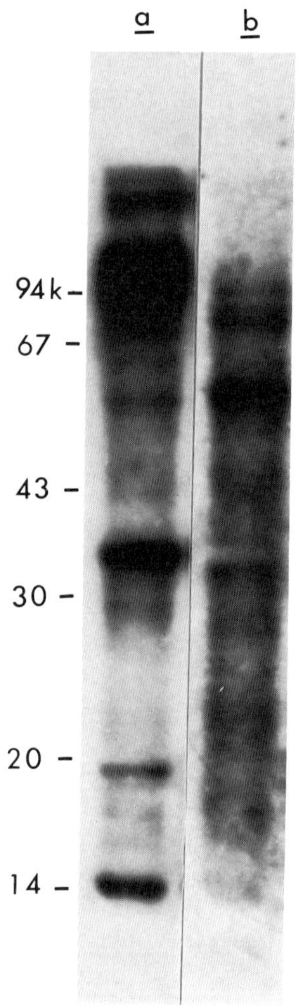

Fig. 2. Immuno-reactivity of anti-nodule and anti-root sera to a fraction of nodule cytoplasmic proteins. The 0%–30% ammonium sulfate fraction was electrophoresed on SDS gels, transferred to nitrocullulose paper and reacted with anti-nodule serum *(a)* and anti-root serum *(b)*

proteins, several proteins were detected that appear to accumulate specifically in nodule tissue (Fig. 2). Nodule-specific polypeptides detected by this technique were also of higher molecular weight.

Molecular Cloning of Nodulin Genes

Analyses of RNA:cDNA hybridization kinetics have revealed the presence of a moderately-abundant, nodule-specific RNA class in addition to the super-abundant class encoding leghemoglobin (Auger et al. 1979, Auger and Verma 1981). Since these sequences are induced in parallel to leghemoglobin, they have been postulated to encode nodulins (Verma et al. 1981a). In order to isolate these sequences, to identify the polypeptides they encode, and to obtain probes for the analysis of nodulin genes, a soybean nodule cDNA library was constructed. Analysis of 5,700 ampicillin sensitive clones showed that 13% of the clones contained sequences that are present in root tissue while 37% contained moderately abundant, nodule-specific sequences. From these, four non-leghemoglobin clones representing three unique, nodule-specific sequences were isolated.

Translation of mRNAs hybrid-selected by these clones followed by immuno-precipitation with antibodies against nodule proteins showed that the cloned sequences select mRNAs encoding polypeptides of 44,000, 27,000 and 24,000 MW (Fig. 3). These polypeptides were also detectable in immuno-precipitates of translation products of total nodule mRNA. In addition, one of the clones (# 60) appeared to hybrid-select several other mRNAs which give rise to translation products that were specifically immuno-precipitated by anti-nodule serum. These results indicate that some nodule mRNAs may share partial homology. However, since this clone hybridizes to only one *Eco*RI genomic fragment (Fuller et al. 1983), any such cross-homologies must be limited.

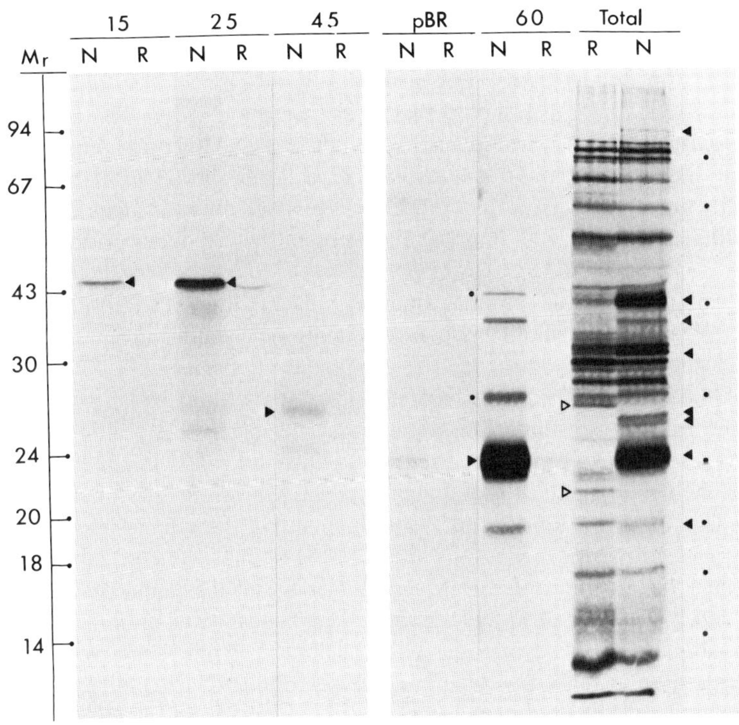

Fig. 3. Immuno-precipitation of translation products of mRNAs hybrid-selected with cloned sequences. Translation products were reacted with antisera prepared against total soluble proteins of root *(R)* or nodule *(N)*, precipitated with Sepharose-ProteinA and electrophoresed. Other immuno-precipitations are: *Total* translation products of total nodule poly(A)$^+$ RNAs; *pBR* translation products of RNA ('selected') by pBR322. *Filled arrowheads* products which react specifically with anti-nodule serum. *Open arrowheads* products which react specifically with anti-root serum

Translation products of total nodule poly(A)$^+$ mRNAs were also reacted with anti-root serum (Fig. 3) which indicated two polypeptides (22,000 and 29,500 MW) that may be more abundant in root tissue. Several polypeptides (120, 44, 41, 36, 27.5 and 27, 24 and 20 thousand MW) reacted with nodule but did not react (or reacted poorly) with root antiserum.

Southern blot hybridizations of soybean genomic DNA with the nodulin clones indicated that, in contrast to leghemoglobin which is represented as a closely related family of sequences, nodulins are encoded by 'single-copy' genes. We have isolated some of these sequences from a soybean genomic library.

Summary and Overview

Both genetic and molecular data strongly suggest the involvement of several nodule-specific genes in legume-*Rhizobium* symbiosis. We have identified, in soybean, a number of host gene products that are induced during infection and development of root nodules. These tissue-specific proteins, nodulins, appear in parallel to leghemoglobins. Their induction occurs prior to, and is independent of the development of nitrogenase activity in nodule tissue. The purification and characterization of one of the nodulins, Nodulin-35, has shown that it has uricase activity. The latter is localized in peroxisomes and is involved in ureide metabolism. Other nodulins represent a relatively smaller fraction of the soluble cytoplasmic proteins of the host and consequently their purification has proven to be more difficult. This may be facilitated by the development of monoclonal antisera against these proteins (in progress).

Molecular cloning of nodule-specific sequences indicated that in addition to leghemoglobin mRNAs, there is another sequence that is very abundant representing almost 6.5% of the total polysomal mRNA in nodules (Fuller et al. 1983). This transcript could be a part of the superabundant class representing 18% by mass of nodule mRNA which has previously been interpreted to consist solely of leghemoglobin RNA (Auger et al. 1979). The occurrence of such an abundant transcript must be important in the development of root nodule symbiosis, however, the 44,000 MW translation product of this mRNA is not very abundant in the soluble fraction of nodule proteins.

Understanding the function of nodule-specific genes and their mode of regulation following infection of the plant by *Rhizobium* is of fundamental importance for improving or manipulating this symbiosis. The availability of individual molecular probes for nodule-specific sequences would allow isolation and characterization of these genes in detail and the determination of specific regulatory elements that are responsible for their induction. Information regarding the organization of nodulin genes on the chromosome and their relative location with respect to leghemoglobin genes would also be valuable in understanding the coordination of their induction following infection of the plant by *Rhizobium*. Isolation of *Rhizobium* and plant mutants which affect a specific stage of nodule development should further facilitate this analysis.

Acknowledgements. This study was supported by Research grants for the Natural Sciences and Engineering Research Council of Canada and the Quebec Ministry of Education. F.F. was supported by a NATO postdoctoral fellowship.

References

Auger S, Verma DPS (1981) Biochemistry 10:1300–1306
Auger S, Baulcombe D, Verma DPS (1979) Biochem Biophys Acta 563:496–509
Bergersen FJ, Nutman PS (1957) Heredity 11:175–184
Bethelnfalvay GJ, Aubu-shakra SS, Phillips DA (1978) Plant Physiol 62:127–130
Caldwell BE, Vest HG (1977) In: Harey RWF, Silver WS (eds) A treatise on dinitrogen fixation. Wiley Interscience, New York, pp 557–576
Christeller JT, Laing WA, Sutton WD (1977) Plant Physiol 60:47–50
Cullimore JV, Lara M, Lea PJ, Miflin BJ (1983) Planta 157:245–253
Francis AJ, Alexander M (1972) Can J Microbiol 18:861–864
Freney JR, Gibson AH (1975) Aust J Plant Physiol 2:663–668
Fuller F, Künstner P, Nguyen T, Verma DPS (1983) Proc Natl Acad Sci USA 80:2594–2598
Holl FB (1973) Can J Genet Cytol 15:659
Legocki RP, Verma DPS (1979) Science 205:190–193
Legocki RP, Verma DPS (1980) Cell 20:153–163
Mulongoy K, Elkan GH (1977) J Bacteriol 131:179–187
Noel DK, Stacey G, Tandon SR, Silver LE, Brill WJ (1982) J Bacteriol 152:485–494
Nutman PS (1981) In: Gibson AH, Newton WE (eds) Current perspectives in nitrogen fixation. Aust Acad Sci, Canberra, pp 194–204
Puppo A, Rigaud J, Job D (1980) Plant Sci Lett 20:1–6
Robertson JG, Farnden KJ, Warbuton MP, Bank JM (1975a) Aust J Plant Physiol 2:265–272
Robertson JG, Warburton MP, Farnden KJF (1975b) FEBS Lett 55:33–37
Scott DB, Farnden KJF, Robertson JG (1976) Nature 263:703–705
Streeter JG (1977) Plant Physiol 60:235–239
Tajima S, Yamamoto Y (1975) Plant Cell Physiol 16:271–282
Verma DPS (1982) In: Smith H, Grierson D (eds) The molecular bioloty of plant development. Blackwell Sci Publ, Oxford, pp 437–466
Verma DPS, Long S (1983) In: Jeon K (ed) International review of cytology, Suppl 14. Academic Pres, London New York, pp 211–245
Verma DPS, Zogbi V, Bal AK (1978) Plant Sci Lett 13:137–142
Verma DPS, Legocki RP, Auger S (1981a) In: Gibson AA, Newton WE (eds) Current perspectives in notrogen fixation. Aust Acad Sci, Canberra, pp 205–208
Verma DPS, Haugland R, Brisson N, Legocki R, Lacroix L (1981b) Biochem Biophys Acta 653: 98–107
Vose, Ruschel (1981) In: Associate notrigen fixation. CRC Press
Werner D, Mörschel E (1978) Planta 141:169–177
Weber CR (1966) Agronoy J 58:43–46

General Organization of Nitrogen Fixation Genes in *Rhizobium phaseoli*

R. PALACIOS, C. QUINTO, H. DE LA VEGA, M. FLORES, L. FERNÁNDEZ, M. HERNÁNDEZ, T. BALLADO, and G. SOBERÓN-CHÁVEZ [1]

To identify and characterize nitrogen fixation genes (*nif* genes) in *Rhizobium phaseoli* we have followed an experimental approach based on the highly conserved homology among the DNA sequences that codify for nitrogenase polypeptides of different species that fix nitrogen (Nuti et al. 1979, Ruvkun and Ausubel 1980, Mazur et al. 1980). In this communication we review experimental evidence from our laboratory that supports the following conclusions: (1) there is a common pattern of organization of *nif* genes in *Rhizobium phaseoli;* (2) DNA sequences related with *nif* genes are reiterated in *Rhizobium phaseoli;* and (3) these sequences are localized in plasmids.

In order to obtain general conclusions in regard to the organization of *nif* genes in *Rhizobium phaseoli,* we decided to work with strains from different geographical origins. Genome Southern blots (Southern 1975) of *Eco*R1 digests from different strains were hybridized with the *Eco*R1 insert from plasmid pSA30. This insert contains DNA sequences that codify for *Klebsiella pneumoniae* nitrogenase polypeptides (Cannon et al. 1979). The different *R. phaseoli* strains that we have tested revealed three hybridization bands (Quinto et al. 1982). Two of the bands, with lengths of 4.7 kb (band a) and 4.1 kb (band b), are common to all the strains. The third band (c) gives a less intense hybridization signal and has a characteristic size of either 5.2 kb (2 strains), 4.5 kb (9 strains) or 3.6 kb (1 strain). Examples of these experiments are presented in Fig. 1A. Moreover, earlier reports from other laboratories (Ruvkun and Ausubel 1980, Prakash et al. 1981) show hybridization patterns of *R. phaseoli* with *Klebsiella pneumoniae nif* genes that are consistent with the ones presented here. In contrast, experiments performed with *Rhizobia* from other cross inoculation groups (Nuti et al. 1979, Ruvkun and Ausubel 1980, Prakash et al. 1981) including *R. meliloti, R. trifolii, R. leguminosarum,* and *R.* sp., give patterns different from that in *R. phaseoli.*

The above presented data suggest a close relation between the *nif* genes organization and the cross inoculation group of *Rhizobium.* This relation could be either physical or functional. A physical relation could mean that both *nif* genes and genes responsible for specificity of nodulation are located in the same replicon (plasmid) and that at the same time that such a replicon does not recombine with replicons from other cross inoculation groups of *Rhizobium.* On the other hand, a functional relation could mean that to fix nitrogen in *Phaseolus vulgaris,* this particular pattern of

1 Centro de Investigación sobre Fijación de Nitrógeno, UNAM. Ap. postal 565-A, Cuernavaca, Mor., Mexico

Molecular Genetics of the Bacteria Plant Interaction
(ed. by A. Pühler)

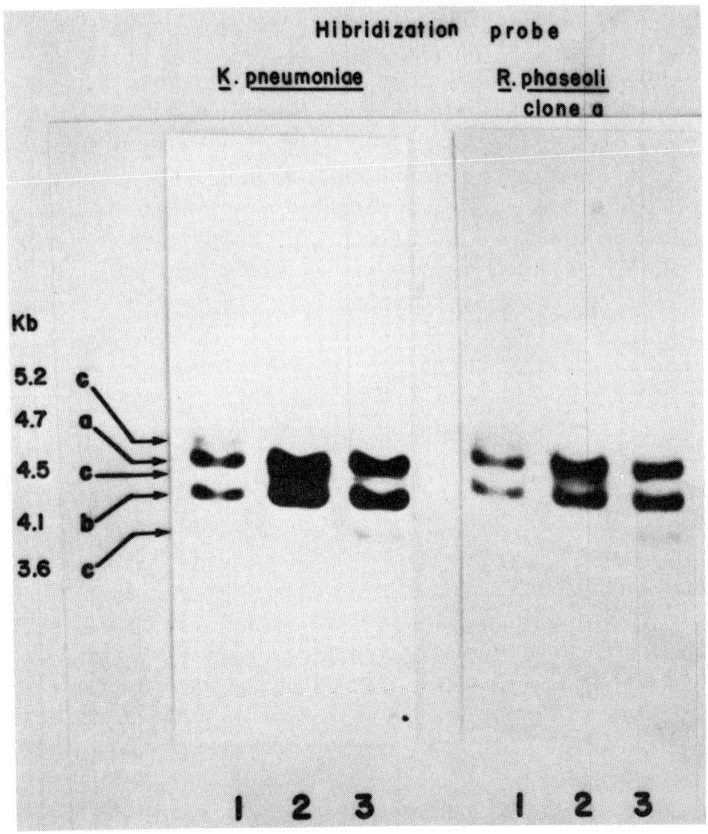

Fig. 1. Reiteration of *nif* DNA sequences in *R. phaseoli.* Genome Southern blots of total DNA digested with *Eco*R1 were hybridized with the indicated probes using 50% formamide at 42°C (*Klebsiella* probe) or sodium phosphate buffer at 65°C (*R. phaseoli* probe) as reported (Quinto et al. 1982). Strains: *1* CFN-Rph-1; *2* CFN-Rph-42; *3* CFN-Rph-215

nif genes organization is a necessary condition. However, to make definitive conclusions, more strains, specially some with a broader host spectrum of infection, must be analyzed.

It has been shown that in *R. meliloti* the DNA sequences that hybridize under stringent conditions with *K. pneumoniae nif* genes consist of a contiguous region of about 1.6–2.2 kb containing genes H and D (Ruvkun and Ausubel 1980, Corbin et al. 1981). The pattern of *R. phaseoli* is not consistent with this simple arrangement of hybridizing sequences. The following possibilities can be suggested: (a) *R. phaseoli* DNA hybridizes with more extensive *Klebsiella* sequences, (b) sequences that hybridize with *Klebsiella* DNA are not contiguous in *R. phaseoli,* as has been reported for the blue-green algae *Anabaena* (Mazur et al. 1980), (c) there are duplications of the *R. phaseoli* sequences that hybridize with *Klebsiella nif* genes.

To gain insight into the general organization of *R. phaseoli* DNA sequences that hybridize with *Klebsiella nif* genes, DNA from one strain was cloned in the *Eco*R1 site of the pBR328 (Sobéron et al. 1980) vector. Recombinant plasmids containing inserts corresponding to the a (4.7 kb), b (4.1 kb) and c (4.5 kb) bands were isolated and used as hybridization progress with total genome digests of different strains of *R. phaseoli* (Quinto et al. 1982). Each of the inserts (a, b, or c) hybridizes with the three bands that give positive reactions with *Klebsiella nif* genes. An example using band a as probe is presented in Fig. 1B. Inserts a and b give strong hybridization signals with bands a and b from different strains and less intense signal with band c. On the other hand, insert c gives a strong signal with band c of the different strains (Quinto et al. 1982).

These results indicate that in the inserts there are DNA sequences that are reiterated in the genome of the analyzed strains. An alternative explanation could be that we are dealing with bacterial populations that have a different arrangement of DNA sequences in different cells. This does not seem likely since experiments performed with cloned bacterial cells give the same results. Accordingly, if polymorphism is proposed as an explanation for these data, one must propose at the same time a dynamic state of gene rearrangements in *R. phaseoli.*

Preliminary experiments (not shown) hybridizing fragments of the *Klebsiella pneumoniae* insert with the different *R. phaseoli* inserts suggest that bands a and b contain genes H and D, while band c contains only gene H. On the other hand, initial DNA sequence data has revealed the presence of gene H coding sequences in each of the three *R. phaseoli* bands. However, it is not clear yet if the different reiterated sequences contain functional genes. It is now a well documented fact that at least some *Rhizobium nif* genes from different cross inoculation groups are located in large plasmids (Prakash et al. 1981, Hombrecker et al. 1981, Banfalvi et al. 1981, Rosenberg et al. 1981). We have found hybridization of each of the three *R. phaseoli* inserts with unique plasmids in different strains (Quinto et al. 1982). Furthermore, as shown in the experiment of Fig. 2, one strain cured by heat treatment loses a plasmid of molecular weight 170×10^6, and no hybridization is now found with either plasmid profiles or total genome digestions using as probe *R. phaseoli nif* sequences. This indicates that at least in this strain, all reiterated *nif* sequences are located in a single plasmid. This cured strain also loses the capacity to nodulate *Phaseolus vulgaris* suggesting that nodulation genes are also located in the same plasmid. The location of nodulation genes in *R. phaseoli* plasmids was initially reported by Lamb et al. (1982).

The results reported here suggest a complex organization of *nif* related DNA sequences in *R. phaseoli,* in contrast to a more simple organization of *nif* genes that is suggested by other laboratories working with different cross inoculation groups of *Rhizobium* (see papers by J. Dénarié, A. Kondorosi et al., F. Ausubel et al., A. Pühler et al., G. Dita et al., A.W. Johnston et al., J. Shine et al., this vol.). Our approach has revealed that some sequences are present three times in the *R. phaseoli* genome. However, reiterations preserving restriction enzyme sites have not been found. Studies searching for microheterogeneity in different clones and their adjacent DNA regions, as well as experiments based on the introduction of genetic elements (transposons) in the different regions, will help to understand the actual number and

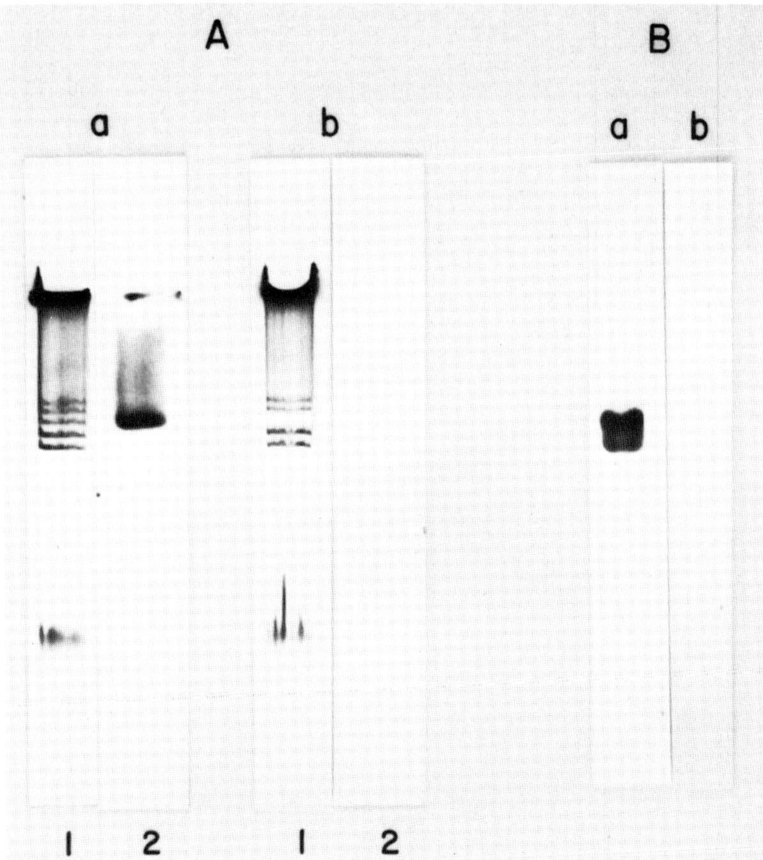

Fig. 2a, b. Location of the different *nif*-related DNA reiterations in a single plasmid. Strain CFN-Rph-42 was heat-treated and a nod$^-$ strain lacking plasmid c (M.W. 170×10^6) was isolated. a wild-type strain; b cured strain. *A* plasmid profiles: *1* ethidium bromide strain; *2* hybridization with *R. phaseoli nif* clone a (5.7 kb). *B* genome Southern blots of total DNA digested with *Eco*R1, hybridized with *nif* clone a

length of the reiterated regions and to establish if this feature is also present in other cross inoculation groups of *Rhizobium*. Stable gene duplications are not frequent in bacterial systems (Riley and Anilionis 1978). The finding of highly conserved *nif*-related DNA reiterations in *R. phaseoli* leads one to search for a significant role of this type of genetic organization.

Acknowledgements. This work was supported in part by grants from Consejo Nacional de Ciencia y Technología and from Fondo de Estudios e Investigaciones Ricardo J. Zevada.

References

Banfalvi Z, Sakanyan V, Koncz C, Kiss A, Dusha I, Kondorosi A (1981) Mol Gen Genet 184: 318–325
Cannon FC, Riedel GE, Ausubel FM (1989) Mol Gen Genet 174:59–66
Corbin D, Ditta G, Helinski DR (1981) J Bacteriol 149:221–229
Hombrecker G, Brewing NJ, Johnston AWB (1981) Mol Gen Genet 182:133–136
Lamb JW, Hombrecker G, Johnston AWB (1982) Mol Gen Genet 186:449–452
Mazur BJ, Rice D, Haselkorn R (1980) Proc Natl Acad Sci USA 77
Nuti MP, Lepidi AA, Praksh RK, Schilperoort RA (1979) Nature 282:533–535
Prakash RK, Schilperoort RA, Nuti MP (1981) J Bacteriol 145:1129–1136
Riley M, Anilionis A (1978) Annu Rev Microbiol 32:519–560
Rosenberg C, Boistard P, Dénarié J, Casse-Delbart F (1981) Mol Gen Genet 184:326–333
Ruvkun GB, Ausubel FM (1980) Proc Natl Acad Sci USA 77:191–195
Sobéron X, Covarrubias L, Bolivar F (1980) Gene 9:287–305
Southern EM (1975) J Mol Biol 98:503–517

Identification of Nitrogen Fixation and Nodulation Genes on a Large Plasmid from a Broad Host Range *Rhizobium* sp.

C.E. PANKHURST[2], W.J. BROUGHTON[1], C. BACHEM[1], E. KONDOROSI[3], and A. KONDOROSI[4]

Introduction

The complexities of the infective characteristics of rhizobia are enormous. Some fast-growing rhizobia, for example strains of *R. trifolii* or *R. meliloti* have a limited host range, while others, for example the slow-growing rhizobia commonly encountered in the nodules of tropical legumes are 'promiscuous' and may nodulate and fix nitrogen with a wide range of different legumes (Vincent 1974). *Rhizobium* strains of the latter type presumably contain genes specifing broad host range.

A fast growing *Rhizobium* strain NGR234 (= MPIK3030) isolated from *Lablab purpereus,* was found to produce effective nitrogen-fixing root nodules on a number of tropical legumes previously believed to have only slow-growing symbionts (Trinck 1980). This strain was also found to form nodules on the non legume *Parasponia andersonii* (Trinck and Galbraith 1980). MPIK3030 therefore appeared to combine the desirable attributes of fast growth rate with broad host range making it an attractive strain for molecular genetic analysis. For these reasons we have begun a programme to study *Rhizobium* sp. strain MPIK3030 with the ultimate aim of characterizing the nodulation *(nod)* genes and genes required for broad host range.

Host Range of MPIK3030

Trinck (1980) examined the ability of MPIK3030 to nodulate and fix nitrogen with several tropical legumes and with temperate legumes representing the major cross inoculation groups. We have extended this study and what is currently known concerning the host range of MPIK3030 is presented in Table 1. In addition to forming Nod^+ Fix^+ nodules on a range of tropical legumes, MPIK3030 formed Nod^+ Fix^+ nodules on *Glycine max,* usually nodulated by *R. japonicum.* MPIK3030 was unable to nodulate hosts belonging to the lotus *(Lotus peduncualtus),* lupin *(Lupinus angustifolius),* clover *(Trifolium pratense),* pea *(Pisum sativum)* and bean *(Phaseolus vulgaris)* cross inoculation groups (Table 1). With *Medicago sativa,* MPIK3030 formed occasional tumour like nodules. These tumour like structures may become very large (3–5 mm dia.; Pankhurst, unpublished observations).

1 Max-Planck-Institut für Züchtungsforschung, Abteilung Schell, 5000 Köln 30, FRG
2 Applied Biochemistry Division, D.S.I.R., Palmerston North, New Zealand
3 Institute of Biochemistry, Biol. Res. Cts., Hung. Acad. Sci., P.O. Box 521, 6701 Szeged, Hungary
4 Institute of Genetics, Biol. Res. Cts., Hung. Acad. Sci., P.O. Box 521, 6701 Szeged, Hungary

Molecular Genetics of the Bacteria-Plant Interaction
(ed. by A. Pühler)

Table 1. Host range of *Rhizobium* sp. strain MPIK3030

Host legume	Nodule phenotype	Reference [a]
Calopogonium caeruleum	Nod^+ Fix^+	(1)
Desmodium intortum	Nod^+ Fix^+	(2)
Flemingia congesta	Nod^+ Fix^+	(1)
Glycine max	Nod^+ Fix^+	(1)
Lablab purpureus	Nod^+ Fix^+	(1)
Macroptilium atropurpureum	Nod^+ Fix^+	(1)
Psophocarpus tetragonolubus	Nod^+ Fix^+	(2)
Stylosanthes hamata	Nod^+ Fix^+	(2)
Tephrosia candida	Nod^+ Fix^+	(1)
Vigna unguiculata	Nod^+ Fix^+	(1)
Arachis hypogaea	Nod^+ Fix^-	(2)
Lotus pedunculatus	Nod^-	(2)
Lupinus angustifolius	Nod^-	(1)
Medicago sativa	Nod^-	(1, 2)
Phaseolus vulgaris	Nod^-	(1)
Pisum sativum	Nod^-	(1)
Trifolium pratense	Nod^-	(1)
Parasponia andersonii (non-legume)	Nod^+ Fix^-	(3)

[a] References: (1) Trinck (1980); (2) this vol.; (3) Trinck and Galbraith (1980)

Plasmid Content of MPIK3030

Using the Eckhardt agarose gel electrophoresis procedure (Banfalvi et al. 1981) MPIK3030 was found to contain a single large plasmid (Fig. 1a). Comparing the mobility of this plasmid with plasmids of known molecular weight electrophoresed in the same gel, the molecular weight of pMPIK3030*a* was estimated as 310 ± 20 Md.

Presence of *nif* on pMPIK3030a

To determine if genes for nitrogen fixation *(nif)* were present on pMPIK3030*a*, a Southern blot of an Eckhardt gel containing pMPIK3030*a* was hybridized with a ^{32}P labelled probe (pID1) which contains the *nifH* and the *nifD* genes of *R. meliloti* strain 41 (Banfalvi et al. 1981). As shown in Fig. 1a strong hybridization between pID1 and pMPIK3030*a* was obtained, demonstrating the presence of *nif* genes on this plasmid.

Isolation of pMPIK3030a

Because of the high molecular weight of pMPIK3030*a* preparative extraction of this plasmid using the alkaline denaturation procedure (Casse et al. 1979) was unsuccessful. A method incorporating the cell lysis procedure of Schwinghamer (1980) and

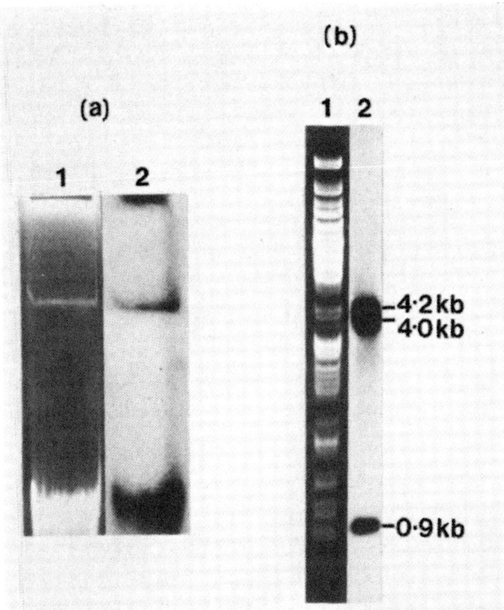

Fig. 1a, b. Detection of a large plasmid in *Rhizobium* sp. strain MPIK3030. **a** Eckhardt gel showing pMPIK3030*a* *(lane 1)* and hybridization of pMPIK-3030*a* with pID1 *(lane 2)*. **b** Agarose gel showing an *Eco*RI restriction digest of pMPIK3030*a* *(lane 1)* and hybridization of the *Eco*RI restriction digest with pID1 *(lane 2)*. The size of the *Eco*RI fragments hybridizing with pID1 is indicated

described in detail by Rosenberg et al. (1982) was therefore used. This method gave yields of pMPIK3030*a* of 5–10 μg DNA per litre of early log phase culture (5×10^7 to 1×10^8 cells/ml).

An *Eco*RI restriction digest of pMPIK3030*a* hybridized strongly with the pID1 probe (Fig. 1b) confirming that *nif* was present on pMPIK3030*a* and showing further that the hybridizing region was contained on 3 *Eco*RI fragments of size 4.2, 4.0 and 0.9 kb (Fig. 1b).

Cloning pMPIK3030a into the Cosmid Vector pJB8

Cosmid pJB8 is a modified derivative of the cosmid vector *Homer* 1 (Ish-Horowicz and Burke 1981). It is 5.4 kb in length and has been constructed in a way to overcome two major problems generally associated with cosmid cloning. These are (1) the formation of cosmids containing noncontiguous DNA inserts, and (2) cosmids containing multiple vectors and multiple inserts (Ish-Horowicz and Burke 1981). Cosmid pJB8 has two *Eco*RI sites separated by a *Bam*HI linker into which DNA fragments generated by partial cleavage with *Mbo*I of *Sau*3A can be inserted. Digestion with *Eco*RI excises the inserted DNA from the hybrid cosmid.

Purified pMPIK3030*a* DNA was cloned into pJB8 according to the methods of Ish-Horowicz and Burke (1981). A cosmid bank containing 120 individual hybrid cosmid clones of pMPIK3030*a* was established in *E. coli* strain HB101 (*hsd*R$^-$ *rec*A$^-$). Using a 'miniprep' procedure for isolating the cosmids from HB101 (Ish-Horowicz and Burke 1981) and digesting the cosmid DNA with *Eco*RI, it was shown that the insert size of pMPIK3030*a* in individual cosmids varied from 32 to 47 kb. As pMPIK-3030*a* has an estimated length of 462 kb, it was calculated that the entire plasmid should be represented in a random selection of 50 cosmid clones.

Cosmid Map of pMPIK3030a

Cosmid clones containing *nif* were identified using the colony hybridization procedure described by Ish-Horowicz and Burke (1981) and pID1 as a hybridization probe. Two clones containing *nif* but otherwise different *Eco*RI restriction patterns were then selected. Cosmid DNA was isolated from these two clones using the 'miniprep' protocol (scaled up by a factor of 40) and purified by CsCl-EtBr centrifugation. The DNA was labelled with ^{32}P by nick translation and used as a hybridization probe against DNA from 50 cosmid clones. The DNA from these 50 clones was prepared using the 'miniprep' procedure, digested with *Eco*RI and contained on two sets of nitrocellulose filters. In this way, cosmid clones hybridizing with the probes could be identified. Cosmids in which only 2 or 3 fragments showed hybridization with the probe were chosen and used for preparing the next hybridization probe. By repeating this process we have been able to order the DNA sequences contained in individual cosmids along 50% of the estimated length of pMPIK3030*a*.

Location of the *nod* Region on pMPIK3030a

The possibility of using cloned regions of the megaplasmid pRme 41*b* of *R. meliloti* strain 41 (AK631) (Banfalvi et al. 1981) as hybridization probes to identify the nodulation *(nod)* region on pMPIK3030*a* was investigated. When purified pMPIK-3030*a* DNA was hybridized to total DNA of AK631 a large number of hybridizing bands were detected (Fig. 2). When hybridized to total DNA of *R. meliloti* AK1329 (a *nod, nif* deletion mutant of AK631), several of the hybridzing bands present in AK631 were missing (Fig. 2 cf. lanes 1 and 2). The bands missing from AK1329 probably contain *nif* and *nod* genes deleted from AK631. Thus it appeared that there would be sufficient homology between pRme41*b* and pMPIK3030*a* to use cloned fragments of pRme41*b* as hybridization probes to locate the *nod* region on pMPIK-3030*a*. Also shown in Fig. 2 is the hybridization obtained with pMPIK3030*a* and total DNA from *Agrobacterium tumefaciens* (LBA288), *R. trifolii* (LPR5001) and various transconjugants of these strains (Hooykaas et al. 1981). Several weak hybridizing bands were found with LBA288 (Fig. 2 lane 3) but a derivative of LBA288 containing pRtr5*a* (a *sym* plasmid from LPR5001) showed several new strong hybridizing bands (Fig. 2 lane 4). *R. trifolii* strain LPR5001 (Fig. 2 lane 5) showed several strong hybridizing bands while a heat cured derivative of LPR5001 which lacks pRtr5*a* showed little hybridization (Fig. 2 lane 6). Re-introduction of pRtr5*a* into the cured derivative (strain LPR5041, Fig. 2 lane 7) restored the hybridization. These data indicate that the strong hybridization obtained in these bacteria with pMPIK-3030*a* is associated with the presence of pRtr5*a*. Of interest is the extensive homology between pMPIK3030*a* and AK631.

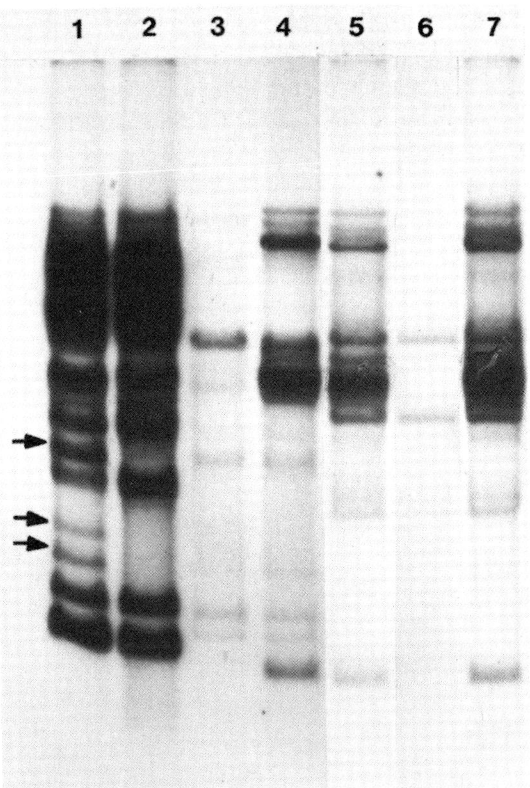

Fig. 2. Hybridization of pMPIK-3030*a* to *Eco*RI restricted total DNA of *R. meliloti* AK631 *(lane 1); R. meliloti* AK1329 *(lane 2); Agrobacterium tumefaciens* LBA288 *(lane 3);* LBA-2702, a LBA288 derivative containing pRtr5*a*, a *sym* plasmid from LPR5001 *(lane 4); R. trifolii* LPR5001 *(lane 5);* LPR5054, a LPR5001 derivative cured of pRtr5*a (lane 6);* LPR5041, a derivative of LPR-5054 into which pRtr5*a* has been reintroduced *(lane 7). Arrows* indicate hybridizing bands present in AK631 but absent in AK1329

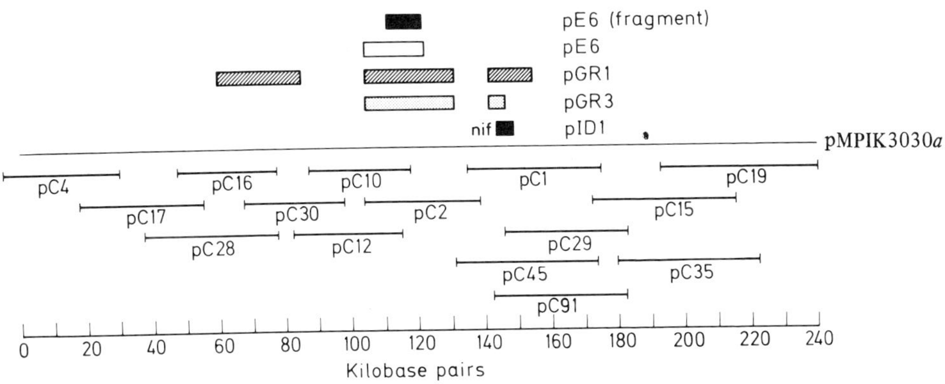

Fig. 3. Map of a 240 kb region of pMPIK3030*a* showing the DNA regions contained by various cosmid clones and the pattern of hybridization obtained with plasmids pID1, pGR3, pGR1, cosmid pE6 and a 9.0 kb fragment of pE6 with these clones

Several R68.45 R prime plasmids of pRme41*b* have been constructed (Kondorosi et al., this vol.). Two of these R primes, pGR1 and pGR3, containing 160 kb and 88 kb of pRme41*b* respectively, have been shown to carry *nif* and *nod* genes by their ability to restore the Nod^+ Fix^+ phenotype to AK1329. When these two R primes were used as hybridization probes against the pMPIK3030*a* cosmid bank, hybridization to clones containing *nif* and also to several other clones was detected (Fig. 3).

Two prominent regions of hybridization were found between pGR1, pGR3 and the pMPIK3030*a* cosmid map. The first region covering approximately 15 kb, contained the *nif* genes defined by pID1 (Fig. 3). A second region, covering approximately 25 kb and defined by clones pC2 and pC10 (Fig. 3) was located about 15 kb to the left of the *nif* region. Additional hybridization was obtained with pGR1 to the left of this region (e.g., with clones pC30 and pC16). The hybridization between pGR3 and cosmid clones pC1, pC2 and pC10 is shown in Fig. 4.

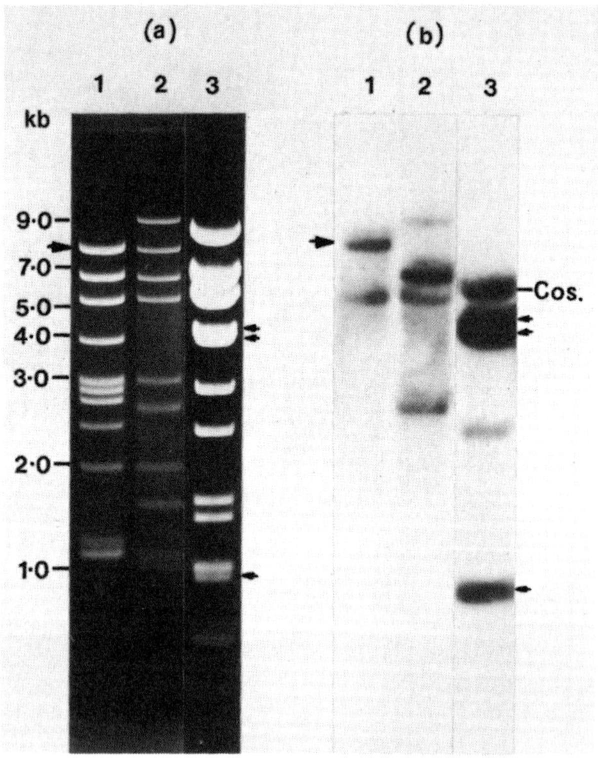

Fig. 4a, b. Hybridization of the *R. meliloti* R prime pGR3 to *Eco*RI restricted DNA of three pMPIK3030*a* cosmid clones. **a** *Eco*RI restricted DNA of pC10 *(lane 1)*, pC2 *(lane 2)* and pC1 *(lane 3)*. *Small arrows* three *Eco*RI restriction fragments in pC1 that contain *nif* genes. *Large arrow* 7.7 kb fragment of pC10 that also hybridizes with a 9.0 kb fragment of pE6 (which possibly contains *R. meliloti nod* genes). The 5.4 kb fragment of the cosmid vector pJB8 contained in each clone is indicated

Although direct evidence is lacking, two lines of indirect evidence support the location of *nod* functions in the region defined by pC2 and pC10. Firstly, the hybridization data obtained with pGR1 and pGR3 indicates that apart from the *nif* region both R primes showed strong hybridization to the region of pMPIK3030*a* contained by pC2 and pC10. Since both R primes restore nodulation as well as nitrogen fixation to AK1329 it is reasonable to suggest that this region could contain *nod* genes. Secondly, hybridization between cosmids pC10 and pC2 and a cosmid clone of *R. meliloti* strain AK631 which contains a 9.0 kb *Eco*RI fragment believed to harbour *nod* genes (Kondorosi et al., this vol.) was obtained (Fig. 3). This cosmid, pE6, contains a 30 kb DNA insert located about 18 kb from the *nif*D gene in pRme41*b* (Kondorosi et al., this vol.). When pC10 was used as a hybridization probe against *Eco*RI restricted pE6 DNA, strong hybridization between pC10 and the 9.0 kb fragment of pE6 was obtained (data not shown). Hybridization of restricted pC10 DNA with the 9.0 kb fragment of pE6 showed strong homology between this fragment and the largest (7.7 kb) fragment of pC10 (Fig. 4).

Further supportive evidence for the location of *nod* genes in the region of the pMPIK3030*a* cosmid map defined by pC10 and pC2 comes from genetic evidence of other studies which demonstrate that *nod* and *nif* genes are closely linked in *R. meliloti* (Banfalvi et al. 1981, Rosenberg et al. 1981, Kondorosi et al., this vol.), *R. leguminosarum* (Johnston et al., this vol.), and *R. trifolii* (Shine et al., this vol.). In addition, DNA sequences contained on a cosmid clone able to complement a Nod^- mutant of *R. meliloti* strain 1021 were located within 30 kb of the *nif* loci on the *nif*D side of the structural genes for nitrogenase (Long and Ausubel 1982).

Characterization of the *nif* Region of pMPIK3030a

The *nif* region of pMPIK3030*a* defined by hybridization with pID1, pGR1 and pGR3 contains three *Eco*RI restriction fragments of 4.2, 4.0 and 0.9 kb (Figs. 1 and 4). Hybridization of pID1 to DNA of *nif* containing clones pC1, pC45 and pC91 restricted with different restriction enzymes indicated further the absence of sites for *Hin*dIII, *Kpn*I, *Sma*I and *Bgl*II and the presence of single sites for *Bam*HI and *Sal*I within this defined region. To determine which of the 3 *Eco*RI fragments contain *nif*D and/or *nif*H, individual fragments of pID1 carrying *nif*D (*Sma*I 2.2 kb fragment) and *nif*H (*Hin*dIII 2.1 kb fragment) (Banfalvi et al. 1981) were isolated and used as hybridzation probes against *Eco*RI restricted pC1 and pC45 DNA. The result of this hybridization was inconclusive as hybridization to all 3 *Eco*RI fragments was obtained with both probes. However, similar hybridization to *Bam*HI digested pC1 and pC45 DNA showed the *nif*D probe to hybridize with two fragments (7.5 kb and 2.1 kb) and the *nif*H probe to hybridze to only the 7.5 kb fragment (data not shown). Also, clone pC29 which contains a DNA insert that ends in the *nif* region (Fig. 3) lacks the 2.1 kb *Bam*HI fragment that hybridizes with *nif*D. From these data we tentatively conclude that *nif*D is situated to the left of *nif*H. The ordering of *nif*D to the left of *nif*H has also been established for the *nif* region of *R. meliloti* (Banfalvi et al. 1981, Ruvkun et al. 1982) and *R. japonicum* (Fuhrman and Hennecke 1982).

Conclusions

The *nif* region and a putative *nod* region have been identified on cosmid clones of the single large plasmid of the broad host range *Rhizobium* sp. MPIK3030. The *nod* region was identified by hybridization with two R prime plasmids and a cosmid of *R. meliloti* DNA containing *nod* functions, and appears to be located about 25 kb to the left of the defined *nif* region on pMPIK3030*a*.

References

Banfalvi Z, Sakanyan V, Koncz C, Kiss A, Dusha I, Kondorosi A (1981) Location of nodulation and nitrogen fixation genes on a high molecular weight plasmid of *R. meliloti*. Mol Gen Genet 184:318–325

Casse F, Boucher C, Julliot JS, Michel H, Dénarié J (1979) Identification and characterization of large plasmids in *Rhizobium meliloti* using agarose gel electrophoresis. J Gen Microbiol 113: 229–242

Fuhrmann M, Hennecke H (1982) Coding properties of cloned nitrogenase structural genes from *Rhizobium japonicum*. Mol Gen Genet 187:419–425

Hooykaas PJJ, Brussel AAN Van, Dulk-Ras H Den, Slogteren GMS Van, Schilperoort RA (1981) Sym-plasmid of *Rhizobium trifolii* expressed in different rhizobial species and in *Agrobacterium tumefaciens*. Nature 291:351–353

Ish-Horowicz D, Burke JF (1981) Rapid and efficient cosmid cloning. Nucleic Acid Res 9:2989–2998

Long SR, Buikema WJ, Ausubel FM (1982) Cloning of *Rhizobium meliloti* nodulation genes by direct complementation of Nod⁻ mutants. Nature 298:485–488

Rosenberg C, Boistard P, Dénarié J, Casse-Delbart F (1981) Genes controlling early and late functions in symbiosis are located on a megaplasmid in *Rhizobium meliloti*. Mol Gen Genet 184: 326–333

Rosenberg C, Casse-Delbart F, Dusha I, David M, Boucher C (1982) Megapalsmids in the plant-associated bacterium *Rhizobium meliloti* and *Pseudomonas solanacearum*. J Bact 150:402–406

Ruvkun GB, Sundaresan V, Ausubel FM (1982) Directed transposon Tn5 mutagenesis and complementation analysis of *Rhizobium meliloti* symbiotic nitrogen fixation genes. Cell 29: 551–559

Schwinghamer EA (1980) A method for improved lysis of some Gram-negative bacteria. FEBS Lett 7:157–162

Trinck MJ (1980) Relationships amongst the fast-growing rhizobia of *Lablab purpureus, Leucaena leucocephala, Mimosa* spp., *Acacia farnesiana* and *Sesbania grandiflora* and their affinities with other rhizobial groups. J Appl Bacteriol 49:39–53

Trinck MJ, Galbraith J (1980) The *Rhizobium* requirements of the non-legume *Parasponia* in relationship to the cross-inoculation group concept of legumes. New Phytol 86:17–26

Vincent JM (1974) Root-nodule symbiosis with *Rhizobium*. In: Quispel A (ed) The biology of nitrogen fixation. Elsevier/North Holland Publ. Comp, Amsterdam, pp 265–341

Identification, Broad Host Range Mobilization and Mutagenesis of a *Rhizobium trifolii* Sym::R68.45 Cointegrate Plasmid

C.W. RONSON [1] and D.B. SCOTT [2]

Introduction

Observations that infectiveness (Nod$^+$) and effectiveness (Fix$^+$) are unstable in certain strains of *Rhizobium trifolii* led to the suggestion that these symbiotic properties were plasmid-borne. Recently several reports have demonstrated the presence of large plasmids in *Rhizobium* strains, and confirmed that some of the plasmids encode symbiotic genes (reviewed by Dénarié et al. 1981).

In this paper we describe the identification of a symbiotic plasmid pRtr514a in an unstable strain of *Rhizobium trifolii.* The plasmid was mobilised in the form of a cointegrate plasmid, pPN1, with the broad host range plasmid R68.45 (Scott and Ronson 1982) and was transferred to strains of *Escherichia coli, Pseudomonas aeruginosa* and various rhizobial species (Ronson et al. 1983). Its effects on the symbiotic properties of the other rhizobial species, and the use of an *E. coli*/pPN1 strain to achieve transposon mutagenesis of the plasmid are also reported.

Material and Methods

Bacterial Strains and Media

Bacterial strains are described in Table 1. *R trifolii* strain PN165 was isolated from strain NZP561 after heat-treatment carried out as described by Scott and Ronson (1982). No differences in the plasmid profile of strains PN165 and NZP561 were observed. TY medium, S10 defined medium, and Oxoid Nutrient medium have been described (Ronson et al. 1981, Scott and Ronson 1982, Ronson et al. 1983).

Crosses

These were done on sterile Millipore filters as described (Scott and Ronson 1982, Ronson et al. 1983). After overnight incubation of the filter on a TY plate, bacteria on the filter were suspended, washed, diluted and then plated on suitable selective

1 Grasslands Division and
2 Applied Biochemistry Division, Department of Scientific and Industrial Research, Palmerston North, New Zealand

Molecular Genetics of the Bacteria-Plant Interaction
(ed. by A. Pühler)

Table 1. Bacterial strains and plasmids

Strain	Relevant characteristics	Source or reference
Rhizobium trifolii		
NZP514	Nod^+ Fix^+	DSIR Culture Collection
NZP597	Nod^-	Spontaneous derivative of NZP514
PN100	Nod^+ Fix^+ *str*-1 *rif*-1	Spontaneous mutant of NZP514
PN104	Nod^-*str*-1 *rif*-1	Heat-cured derivative of PN100
PN109	Nod^+ Fix^+ *kan*-1	Spontaneous mutant of NZP514
PN116	Nod^+ Fix^+ R68.45	Derivative of NZP514
PN118	Nod^-5*fur*-1	Spontaneous mutant of NZP597
PN150	Nod^+ Fix^+ 5*fur*-1/pPN1	This study
NZP561	Nod^+ Fix^+	DSIR Culture Collection
PN165	Nod^-	Heat-cured derivative of NZP561
NZP596	Nod^+ Fix^+	DSIR Culture Collection
R. leguminosarum		
3841	Nod^+ Fix^+ *str*-277	Brewin et al. (1980)
NZP5471	Nod^+ Fix^+	DSIR Culture Collection
R. phaseoli		
NZP5097	Nod^+ Fix^+	DSIR Culture Collection
R. loti		
NZP2037	Nod^+ Fix^+ *(L. pedunculatus)*	DSIR Culture Collection
NZP2213	Nod^+ Fix^+ *(L. tenuis)*	DSIR Culture Collection
Escherichia coli		
HB101	*rec*A r^- m^-	G. Ditta
PN200	HB101/pPN1	This work
PN205	HB101: : Tn7	Derivative of HB101
PN206	HB101: :Tn7/pPN1	Derivative of PN205
Pseudomonas aeruginosa		
PAO2003	*rec*-2	Chandler and Krishnapillai (1974)
PN203, PN204	PAO2003/pPN1	This work
PAO1162	r^- m^+	
Plasmids		
pRtr514a	Nod^+	Scott and Ronson (1982)
R68.45	KmTcCbIncP	Haas and Holloway (1976)
pPN1	pRtr514a: :R68.45	This vol, Scott and Ronson (1982)

media (S10 agar containing tetracycline for *Rhizobium* recipients; Oxoid nutrient agar containing kanamycin for *E. coli* recipients; Oxoid nutrient agar containing carbenicillin for *P. aeruginosa* recipients). Transconjugants were purified by 4 cycles of single colony streaking, unless otherwise stated. The transfer frequency was calculated as the number of transconjugants per total recipients.

Plant Tests

Plant tests using *Trifolium pratense, T. repens, Lotus pedunculatus, L. tenuis* and *Medicago sativa* were done using seedlings grown in 150 X 19 mm test tubes on slopes of Jensen's seedling agar as described (Vincent 1970, Ronson et al. 1983). Plant tests using *Pisum sativum* and *Phaseolus vulgaris* were done in 500 ml glass jars containing sterile moistened pumice as described (Ronson et al. 1983). The tubes and jars were placed in a growth cabinet after inoculation. *Rhizobium* strain effectiveness was monitored by comparing the growth response of inoculated plants with that of uninoculated control plants, and by the acetylene reduction assay for nitrogen fixation. Rhizobia were isolated from nodules as described (Scott and Ronson 1982).

Plasmid Isolation and Agarose Gel Electrophoresis

Bacteria were grown in TY broth to an absorbance at 600 nm of about 0.4. Bacteria from 150 μl of culture (*E. coli*/pPN1 strains) or 1 ml of culture (other strains) were washed once with 0.1% (w/v) sarkosyl in 50 mM Tris-20 mM EDTA buffer, pH 8.0, and crude cell lysates prepared and subjected to electrophoresis by the Eckhardt (1978) procedure as modified by Rosenberg et al. (1982), except that the procedure was carried out at 14°C. Gels were stained and photographed as described (Scott and Ronson 1982).

Other Methods

The methods used for plasmid-curing, the study of root-hair curling, and light and electron microscopy of nodules have been described (Ronson et al. 1981, Scott and Ronson 1982).

Results and Discussion

Derivation of pPN1

Rhizobium trifolii strain NZP514 was known to segregate Nod$^-$ derivatives upon long-term storage (R.M. Greenwood, personal communication). Comparison of the plasmid profile of one Nod$^-$ derivative, strain NZP597, with that of NZP514 revealed that the Nod$^-$ strain lacked the smallest plasmid found in the wild-type strain (Fig. 1a, lanes 1 and 2). This 180 million molecular weight plasmid, termed pRtr514a, was also missing from Nod$^-$ derivatives of NZP514 which were obtained by heat-treatment. The Nod$^-$ derivatives were unable to induce root hair curling of white clover, suggesting that pRtr514a controlled an early step in root nodule formation.

To confirm that pRtr514a indeed carried symbiotic genes, attempts were made using various donors to transfer Nod$^+$ to strain PN104, a Str^R Rif^R derivative of NZP514 heat-cured of pRtr514a. No nodules formed by recipient rhizobia were obtained when a genetically-marked (Kan^R) derivative of NZP514, strain PN109, was used as donor. Control experiments showed that as few as 10 colony-forming units of PN109 formed nodules when inoculated in the presence of 10^8 cfu of PN104. Hence the frequency of transfer of Nod$^+$ from PN109 to PN104 was less than 10^{-7},

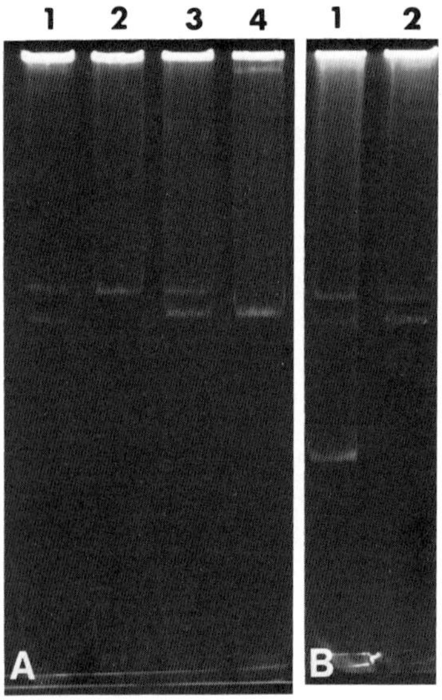

Fig. 1A, B. Agarose gel electrophoresis of plasmids in *R. trifolii* strains and *E. coli* strain PN200. **A** *lane 1, R. trifolii* strain NZP514; *lane 2,* strain NZP597; *lane 3,* Nod$^+$ transconjugant from cross PN116 × PN-118; *lane 4, E. coli* strain PN200. **B** *lane 1, R. trifolii* strain obtained after storage of PN116; *lane 2,* strain PN150

suggesting that pRtr514a was not self-transmissable. However when strain PN116, a derivative of NZP514 containing R68.45, was used as donor, nodules formed by recipient rhizobia were obtained on plants inoculated with bacteria down to the 10^{-5} dilution of the cross, which had been counterselected on streptomycin plus kanamycin. In these experiments, Nod$^+$ was mobilised at a frequency of 4×10^{-3} per Km^R recipient.

Plasmid analysis of recipient rhizobia isolated from the nodules revealed that most did not contain pRtr514a or R68.45, but instead contained a new plasmid of molecular weight ~ 220 million (Fig. 1a, lane 3). However the nodule isolates were Km^{R}Tc^{R}Cb^R, indicating that they contained R68.45 markers, and one isolate tested transferred Km^R at a frequency of 10^{-3}–10^{-4} to the Nod$^-$ strain PN118. All of the 10 Km^R transconjugants tested were Nod$^+$ and had received only the 220 million MW plasmid. These results suggested that the 220 million MW plasmid, which was termed pPN1, was a cointegrate comprised of pRtr514a (180 million MW) and R68.45 (39 million MW). One strain, termed PN150, containing pPN1 was used for further work. Upon continued subculturing and storage of strain PN150, occasional isolates have been found that lack pPN1 but contain plasmids with identical mobilities to pRtr514a and R68.45 confirming the cointegrate nature of pPN1. Furthermore strains containing a plasmid of the mobility of pPN1 in place of pRtr514a have been found following storage of derivatives of NZP514 containing R68.45, suggesting that cointegrate molecules form spontaneously at high frequency during storage. Such strains nevertheless retained R68.45 (Fig. 1b).

Transfer of pPN1 to *Escherichia coli* and *Pseudomonas aeruginosa*

To determine whether pPN1 retained the broad host range of R68.45, attempts were made to transfer it to strains of *E. coli* and *P. aeruginaosa.* Strain PN150 donated Km^R to *E. coli* strain HB101 at a frequency of 10^{-3}, and the Km^R HB101 transconjugants analysed for plasmids had received pPN1 (Fig. 1a, lane 4). One such transconjugant, strain PN200, was used for further work. pPN1 was maintained stably in PN200, even after several subcultures at 37°C in the absence of selective drugs. However the stability of pPN1 was dependent on the *rec*A mutation in HB101, since PN200 donated pPN1 to other *rec*A *E. coli* strains, but only donated a plasmid of identical mobility to R68.45 to rec^+ strains.

In crosses to the rec^- *P. aeruginosa* strain PAO2003, strain PN150 donated Cb^R at a frequency of about 10^{-4}. Only 2 out of 20 Cb^R transconjugants analysed for plasmids contained pPN1; the other 18 contained a plasmid of identical mobility to R68.45. However both PAO2003/pPN1 strains (PN203 and PN204) maintained pPN1 stably through several subcultures. No strains containing pPN1 were obtained when PN150 was crossed to the rec^+ *P. aeruginosa* strain PAO1162.

E. coli strain PN200, and *P. aeruginosa* strains PN203 and PN204 did not nodulate red clover, and strain PN200 had no effect when assayed for root-hair curling of white clover. Strain PAO2003 caused dramatic distortions to white clover root hairs, and induced callus-like growths at the sites of lateral root initiation. Strains PN203 and PN204 caused similar effects, and no specific effect of pPN1 was detected.

Transfer of pPN1 to Other Rhizobial Species

E. coli strain PN200 was chosen in preference to *R. trifolii* strain PN150 as donor of pPN1 in crosses to several rhizobial strains, because it could be easily counterselected on the basis of its auxotrophy. In addition any donors which escaped counterselection should be easily distinguished from *Rhizobium* recipients on the basis of growth rate, growth on Oxoid nutrient agar and colony morphology, and would not form nodules on clover.

E. coli strain PN200 donated pPN1 to various strains of *R. leguminosarum, R. phaseoli, R. loti* and *R. meliloti* at high frequency (ca. 10^{-3}). pPN1 affected the symbiotic response of the recipient strains, both with respect to clover and their original host legume (Table 2). The responses observed on clover are consistent with the proposed classification of *R. leguminosarum, R. phaseoli* and *R. trifolii* as a single species and *R. loti* and *R. meliloti* as separate species (Crow et al. 1981, Jarvis et al. 1982, C. Jordan, pers. commun.). However the possibility that *R. loti* and *R. meliloti* symbiotic genes interfered with the expression of the clover nodulation genes on pPN1 cannot be discounted.

The alteration in the symbiotic responses of the *R. leguminosarum* and *R. phaseoli* transconjugants on peas and beans respectively was associated with the loss of a native plasmid from the transconjugants. The *R. leguminosarum* strain 3841 transconjugants (Fig. 2a, lane 2), which contained a plasmid band of the same mobility as pRtr514a (Fig. 2b) in addition to pPN1, had lost the plasmid band corresponding to

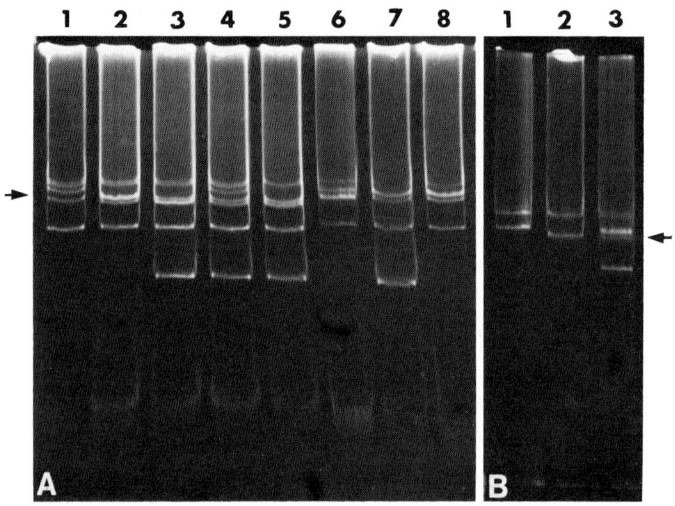

Fig. 2A. Agarose gel electrophoresis of plasmids in *R. leguminosarum* parent strains, pPN1-containing transconjugants and isolates from nodules from pea plants inoculated with pPN1-containing transconjugants. The position of pPN1 is marked by an *arrow. Lane 1, R. leguminosarum* strain JB3841, *lane 2,* transconjugant from cross PN200 × JB3841; *lanes 3, 4, 5,* isolates from nodules obtained from pea plants inoculated with transconjugants from cross PN200 × JB3841; *lane 6,* transconjugant from cross PN200 × NZP5471; *lane 7,* isolate from nodule obtained from pea plant inoculated with transconjugant from cross PN200 × NZP5471; *lane 8, R. leguminosarum* strain NZP5471. **B** Agarose gel electrophoresis of plasmids in *R. trifolii* strain PN150 *(lane 1), R. trifolii* strain PN100 *(lane 2),* and a transconjugant from the cross PN200 × JB3841 *(lane 3).* The position of pRtr514a is marked by an *arrow*

pRL10JI, a plasmid previously identified as essential for the nodulation of peas by that strain (Buchanan-Wollaston et al. 1980, Hirsch et al. 1980, Hombrecher et al. 1981). The loss of a plasmid from the NZP5471 transconjugants was indicated by the marked reduction in the intensity of the second slowest migrating plasmid band in the transconjugants compared to that in the wild-type (Fig. 2a, lanes 6 and 8). The concomitant loss of pea nodulation ability suggests that this plasmid was also a symbiotic plasmid. The *R. phaseoli* strain NZP5097 transconjugants had apparently lost a native plasmid of mobility slightly less than pPN1 (Fig. 4). Their ability to form ineffective nodules on beans may have been due to pPN1, since *R. trifolii* strain PN150 also nodulated beans, albeit poorly. pPN1 also eliminated plasmids from 2 out of the 4 Nod$^+$ strains of *R. trifolii* to which it was transferrred; the plasmid lost from strain PN100 transconjugants (Fig. 3, lanes 1 and 2) was the symbiotic plasmid pRtr514a (Scott and Ronson 1982), while the plasmid lost from strain NZP596 transconjugants (Fig. 3, lanes 3 and 4) hybridised to the *Klebsiella pneumoniae nif* HDKY *Eco*R1 fragment from pSA30, suggesting that it too was a symbiotic plasmid. Hence pPN1 was incompatible with and thus allowed the identification of symbiotic plasmids in several strains of *R. leguminosarum, R. phaseoli* and *R. trifolii.*

To test whether *R. leguminosarum* transconjugants, which retained pea nodulation ability could be isolated, transconjugants which had undergone only one cycle of

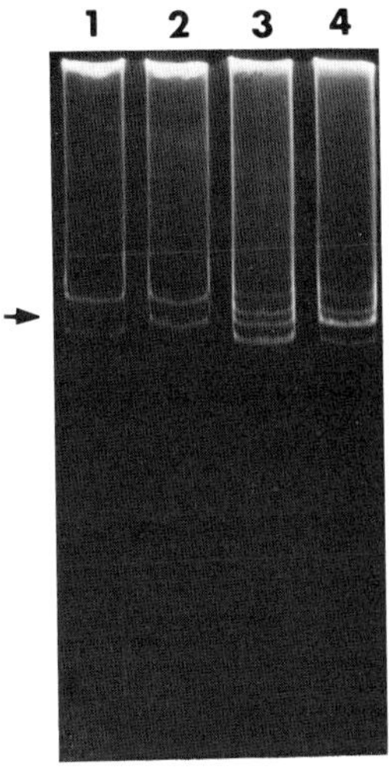

Fig. 3

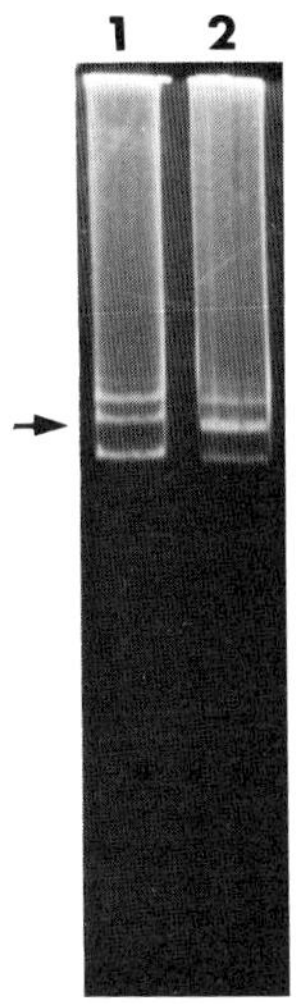

Fig. 4

Fig. 3. Agarose gel electrophoresis of plasmids in *R. trifolii* parent strains and transconjugants. The position of pPN1 is marked by an *arrow*. *Lane 1, R. trifolii* strain PN100; *lane 2,* transconjugant from cross PN200 × PN100; *lane 3, R. trifolii* strain NZP596; *lane 4,* transconjugant from cross PN200 × NZP596

Fig. 4. Agarose gel electrophoresis of plasmids in *R. phaseoli* strain NZP5097 *(lane 1)* and a transconjugant from the cross PN200 × NZP5097 *(lane 2)*. The position of pPN1 is marked by an *arrow*

single colony purification were used to inoculate peas. Three of 4 3841 transconjugants and one of 4 NZP5471 transconjugants formed nodules. Rhizobia were isolated from 2 nodules formed by each of these once-purified Nod$^+$ transconjugants and 4 single colony isolates from each nodule were analysed for plasmids and the ability to nodulate peas and clover. The 3841 nodule isolates had lost pPN1, but contained plasmids of the same mobility as R68.45 and as pRtr514a and nodulated clover. In addition, several of the isolates from nodules formed by 2 of the once-purified transconjugants contained a plasmid of the same mobility as pRL10JI (e.g., Fig. 2a, lane 4) and nodulated peas. However other isolates from the same nodules did not contain pRL10JI (e.g., Fig. 2a, lane 3) and were unable to nodulate peas. Isolates from nodules formed by the third once-purified transconjugant also did not contain pRL10JI, but nevertheless nodulated peas. Examination of the plasmid profile of these isolates over several gels (Fig. 2, lane 5 and not shown) revealed that the second largest plasmid in the isolates had decreased slightly in mobility compared to the analogous plasmid in strain 3841, indicating a large increase in its size. These isolates and the pea nodule isolates which contained both pRL10JI and pRtr514a maintained the ability to nodulate both peas and clover stably over several subcultures. The NZP-5471 nodule isolates (Fig. 2a, lane 7) had lost pPN1 and pRtr514a, but contained R68.45. In addition their native plasmid bands were of the same intensity as in the wild-type, suggesting that they contained the native plasmid missing from the purified transconjugants, and they nodulated peas but not clover. These results suggest that the native symbiotic plasmid had not segregated completely from the once-purified transconjugants and that the pea plant was able to select out rare rhizobia which maintained the symbiotic plasmid instead of pPN1. The observation that pRtr514a and/or R68.45 did not express incompatibility toward pRL10JI is consistent with previous studies which have suggested that symbiotic plasmids from *R. trifolii, R. leguminosarum* and *R. phaseoli* can coexist stably (Brewin et al. 1980, Beynon et al. 1980, Hooykaas et al. 1981), although they share extensive deoxyribonucleic acid homology (Prakash et al. 1981). The reason why pPN1 but not pRtr514a expressed incompatibility is unclear but may be related to pPN1 existing at a higher copy number than pRtr514a due to its R68.45 replicon.

The effect of pPN1 on the symbiotic properties of *R. loti* and *R. meliloti* transconjugants on lotus and lucerne respectively (Table 2) was not due to loss of genetic information, since strains which had lost pPN1 were readily recovered from the pPN1-containing transconjugants and were found to be symbiotically proficient. *R. loti* strain NZP2037 and NZP2213 transconjugants containing pPN1 (Fig. 5, lanes 4 and 7) were very slow-growing, taking 8 days to form 1 mm colonies on S10 plates containing tetracycline, but gave rise to fast-growing "revertants". Such revertants had lost pPN1 but contained R68.45 and occasionally a plasmid the size of pRtr514a (Fig. 5, lanes 3, 5 and 8), and nodulated lotus effectively. *R. meliloti* strain NZP4015 transconjugants containing pPN1 (Fig. 6, lane 3) did not nodulate lucerne within two weeks of inoculation, but effective nodules appeared upon subsequent incubation. Rhizobia isolated from such nodules had lost pPN1 but contained R68.45 (Fig. 5, lane 5). In addition, effective *R. meliloti* strains containing R68.45, with (Fig. 5, lane 4) or without a plasmid the mobility of pRtr514a, were recovered after continued subculturing of the pPN1-containing transconjugants. Therefore the effect of pPN1 seemed to be due to interference with the expression of the symbiotic genes or function of their products. A similar effect has been observed with a *R. leguminosarum* plasmid in *R. phaseoli* (Beynon et al. 1980), and the effect has been termed physiological incompatibility (Beringer et al. 1980).

Table 2. Effect of pPN1 on the symbiotic responses of *Rhizobium* strains on their original host legume and clover

Recipient strain	Normal host	Symbiotic response of normal host	pPN1-containing strain on: clover
3841	*Pisum sativum*	Nod^-	$Nod^+ Fix^+$
NZP5471	*Pisum sativum*	Nod^-	$Nod^+ Fix^+$
NZP5097	*Phaseolus vulgaris*	$Nod^+ Fix^-$ (large, bacteroid-containing nodules)	$Nod^+ Fix^-$ (large, bacteroid-containing nodules)
NZP2037	*Lotus pedunculatus*	$Nod^+ Fix^-$ (small nodules)	$Nod^+ Fix^-$ (sparse late-appearing nodules, which apparently did not contain intracellular rhizobia)
NZP2213	*Lotus tenuis*	$Nod^+ Fix^-$ (sparse, small nodules)	as for NZP2037
NZP4015	*Medicago sativa*	Nod^-	Nod^- (caused root hair curling of white clover)

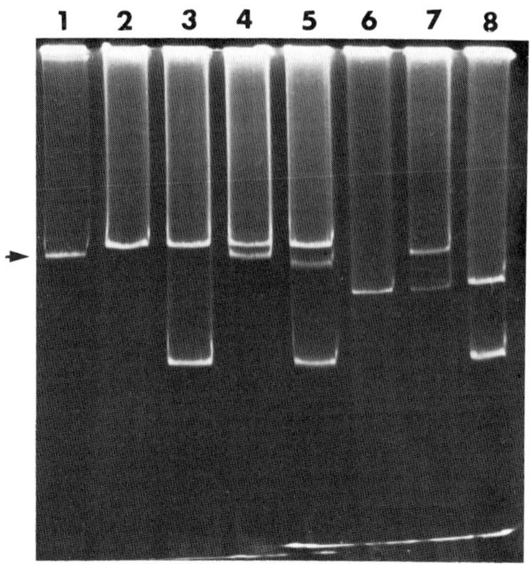

Fig. 5

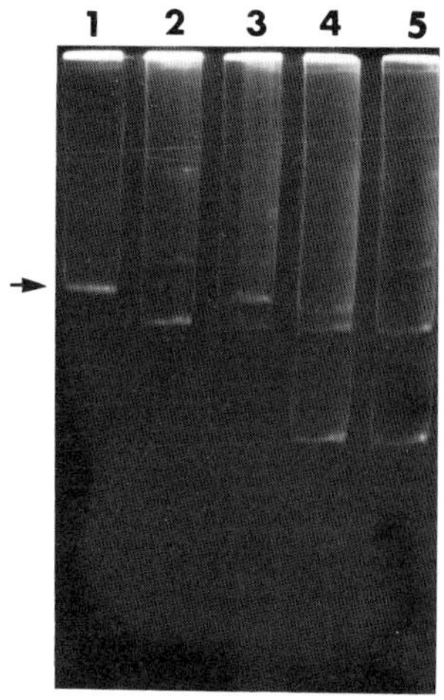

Fig. 6

Fig. 5. Agarose gel electrophoresis of plasmids in *E. coli* strain PN200, and *R. loti* parent strains and transconjugants. The position of pPN1 is marked by an *arrow*. *Lane 1, E. coli* strain PN200; *lane 2, R. loti* strain NZP2037; *lane 3,* fast-growing transconjugant from cross PN200 × NZP2037; *lane 4,* slow-growing transconjugant from cross PN200 × NZP2037; *lane 5,* fast-growing "revertant" from NZP2037/pPN1 transconjugant; *lane 6, R. loti* strain NZP2213; *lane 7,* slow-growing transconjugant from cross PN200 × NZP2213; *lane 8,* fast-growing transconjugant from cross PN200 × NZP2213

Fig. 6. Agarose gel electrophoresis of plasmids in *E. coli* strain PN200 *(lane 1), R. meliloti* strain NZP4015 *(lane 2),* a transconjugant from cross PN200 × NZP4015 *(lane 3),* and derivatives from that transconjugant *(lanes 4 and 5).* The position of pPN1 is marked by an *arrow*

Transposon Mutagenesis of pPN1

pPN1 was transferred to strain PN205, a derivative of *E. coli* strain HB101 containing the transposon Tn7 in its chromosome. When the resultant strain PN206 was used as donor in crosses to the Nod⁻ *R. trifolii* strain PN165, *R. trifolii* transconjugants resistant to streptomycin (selective for Tn7) and tetracycline (selective for pPN1) were obtained at a frequency of about 10^{-6}, compared to 10^{-3} for the transfer of pPN1 alone. The transconjugants were also resistant to spectinomycin, another Tn7 marker, indicating that they were not spontaneous streptomycin-resistant mutants. Of about 500 $Str^R Tc^R$ PN165 transconjugants assayed on plants, four were Nod^-. Three of these had received a plasmid slightly larger than R68.45, which was probably R68.45::Tn7, and the fourth had received a plasmid of molecular weight 120 million, which was presumably a deleted derivative of pPN1. A further 8 transconjugants formed ineffective nodules and had received a plasmid of the mobility of pPN1, or, in one case, slightly faster. These results indicate both that strain PN165 had lost Fix genes as well as Nod genes and that pPN1 encodes both Fix and Nod genes. The sites of the Tn7 insertions in pPN1 are presently being physically mapped, using plasmids purified from *E. coli* strains.

Summary

A *Rhizobium trifolii* symbiotic plasmid pRtr514a was mobilized by the broad host range plasmid R68.45 in the form of a cointegrate plasmid pPN1. pPN1 was stable in *rec*$^-$ strains of *E. coli* and *P. aeruginosa,* and in most fast-growing *Rhizobium* strains, suggesting that most *Rhizobium* strains may be naturally inefficient in genetic recombination.

pPN1 affected the symbiotic properties, both with respect to clover and their original host legume, of the *Rhizobium* strains to which it was transferred (Table 2). It apparently eliminated native symbiotic plasmids from strains of *R. leguminosarum* and *R. phaseoli,* and also from some Nod$^+$ strains of *R. trifolii.* However its effect on the symbiotic properties of *R. loti* and *R. meliloti* strains on lotus and lucerne respectively was due to interference with the expression of the native symbiotic genes rather than loss of genetic information, since symbiotically proficient strains which had lost pPN1 were readily recovered from the pPN1-containing transconjugants.

Transposon mutagenesis of pPN1, achieved using an *E. coli*::Tn7/pPN1 strain as donor in crosses to a Nod$^-$ *R. trifolii* strain, indicated that the plasmid encoded genes required for several of the steps leading to the formation of a root nodule on clover.

References

Beringer JE, Brewin NJ, Johnston AWB (1980) The genetic analysis of *Rhizobium* in relation to symbiotic nitrogen fixation. Heredity 45:161–186

Beynon JL, Beringer JE, Johnston AWB (1980) Plasmids and host-range in *Rhizobium leguminosarum* and *Rhizobium phaseoli.* J Gen Microbiol 120:421–429

Brewin NJ, Beringer JE, Johnston AWB (1980) Plasmid-mediated transfer of host-range specificity between two strains of *Rhizobium leguminosarum.* J Gen Microbiol 120:413–420

Buchanan-Wollaston AV, Beringer JE, Brewin NJ, Hirsch PR, Johnston AWB (1980) Isolation of symbiotically defective mutants in *Rhizobium leguminosarum* by insertion of the transposon Tn5 into a transmissible plasmid. Mol Gen Genet 178:185–190

Chandler PM, Krishnapillai V (1974) Isolation and properties of recombination-deficient mutants of *Pseudomonas aeruginosa.* Mut Res 23:15–23

Crow VL, Jarvis BDW, Greenwood RM (1981) Deoxyribonucleic acid homologies among acid-producing strains of *Rhizobium.* Int J Syst Bacteriol 31:152–172

Dénarié J, Boistard P, Casse-Delbart F, Atherly AG, Berry JO, Russel P (1981) Indigenous plasmids of *Rhizobium.* In: Giles KL, Atherly AG (eds) Int Rev Cytol Suppl 13. Academic Press, London New York, pp 225–246

Eckhardt T (1978) A rapid method for the identification of plasmid deoxyribonucleic acid in bacteria. Plasmid 1:584–588

Haas D, Holloway BW (1976) R factor variants with enhanced sex factor activity in *Pseudomonas aeruginosa.* Mol Gen Genet 144:243–251

Hirsch PR, Montagu M Van, Johnston AWB, Brewin NJ, Schell J (1980) Physical identification of bacteriocinogenic, nodulation and other plasmids in strains of *Rhizobium leguminosarum.* J Gen Microbiol 120:403–412

Hombrecher G, Brewin NJ, Johnston AWB (1981) Linkage of genes for nitrogenase and nodulation ability on plasmids in *Rhizobium leguminosarum* and *R. phaseoli.* Mol Gen Genet 182: 133–136

Hooykaas PJJ, Brussel AAN Van, Dulk-Ras H Den, Slogteren GMS Van, Schilperoort RA (1981) Sym plasmid of *Rhizobium trifolii* expressed in different rhizobial species and *Agrobacterium tumefaciens.* Nature 291:351–353

Jarvis BDW, Pankhurst CE, Patel JJ (1982) *Rhizobium loti:* a new species of legume root nodule bacteria. Int J Syste Bacteriol 32:378–380

Prakash RK, Schilperoort RA, Nuti MP (1981) Large plasmids of fast-growing rhizobia: homology studies and location of structural nitrogen fixation *(nif)* genes. J Bacteriol 145:1129–1136

Ronson CW, Lyttleton P, Robertson JG (1981) C4-dicarboxylate transport mutants of *Rhizobium trifolii* form ineffective nodules on *Trifolium repens.* Proc Natl Acad Sci USA 78:4284–4288

Ronson CW, Riddiford AHH, Scott DB (1983) Transfer of *Rhizobium trifolii* symbiotic plasmid::R68.45 cointegrate to strains of *Escherichia coli, Pseudomonas aeruginosa* and fast-growing *Rhizobium* species. (submitted)

Rosenberg C, Casse-Delbart F, Dusha I, David M, Boucher C (1982) Megaplasmids in the plant-associated bacteria *Rhizobium meliloti* and *Pseudomonas solanacearum.* J Bacteriol 150: 402–406

Scott DB, Ronson CW (1982) Identification and mobilization by cointegrate formation of a nodulation plasmid in *Rhizobium trifolii.* J Bacteriol 151:36–43

Vincent JM (1970) International biological programme handbook, No 15. A manual for the practical study of root-nodule bacteria. Blackwell, Oxford

Genetic Analysis of the Symbiotic Regions in *Rhizobium trifolii* and *Rhizobium parasponia*

B.G. ROLFE, M.A. DJORDJEVIC, N.A. MORRISON, J. PLAZINSKI, G.L. BENDER, R. RIDGE, W. ZURKOWSKI, J.T. TELLAM, P.M. GRESSHOFF, and J. SHINE [1]

Rhizobium Strains Which Can Infect Legumes and the Non-Legume *Parasponia*

Rhizobium bacteria which can form nitrogen-fixing symbiosis with leguminous plants have been divided into two main groups – the so-called fast-growers and the slow-growers on the basis of their growth on laboratory media (Vincent 1982). However, it has been shown recently that a number of slow-growing rhizobia isolated from cowpea type plants are able to nodulate the non-legume, *Parasponia* (a member of the Ulmaceae). Similarly, a number of isolates from *Parasponia* nodules proved to be typical slow-growing organisms which had a variable host range amongst the cowpea plants (Trinick and Galbraith 1980).

Of over two hundred isolations made in New Guinea from nodules of cowpea group plants (which are normally nodulated by slow-growers), Trinick (1980) found only one fast-growing bacterium. This strain, NGR234, resembles fast-growing rhizobia from the tropical tree *Leucaena leucocephala,* and is able to nodulate a broad range of cowpea group plants, as well as lucerne *(Medicago sativa), Leucaena* and *Acacia farnesiana* (Trinick 1980). In addition, this fast-growing *Rhizobium* is able to nodulate the non-legume, *Parasponia* (Trinick and Galbraith 1980). In contrast to the slow-growers from *Parasponia,* the symbiosis between *Parasponia* and strain NGR234 is not effective in nitrogen fixation. Clearly the genotype of this organism allows it to nodulate a wide range of plants, crossing the boundaries that separate not only the fast- and slow-growing rhizobia but also the legume and *Parasponia.*

It has been shown that the symbiotic properties of many fast-growing rhizobia from temperate regions can be lost through a process of prolonged exposure to elevated temperatures. The eradication of the symbiotic capability is attributed to the loss of large endogenous plasmids, or to internal deletion in these plasmids (Zurkowski and Lorkiewicz 1978, Zurkowski 1982). A log phase culture of strain ANU240 (a spontaneous streptomycin resistant mutant of NGR234) was evenly spread on the surface of solid YM medium and the plates incubated at 37°*C* for 7 days during which no growth occurred. The YM plates were placed at room temperature and after 5 days about 100 single colonies arose per plate. The single colonies from the cultures treated at 37°C were purified and checked for the antibiotic

1 Genetics Department, Research School of Biological Sciences, Australian National University, Canberra City, A.C.T. 2601, Australia

Molecular Genetics of the Bacteria-Plant Interaction
(ed. by A. Pühler)

resistance pattern and growth characteristic of the strain to ensure that they were not contaminants. The purified clones were tested for their ability to nodulate the tropical legume siratro *(Macroptilium atropurpureum)* by the plate method (Rolfe et al. 1980, Cen et al. 1982). Approximately 25% of the tested clones failed to nodulate. The nodulating (Nod$^+$) and non-nodulating (Nod$^-$) phenotypes were found to be stable and non-reverting through subculture. Plasmid profiles of these Nod$^-$ strains were examined by a modified gel technique (Djordjevic et al. 1982) and in all cases the Nod$^+$ strains possessed a plasmid of approximately 300 Mdals (designated pNM3AN) (Fig. 1A). Invariably the 200 plasmid was not detected in the Nod$^-$ strains nor was a deleted form of this large plasmid observed.

One heat cured Nod$^-$ strain, ANU264, was tested for the ability to nodulate a number of plants, usually nodulated by NGR234. Strain ANU264 was found to be unable to nodulate the following plants: *Vigna unguiculata, Lablab purpureus, Macroptilium lathyroides, Leucaena leucocephala* and *Parasponia andersonii;* while strain ANU240 was able to nodulate all these plants.

The slide method of Fahraeus (1957) was used to examine the roots of siratro plants inoculated with either strain ANU240 or ANU264. While strain ANU240 was able to induce curling, extreme distortions, and infection threads on root hairs, the mutant strain had no observable effect on root hair morphology.

Southern hybridization techniques demonstrated that radioactively labeled DNA from a cloned fragment of the *R. trifolii nif*HD gene hybridized to the DNA of strain ANU240 but not strain ANU264. The *nif*HD gene region of strain ANU240 appears to

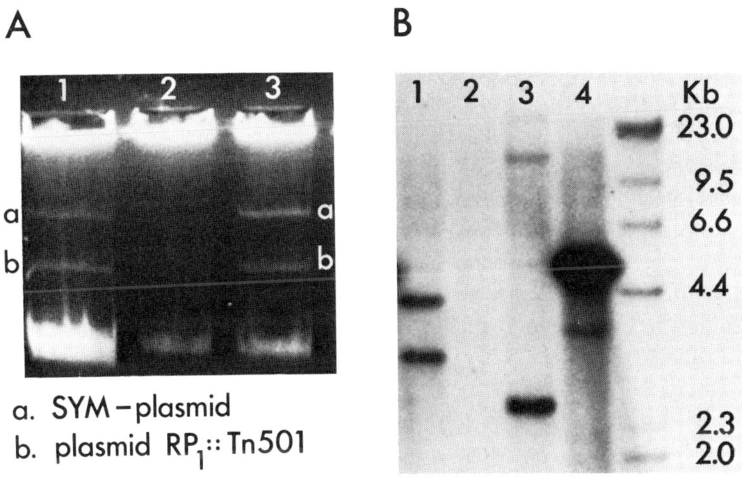

a. SYM – plasmid
b. plasmid RP_1 :: Tn501

1. ANU 239 (RP_1::Tn501)
2. ANU 265
3. Nod$^+$ transconjugant

1. ANU 237
2. ANU 265
3. ANU 265 (p JB5 JI)
4. ANU 265 (p BRIAN)

Probe: *R. trifollii nif* HD

Fig. 1A, B

be contained on two *Eco*RI generated fragments of approximately 3 and 4 kilobases (Fig. 1B). DNA from strains ANU264 (pBR1AN) and ANU264 (pJB5JI) were included in the analysis as controls, since these plasmids have previously been shown to encode nitrogenase genes (Johnston et al. 1978, Rolfe et al. 1981).

The ability to nodulate siratro effectively was restored to the heat cured strain by reintroducing pNM3AN. This was achieved by mobilizing the Sym-plasmid with the P1 group plasmid RP1::Tn501 (Pemberton and Bowen 1981), from strain ANU239 (a rifampicin resistant derivative of NGR234) to ANU265 (a spectinomycin resistant ANU264 derivative). Plasmid gels showed that the nodulating transconjugants contained the Sym-plasmid pNM3AN and RP1::Tn501 (Fig. 1A). No evidence of stable cointegration of the two plasmids was found.

These results indicate that the 300 plasmid of ANU240 is required for the formation of infection threads and root hair curling on siratro and for the nodulation of a range of different legumes and the non-legume, *Parasponia*. The pattern of hybridization of cloned *nif*HD DNA to total genomic DNA plots indicates that the genes for the nitrogenase enzyme complex also reside on the 300 plasmid PNM3AN.

Transposon Mutagenesis of Strain ANU237 and its Derivatives

The suicide plasmids pJB4JI and pSP601 (which contain Tn5 and phage Mu) transferred kanamycin-resistance (Km^R) to strain ANU237 at about 10^{-3} frequency. However, the co-inheritance of other antibiotic resistance markers of these plasmids was also found. Furthermore, when presumptive Tn5 transposition mutants were mated with *E. coli* strain RR1, about 30%–50% of the clones were able to retransfer Km^R at a high frequency to *E. coli*. An alternative Tn5 mutagenic system provided by Drs. A. Pühler and R. Simon, was then tested with strain ANU237 and its derivatives. This involved use of the constructed plasmid pSUP101 which was made by cloning the mobilization site (*mob* site or *ori*-T) of plasmid RP4 into the tetracycline resistance gene of the multicopy vector pACYC184. This allows pSUP101 to be mobilized in *trans* by *Inc*P1 group plasmids, since the *mob* site (*ori*T) is the origin of transfer of the *Inc*P1 transfer system. Tn5 was inserted into pSUP101 to give pSUP1011. Both pSUP101 and pSUP1011 can be mobilised between *E. coli* strains at up to 100% frequency. Since the origin of vegetative replication (*ori*-V) of pACYC184 (a *col*-E1 like vector) is primarily suited to *E. coli* cells, these plasmids cannot be maintained in bacteria other than *E. coli*. The *Inc*P1 transfer system however, has broad host range capability so that while the plasmids can be mobilised to bacteria other than *E. coli* they cannot be maintained. pSUP1011 is therefore an excellent vehicle for suicide mutagenesis in *Rhizobium*.

When the suicide plasmid pSUP1011 was used in filter matings with ANU240, Km^R was transferred at about 10^{-5} frequency. Kanamycin-resistant colonies isolated from this transposon mutagenic system did not show retransfer of Km^R. However, about 20% of the Km^R colonies were also Cm^R (the other marker on pSUP1011). Clones which were Km^R Cm^S were considered to be potential bona fide Tn5 mutants. Over 3000 Km^R colonies isolated from two separate mating

Table 1. Tn5 mutagenesis of strain ANU237

Mutant type	Number	Comments
Auxotrophs		
ADE	3	
URA	3	Reversion frequencies: $< 10^{-8}$
ARG	2	
MET	1	
CYS	2	
TRP	4	Tn5 insertions: 21 mutants analysed,
LEU	1	18 single Tn5 insertions
ASP	2	2 double Tn5 insertions
TYR	1	1 triple Tn5 insertion
ADE URA	2	
ASP GLU	4	Tn5 insertions appear to be random
MET CYS	1	
ISO VAL	3	
Unknown requirement	7	
Symbiotic		
Nod^-	5	1 has single Tn5 on Sym-plasmid 4 are chromosomally located
Delayed nodulation	9	
Defective nodulation	1	
Fix^-	1	

experiments were purified and tested for possible auxotrophic and symbiotic mutants using TM minimal agar and siratro plants. A range of auxotrophic mutants at about 1.3% were isolated as well as various symbiotic mutants (Table 1). The chromosomal or plasmid location of Tn5 in the different Nod^- mutants was assayed by hybridization of the cloned *Eco*RI fragment (containing Tn5 and associated *Rhizobium* DNA isolated from the various Nod^- mutants) to gels containing DNA from the parent strain ANU240 and ANU264 (which lacks the endogenous Sym-plasmid pNM3AN).

Construction of a Mobilizable Sym-Plasmid pNM4AN

The $Cm^R Km^R$ clones which occurred at 20% of selected Km^R ANU240 clones were probably due to the integration of pSUP1011 into a site somewhere within the genome. Integration into a replicon would rescue both the Km^R and Cm^R markers, as well as inserting the *mob* site. If such insertions were in the chromosome it could provide a site for mobilizing the chromosome at high frequency, in *trans,* using the *tra* functions of an *Inc*P1 group plasmid. If the insertion was in the symbiotic plasmid, pNM3AN, then this also should be mobilized at high frequency.

Models of cointegration or replicon fusion promoted by transposons, predict that two copies of Tn5 will be present in a fusion between pNM3AN and PSUP1011. These copies of Tn5 will be directly repeated, flanking the remainder of pSUP1011.

The *mob* site should reside between the transposons. Selective pressure can be applied to maintain the cointegrate by selecting for Cm^R. If homologous recombination occurs between the two Tn5 copies it would give pNM3AN::Tn5 and a copy of pSUP1011. The pSUP1011 cannot replicate independently in *Rhizobium* and Cm^R would be lost. Any cells which resolved the cointegrate would therefore be susceptible to Cm selection.

In order to test this hypothesis pSUP1011 was mated into ANU239 (R6845 ΔKm). Plasmid R6845 ΔKm was used to provide *tra*$^+$ functions for subsequent mobilization of Km^R or Cm^R. Recipient transconjugants were selected on TY Rif with Km, Km and Cm, and Cm alone to measure the transfer rates for these various markers. The transfer rate of Cm^R was the same as for Km^R Cm^R (5×10^{-7}) while the transfer rate for Km^R was about 10^{-6}. When Km^R clones were spot mated with *E. coli* RR1 to check for transferability of Km^R and Cm^R, only those clones which were Cm^R were able to transfer both Km^R and Cm^R. Those clones which were Km^R Cm^S did not transfer Km^R to *E. coli*. This is further evidence that true Tn5 suicide has occurred in Km^R and Cm^S clones since they appear to have lost both Cm^R and the *mob* site.

In all the above crosses with ANU239 (R6845 ΔKm) derivatives the plasmid R6845 ΔKm was also transferred to *E. coli* as indicated by Tc^R and Ap^R. The ANU239

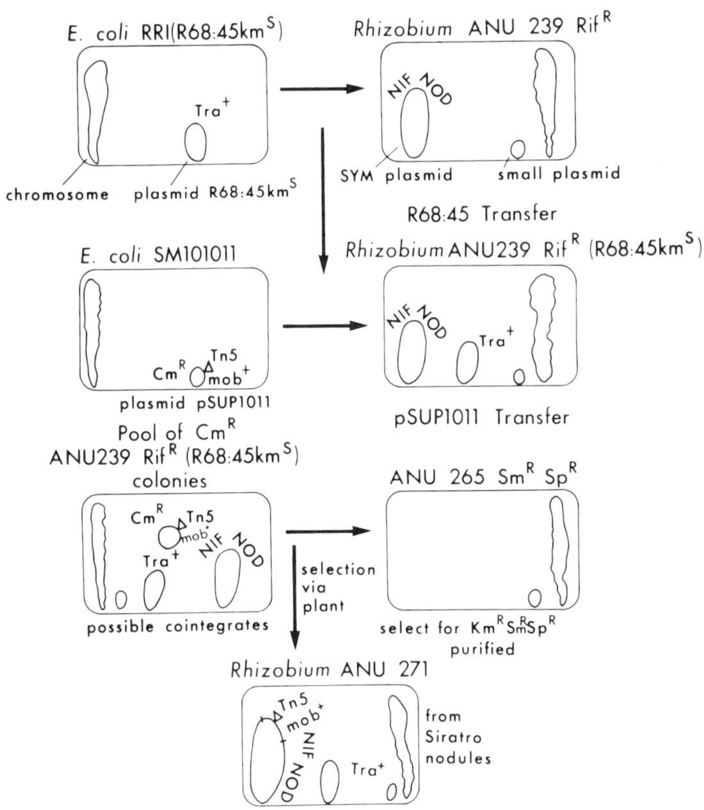

Fig. 2. Construction of mobilizable Sym-plasmid

(R6845 ΔKm) derivatives were all free of contamination by the donor *E. coli* (Fig. 2). The various ANU239 (R68:45 Km^S) (pSUP1011) colonies were pooled and then mated with strain ANU265 selecting for Km^R Sm^R Sp^R colonies. The resulting plate of colonies was replica plated twice on selection plates and then the cell cultures were used to inoculate siratro plants. After 20 days 14 nodules had formed (Fig. 2). The bacteria isolated from such nodules on siratro were shown to be Km^R Cm^R Tc^R Sm^R Sp^R Rif^s and able to renodulate plants and retransfer Km^R Cm^R at high frequency (about 10^{-2}). The original donor strain was not found in the inoculum nor in the nodules formed.

Nodulation Properties of Different Transconjugants

The strain ANU271 [ANU265 Sm^R Sp^R (R68.45 Km^S) (pSUP1011-Sym-plasmid)] isolated from siratro nodules was used as a donor strain to transfer the newly constructed mobilizable Sym-plasmid (pNM4AN) into a variety of bacteria and test whether broad host range nodulation could be expressed in other strains (Fig. 3). The fast-growing rhizobia which normally infect temperate legumes *(R. leguminosarum, R. meliloti, R. trifolii)* could readily nodulate siratro and cowpeas when they contained Sym-plasmid pNM4AN. These nodules are found along the tap root of the siratro plants and form slightly more slowly than strain ANU240 nodules. In contrast, *Agrobacterium tumefaciens* strains carrying the pNM4AN Sym-plasmid do not form nodules on the tap roots of siratro plants but form poorly developed nodules on the lateral roots. *E. coli* strains containing plasmid pNM4AN do not nodulate any test plants.

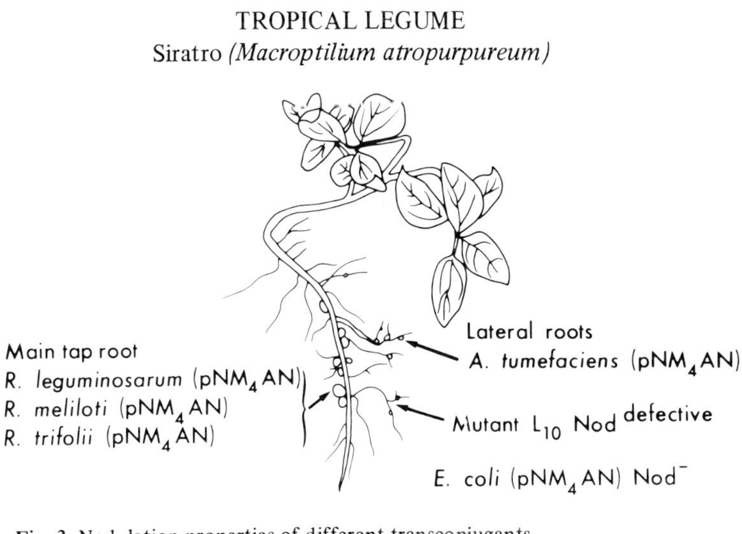

Fig. 3. Nodulation properties of different transconjugants

Properties of *Rhizobium* Symbiotic Plasmids

Sym-plasmid pBR1AN is a plasmid constructed in vivo which encodes clover-specific nodulation and nitrogenase genes (Fig. 4). It was originally derived from the self-transmissible *R. leguminosarum* Sym-plasmid pJB5JI via an in vivo recombination event which involved at least 60 kb of DNA. The approximate extent of the area involved in this recombination is shown. This recombination event resulted in the pea-specific nodulation and nitrogen fixation functions being exchanged for clover-specific nodulation and nitrogen fixation functions. As a result the two plasmid molecules have retained identical sizes (130 Mdal) as well as a number of non-symbiotic gene markers that were originally encoded on pJB5JI (Table 2).

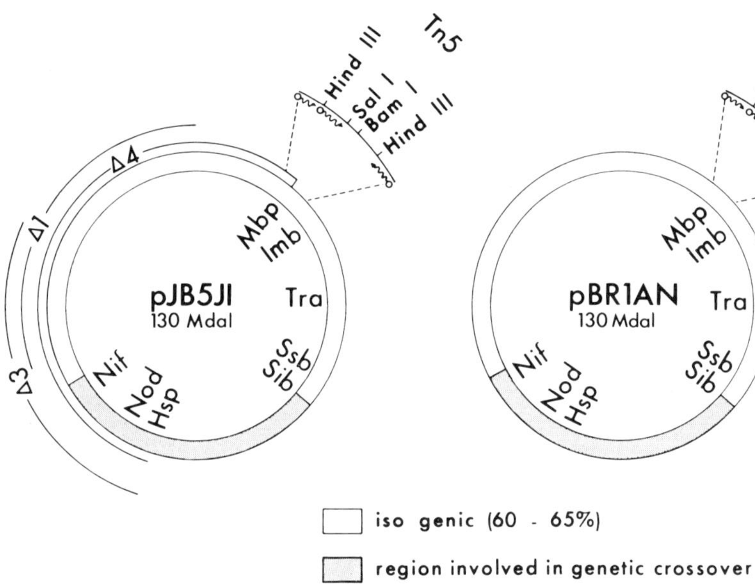

Fig. 4. Construction and characterization of a clover Sym-plasmid

Both plasmids pBR1AN and pJB5JI encode: common transfer function *(Tra⁺)*, kanamycin resistance (due to transposon Tn5 insertion into a gene encoding "medium" bacteriocin production, *Mbp*: :Tn5), the immunity to killing by medium bacteriocins *(Imb⁺)*, the suppression of "small" bacteriocin production *(Ssb⁺)* and the suppression of the immunity to "small" bacteriocins *(Sib⁺)*. The expression of the *Ssb⁺* and *Sib⁺* markers are apparent when either pJB5JI or pBR1AN is transferred to a *Rhizobium* strain producing "small" bacteriocins. Hybridization data obtained using plasmid pJB5JI and various deleted derivatives of pJB5JI as hybridization probes, coupled with an analysis of the various non-symbiotic markers retained by the deleted pJB5JI derivatives, suggests that these plasmid molecules are up to 65% isogenic. The region

Table 2. Comparison of the known genetic markers on Sym-plasmids pBR1AN and pJB5JI

Genetic markers	Plasmid pJB5JI	Plasmid pBR1AN
Nodulation ability:		
Root adhesion *(Roa⁺)*	+	+
Hair curling *(Hac⁺)*	+	+
Nodule initiation *(Noi⁺)*	+	+
Host specificity:		
Peas	+	–
White clover	–	+
Subterranean	+	+
Nodule function *(Fix⁺)*	+	+
Self-transmissibility	+	+
Presence of Tn5	+	+
Medium bacteriocin production *(Mbp⁺)*	–	–
Immunity to medium bacteriocins *(Imb⁺)*	+	+
Suppression of small bacteriocins *(Ssb⁺)*	+	+
Suppression of small bacteriocin immunity *(Sib⁺)*	+	+
Hybridzation to:		
R. trifolii nod probe	±	+
*R. trifolii nif*HD probe [a]	+	+

[a] The *R. trifolii nif*HD sequence cross-hybridizes to the *R. leguminosarum nif*H sequence encoded on plasmid pJB5JI

of difference between the two plasmids includes the genes that control the host-specific nodulation and the structural nitrogenase genes which are some 25 kb apart, but also extends at least another 40 kb out from this cluster of symbiotic genes.

Transfer of Sym-Plasmids to Both Fast- and Slow-Growing Rhizobia

The transfer of Sym-plasmids pBR1AN and pJB5JI to various symbiotic mutants of the Rhizobiaceae resulted in the variable expression of the plasmid-borne symbiotic functions. For example, the transfer of pJB5JI or pBR1AN to various *R. meliloti* megaplasmid deletion mutants did not result in the ability of these strains to nodulate peas or clovers. The transfer of pBR1AN to an *Agrobacterium tumefaciens* strain conferred clover nodulation to this strain, whereas pJB5JI could not induce this *Agrobacterium* strain to nodulate peas. The transfer of pBR1AN to various Sym-plasmid deleted or Sym-plasmid cured *R. leguminosarum* or *R. trifolii* strains resulted in the ability to nodulate clovers, whereas the reverse was true when pJB5JI was transferred to these strains.

The transfer of these Sym-plasmids to various cell surface defective *(Muc⁻) R. trifolii* and *R. leguminosarum* strains resulted in the restoration of the ability to nodulate in the non-nodulating strains, however the cell surface polysaccharide production was not corrected and only poor, ineffective nodulation occurred. When these plasmids were transferred to the Muc$^-$ derivative (which had not lost the ability to nodulate) no enhanced nodulation capacity could be transferred and the Muc$^-$ phenotype remained unaltered. These results implied that a Muc$^+$ phenotype is required for effective nodulation and that mucoidy production is not Sym-plasmid controlled. One particular Muc$^-$ mutant, ANU14, demands special attention. This Nod$^-$ Muc$^-$ mutant was derived from ANU13, a heat-cured mutant of ANU12, via a second round of heat-curing of this strain. The Muc$^-$ phenotype is conditional as it is only apparent when cells are cultured on media where either mannitol, sorbitol, or fructose is used as a carbon-source. The transfer of pBR1AN to this strain does not confer clover nodulation, whereas the transfer of pJB5JI to this strain confers pea nodulation (Table 3). Kanamycin resistance and the ability to nodulate clovers can be transferred from ANU14 (pBR1AN) cells indicating that the plasmid molecule is intact. Plasmid pBR1AN expressed clover nodulation in ANU13, the mutant from which ANU14 was derived. Therefore it seems clear that this mutation in ANU14 has somehow masked the phenotypic expression of the clover nodulation genes carried by pBR1AN but has not affected the phenotypic expression of pJB5JI. The precise genetic location of the Muc$^-$ mutation of strain ANU14 is still unknown.

Table 3. Interaction of a non-mucoid mutation and Sym-plasmid phenotypic expression

Strain	Nodulation response on			
	Red clovers	White clovers	Subterranean clovers	Peas
R. trifolii				
ANU13	Nod$^-$	Nod$^-$	Nod$^-$	Nod$^-$
ANU13 (pJB5JI)	Root distortions	Root distortions	Nod$^+$ Fix$^+$	Nod$^-$
ANU13 (pBR1AN)	Nod$^+$ Fix$^+$	Nod$^+$ Fix$^+$	Nod$^+$ Fix$^+$	Nod$^-$
ANU14 (pJB5JI)	Nod$^-$	Nod$^-$	Nod$^+$ Fix$^-$	Nod$^+$ Fix$^+$
ANU14 (pBR1AN)	Nod$^-$	Nod$^-$	Nod$^-$	Nod$^-$
ANU14 Muc$^+$ revertant				
ANU14R (pBR1AN)	Nod$^+$ Fix$^+$	Nod$^+$ Fix$^+$	Nod$^+$ Fix$^+$	Nod$^-$

Strain ANU13 is T12 Nod$^-$16; ANU14 is T12 Nod$^-$ 16 Muc$^-$ 2; ANU14 is the Muc$^+$ revertant of the non-mucoid parent strain

Fast-Growing Broad Host Range Strain ANU237

When plasmids pJB5JI and pBR1AN were introduced into strains ANU237 or ANU264 (a streptomycin-resistant derivative of ANU237 which has lost the endogenous Sym-plasmid pNM3AN) the resulting transconjugants showed a change in the spectrum of plants that could be nodulated (Table 4).

On white clover, strain ANU237 (pJB5JI) produced nodules of a similar type to ANU237 (pBR1AN). This is a surprising result since plasmid PJB5JI codes for pea nodulation and strain ANU237 is unable to nodulate clovers. This indicates that

Table 4. Sym-plasmid interactions

Strains	White clovers	Subterranean clovers	Peas
ANU237	Nod$^-$ (root distortions)	Nod$^-$	Nod$^-$ (some pigmentation of roots)
ANU237 (pBR1AN)	Nod$^+$	Nod$^+$	Nod$^-$
ANU237 (pJB5JI)	Nod$^+$	Nod$^-$	Nod$^+$
ANU264	Nod$^-$	Nod$^-$	Nod$^-$
ANU264 (pBR1AN)	Nod$^+$	Nod$^+$	Nod$^-$
ANU264 (pJB5JI)	Nod$^-$ (Root distortions)	Nod$^-$	Nod$^+$

All nodules are non-nitrogen-fixing (Fix$^-$)

there is some kind of interaction between the endogenous Sym-plasmid pNM3AN and the introduced pJB5JI which now leads to a change in the spectrum of plants that the constructed *Rhizobium* strain can nodulate. In the heat cured strain ANU264 the plasmid pJB5JI behaves as expected (Table 4).

Strain ANU264 (pBR1AN) forms nodules similar to a stage reached after 10 days by a wild type *R. trifolii* strain on clovers. Nodules formed by strains ANU237 (pBR1AN) stop their development at an earlier stage. Strain ANU264 (pBR1AN) also forms more nodules on clovers than strain ANU237 (pBR1AN) under similar inoculation and growth conditions. This again indicates that Sym-plasmid interactions influence nodule induction and the degree of nodule development.

Transfer to Slow-Growing Rhizobia: Effects on Siratro Nodulation

When the pea plasmid pJB5JI was transferred to the slow-growing cowpea rhizobia strains ANU289, CB756 and 32H1, no delay in the onset and early development of nodulation was observed. However, the plants had an ineffective phenotype and no nitrogen was fixed. Examination of these nodules by both light and electron microscopy showed that the strains containing plasmid pJB5JI had faulty bacterial release and could be described phenotypically as Bar$^-$ (defective bacterial release from the ramifying infection thread) (Figs. 5–7).

Biological Consequences of in Vitro Constructed Sym-Plasmid

The construction of plasmid pSO32, which contains a 14 kb fragment of the nodulation region of strain ANU843, will be described elsewhere (Shine et al., this vol.). This region of the Sym-plasmid from *R. trifolii* is sufficient to induce white clover nodules when introduced into Nod$^-$ *R. trifolii* strains, *R. leguminosarum, Agrobacterium tumefaciens,* and broad host range strain ANU237.

When nodules from both white and subterranean clovers were examined by microscopy marked differences could be observed (Figs. 8–12). In white clover nodules poor bacterial release (Bar$^-$ phenotype) occurs and very little nodule development

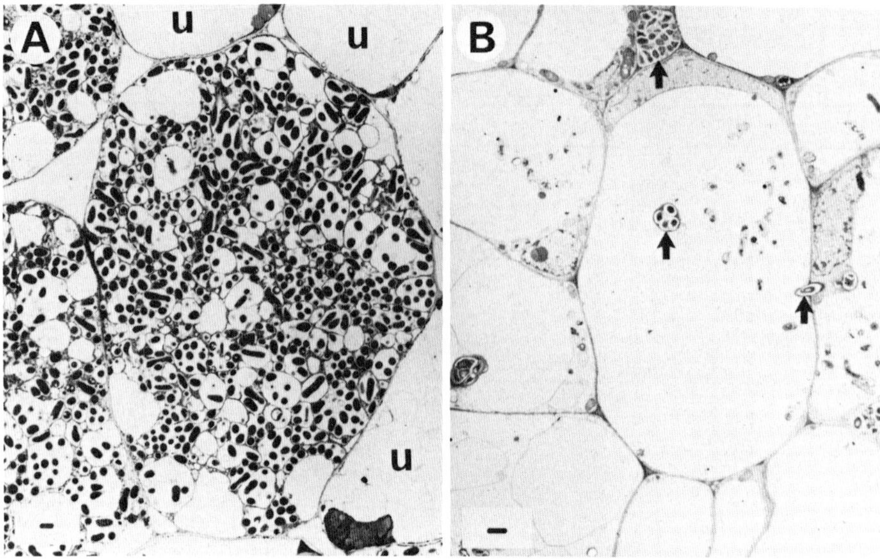

Fig. 5A, B. Electron micrographs of single cells of siratro *(Macroptilium atropurpureum)* nodule after inoculation with either **A** ANU289 (wild type) or **B** ANU289 (pJB5JI) (transconjugant). The effect of the introduced plasmid into the wild type is obvious; there is a small amount of bacterial release from the infection threads *(arrows)*, the bacteria rapidly degenerate after release. *U* uninfected cells; *bars* 1 micron

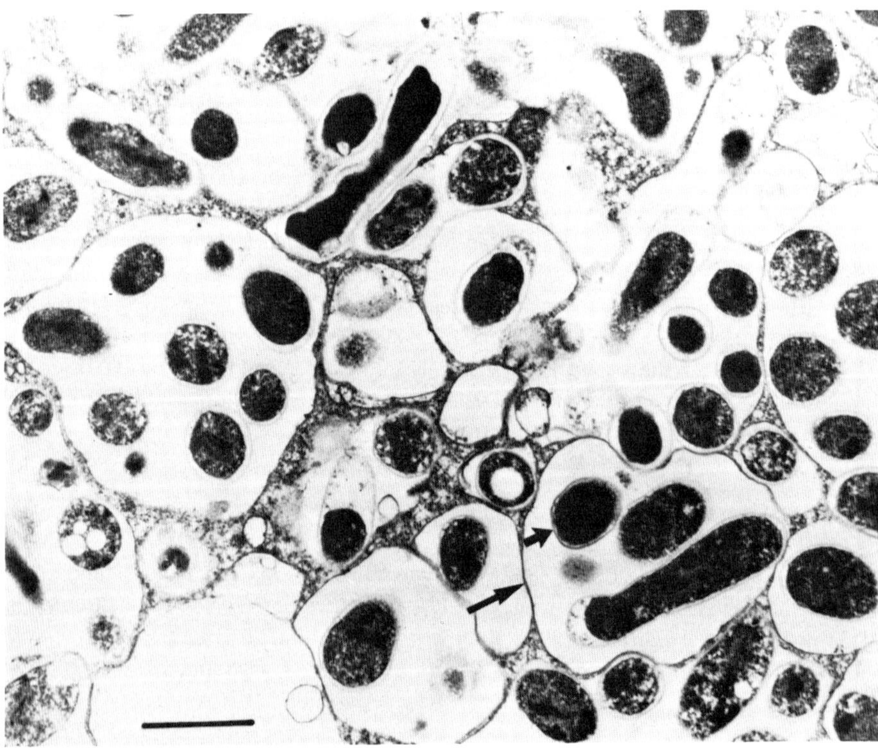

Fig. 6. Enlargement of Fig. 5A showing the "packets" *(large arrow)* of bacteroids *(small arrow)*, numbering up to 15 per packet. *Bar* 1 micron

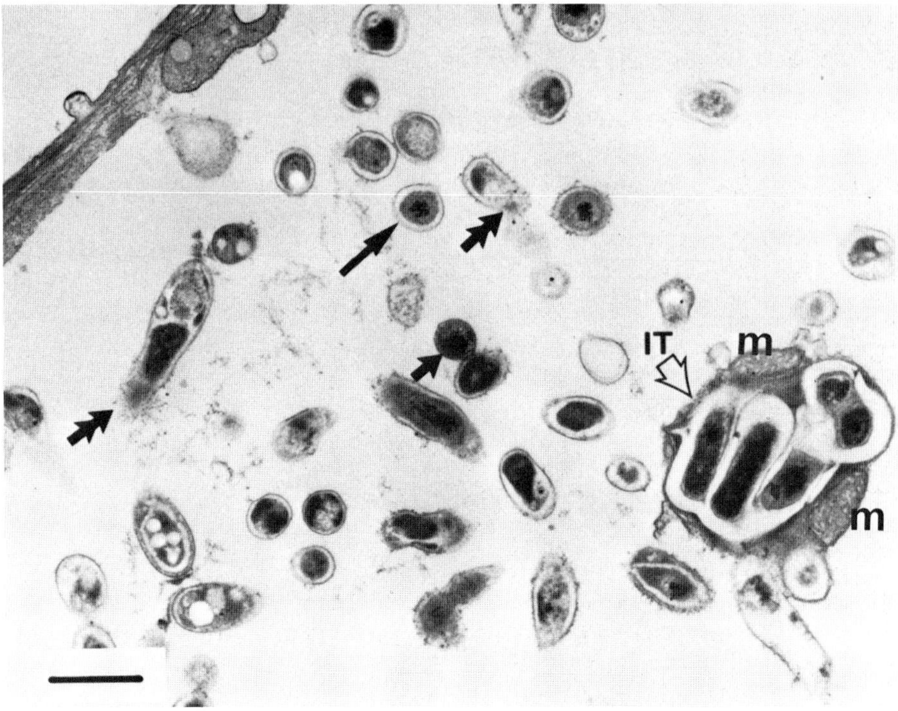

Fig. 7. Enlargement of Fig. 5B showing lack of bacteroid and peri-bacteroid membrane development. The bacteria lyse *(large arrow)* soon after release and degeneration follows *(double arrows)*. Some bacteria have large cell inclusions *(small arrow)*. *IT* infection thread; *M* mitochondria; *bar* 1 micron

takes place (Figs. 8–10). In marked contrast, nodules formed on subterranean clovers by the mutant carrying the reconstructed plasmid pSO32 (ANU845/32) showed more areas of bacterial release and the beginnings of a poorly developed bacteroid zone (Figs. 11, 12). In fact, many of the released bacteria had changed morphology and started swelling in some cases to the size of wild type bacteroids. However, the release of bacteria in ANU845/32-induced nodules occurred not as separately packaged bacteria within single peribacteroid membranes, but as three to six bacteria per membrane envelope. Clearly the in vitro constructed Sym-plasmid can be used to not only determine the minimum amount of *Rhizobium* DNA required for infection and nodule induction, but also to probe different host plant responses to the same set of bacterial genes.

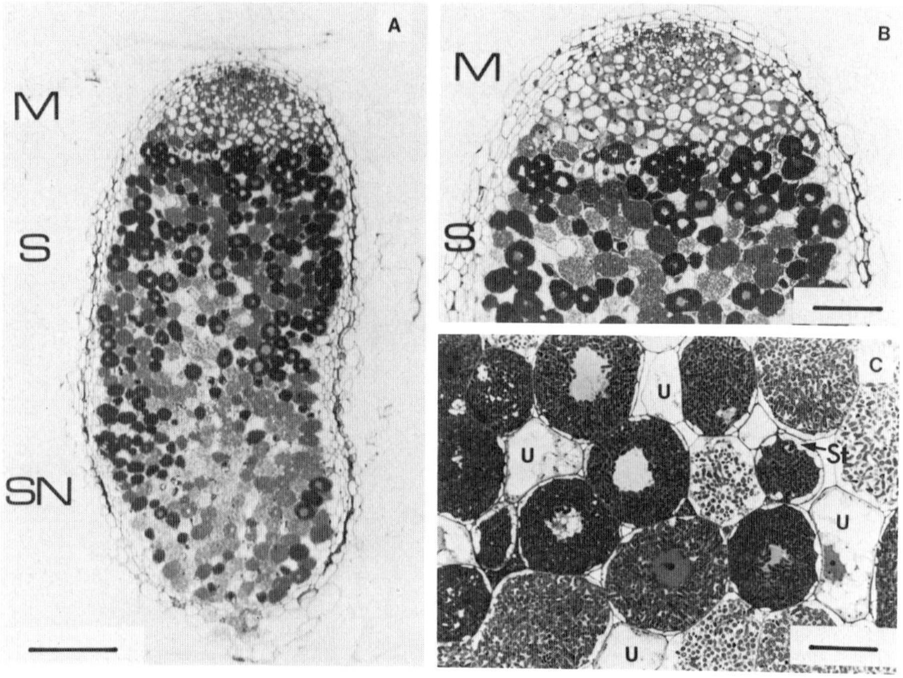

Fig. 8A–C. Light micrographs of a white clover nodule containing ANU843 (wild type) bacteroids, 18 days after inoculation. **A** Longitudinal section of whole nodule; note the prominent meristematic zone *(M)*, the large symbiotic region *(S)* containing mostly bacteroid-filled cells, and the region of senescence *(SN)* where the majority of cells contain degenerating bacteroids. There are few uninfected cells. *Bar* 0.2 mm. **B** Enlargement of the active zone of the nodule; *bar* 0.1 mm. C Bacteroid-filled cells of the symbiotic region; the relatively high osmotic potential of the infected cells crush the uninfected cells *(U)*. *St* starch; *bar* 25 microns. 0.5 μm Spurr's sections stained with toluidine blue

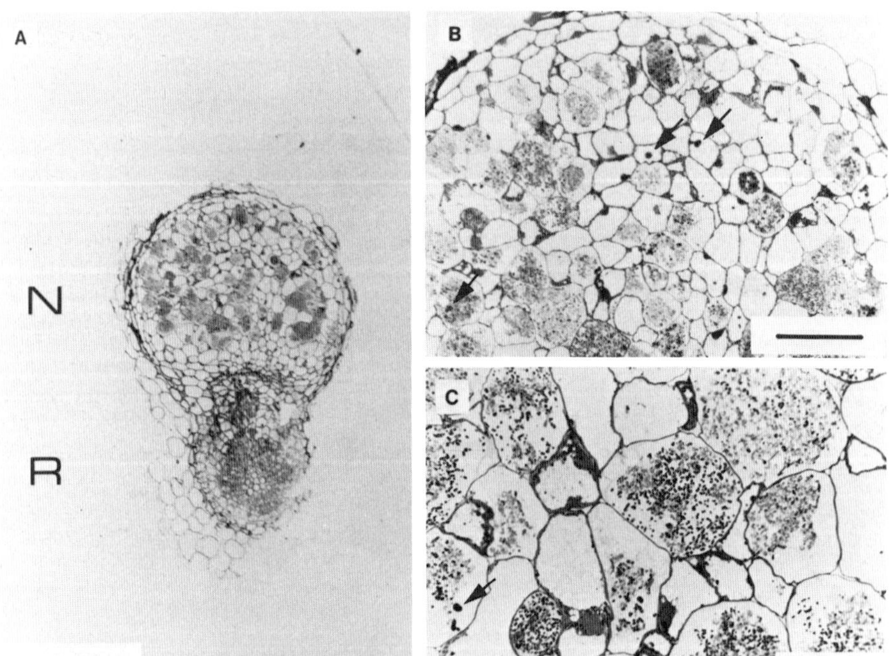

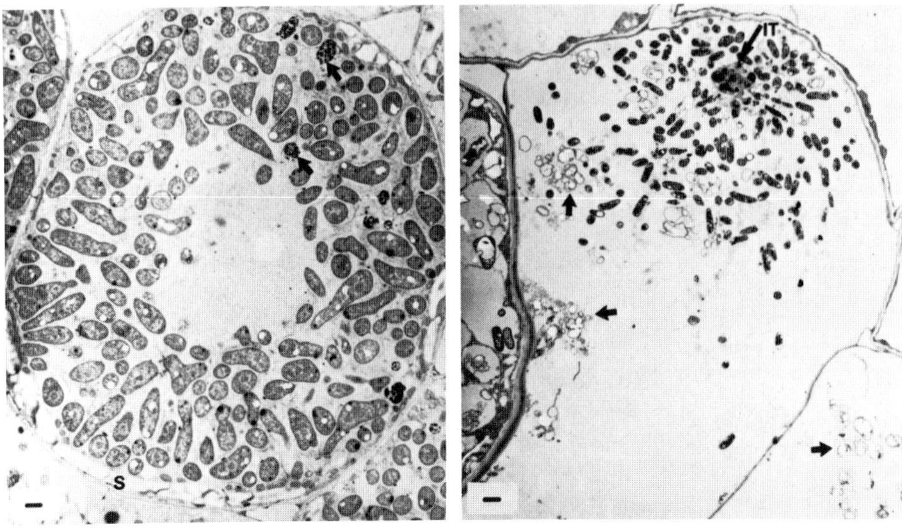

A B

Fig. 10. A Electron micrograph of a white clover nodule containing ANU843 (wild type) showing bacteroids in one cell. Note several degenerating bacteroids (*arrows*). **B** A similar micrograph of a nodule containing ANU32/845 (mutant) showing release of bacteria but no bacteroid formation. Membrane synthesis appears to be misdirected *(arrows)*. *IT* infection thread; *S* starch; *bars* 1 micron

---

◀ Fig. 9 A–C. Light micrographs of a white clover nodule containing ANU845/32 (mutant) bacteria, 18 days after inoculation. **A** Longitudinal section of whole nodule *(N)* and transverse section of root *(R)* (cf. Fig. 8A), no clear meristematic zone or symbiotic region have developed. *Bar* 0.2 mm. **B** and **C** enlarged views of the nodule showing released bacteria but no bacteroid development; infection threads *(arrows)* can be seen. *Bars* **B** 0.2 mm, **C** 25 microns. 0.5 μm Spurr's sections stained with toluidine blue

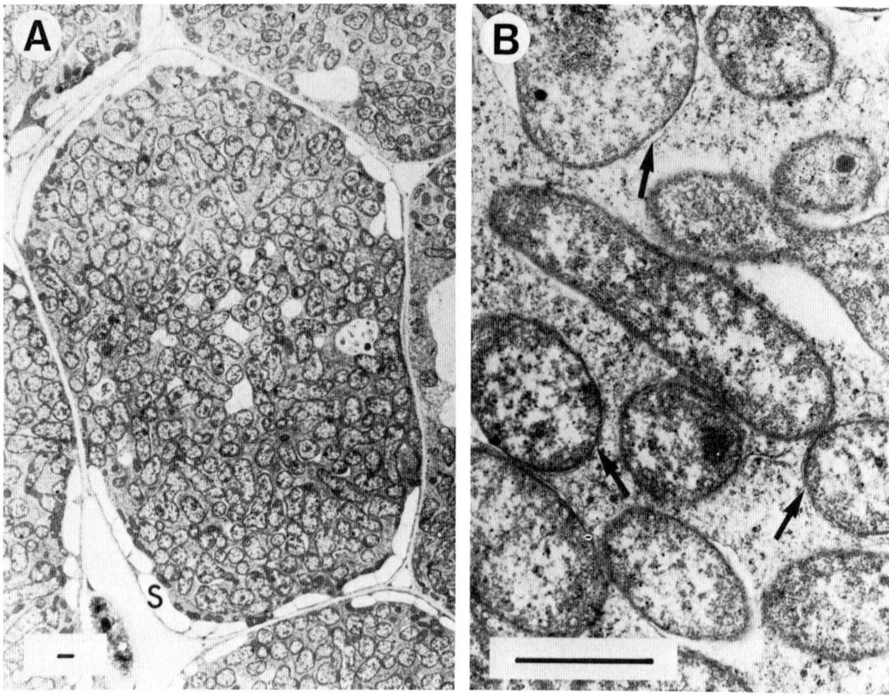

Fig. 11. A Electron micrograph of a subterranean clover nodule cell infected with ANU843 (wild type), 18 days after infection. *S* starch. **B** Enlargement showing bacteroids with individual peribacteroid membranes *(arrows)*. *Bars* 1 micron

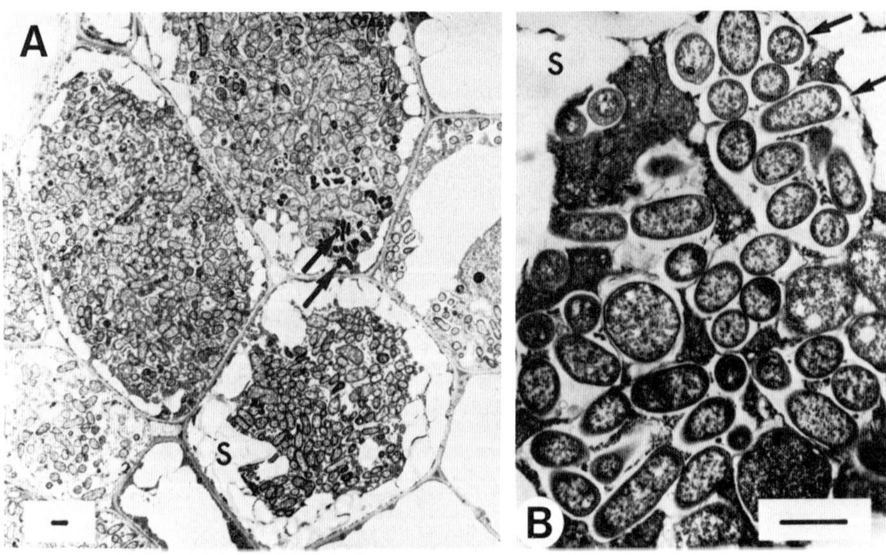

Fig. 12. A Electron micrograph of a subterranean clover nodule cell infected with ANU845/32 (mutant), 18 days after inoculation. There were more areas of bacterial release giving an appearance of a poorly developed bacteroid zone. "Bacteroids" show some degeneration *(arrows)*. *S* starch. **B** Enlargement showing incomplete membrane formation, the "bacteroids" being in "packets" *(arrows)* which is an early stage of development in the wild type. The "bacteroids" are also generally half the size of the wild type (cf. Fig. 11B). *Bars* 1 micron

References

Cen Y, Bender GL, Trinick MJ, Morrison NA, Scott KF, Gresshoff PM, Shine J, Rolfe BG (1982) Transposon mutagenesis in rhizobia which can nodulate both legumes and the non-legume *Parasponia*. Appl Environ Microbiol 43:233–236

Djordjevic MA, Zurkowski W, Rolfe BG (1982) Plasmids and stability of symbiotic properties of *Rhizobium trifolii*. J Bacteriol 151:560–568

Fåhraeus G (1957) The infection of clover root hairs by nodule bacteria, studied by a simple slide technique. J Gen Microbiol 16:374–381

Johnston AWB, Beynon JL, Buchanan-Wollaston AV, Setchell SM, Hirsch PR, Beringer JE (1978) High frequency transfer of nodulating ability between strains and species of *Rhizobium*. Nature 276:635–636

Pemberton JM, Bowen AR, Bowen AR StG (1981) High frequency chromosome transfer in *Rhodopseudomonas sphaeroides* promoted by broad host range plasmid RP1 carrying mercury transposon Tn501. J Bacteriol 147:110–117

Rolfe BG, Gresshoff PM, Shine J (1981) Rapid screening for symbiotic mutants of *Rhizobium* and white clover. Plant Sci Lett 19:277–284

Trinick MJ (1980) Relationships amongst the fast-growing rhizobia of *Lablab purpureus, Leucaena leucocephala, Mimosa* spp., *Acacia farnesiana* and *Sesbania grandioflora*, and their affinities with other rhizobial groups. J Appl Bacteriol 49:39–53

Trinick MJ, Galbraith J (1980) The *Rhizobium* requirements of the non-legume *Parasponia* in relationship to the cross-inoculation group concept of legumes. New Phytol 88:17–26

Vincent JM (1982) Nature and basic properties of the rhizobia. In: Vincent JM (ed) Nitrogen fixation in legumes. Academic Press, London New York, p 5

Zurkowski W (1982) Molecular mechanism for the loss of nodulation properties of *Rhizobium trifolii*. J Bacteriol 150:999–1007

Zurkowski W, Lorkiewicz Z (1978) Effective method for the isolation of non-nodulating mutants of *Rhizobium trifolii*. Genet Res 32:311–314

Molecular Anatomy of the Symbiotic Region in *Rhizobium trifolii* and *Rhizobium parasponia*

J. SHINE, K.F. SCOTT, F.F. FELLOWS, M.A. DJORDJEVIC, P. SCHOFIELD, J.M. WATSON, and B.G. ROLFE [1]

Biological nitrogen fixation occurs in prokaryotic organisms, through induction of the enzyme complex nitrogenase, which converts atmospheric nitrogen to ammonia. Whilst this enzyme complex is common to all characterised nitrogen-fixing organisms, the conditions under which it is expressed vary considerably between species. In bacteria of the genus *Rhizobium,* nitrogenase synthesis is normally repressed under free-living conditions and is induced only within a complex symbiosis formed with leguminous plants. As a first step towards characterising the factors which control nitrogen fixation gene expression within this agriculturally-important symbiosis, we have undertaken a molecular analysis of the genes encoding nodulation *(nod)* and nitrogen fixation *(nif)* in two species of *Rhizobium* (viz. *Rhizobium trifolii* and *Parasponia Rhizobium* sp. ANU289). *R. trifolii* is a classic example of a fast-growing *Rhizobium* with a narrow host range which normally cannot be induced to fix nitrogen in culture. On the other hand, *Parasponia Rhizobium* sp. ANU289 is a slow-growing organism with a very broad host range capable of the effective nodulation of a wide variety of tropical legumes as well as the non-legume *Parasponia* (Trinick 1980). This organism can be induced to fix nitrogen in culture, although the level of this fixation is some 100-fold less than can be obtained from the free-living bacterium *Klebsiella pneumoniae.*

Our approach to the analysis of the genetic control of nodulation and nitrogen fixation in these organisms has been initially to isolate the appropriate genes involved in these functions. "Nodulation" genes have been identified by transposon mutagenesis and plant-screening assays (Rolfe et al. 1980, Scott et al. 1982), while the *Rhizobium nif* genes have been isolated by direct hybridization to homologous *nif* sequences cloned from *K. pneumoniae* (Ruvkun and Ausubel 1980, Scott et al. 1981). Having isolated the appropriate DNA sequences, we have then attempted to both map the various genes by in vitro mutagenesis and to determine their primary nucleotide sequence. To elucidate the function of each of these symbiotic genes, we have then introduced them into other *Rhizobia* and related bacteria to assay the effect that the presence of such genes has on the symbiotic or nitrogen fixation phenotype of different micro-organisms (Rolfe et al., this vol.).

1 Centre for Recombinant DNA Research and Department of Genetics, Research School of Biological Sciences, Australian National University, Canberra, A.C.T. 2601, Australia

Molecular Genetics of the Bacteria-Plant Interaction
(ed. by A. Pühler)

R. trifolii Nitrogen Fixation and Nodulation Genes

The *R. trifolii nif* and *nod* genes have been isolated by molecular cloning and used as probes to examine their organisation of the *Rhizobium trifolii* genome. Such an analysis demonstrates that both sets of genes are carried on the same large, so-called *sym*biotic (Sym) plasmid (Fig. 1). In addition, the *nif* and *nod* genes were found to be very *closely* linked on this large plasmid, since a particular heat-induced deletion of the Sym plasmid (strain ANU871), resulting in the loss of about 40 kilobases (kb) of plasmid DNA, was shown to have lost both *nod* and *nif* gene sequences. To investigate this linkage more closely, we have isolated overlapping clones of the *nif* and *nod* regions and determined a more detailed restriction map of this portion of the Sym plasmid. As shown in Fig. 2, analysis of such clones demonstrated that in the *R. trifolii* strain ANU843, the *nif*HD region was located some 20 kb from the nodulation region.

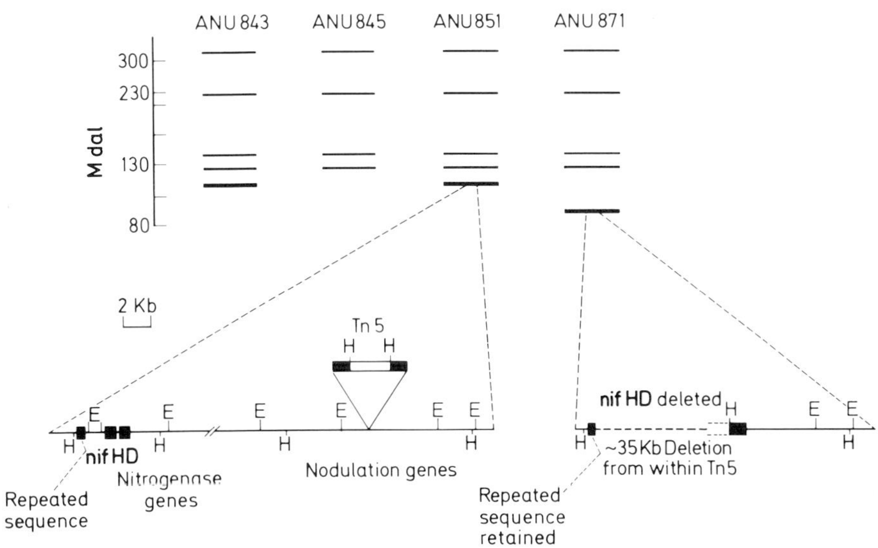

Fig. 1. Plasmid profiles and partial restriction maps of the symbiotic regions from *R. trifolii* strains. *ANU843* parent strain, *ANU845* Sym plasmid cured derivative of *ANU843*, *ANU851* Tn5-induced nodulation-deficient mutant of *ANU843*, *ANU871* heat-induced deletion mutant of *ANU843* deficient in nodulation

We have also examined, in some detail, the *nif* region of a closely related *R. trifolii* strain (ANU329) which was isolated from a clone 'bank' of *R. trifolii* DNA in bacteriophage lambda. Sequences hybridising to the *K. pneumoniae nif*H gene were subcloned into plasmids in order to determine the complete DNA sequence of the *R. trifolii* nitrogenase structural genes. Examination of this sequence demonstrated that the *nif*H and *nif*D genes were contiguous, as in the free-living bacterium *K. pneu-*

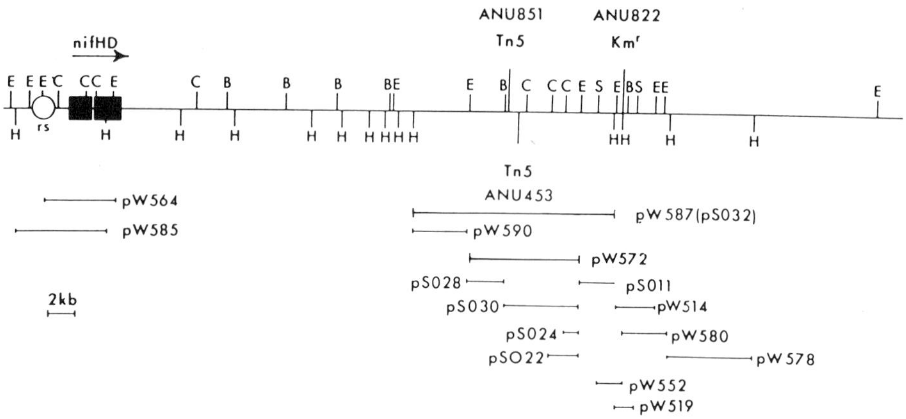

Fig. 2. Restriction map of the nodulation-nitrogen fixation region of the Sym-plasmid in *R. trifolii* ANU843. *C ClaI*; *E EcoRI*; *H HindIII*; *B BamHI*; *S SalI*; *rs* repeated sequences

moniae (Scott et al. 1981) and in the fast-growing legume symbiont, *Rhizobium meliloti* (Torok and Kondorosi 1981, Ruvkun et al. 1982). Furthermore, the primary structure of the Fe-protein encoded by the *R. trifolii nif*H gene is extremely homologous to that of the corresponding protein from several other micro-organisms. This sequence homology is clustered in specific regions of the protein, especially in the N-terminal region and around five cysteine residues which are conserved in all nitrogen-fixing organisms examined to date. The constraints placed on the protein sequence in these regions, presumably implicate them in the maintenance of Fe-protein function.

One other interesting observation was made during the examination of the *nif* region of *R. trifolii*. When sequences lying on the 5′-side of the *nif*H gene were used to probe genomic blots of *R. trifolii* DNA, we consistently observed several other restriction fragments which hybridised to this sequence, demonstrating that a sequence on the 5′-side of the *nif*H gene is repeated several times in the genome. This has proven to be a general phenomenon in all of the *R. trifolii* strains so far investigated regardless of their geographical origin. Furthermore, all copies of this sequence occur on the Sym plasmid in each strain. We are currently determining the size and the sequence of this particular class of repeated DNA.

The nodulation region of *R. trifolii* was initially identified by Tn5 mutagenesis resulting in the isolation of two independent Tn5-induced mutants (ANU851 and ANU453) which gave rise to a nodulation-defective phenotype. Analysis of cloned Tn5-containing fragments derived from these mutants demonstrated that in each strain the transposon insertion was located in the same region of the Sym plasmid. In fact, precise restriction analysis showed that the site of insertion of Tn5 in the two strains mapped to within 1 kb of each other (Fig. 2). To determine the boundaries of the nodulation region, we have subcloned a 14 kb *Hin*dIII fragment (pW587, pSO32, Fig. 2) from this region of the wild-type *R. trifolii* Sym plasmid. Reintroduction of this DNA into a mutant of ANU843 (ANU845), which had been cured of the large

Sym plasmid and hence was unable to nodulate clover, resulted in the restoration of nodulation. It thus appears that the genetic information required for bacterial attachment, root hair curling, initiation and early development of nodules is present on this 14 kb fragment. In addition, the introduction of this fragment into *Agrobacterium tumefaciens* resulted in the ability of this bacterium to nodulate white clover. These results imply that host specificity is at least partially determined by functions encoded on this 14 kb fragment. We have now extended these studies to examine smaller portions of the 14 kb DNA in order to further define the limits of the 'nodulation' genes. Restriction analysis of the 14 kb fragment showed that it contained an internal 7 kb *Eco*R1 fragment (Fig. 2) and when this smaller fragment is reintroduced into ANU845, it does not restore nodulation capacity. However, upon microscopic examination this strain was found to have a dramatic effect on root hairs in that they were invaded by the bacteria (Figs. 3 and 4) although no root hair curling or infection thread formation was apparent. From these observations, it is clear that the 7 kb DNA fragment encodes functions that enable the bacteria to invade and grow in a root hair, since no root hair curling or infection threads are observed, it seems likely that additional functions encoded on the 14 kb fragment are required for the controlled infection of a root hair by *R. trifolii.* We are currently attempting to define more exactly the sequences involved in the different steps leading to nodulation encoded by this region of DNA. Already we know that insertion mutations created to the 3′-side of the 14 kb *Hin*dIII fragment (e.g., ANU822, Fig. 2) do not have any effect on the symbiotic phenotype. This is, of course, consistent with the view that this fragment contains the Sym plasmid-encoded information necessary for nodulation.

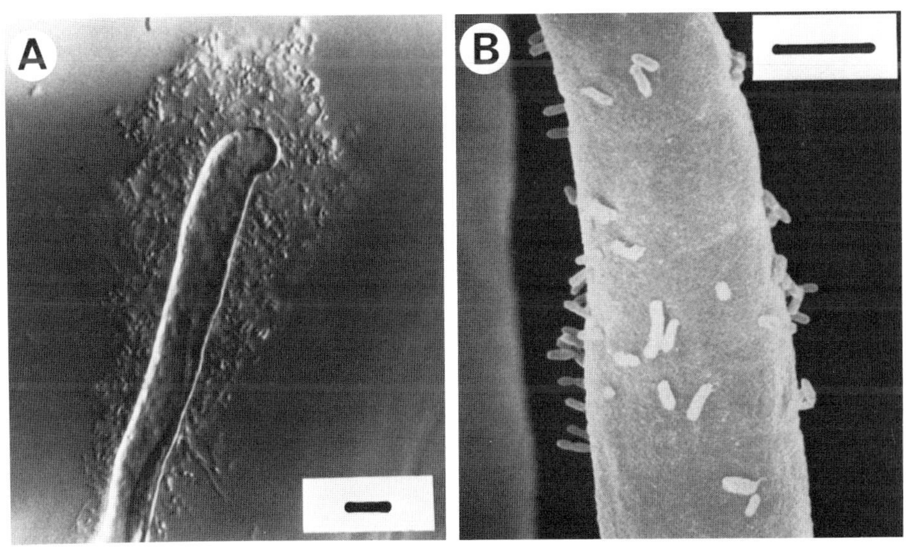

Fig. 3. A ANU843 (wild type) colonising the root hair tip of white clover; curling of the hair has just started. *Bar* 10 microns. Observed through differential interference contast (Nomarksi) optics. **B** Scanning electron micrograph of ANU843 on white clover root hair; note polar attachment of some bacteria. *Bar* 5 microns

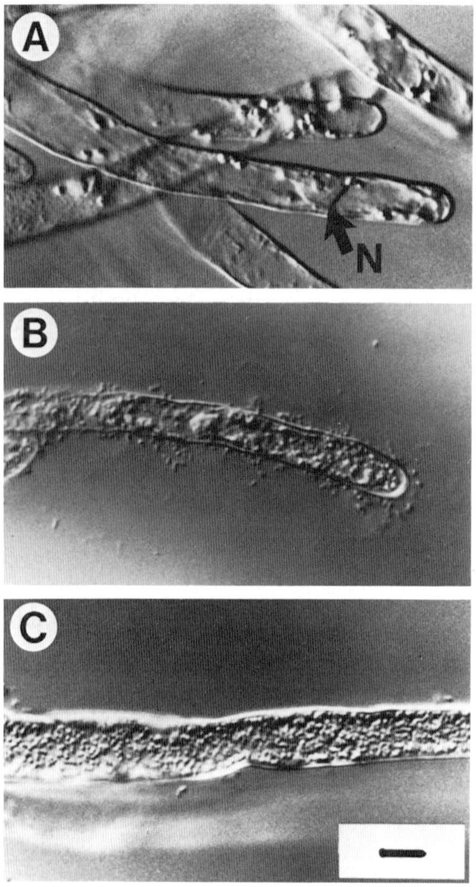

Fig. 4A–C. Strain ANU845/33 (mutant) acts like a pathogen. **A** Swollen bacteria can be observed swimming inside the root hair within 24 h of inoculation; note the proximity of the nucleus *(N)* which is in the position for normal infection thread initiation. **B** 36 h after inoculation bacteria in the root hair are less swollen and more numerous. **C** By 48 h normal-sized bacteria form a colony inside the hair. *Bar* 10 microns; micrographs taken using differential interference contrast (Nomarski) optics

Nitrogenase Structural Genes in *Parasponia Rhizobium* sp. ANU289

The *nif* genes from the slow-growing *Parasponia Rhizobium* sp. ANU289 have also been isolated from a lambda bank of genomic DNA by virtue of their homology to *Klebsiella nif* sequences. One of these lambda clones has been characterised in detail and the complete DNA sequence of *nif*H and its flanking regions determined. Several major observations arose from these studies. (1) As with the *R. trifolii* sequence, the Fe-protein encoded by the *nif*H gene is extremely homologous to products of the other Fe-protein genes. (2) When we examined the sequences to the 5′-side of the *nif*H gene (i.e., the potential regulatory sequences), we found a sequence which strongly resembles a prokaryotic transcription termination signal very close to the beginning of the coding region of the *nif*H gene. Since this sequence appeared too close to the *nif*H gene to allow for a separate promoter between it and the beginning of the *nif*H coding sequence, we immediately suspected, by analogy to similar situations in the *E. coli* biosynthetic operons (Yanofsky 1981), that some sort of attenuation mechanism may be operating. If we examine the sequence more closely, we can see that it does in fact contain all the features necessary to fulfill a role as an attenuator of transcription termination. There is a short peptide-coding sequence containing two contiguous glutamine residues. This is immediately followed by a region of extensive secondary structure which could encode either an antiterminator or a terminator stem. A possible role for glutamine attenuation in the regulation of *nif*H gene expression is consistent with the observation that in *Parasponia Rhizobium* sp. ANU289, addition of glutamine to the nitrogen-fixing culture results in a very rapid inhibition of nitrogenase activity. (3) Examination of the DNA sequence of the *Parasponia Rhizobium nif* region shows that, unlike all the other nitrogen-fixing organisms examined to date, the *nif*D gene is not closely linked to the *nif*H gene. In addition, examination of genome blots of ANU289 DNA, probed with either *nif*H or *nif*D-specific gene sequences from *K. pneumoniae,* shows that in every case different restriction fragments hybridise specifically with the two different probes. The observation that *nif*H and *nif*D genes are not closely linked in this *Rhizobium* strain has been confirmed by the isolation and characterisation of several *nif*D clones. We now know that the *nif*H gene is separated by at least 10 kb from the *nif*D gene and would thus appear to lie on a separate operon. We are currently generating a *nif*H mutant of ANU289 in order to examine the effect of specific mutations on the regulation of the *nif*H gene and to elucidate the mechanism used for the co-ordinate regulation of the *nif*H and *nif*D genes in this organism.

Acknowledgements. The authors wish to acknowledge in a special way the contribution made to this work by Dr. Florence Fellows, who was recently killed in a motor accident. Her work on the isolation and characterisation of the *Parasponia Rhizobium* sp. ANU289 *nif*D gene has been a significant contribution to the field.

References

Rolfe BG, Gresshoff PM, Shine J (1980) Rapid screening for symbiotic mutants of *Rhizobium* and white clover. Plant Sci Lett 19:277–284

Rolfe BG, Djordjevic MA, Morrison NA, Plazinski J, Bender GL, Ridge R, Tellam JT, Gresshoff PM, Shine J (this vol) Genetic analysis of the symbiotic regions in *Rhizobium trifolii* and *Rhizobium parasponia*

Ruvkun GB, Ausubel FM (1980) Interspecies homology of nitrogenase genes. Proc Natl Acad Sci USA 77:191–195

Ruvkun GB, Sunderasan V, Ausubel FM (1982) Directed transposon Tn5 mutagenesis and complementation analysis of *R. meliloti* symbiotic nitrogen fixation genes. Cell 29:551–559

Scott KF, Rolfe BG, Shine J (1981) Biological nitrogen fixation: primary structure of the *Klebsiella pneumoniae nif*H and *nif*D genes. J Mol Appl Genet 1:71–81

Scott KF, Hughes JE, Gresshoff PM, Beringer JE, Rolfe BG, Shine J (1982) Molecular cloning of *Rhizobium trifolii* genes involved in symbiotic nitrogen fixation. J Mol Appl Genet 1: 315–326

Torok I, Kondorosi A (1981) Sequence of the *R. meliloti* nitrogenase reductase (*nif*H) gene. Nucleic Acids Res 9:5711–5723

Trinick MJ (1980) Relationships amongst the fast-growing rhizobia of *Lablab purpureus, Leucaena leucocephala, Mimosa* spp., *Acacia farnesiana* and *Sesbania grandioflora,* and their affinities with other rhizobial groups. J Appl Bacteriol 49:39–53

Yanofsky C (1981) Attenuation in the control of exprcssion of bacterial operons. Nature 289: 751–758

Stem Nodulation in *Aeschynomene:* A Model System for *Bacterium*-Plant Interactions

R.P. LEGOCKI, A.R.J. EAGLESHAM, and A.A. SZALAY [1]

Formation of stem nodules on *Aeschynomene* plants was first communicated in 1928 by Hagerup. Although several other reports followed (von Suessenguth and Beyerle 1936, Arora 1954, Jenik and Kubikova 1969, Barrios and Gonzales 1971), the development and structure of stem nodules have not been investigated in detail. Stem nodules are usually formed under waterlogged conditions, however stem rhizobia are also capable of forming both stem and root nodules in non-flooded soil. Eaglesham and Szalay (1983) have shown that waterlogging is not a prerequisite for the formation of stem nodules.

Due to the presence of chloroplasts in the cortical cells of stem nodules, their morphology is considerably different from that of root nodules. Since the same *Rhizobium* species is able to form effective nodules on both roots and stems of the host plant, this *Rhizobium*-legume association may represent a unique system for the comparative study of differentation and gene expression in root and stem tissues.

We report here the presence of two leghemoglobin species, designated Lbα and Lbβ, whose expression appears to be different in stem and root nodules. We also report the isolation and cloning of *nif* genes from stem *Rhizobium,* as well as the isolation of at least two *nif*-associated promoters of different promoter strengths in *E. coli.* In addition, we report construction of a promoter library of stem *Rhizobium,* from which about 180 individual promoter clones were isolated and partially characterized. Selected promoter-containing clones, including those of *nif,* are as small as 200–400 bp in size.

Results and Discussion

Characterization of Stem *Rhizobium*

A strain of stem *Rhizobium* was isolated from submerged stem nodules of spontaneously nodulated *Aeschynomene indica,* grown under waterlogged conditions. The colony-purified *Rhizobium* culture, referred to as stem *Rhizobium* strain BTAil, grows in yeast-mannitol medium at 28°–30°C with a generation time of about 4 h (M.D. Stowers and A.R.J. Eaglesham, pers. commun.). Two EcoRI fragments of a total DNA digest from stem *Rhizobium* were found to hybridize with the KDH region

1 Boyce Thompson Institute for Plant Research, Cornell University, Ithaca NY 14853, USA

Molecular Genetics of the Bacteria-Plant Interaction
(ed. by A. Pühler)

of *Klebsiella pneumoniae* (Ray G. Hadley, pers. commun.). The limited number of DNA fragments containing *nif* structural genes observed in stem *Rhizobium,* along with its rapid growth rate, resemble the characteristics of fast-growing rhizobia. Broth cultures of stem *Rhizobium* are capable of nodulating the host plant after a single application of rhizobia onto the stem.

Structure of Stem Nodules

Stem nodules of *Aeschynomene* are about 3–7 mm in diameter (Fig. 1) and their cortical cells contain large numbers of chloroplasts. Nodules are detectable about 7 days following inoculation, and 3 weeks later are fully differentiated, reaching maximal levels of nitrogen fixation. Cross-sections of nodules show the presence of a red pigment (leghemoglobin) about 9–11 days after inoculation. Scanning electron micrographs of stem nodules show the presence of bacteroids in infected host cells, which are localized in the central portion of the nodule (Fig. 2). Transmission electron microscopy has also shown the presence of membrane envelope in stem nodules of *Aeschynomene indica* (not shown).

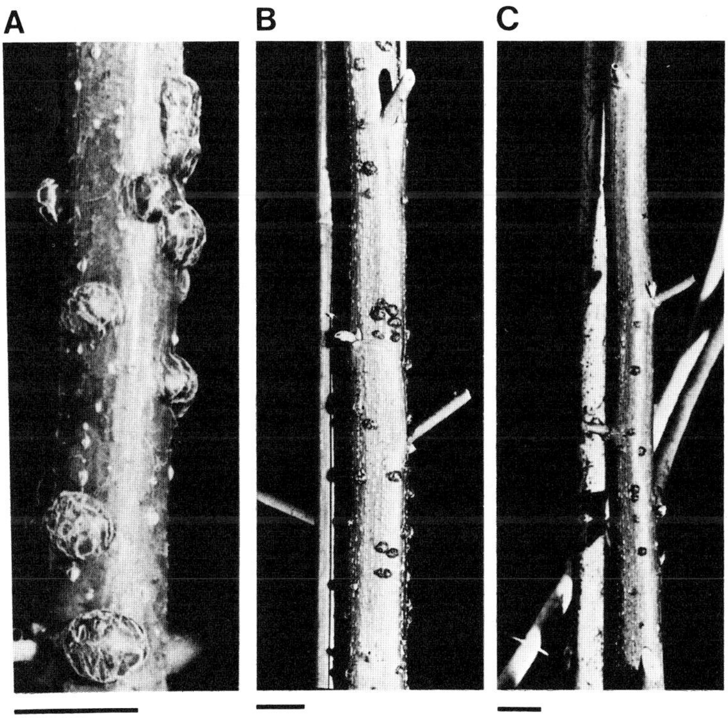

Fig. 1A–C. Stem nodules of *Aeschynomene.* **A** *A. indica* grown without N supplied; **B** *A. scabra* grown without N supplied; **C** *A. scabra* grown with 17 mM N supplied. *Scale bars* indicate 1 cm

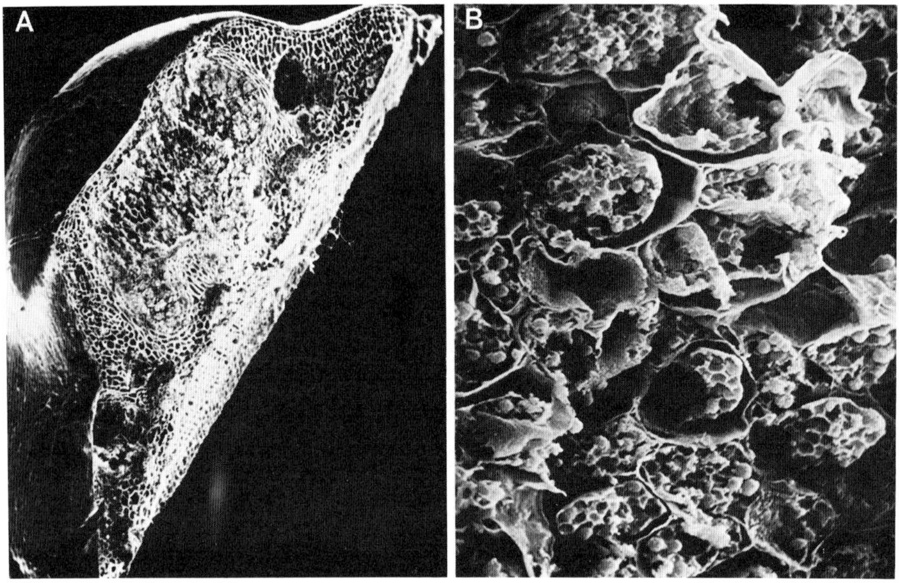

Fig. 2A, B. Scanning electron micrographs of stem nodules of *A. pratensis* showing **A** a cross-section of a nodule, and **B** presence of bacteroids in infected plant cells

Although the stem nodule represents a unique symbiotic organ (e.g., its location on the stem and the presence of chloroplasts), the time lengths required for nodulation to occur and for detection of leghemoglobin are similar to those observed in the soybean root symbiosis.

Nodulation Studies of *Aeschynomene*

To determine the host specificity of stem *Rhizobium,* 16 different species of *Aeschynomene* and 10 other legumes were inoculated with stem rhizobia and examined for the presence of nodules. Results, shown in Table 1, indicated that the ability of stem *Rhizobium* to form nodules is restricted to only certain species of *Aeschynomene.* It is interesting to note that the nodulating species of *Aeschynomene* were able to form both stem and root nodules, and none of the plants examined showed the presence of stem or root nodules only (Table 1). This result suggests the involvement of specific plant genes in the nodulation process, whose expression in *Aeschynomene* is not restricted to the root. Genetic studies of the nodulating and non-nodulating species of *Aeschynomene* may therefore represent an important approach to the identification of plant genes involved in nodulation. It is not clear, however, if the phenomenon of stem nodulation in *Aeschynomene* is determined by specific plant or bacterial genes, or both, in addition to genes involved in root nodulation.

In contrast to root-nodulating legumes, the formation of stem nodules on *A. scabra* was not significantly inhibited at high concentrations of applied nitrogen (17 mM). Nitrogen at 17 mM inhibited stem nodule development rather than the

Table 1. Nodulation specificity of stem *Rhizobium*

Nodulation response tested	Result	Species
Root and stem nodulation	Positive	*A. indica. A. scabra, A. rudis, A. evenia, A. denticulata, A. pratensis, A. sensitivia*
Root and stem nodulation	Negative	*A. brasiliana, A. elegans*
Root nodulation only	Negative	*A. afraspera, A. americana, A. falcata, A. fasicularia, A. histrix, A. paniculata, A. villosa, Trifolium pratense* (unknown cv.), *Medicago sativa* (cv. Saranak), *Pisum sativum* (cv. Sparkle), *Phaseolus vulgaris* (cv. Red Cloud), *Glycine max* (cv. Wilkin and TGm 344), *Vigna unguiculata* (cv. VITA-3 and VITA-5), *Arachis hypogaea* (cv. NC7), *Lupinus angustifolius* (cv. Unicrop), *Cajanus cajan* (cv. CITA-1), *Vigna radiata* (cv. MB33)

infection process per se, i.e., although nodule number was not significantly affected, nodule dry weight was reduced by about 85% (Eaglesham and Szalay 1983).

Nitrogen Fixation in Stem Nodules

Acetylene reduction assays (Hardy et al. 1968) resulted in stem nodule activity of 270 μmol/h/g dry nodule, an efficiency which is comparable to or higher than that found in soybean root nodules. Repeated inoculation of *Aeschynomene* with stem *Rhizobium* resulted in an increased number of nodules (450 per plant), and thus in higher amounts of nitrogen fixed per plant (see also Eaglesham and Szalay 1983).

Identification of Leghemoglobin

Analysis of total plant cytoplasmic proteins from stem and root nodules of *Aeschynomene scabra* by SDS-PAGE (Laemmli 1970) indicated the presence of an abundant polypeptide of about 14,500 molecular weight in both tissues, which was identified as leghemoglobin (Fig. 3, arrowhead). Figure 3 illustrates that the majority of proteins present in stem nodules are also found in root nodules. Two-dimensional PAGE (O'Farrell 1975) also indicated that several polypeptides are unique to either stem or nodule tissue (data not shown).

Scanning of electropherograms similar to that in Fig. 3 indicated that leghemoglobin represents about 11% and 7% (w/w) of total cytoplasm from stem and root nodules, respectively. Since stem nodules of *Aeschynomene* contain chloroplasts, the apparently higher amount of leghemoglobin may be necessary to compensate for an increased concentration of oxygen in this tissue. Electrophoretic analysis under non-denaturing conditions (Davis 1964) of fresh protein extracts from stem and root nodules of *A. scabra,* indicated the presence of only one distinguishable component in both tissues. Its R_f values with respect to electrophoretic mobility of LbS[2] and LbF[3] of soybean leghemoglobins (Ellfolk 1972) are approximately 0.8 and 0.6, respectively.

2 Electrophoretically slow-moving component of leghemoglobin

3 Electrophoretically fast-moving component of leghemoglobin

Isoelectric focusing under non-denaturing conditions (Fuchsman and Appleby 1979) in a pH range 5–8, indicated the presence of 4 molecular species of *Aeschynomene* leghemoglobin, designated α, β_1, β_2 and β_3 (Fig. 4). Each leghemoglobin band was isolated on a preparative scale, treated with a lysis buffer containing 9 M urea (O'Farrell 1975) to dissociate heme, and subjected to isoelectric focusing under denaturing conditions (O'Farrell 1975). The results indicated that the components β_1, β_2 and β_3 represent the same leghemoglobin (pI of about 7.6) whereas the component α remains distinctly different (pI of about 8.4). The observed resolution of the β-leghemoglobin into 3 components under non-denaturing conditions is believed to be due to a differential charge of heme in its aquo-ferric complexes, which has previously been reported in soybean (Fuchsman and Appleby 1979). Thus, we report here the presence in stem and root nodules of *Aeschynomene scabra* of 2 distinct molecular species of leghemoglobin, designated Lbα and Lbβ. Analysis of purified leghemoglobins by a variety of electrophoretic techniques, including two-dimensional PAGE, indicated similar amounts of Lbβ in stem and root nodules, however Lbα was found to be substantially more abundant in the stem nodule tissue (see also Fig. 4). Whether the increased level of Lbα in stem nodules is due to elevated amounts of oxygen in this photosynthetic tissue, and whether or not the *Aeschynomene* leghemoglobins are developmentally regulated (Verma et al. 1979, 1981), yet differently in stem and root nodules, remain subjects for further investigation. The sequencing of leghemoglobins Lbα and Lbβ from stem nodules, which is now in progress, may also contribute to our understanding of the stem nodule symbiosis.

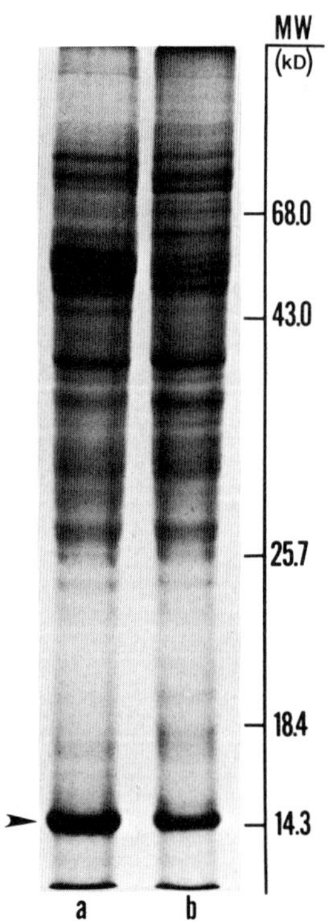

Fig. 3

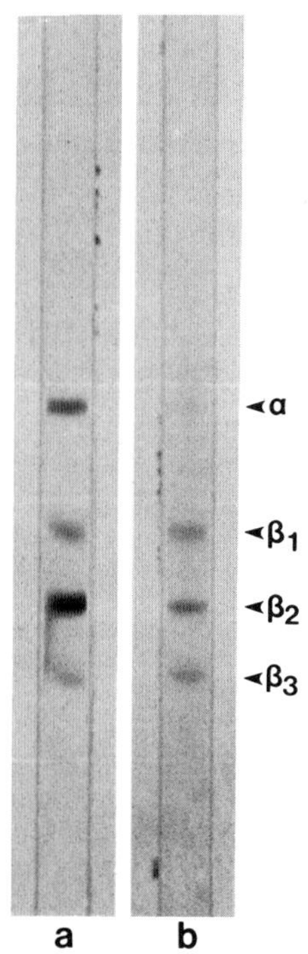

Fig. 4

Fig. 3a, b. SDS-polyacrylamide gel electrophoresis of soluble cytoplasmic proteins from **a** stem and **b** root nodules of *A. scabra* (Coomassie blue stain). *Arrowhead* indicates leghemoglobin

Fig. 4a, b. Isoelectring focusing of leghemoglobins isolated from **a** stem and **b** root nodules of *A. scabra*. Components β_1, β_2 and β_3 were shown to represent one leghemoglobin only (see text for details). Gels were photographed without staining

Cloning of *nif* Genes from Stem *Rhizobium*

Electrophoresis of a total EcoRI digest of the stem *Rhizobium* DNA, followed by hybridization with KDH probe of *Klebsiella pneumoniae,* indicated the presence of *nif* genes on two DNA fragments, about 28 kb and 6 kb in size (Ray G. Hadley, pers. commun.). The 28 kb DNA region was isolated by electroelution, ligated (Dugaiczyk et al. 1975) into EcoRI site of the plasmid pSUP201, and used to transform *E. coli* strain JA221. Of 329 transformants examined, one showed a strong hybridization to the KDH probe. Plasmid analysis of this *nif*-containing clone (pRL4) indicated the presence of a 6.7 kb DNA insert, suggesting that a significant deletion of the 28 kb target DNA had occurred upon cloning.

Construction of an Expression Vector for Monitoring Promoter Activity

The cloning and identification of *nif*-associated promoters from stem *Rhizobium,* described below, was carried out in a β-galactosidase expression vector pRL11H, shown in Fig. 5. A 48 bp synthetic *lac* promoter (Marians et al. 1976, Heyneker et al. 1976) of the expression vector pXJ003 (John Rossi, pers. commun.) was removed and replaced with a 3.4 kb foreign DNA fragment having no promoter activity, to eliminate possible read-through from the β-lactamase (penicillinase) promoter. Plasmid pRL11H was tested in *E. coli* strain MC1061 on agar plates containing X-gal (white colonies), as well as by β-glactosidase enzyme assay (see below), and was shown to have no detectable promoter activity. This plasmid vector contains a unique promoter-cloning site (HindIII), followed by a 180 bp ribosome-binding site of the chloramphenicol acetyltransferase gene, a *lac* Z gene encoding β-galactosidase, a portion of pBR322 specifying resistance to ampicillin, and the 3.4 kb foreign DNA fragment (Fig. 5).

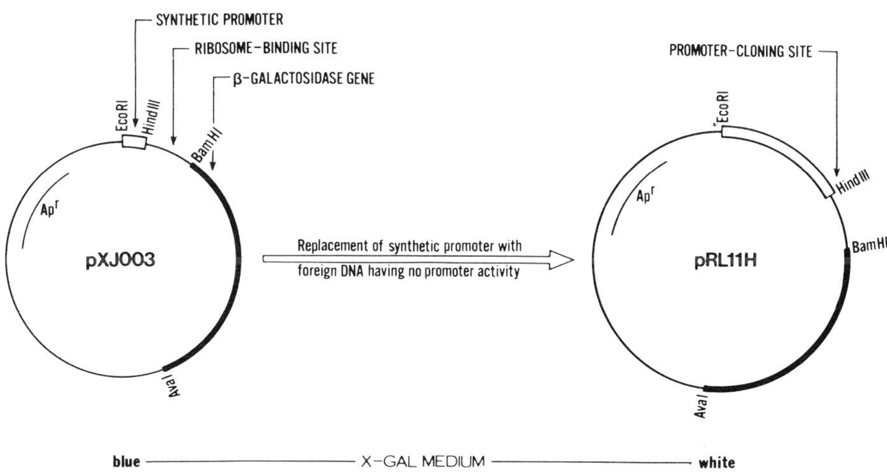

Fig. 5. Construction of a β-galactosidase expression vector (pRL11H) for monitoring promoter activity in *E. coli*

Isolation of *nif*-Associated Promoters from Stem *Rhizobium*

Plasmid pRL4, containing *nif* sequences of stem *Rhizobium* was subjected to restriction analysis using 14 different restriction endonucleases, followed by Southern blotting and hybridization of the resulting DNA fragments with KDH probe. Based on their small size and intense hybridization with KDH (not shown), two restriction fragments were selected for promoter isolation: a 1.5 kb AvaI fragment, and a 1.0 kb MboII fragment (see Fig. 6).

To further reduce the size of the two *nif*-associated DNA fragments, they were resected with exonuclease Bal31 (Gray Jr. et al. 1975) at 10 different time lengths. Resected fragments were treated with AMV reverse transcriptase (Kacian and Myers 1976) to blunt-end the molecules by "filling-in", and following addition of synthetic HindIII linkers, the DNA mixture was ligated into the expression vector pRL11H, linearized with HindIII (see above; see also Fig. 6). The ligation product was used

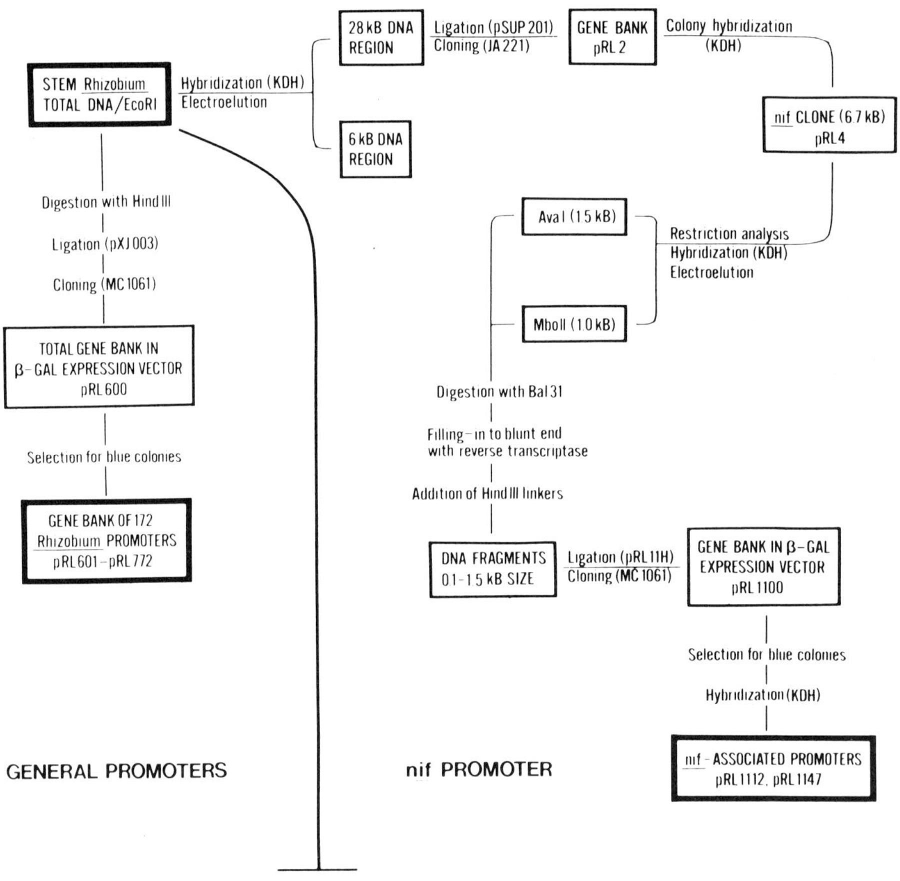

Fig. 6. A scheme illustrating cloning and isolation of two *nif*-associated promoters from stem *Rhizobium* (pRL1112 and pRL1147), as well as the construction of a large number of other stem *Rhizobium* promoters (pRL601–pRL772) active in *E. coli*

to transform a *lac*⁻ strain of *E. coli* (MC1061), and ampicillin-resistant transformants were selected.

Screening for promoter activity was carried out by replica-plating ampicillin-resistant colonies onto agar plates containing M9 glucose minimal medium (Miller 1972) supplemented with 0.2% casamino acids, leucine at 49 μg/ml, X-gal[4] at 40 μg/ml, and ampicillin at 100 μg/ml. Out of about 4000 colonies examined, approximately 70 were blue after 20–24 h incubation, of which 48 were chosen for plasmid isolation. Following further selection for the smallest size of promoter DNA cloned, as judged by restriction analysis, 7 of the original transformants were studied in more detail. Figure 7 shows HindIII digestion profiles of plasmid DNA isolated from the selected clones. Size of the promoter-containing DNA fragments varies from about 250 to 400 bp. Hybridization of the KDH probe to DNA shown in Fig. 7 indicated the presence of *nif* genes on two promoter-containing fragments (lanes f and g), designated pRL1112 and pRL1147, respectively (data not shown). The strength of the cloned promoters, including those which do not hybridize with KDH, was determined by the β-galactosidase assay (Miller 1972) in *E. coli.* Table 2, in conjunction with Fig. 7, shows that there is no apparent relationship between the fragment size and its promoter activity.

Sequence analysis of the *nif*-associated promoters pRL1112 and pRL1147 from stem *Rhizobium*, as well as their activity in *E. coli* in the presence and absence of *nif* A gene product, will be published elsewhere (Legocki et al. 1983).

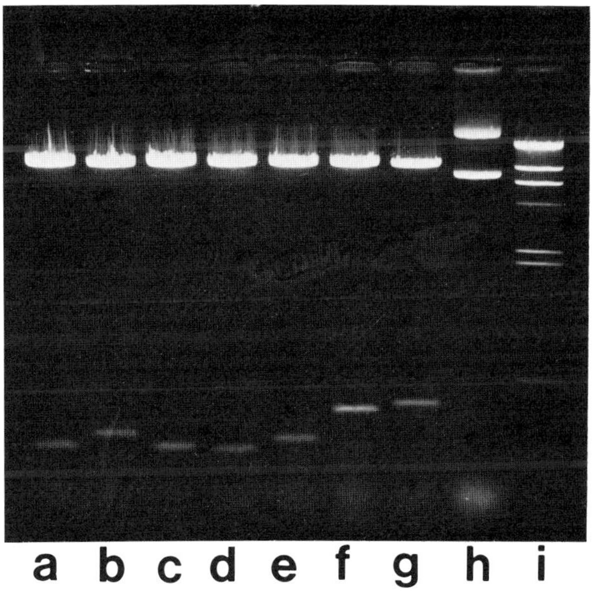

Fig. 7. Agarose gel electrophoresis of selected promoter clones from stem *Rhizobium* (HindIII digest). Plasmids: *a* pRL1107, *b* PRL1109, *c* pRL1111, *d* pRL1120, *e* pRL1129, *f* pRL1112, *g* pRL1147. Lanes *h* and *i* contain size markers

4 5-bromo-4-chloro-3-indolyl-β-D-galactopyranoside (Sigma), dissolved in dimethyl formamide

Table 2. Activity of stem *Rhizobium* promoters (β-galactosidase assay)

Plasmid	Elapsed time of reaction (min)	Absorbance at 600 nm	Absorbance at 550 nm	Absorbance at 420 nm	Units per 0.5 ml culture	Hybridization to KDH probe
pRL11H	51.0	0.645	0.183	0.317	– 0.2	–
pRL1100	71.5	0.537	0.176	0.556	12.0	ND
pRL1107	73.3	0.777	0.223	0.906	18.1	–
pRL1109	75.5	0.742	0.237	1.449	36.9	–
pRL1111	78.3	0.699	0.203	0.995	23.4	–
pRL1112	20.3	0.632	0.180	1.656	208.7	+
pRL1120	5.0	0.696	0.197	1.694	775.4	–
pRL1129	6.3	0.570	0.154	1.381	615.8	–
pRL1147	49.5	0.708	0.201	1.375	58.4	+

Note Added in Proof. The pRL1112 promoter has been sequenced and identified as that of *nif*H. The transcription initiation site of the *nif*H promoter-*lac*Z gene fusion is different in *E. coli* than in stem *Rhizobium* bacteroids, as judged by S1-mapping. Following insertion of this transcriptional fusion into the stem *Rhizobium* chromosome, nodules induced by the transformed rhizobia showed substantially higher β-galactosidase activity than wild-type nodules.

Acknowledgements. Plasmid pSUP201 was kindly provided by Drs. R. Simon and A. Pühler; plasmid pXJ003 was a generous gift from Dr. John Rossi. We wish to thank Dr. Adrienne Koermendy for providing us with plant material used in this study. Roman Legocki wishes to thank Dr. Ray G. Hadley for his valuable help and advice regarding techniques used. We thank Dr. William Langridge for reading the manuscript.

References

Arora N (1954) Morphological development of the root and stem nodules of *Aeschynomene indica* L. Phytomorphology 4:211–216

Barrios S, Gonzalez V (1971) Rhizobial symbiosis on Venezuelan savannas. Plant and Soil 34: 707–719

Davis BJ (1964) Disc electrophoresis, method and application to human serum proteins. Ann NY Acad Sci 121:404–427

Dugaiczyk A, Boyer HW, Goodman HM (1975) Ligation of EcoRI endonuclease-generated DNA fragments into linear and circular structures. J Mol Biol 96:171–184

Eaglesham ARJ, Szalay AA (1983) Aerial stem nodules on *Aechynomene* spp. Plant Sci Lett 29:265–272

Ellfolk N (1972) Leghaemoglobin, a plant haemoglobin. Endeavour 31:139–142

Fuchsman WH, Appleby CA (1979) Separation and determination of the relative concentrations of the homogeneous components of soybean leghemoglobin by isoelectric focusing. Biochem Biophys Acta 579:314–324

Gray HB Jr, Ostrander DA, Hodnett JL, Legerski RJ, Robberson DL (1975) Extracellular nucleases of *Pseudomonas* Bal 31. Nucleic Acids Res 2:1459–1492

Hagerup O (1928) En hygrofil baelgplante *(Aechynomene aspera* L.) med bakterieknolde paa stangelen. Dan Bot Ark 15:1–9

Hardy RWF, Holsten RD, Jackson EK, Burns RC (1968) The acetylene-ethylene assay for N_2-fixation: laboratory and field evaluation. Plant Physiol 43:1185–1207

Heyneker HL, Shine J, Goodman HM, Boyer HW, Rosenberg J, Dickerson RE, Narang SA, Itakura K, Lin S, Riggs AD (1976) Synthetic *lac* operator DNA is functional in vivo. Nature 263:748–752

Jenik J, Kubikova J (1969) Root system of tropical trees. 3. The heterorhizis of *Aseschynomene elaphroxylon* (Guill. et. Perr.) Taub. Preslia 41:220–226

Kacian DL, Myers JC (1976) Synthesis of extensive, possibly complete, DNA copies of poliovirus RNA in high yields and at high specific activities. Proc Natl Acad Sci USA 73:2191–2195

Laemmli UK (1970) Cleavage of structural proteins during the assembly of the head of bacteriophasity. Nat New Biol 227:680–685

Legocki RP, Yun A, Szalay AA (1983) (in preparation)

Marians KJ, Wu R, Stawinski J, Hozumi T, Narang SA (1976) Cloned synthetic *lac* operator DNA is biologically active. Nature 263:744–748

Miller JH (1972) Experiments in molecular genetics. Cold Spring Harbor Laboratory, New York

O'Farrell PH (1975) High resolution two-dimensional electrophoresis of proteins. J Biol Chem 25 250:4007–4021

Suessenguth K von, Beyerle R (1936) Über Bakterienknöllchen am Sproß von *Aeschynomene paniculata* Willd. Hedwigia 75:234–237

Verma DPS, Ball S, Guirin C, Wanamaker L (1979) Leghemoglobin biosynthesis in soybean root nodules. Characterization of the nascent and released peptides and the relative rate of synthesis of the major leghemoglobins. Biochemistry 18:476–483

Verma DPS, Haugland R, Brisson N, Legocki RP, Lacroix L (1981) Regulation of the expression of leghaemoglobin genes in effective and ineffective root nodules of soybean. Biochem Biophys Acta 653:98–107

C. The *Agrobacterium*-Plant Interaction

Onc- and *vir*-Genes Located on the Ti-Plasmid of *Agrobacterium tumefaciens*

J. HILLE, J. VAN KAN, and R.A. SCHILPEROORT [1]

Introduction

The capacity of *Agrobacterium tumefaciens* to induce tumors is correlated with the presence of a large plasmid that is called the tumor-inducing (Ti) plasmid (Van Larebeke et al. 1974, Watson et al. 1975, Zaenen et al. 1974). In the absence of the Ti-plasmid the bacterium is no longer able to induce tumors, whereas this property is acquired again upon re-introduction of this plasmid. A defined fragment of this plasmid, the T-region, is found to be stably integrated in the nuclear DNA of the tumor cell (Chilton et al. 1977, 1980). As part of the plant genome, this DNA is called T-DNA. It is transcribed into 7 poly A containing mRNA's (Willmitzer et al. 1982), of which at least some are capped and translated into protein (Schröder and Schröder 1982). At a distance from the T-region, a segment has been found on the Ti-plasmid, which is also essential for the bacterium to be oncogenic (Garfinkel et al. 1980, Ooms et al. 1980). This region has never been detected in the plant tumor cell, which makes it attractive to speculate that this region is expressed inside the bacterium. That this is likely indeed came from genetic complementation studies (Hille et al. 1982, Klee et al. 1982). Therefore, it is proposed to call it *vir*-region, in order to distinguish it from the T-region carrying genes for oncogenicity that have to be expressed in the plant tumor cell.

Ti-plasmids have a host range that is limited to the *Rhizobiaceae* (Hooykaas 1979). This can be enlarged by the construction of cointegrate plasmids composed of a Ti-plasmid and a wide host range plasmid. Upon transfer of such cointegrates to *E. coli,* they are usually unstable (De Picker et al. 1980), and dissociate again into the Ti-plasmid and the wide host range plasmid. Since the Ti-plasmid is not able to replicate in *E. coli,* this plasmid is subsequently lost. We have isolated a cointegrate plasmid, consisting of the wide host range *Inc* P-1 plasmid R772 and an octopine Ti-plasmid (pTiB6), which is perfectly stable in both *A. tumefaciens* and *E. coli.* This cointegrate has been used to construct in vivo R-prime plasmids, which have been used in the complementation experiments with Ti-plasmid *vir*-region mutants. The cointegrate has also been used in *E. coli* for site-directed mutagenesis of the T-region. Results will be discussed in relation to the use of Ti-plasmids as a vector for the introduction of foreign genes in plants.

1 Department of Biochemistry, State University of Leiden, Wassenaarseweg 64, 2333 AL Leiden, The Netherlands

Molecular Genetics of the Bacteria-Plant Interaction
(ed. by A. Pühler)

Stability of an R772::Ti Cointegrate Plasmid

Several R::Ti cointegrates have been isolated by our group. One of them, an R772::TiB6 cointegrate plasmid (named pAL969; Hooykaas et al. 1980), is completely stable in both *A. tumefaciens* and *E. coli,* and when it is present in *A. tumefaciens* normal tumor induction is obtained on several plant species. To understand the remarkable stability of this cointegrate plasmid we constructed a physical map of it. At first, a map has been made with 10 different restriction endonucleases for the wide host range *Inc* P-1 plasmid R772 (molecular weight of 40,500,000) using double digestions, Southern blot hybridization and molecular cloning. By this, a transposable element, called IS70, has been identified in R772, and after its isolation is characterized in detail. The molecular weight of IS70 is 2,500,000 and homoduplex studies showed that at its ends it carries an inverted repeat of approximately 50 base pairs. Subsequently a physical map of the cointegrate plasmid pAL969 has been made, using the maps of R772 and pTiB6. This has demonstrated that on R772 cointegration had occurred at the position of IS70, suggesting that transposition of IS70 was responsible for cointegrate formation between R772 and pTiB6, as was assumed earlier (Hooykaas et al. 1980). To study this in more detail, IS70 has been used as a probe in a Southern blot hybridization with plasmid pAL969. This experiment showed that this element is present twice, i.e., on both fragments that carry the sites where R772 and the Ti-plasmid are fused, indicating that cointegration was mediated indeed by IS70. Moreover, it was found that only one of the IS70 copies is of full size, whereas the other one is seriously deleted; at most only a molecular weight of 500,000 of IS70 is detected (see Fig. 1). This observation is taken as a likely explanation for the stability of the cointegrate plasmid, assuming that the homology between the complete and the deleted IS70 copies is not enough for efficient homologous recombination to occur.

Complementation of *vir*-Genes

Ti-plasmid mutations, in a 20,000,000 molecular weight region left from the T-region, have been described to cause avirulence of the bacterium, although the T-region is not affected. If it is assumed that this region for virulence needs to be expressed in the bacterium, the mutations in that segment should be complementable in trans. To test this assumption one has to clone fragments of that region on a plasmid that both replicates in *Agrobacterium* and is compatible with Ti-plasmids. Moreover, it is required to clone large fragments because nothing is known about operon structure or regulation in that region. Recently, we described a procedure by which it is possible to clone in vivo large segments of plasmid DNA (Hille and Schilperoort 1981). By enclosing the R772 replicator, in the R772::pTiB6 cointegrate plasmid (pAL969), between two copies of the transposon Tn1831 in the same orientation, several R-prime plasmids are obtained. These carry different parts of the octopine Ti-plasmid, and are compatible with Ti-plasmids. The largest R-prime carries almost the entire Ti-plasmid, except a 7,000,000 molecular weight region, in which the Ti-plasmid

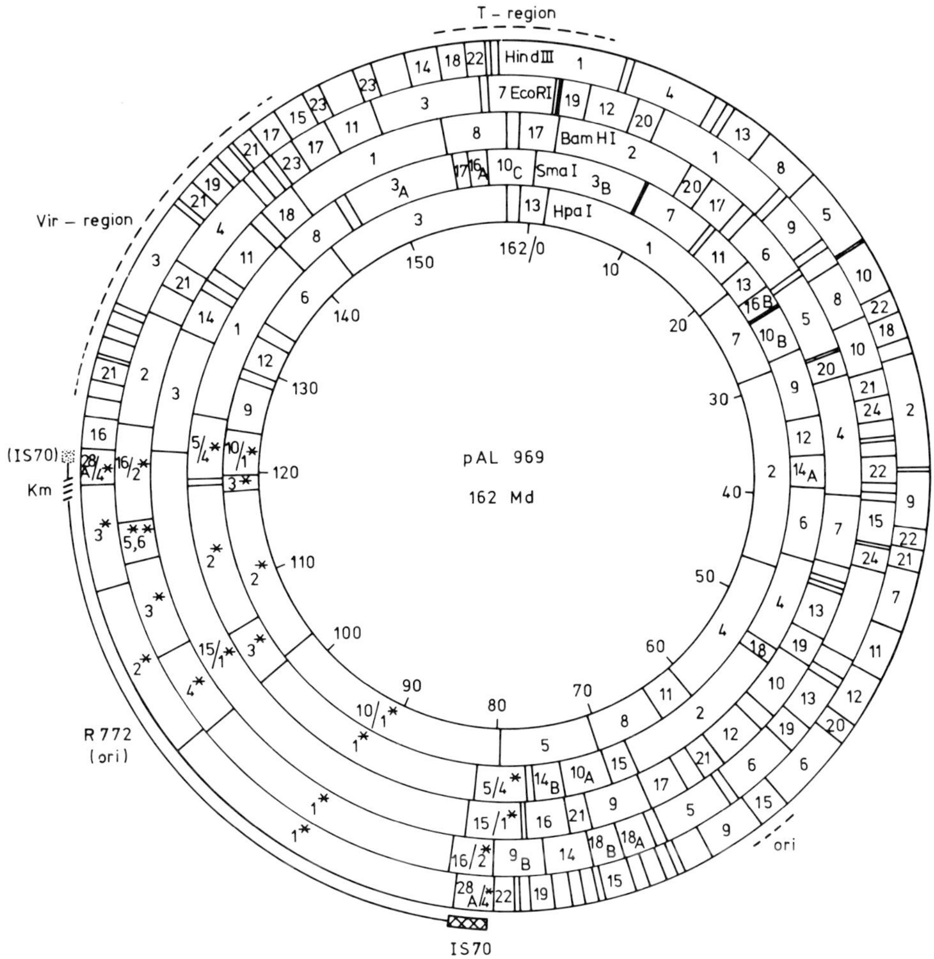

Fig. 1. Map of the R772::Ti cointegrate plasmid pAL969. On the map, the R772 fragments are indicated with a *star*. The R772::Ti fusion fragments have been given two numbers, the R772 and the Ti-number. The map of the Ti-part is shown as described by De Vos et al. (1981)

replicator is located. *Agrobacteria,* harbouring this R-prime, induce tumors normally, indicating that the Ti-plasmid replicator region is not essential for tumor induction (Hille et al. 1982). By the introduction of smaller R-primes, *Agrobacteria* were made diploid for genes located at the left half of the Ti-plasmid. All described Tn904 insertions (Ooms et al. 1980) and Tn5 insertions (Garfinkel et al. 1981) in the virulence region on the left half of the octopine Ti-plasmid, that cause avirulence or attenuated virulence, were shown to be complementable in trans (Hille et al. 1982, Klee et al. 1982). Therefore, it is proposed to make a distinction between *onc*- and *vir*-genes. *Onc* (o)-genes are located on T-DNA and need to be expressed in the plant tumor cell. *Vir* (ulence)-genes, however, can be complemented in trans, indicating that these are expressed inside the bacterium.

Site-Directed Mutagenesis of *onc*-Genes

The cointegrate plasmid pAL969 is stably present in *E. coli.* Therefore, site-directed mutagenesis of this Ti-plasmid can directly be performed in *E. coli* (Hille et al. 1983). T-region fragment *Eco*RI-7 was cloned into the vector pACYC184. The *Pst*I fragment located in the T-region was replaced by a *Pst*I fragment carrying Cm^r. The cointegrate plasmid and the mutated T-DNA clone were put together in one bacterial cell of *E. coli,* which was subsequently used as a donor in a cross to an *E. coli* recipient having no plasmids. Selection was made for transfer of the Cm^r. This marker can only be transferred by mobilization of the whole mutated T-DNA clone through pAL969, or by double crossing-over, between the homologous T-region parts of the T-DNA clone and pAL969, resulting in site-directed mutagenesis of plasmid pAL969, i.e., the introduction of Cm^r in its T-region. Mobilization occurred at a frequency of 2 out of 3, and in 1 out of 3 it was due to site-directed mutagenesis. One of these site-directedly mutated pAL969 plasmids was transferred to *Agrobacterium,* and the transconjugant tested for tumor induction. On tobacco and kalanchoë mainly roots developed, which is in agreement with the position of the mutation on the genetic map of the T-region (Ooms et al. 1981, Garfinkel et al. 1981), on which the *Pst*I substitution takes off part of the tumor morphology rooting locus. On the transcription map (Willmitzer et al. 1982) transcript 4 is mutated and possibly also 6. This Ti-plasmid was further mutated by introducing a Tn1831 insertion in transcript 2 (located in the tumor morphology shooting locus), resulting in a plasmid which now carries T-region mutations at the positions of transcripts 2, 4 and possibly 6. Also a plasmid was constructed which was mutated in transcripts 1, 4 and possibly 6. When *Agrobacterium* strains, carrying these plasmids, were tested, they turned out to be no longer able to induce tumors on tobacco, tomato and kalanchoë. However, the important question has then to be put forward: such mutant strains cannot induce tumors, but do they still incorporate T-DNA into the plant cell? In principle one would test this by looking for expression of the T-DNA encoded enzyme octopine synthase, since in these mutated plasmids this locus is not affected. However, there is no tumor formation after infection with these bacteria, and therefore it is rather difficult to test the expression of this locus. To overcome this problem a 1:1 ratio mixed infection was carried out using one of the mutants and a strain which still had a normal tumor inducing capacity but which carried a deletion eliminating the octopine synthase locus. The mixed infection on tomato resulted in tumors with octopine synthase activity. This activity can only be brought about by the mutant and therefore strongly suggests that although several *onc*-genes are mutated still T-DNA transfer and integration occurs in the plant. The mixed infection was performed with bacteria carrying a Rec mutation, excluding the possibility that after bacterial conjugation homologous recombination could have resulted in bacteria carrying intact *onc*-genes and an intact octopine synthase locus, i.e., having a wild-type T-region. It can however still be envisaged that illegitimate recombination, due to transposable elements, might have occurred. If so, it will have taken place at a very low frequency and therefore certainly not have contributed significantly to the induction of tumor cells in which octopine synthase activity is expressed. Since *Agrobacteria* compete for a limited number of attachment sites on plant cell walls (Lippincott and Lippincott

1969, Schilperoort 1969), that have to be occupied in order to induce tumor cells, it is even not likely that a few *Agrobacteria,* carrying a wild-type T-region, in the presence of millions of mutant *Agrobacteria* can contribute to the induction of any tumor cells.

In view of using the intact Ti-plasmid as a vector for the introduction of new genes into plant cells, the application of a stable R::Ti cointegrate plasmid can be advantageous. The cointegrate plasmid can be transferred reciprocally between *A. tumefaciens* and *E. coli.* The advantages of using *E. coli* as a host seem obvious. *E. coli* has a much faster growth rate, a lot of mutants are available and special cloning vehicles like cosmids and insertional activation vectors have been developed. We have demonstrated site-directed mutagenesis of an R772::Ti cointegrate by conjugative transfer of this plasmid and testing for replacement of a selectable marker. The procedure probably can even further be simplified by the use of *pol* A ts strains. In such strains, col E1 derived plasmids can replicate at 32°C, but not at 42°C, although other plasmids, like the described cointegrate are stably maintained at both temperatures. Therefore, simply selecting for e.g. an antibiotic resistance marker at 42°C would directly give strains, that only carry the site-directed mutation in the cointegrate plasmid.

References

Chilton MD, Drummond MH, Merlo DJ, Sciaky D, Montoya AL, Gordon MP, Nester EW (1977) Stable incorporation of plasmid DNA into higher plant cells: The molecular basis of crown gall tumorigenesis. Cell 11:263–271

Chilton MD, Saiki RK, Yadav N, Gordon MP, Quetier F (1980) T-DNA from *Agrobacterium* Ti-plasmid is in the nuclear DNA fraction of crown gall tumor cells. Proc Natl Acad Sci USA 77:4060–4064

Garfinkel DJ, Nester EW (1980) *Agrobacterium tumefaciens* mutants affected in crown gall tumorigenesis and octopine catabolism. J Bacteriol 144:732–743

Garfinkel DJ, Simpson RB, Ream LW, White FF, Gordon MP, Nester EW (1981) Genetic analysis of crown gall: Fine structure map of the T-DNA by site-directed mutagenesis. Cell 27:143–153

Hille J, Schilperoort RA (1981) The use of transposons to introduce well-defined deletions in plasmids: Possibilities for in vivo cloning. Plasmid 6:151–154

Hille J, Klasen I, Schilperoort RA (1982) Construction and application of R-prime plasmids, carrying different segments of an octopine Ti-plasmid from *Agrobacterium tumefaciens,* for complementation of *vir*-genes. Plasmid 7:107–118

Hille J, Kan J van, Schilperoort R (1983) Site-directed mutagenesis in *Escherichia coli* of a stable R772::Ti cointegrate plasmid from *Agrobacterium tumefaciens.* J Bacteriol 154:693–701

Hooykaas PJJ (1979) The role of plasmid determined functions in the interactions of *Rhizobiaceae* with plant cells. A genetic approach. Thesis, Univ Leiden, Netherlands

Hooykaas PJJ, Dulk-Ras H Den, Schilperoort RA (1980) Molecular mechanism of Ti-plasmid mobilization by R-plasmids: isolation of Ti-plasmids with transposon-insertions in *Agrobacterium tumefaciens.* Plasmid 4:64–75

Klee HJ, Gordon MP, Nester EW (1982) Complementation analysis of *Agrobacterium tumefaciens* Ti-plasmid mutations affecting oncogenicity. J Bacteriol 150:327–331

Larebeke N Van, Engler G, Holsters M, Elsacker S Van Der, Zaenen I, Schilperoort RA, Schell J (1974) Large plasmids in *Agrobacterium tumefaciens* essential for crown gall inducing ability. Nature 252:169–170

Lippincott BB, Lippincott JA (1969) Bacterial attachment to a specific wound site as an essential stage in tumor initiation by *Agrobacterium tumefaciens.* J Bacteriol 97:620–628

Ooms G, Klapwijk PM, Poulis JA, Schilperoort RA (1980) Characterization of Tn904 insertions in octopine-Ti-plasmid mutants of *Agrobacterium tumefaciens.* J Bacteriol 144:82–91

Ooms G, Hooykaas PJJ, Moolenaar G, Schilperoort RA (1981) Crown gall plant tumors of abnormal morphology, induced by *Agrobacterium tumefaciens* carrying mutated octopine Ti-plasmids; analysis of T-DNA functions. Gene 14:33–50

Picker A De, Block M De, Inzé D, Montagu M Van, Schell J (1980) IS-like element IS8 in RP4 and its involvement in cointegration. Gene 19:329–338

Schilperoort RA (1969) Investigations on plant tumors: Crown gall. Thesis, Univ Leiden, Netherlands

Schröder G, Schröder J (1982) Hybridization selection and translation of T-DNA encoded mRNAs from octopine tumors. Mol Gen Genet 185:51–55

Vos G De, Beuckeleer M De, Montagu M Van, Schell J (1981) Restriction endonuclease mapping of the octopine tumor-inducing plasmid pTi Ach5 of *Agrobacterium tumefaciens.* Plasmid 6:249–253

Watson B, Currier TC, Gordon MP, Chilton MD, Nester EW (1975) Plasmid required for virulence of *Agrobacterium tumefaciens.* J Bacteriol 123:255–264

Willmitzer L, Simons G, Schell J (1982) The T_L-DNA in octopine crown gall tumours codes for seven well-defined polyadenylated transcripts. EMBO J 1:139–146

Zaenen I, Larebeke N van, Teuchy H, Montagu M Van, Schell J (1974) Supercoiled circular DNA in crown gall inducing *Agrobacterium* strains. J Mol Biol 86:109–127

Plasmid Genes Essential for the Interactions of Agrobacteria and Rhizobia with Plant Cells

P.J.J. HOOYKAAS[1]

Introduction

Agrobacteria and rhizobia are gram-negative soil bacteria that belong to the bacterial family of *Rhizobiaceae*. They have characteristic interactions with plants, and based on these interactions they are divided into species. The species *Agrobacterium tumefaciens* forms tumors called crown galls on many dicotyledonous plant species, whereas the species *A. rhizogenes* induces root proliferations (hairy root) in such plants. Rhizobia induce the formation of nitrogen-fixing root nodules on legumes and a few non-legumes (*Parasponia* species) (Trinick 1979), and species are distinguished on the basis of their host range. The species *Rhizobium trifolii* induces root nodules on clovers, *R. phaseoli* on beans, *R. leguminosarum* on peas and vetches, and *R. meliloti* on honeyclovers and alfalfa.

Agrobacteria induce tumors on many dicotyledonous plant species (De Cleene and De Ley 1977), but not all strains cover the same wide host range (Panagopoulos and Psallidas 1973). Wound-sites are necessary for infection by agrobacteria. Here the bacteria penetrate the plant, and, subsequently, grow, and multiply in the plant's wound sap. The agrobacteria also attach to healthy plant cells (Schilperoort 1969, Lippincott and Lippincott 1969), but they do not penetrate into these. Some time after wounding at least some of these healthy plant cells will start to divide in order to form wound tissue. In a small period of time before cell divisions start they are sensitive to the tumorigenic stimulus transmitted by attached tumorigenic bacteria (reviewed by Braun 1978). The exact nature of the transmitted material is unknown, but it consists at least partly of nucleic acids, since transformed plant cells contain a DNA sequence – the T-DNA – integrated in their nuclear DNA which originates from a plasmid carried by the tumorigenic agrobacteria (Chilton et al. 1977, Thomashow et al. 1980, De Beuckeleer et al. 1981, Ooms 1981). Therefore, tumor induction is probably due to transfer to and subsequent integration of T-DNA into plant DNA. Crown gall cells have two characteristics in which they can be distinguished from normal plant cells, viz. phytohormone autotrophy (Braun 1958) and production of tumorspecific compounds called opines (reviewed by Tempé and Goldmann 1982).

Rhizobia interact with plants in a completely different way (reviewed by Vincent 1980). They attach to plant root hairs, which they induce to form a marked curling, whereafter they penetrate via an infection thread which grows into the root cortex.

1 Biochemical Laboratory, University of Leiden, Wassenaarseweg 64, 2333 AL Leiden, The Netherlands

Molecular Genetics of the Bacteria-Plant Interaction
(ed. by A. Pühler)

The bacteria multiplying in the infection thread trigger certain cortex cells to divide which results in the formation of a root nodule. Until now the nature of this trigger is unknown. The bacteria are eventually released from the infection thread into the plant cells, whereafter they change shape becoming bacteroids capable of fixing molecular nitrogen. In root nodule plant cells new products are being made, such as the enzyme leghemoglobin which has a function as an oxygen carrier in the root nodule cells, and which protects nitrogenase from being inactivated by oxygen.

Plasmids

Plasmids have been detected in most, if not all agrobacterium and rhizobium strains (e.g., Zaenen et al. 1974, Nuti et al. 1977, Casse et al. 1979). They usually carry a number of large plasmids (> 100 megadaltons) and incidentally they may also harbour one or more smaller plasmids. The role of plasmid genes in the interaction of these bacteria with plant cells has been determined by molecular-genetic experiments. From some *A. tumefaciens* strains avirulent derivatives have been obtained by growth at a higher than the normal growth temperature (Hamilton and Fall 1971). These turned out to have lost a plasmid of 120 Mdal (Van Larebeke et al. 1974, Watson et al. 1975), which later on was found to have a thermosensitive replicator (Rogler 1980). The cured strains could be converted again into pathogens via in planta and ex planta crosses with tumorigenic donors (e.g., Kerr et al. 1977, Genetello et al. 1977). The pathogenic derivatives turned out to have acquired a plasmid from the donor. This plasmid was therefore called Ti-plasmid, tumor-inducing plasmid. It was found that Ti-plasmids not only control tumor induction, but also determine the morphology of the tumors that will be formed and the types of opines that will be synthesized in the tumors (Bomhoff et al. 1976, Hooykaas et al. 1977). In this way octopine, nopaline, and null-type Ti-plasmids are distinguished. Octopine Ti-plasmids and null-type Ti-plasmids determine rough tumors on kalanchoë, but those induced by octopine strains are surrounded by roots. Nopaline strains induce smooth tumors on kalanchoë (Hooykaas et al. 1977, Hooykaas 1979). In octopine tumors the opines of the octopine family (e.g., octopine, lysopine, ornopine, histopine) are present as well as agropine and mannopine (Petit et al. 1970, Firmin and Fenwick 1978). In nopaline tumors opines of the nopaline family (nopaline, ornaline) are there and in addition agrocinopines A and B (Petit et al. 1970, Ellis and Murphy 1981), whereas in null-type tumors (neither octopine nor nopaline) agropine, mannopine, and agrocinopines C and D are found (Guyon et al. 1980, Ellis and Murphy 1981). The Ti-plasmids not only determine the synthesis of specific opines in the tumors, but also encode a catabolic system for the degradation of the opines synthesized in the tumors.

Similarly to the work done for *A. tumefaciens* it has been demonstrated for certain *A. rhizogenes* strains that a plasmid is responsible for hairy root induction in plants. This plasmid, which has been introduced into Ti-plasmid free *A. tumefaciens* strains both by in planta (Albinger and Beiderbeck 1977, Moore et al. 1979, White and Nester 1980a) and ex planta crosses (Chilton et al. 1982, Hooykaas et al. 1982a), has been called Ri-plasmid, root-inducing plasmid. During hairy root induction in plants part of the Ri-plasmid becomes integrated into plant DNA (Chilton et al. 1982, Spano et al. 1982) and is expressed into RNA there (Willmitzer et al. 1982a).

Molecular genetic experiments have also been carried out with different *Rhizobium* species showing that in strains of *R. trifolii, R. leguminosarum,* and *R. phaseoli* one large plasmid is present which controls both root nodulation and nitrogen fixation (Hooykaas et al. 1981, 1982b, unpubl., Hombrecher et al. 1981). It has therefore been proposed to call these plasmids Sym-plasmids (symbiosis-inducing plasmids). In strains of *R. meliloti* the genes controlling nodulation and nitrogen fixation have been found to be located on exceptionally large plasmids (Banfalvi et al. 1981, Rosenberg et al. 1981). Southern blotting experiments with the *Klebsiella pneumoniae* structural genes for nitrogenase as a probe have shown that in *Rhizobia* the nitrogenase structural genes are located in the Sym-plasmids (Nuti et al. 1979, Prakash et al. 1981, Hombrecher et al. 1981, Krol et al. 1982).

Expression of Ti-, Ri-, and Sym-Plasmids in New Hosts

Agrobacteria and rhizobia are taxonomically closely related. This becomes the more clear if one compares the preliminary chromosomal linkage maps that are available for *A. tumefaciens* strain C58, for *R. leguminosarum* strain 300, and for some *R. meliloti* strains (Kondorosi et al. 1980, Hooykaas et al. 1982c). R-primes with *R. meliloti* genes can be used for complementation of Rec^+ as well as of Rec^- *A. tumefaciens* auxotrophs (Hooykaas et al. 1982c, unpubl.). To investigate whether a bacterium is to be called *Rhizobium* or *Agrobacterium* solely because of its plasmid content we have transferred Ti- and Ri-plasmids from agrobacteria into rhizobia (Hooykaas et al. 1977, Hooykaas 1979, unpubl.), and Sym-plasmids from rhizobia into agrobacteria (Hooykaas et al. 1981, 1982b, unpubl.). Transfer of Ti-plasmids (of the octopine or nopaline type) into any of the *Rhizobium* species *R. trifolii, R. leguminosarum,* or *R. meliloti* resulted in transconjugant strains capable of degrading opines. However, only *R. trifolii* or *R. leguminosarum* hosts became tumorigenic upon introduction of a Ti-plasmid, although induced tumors were generally smaller than those induced by *Agrobacterium* hosts. *R. meliloti* hosts remained non-pathogenic after receipt of a Ti-plasmid. Introduction of an Ri-plasmid into a *R. trifolii* strain resulted in a strain capable of inducing hairy root in plants. Reversely, Sym-plasmids from *R. trifolii, R. leguminosarum,* and *R. phaseoli* were introduced into *A. tumefaciens* and *A. rhizogenes* strains. These became able to induce root nodules on leguminous plants, but no nitrogen was fixed in such nodules, possibly reflecting a lack of full expression of Sym-plasmid genes in agrobacterium hosts. The host range for root nodule induction by agrobacteria with a Sym-plasmid, or by rhizobial strains carrying a Ti-plasmid was found to be determined solely by the Sym-plasmid, and was not extended by the presence of the (wide host range) Ti-plasmid.

Strains have also been constructed which carry a combination of two different types of plasmids, viz. a Ti-plasmid and an Ri-plasmid (Hooykaas 1979, Costantino et al. 1980), a Sym-plasmid and a Ti-plasmid or a Sym-plasmid and an Ri-plasmid

(Hooykaas 1979, unpubl.). The presence of a plasmid of one type did not interfere with the expression of a plasmid of another type. Agrobacteria or rhizobia with a Sym-plasmid plus a Ti- or a Ri-plasmid induce root nodules on plants forming part of the cross-inoculation group determined by the Sym-plasmid and also induce crown gall or hairy root (in case of an Ri-plasmid) at wound sites on dicotyledonous plants. However, one exception to this rule has been found. *R. leguminosarum* strains are severely hampered in nodulating peas by the carriage of a Ti-plasmid. Agrobacteria with an Ri- and a Ti-plasmid form aberrant tumors from which many root structures appear showing that both plasmids are active in transforming plant cells.

Genetic Map of Ti-Plasmids

Physical (restriction enzyme) maps have been constructed for an octopine Ti-plasmid (Chilton et al. 1978a, Ooms et al. 1980, De Vos et al. 1981), a nopaline Ti-plasmid (De Picker et al. 1980), and a *Rhizobium leguminosarum* Sym-plasmid (Prakash et al. 1982). From the above Ti-plasmids many transposon-insertion and deletion mutants have been isolated by various methods (e.g., Hernalsteens et al. 1978, Klapwijk et al. 1980, Hooykaas et al. 1980a, Ooms et al. 1982). The position of the mutations have been mapped with the aid of the available physical maps of the plasmids, and by correlating phenotype and genotype genetic maps have been obtained for octopine plasmid pTiB6 (Koekman et al. 1979, 1980, 1982, Ooms et al. 1980, 1981, 1982, De Greve et al. 1981) and for nopaline plasmid pTiC58 (Holsters et al. 1980). Genes involved in replication and incompatibility, conjugative transfer, catabolism of opines, virulence and oncogenicity, and phage (AP1) exclusion have been localized. All virulence and oncogenicity genes have been found to be located in roughly one third of the Ti-plasmid, whereas the other genes are located in the remaining part of the plasmid.

The oncogenicity genes have been localized in the T-region, the part of the Ti-plasmid that is introduced into plant cells during tumor induction. Mutagenesis of the T-region revealed that a number of different loci are present in this region (Garfinkel et al. 1981, Ooms et al. 1981). and furthermore that a sequence is bordering the T-region deletion of which leads to strongly diminished virulence (Koekman et al. 1979, Ooms et al. 1982). The T-DNA is transcribed into (m)RNAs in the plant cells by Pol II (Willmitzer et al. 1981, 1982b). For the octopine Ti-plasmid it has been found that the T-DNA carries the gene for octopine synthase, the enzyme responsible for the synthesis of opines of the octopine family in tumor cells (Schröder et al. 1981, Murai and Kemp 1982), and at least two loci that control differentiation in plant cells (Ooms et al. 1981, Garfinkel et al. 1981). Tumors induced by one type of mutant showed extra shoot formation, while tumors induced by the second type of mutant showed extra root formation on kalanchoë and tobacco. Both types of mutants were avirulent on tomato. As it has been found that the balance between auxins and cytokinins controls differentiation in plant tissues (Skoog and Miller 1957) it has been tested whether addition of plant hormones to the 'shooter' and 'rooter' mutants during tumor induction would stimulate tumor formation by these mutants on tomato (Ooms et al. 1981). Tumor formation by the 'rooter' mutants was indeed stimulated if a cytokinin was sprayed daily, and the 'shooter' mutants responded positively to the addition of an auxin during tumor induction. This supports the view that T-DNA loci are directly responsible for the auxin- and cytokinin-autotrophic growth of crown gall cells.

Virulence genes have been mapped in a region of the Ti-plasmid immediately adjacent to the T-region (Ooms et al. 1980, Garfinkel et al. 1981, De Greve et al. 1981, Hooykaas, unpubl.). This region, the Vir-region, has never been found to be present in transformed cells, and the genes in this region probably are expressed in the bacterium, since most of the mutants in this region can be complemented in trans by R-primes carrying segments of the Vir-region (Hille et al. 1982, Klee et al. 1982, Hooykaas, unpubl.). It can be envisaged that genes in this region play a role in the processing of the Ti-plasmid DNA that is eventually found as T-DNA in transformed plant cells, in the transfer of this material into plant cells, and possibly also in the conditioning of plant cells for the conversion into tumor cells.

Complementation Experiments

In order to be able to perform complementation experiments it is necessary to construct Rec^- strains diploid for a specific DNA segment. For *A. tumefaciens* Rec^- derivatives of some strains have been constructed (Klapwijk et al. 1979, Molbas, unpubl.), but the construction of strains diploid for a Ti-plasmid segment has been hampered by incompatibility between plasmids. This has been solved by the construction of R-primes carrying a segment of the Ti-plasmid, but lacking the Ti replication and incompatibility region (Klapwijk and Schilperoort 1979, Hille and Schilperoort 1982). R primes with segments of the Vir-region were found to be able to restore virulence in strains with mutations in the homologous part of the Vir-region. The virulence region of nopaline and null-type Ti-plasmids has strong DNA homology with that of the octopine Ti-plasmid (e.g., Drummond and Chilton 1978, Hepburn and Hindley 1979, Engler et al. 1981). In agreement three avirulent mutants with a mutation in the Vir-region of their nopaline Ti-plasmid were restored in virulence by R-primes with parts of the octopine Ti-plasmid Vir-region (Hooykaas, unpubl.). Octopine and nopaline Ti-plasmids are incompatible, but strains can be constructed in which an octopine and a nopaline Ti-plasmid have formed a stable cointegrate (Hooykaas et al. 1980b). Such strains form tumors in which both octopine and nopaline are synthesized, showing the transfer of both types of T-DNA into plant cells from such bacteria. Cointegrates have also been constructed from octopine and nopaline Ti-plasmids either or both having a mutation in the Vir-region rendering it avirulent (Hooykaas, unpubl.). Strains with such cointegrates also form tumors in which octopine and nopaline are synthesized, showing that also in these cases both types of T-DNA are transferred to plant cells. Strains with cointegrates were avirulent if in the octopine and the nopaline Ti-plasmid part the same virulence function had been inactivated. Null-type Ti-plasmid pTi542 belongs to a different incompatibility group than octopine and nopaline Ti-plasmids (Hooykaas 1979, unpubl.). After introduction of this plasmid into avirulent nopaline Ti-plasmid carrying strains tumors were formed in which nopaline was present, showing that the null-type Ti-plasmid has at least some virulence functions analogous to those of octopine and nopaline Ti-plasmids.

Ri-plasmids belong to an incompatibility group (*inc* Rh-3) different from that of octopine-nopaline Ti-plasmids (*inc* Rh-1) and that of null-type Ti-plasmids (*inc*Rh-2) (Costantino et al. 1980, Hooykaas 1979, unpubl.). It has been found that the Ri-plasmid shows strong DNA homology with some parts of the virulence region of Ti-plasmids and less DNA homology with other parts (White and Nester 1980b, Risuleo et al. 1982). In agreement, when Ri-plasmids were assayed for their ability to complement Vir^- mutants of octopine and nopaline Ti-plasmids it was found that some of the Vir-region loci were complementable, whereas others were not (Hooykaas, unpubl.) Some Sym-plasmids have been assayed for their sequence homology with Ti-plasmids, and have been found to share DNA homology with small parts of the virulence region (Prakash and Schilperoort 1982). However, at the moment, unfortunately, no results from complementation tests are available.

The T-regions of wide host range octopine, nopaline, and null-type Ti-plasmids share a segment of strong DNA homology, the 'common sequence' (Chilton et al. 1978b, De Picker et al. 1978), which covers the loci that have an auxin-like and a cytokinin-like effect in plant cells. Besides this segment of DNA homology regions of non-homology are present which carry loci at least some of which are involved in opine synthesis. Limited host range Ti-plasmids, Ri-plasmids, and Sym-plasmids lack regions that have strong DNA homology with the 'common sequence', although regions with faint DNA homology may be present (Thomashow et al. 1981, White and Nester 1980b, Risuleo et al. 1982, Willmitzer et al. 1982a, Prakash and Schilperoort 1982). Ri-plasmids can functionally complement shooter mutants derived from octopine strains in tumor induction tests on tomato. Whereas strains with solely a Ri-plasmid or with an octopine Ti-plasmid with a shooter mutation are only weakly oncogenic on tomato, strains with a combination of these plasmids are normally oncogenic (Hooykaas, unpubl.).

Different *Rhizobium* strains may carry Sym-plasmids of different sizes and of different incompatibility groups. The Sym-plasmids pSymRtr5 *(R. trifolii)*, pSymRle1 *(R. leguminosarum)* and pSymRph9 *(R. phaseoli)* have been found to be compatible with each other and to be also compatible with the *inc* Rh-1 and *inc* Rh-2 Ti-plasmids and with Ri-plasmids (Hooykaas et al. 1981, 1982b, unpubl.). Therefore, it is probable that complementation experiments with Sym-plasmid mutants can be carried out easily as soon as mutants are available. So far it has been found that strains carrying two different Sym-plasmids have a larger host range for nodulation and nitrogen fixation covering the respective host ranges determined by either of the two Sym-plasmids (Djordjevic et al. 1982, Hooykaas et al. 1981, 1982b, Johnston et al. 1978). Sym-plasmids share extensive regions of DNA homology and these have been mapped on the physical map of the *R. leguminosarum* Sym-plasmid pSymRle1 (Prakash et al. 1982). One large region of strong DNA homology between different Sym-plasmids covers the structural genes for nitrogenase.

Concluding Remarks

In conclusion it can be stated that in *Agrobacterium tumefaciens, A. rhizogenes, Rhizobium leguminosarum, R. trifolii* and *R. phaseoli* strains plasmids largely determine the phytopathogenic and symbiotic properties of their host bacteria. However, small chromosomal differences between these bacteria must exist because Ti-plasmids induce tumors more efficiently when present in the agrobacteria than in these rhizobia, and Sym-plasmids more efficiently induce root nodules when present in the rhizobia than in the agrobacteria. Different types of Ti-plasmids and Ri-plasmids share (large) regions of DNA homology, most of which contain loci essential for tumor formation, and complementation experiments have shown that at least some of the virulence functions determined by Ti- and Ri-plasmids are analogous. Incompatibility studies showed that Ti- and Ri-plasmids can be divided into at least 3 different classes indicating the possiblity that the virulence and oncogenicity genes became linked to different types of plasmids by (illegitimate?) recombination during evolution. For Sym-plasmids the same is true: they share large regions of DNA homology, probably encoding functions involved in the establishment of symbiotic nitrogen fixation, but they are compatible with one another showing that also symbiotic genes probably have been shuffled around different replicons in evolution.

Acknowledgement. I thank Dr. R.J.M. van Veen for critically reading the manuscript.

References

Albinger G, Beiderbeck R (1977) Übertragung der Fähigkeit zur Wurzelinduktion von *Agrobacterium rhizogenes* auf *A. tumefaciens*. Phytopathol Z 90:306–310

Banfalvi Z, Sakanyan V, Koncz C, Kiss A, Dusha I, Kondorosi A (1981) Location of nodulation and nitrogen fixation genes on a high molecular weight plasmid of *R. meliloti*. Mol Gen Genet 184:318–325

Beuckeleer M De, Lemmers M, Vos G De, Willmitzer L, Montagu M Van, Schell J (1981) Further insight on the transferred-DNA of octopine crown gall. Mol Gen Genet 183:283–288

Bomhoff G, Klapwijk PM, Kester HCM, Schilperoort RA, Hernalsteens JP, Schell J (1976) Octopine and nopaline synthesis and breakdown genetically controlled by a plasmid of *Agrobacterium tumefaciens*. Mol Gen Genet 145:177–181

Braun AC (1958) A physiological basis for autonomous growth of the crown-gall tumor cell. Proc Natl Acad Sci USA 44:344–349

Braun AC (1978) Plant tumors. Biochim Biophys Acta 516:167–191

Casse F, Boucher C, Julliot JS, Michel M, Dénarié J (1979) Identification and characterization of large plasmids in *Rhizobium meliloti* using agarose gel electrophoresis. J Gen Microbiol 113: 229–242

Chilton MD, Drummond MH, Merlo DJ, Sciaky D, Montoya AL, Gordon MP, Nester EW (1977) Stable incorporation of plasmid DNA into higher plant cells: the molecular basis of crown gall tumorigenesis. Cell 11:263–271

Chilton MD, Montoya AL, Merlo DJ, Drummond MH, Nutter R, Gordon MP, Nester EW (1978a) Restriction endonuclease mapping of a plasmid that confers oncogenicity upon *Agrobacterium tumefaciens* strain B6-806. Plasmid 1:254–269

Chilton MD, Drummond MH, Merlo DJ, Sciaky D (1978b) Highly conserved DNA of Ti-plasmids overlaps T-DNA maintained in plant tumors. Nature 275:147–149

Chilton MD, Tepfer DA, Petit A, David C, Tempé J (1982) *Agrobacterium rhizogenes* inserts T-DNA into plant roots. Nature 295:432–434

Cleene M De, Ley J De (1977) The host range of crown gall. Bot Rev 42:389–466

Costantino P, Hooykaas PJJ, Dulk-Ras H Den, Schilperoort RA (1980) Tumor formation and rhizogenicity of *Agrobacterium rhizogenes* carrying Ti-plasmids. Gene 11:79–87

Djordjevic MA, Zurkowski W, Rolfe BG (1982) Plasmids and stability of symbiotic properties of *Rhizobium trifolii*. J Bacteriol 151:560–568

Drummond MH, Chilton M-D (1978) Tumor-inducing (Ti-) plasmids of *Agrobacterium* share extensive regions of DNA homology. J Bacteriol 136:1178–1183

Ellis JG, Murphy PJ (1981) Four new opines from crown gall tumours – their detection and properties. Mol Gen Genet 181:36–43

Engler R, Picker A De, Maenhaut R, Villarroel R, Montagu M Van, Schell J (1981) Physical mapping of DNA base sequence homologies between an octopine and a nopaline Ti-plasmid of *Agrobacterium tumefaciens*. J Mol Biol 152:183–208

Firmin JL, Fenwick GR (1978) Agropine – a major new plasmid-determined metabolite in crown gall tumours. Nature 276:842–844

Garfinkel DJ, Nester EW (1980) *Agrobacterium tumefaciens* mutants affected in crown gall tumorigenesis and octopine catabolism. J Bacteriol 144:732–743

Garfinkel DJ, Simpson RB, Ream LW, White FF, Gordon MP, Nester EW (1981) Genetic analysis of crown gall: fine structure map of the T-DNA by site-directed mutagenesis. Cell 27:143–153

Genetello C, Larebeke N Van, Holsters M, Picker A De, Montagu M Van, Schell J (1977) The Ti-plasmids of *Agrobacterium* as conjugative plasmids. Nature 265:561–563

Greve H De, Decraemer H, Seurinck J, Montagu M Van, Schell J (1981) The functional organization of the octopine *Agrobacterium tumefaciens* plasmid pTiB6S3. Plasmid 6:235–248

Guyon P, Chilton M-D, Petit A, Tempé J (1980) Agropine in "null-type" crown gall tumors: evidence for generality of the opine concept. Proc Natl Acad Sci USA 77:2693–2697

Hamilton RH, Fall MZ (1971) The loss of tumor-initiating ability in *Agrobacterium tumefaciens* by incubation at high temperature. Experientia 27:229–230

Hepburn AG, Hindley J (1979) Region of DNA sequence homology between an octopine and nopaline Ti-plasmid of *Agrobacterium tumefaciens*. Mol Gen Genet 169:163–172

Hernalsteens JP, Greve H De, Montagu M Van, Schell J (1978) Mutagenesis by insertion of the drug resistance transposon Tn7 applied to the Ti-plasmid of *Agrobacterium tumefaciens*. Plasmid 1:218–225

Hille J, Schilperoort RA (1981) The use of transposons to introduce well-defined deletions in plasmids: possibilities for in vivo cloning. Plasmid 6:151–154

Hille J, Klasen I, Schilperoort RA (1982) Construction and application of R-prime plasmids, carrying different segments of an octopine Ti-plasmid from *Agrobacterium tumefaciens* for complementation of *vir* genes. Plasmid 7:107–118

Holsters M, Silva B, Vliet F Van, Genetello C, Block M De, Dhase P, Picker A De, Inzé D, Engler G, Villarroel R, Montagu M Van, Schell J (1980) The functional organization of the nopaline *A. tumefaciens* plasmid pTiC58. Plasmid 3:212–230

Hombrecher G, Brewin NJ, Johnston AWB (1981) Linkage of genes for nitrogenase and nodulation ability on plasmids in *Rhizobium leguminosarum* and *R. phaseoli*. Mol Gen Genet 182:133–136

Hooykaas PJJ (1979) The role of plasmid determined functions in the interactions of *Rhizobiaceae* with plant cells. A genetic approach. Thesis, Univ Leiden, Netherlands

Hooykaas PJJ, Klapwijk PM, Nuti MP, Schilperoort RA, Rörsch A (1977) Transfer of the *Agrobacterium tumefaciens* Ti-plasmid to avirulent agrobacteria and to *Rhizobium* ex planta. J Gen Microbiol 98:477–484

Hooykaas PJJ, Dulk-Ras H Den, Schilperoort RA (1980a) Molecular mechanism of Ti-plasmid mobilization by R-plasmids. Isolation of Ti-plasmids with transposon-insertions in *Agrobacterium tumefaciens*. Plasmid 4:65–74

Hooykaas PJJ, Dulk-Ras H Den, Ooms G, Schilperoort RA (1980b) Interactions between octopine and nopaline plasmids in *Agrobacterium tumefaciens*. J Bacteriol 143:1295–1306

Hooykaas PJJ, Brussel AAN Van, Dulk-Ras H Den, Slogteren GMS Van, Schilperoort RA (1981) Sym-plasmid of *Rhizobium trifolii* expressed in different rhizobial species and *Agrobacterium tumefaciens*. Nature 291:351–353

Hooykaas PJJ, Dulk-Ras H Den, Schilperoort RA (1982a) Method for the transfer of large cryptic, non-self-transmissible plasmids: ex planta transfer of the virulence plasmid of *Agrobacterium rhizogenes*. Plasmid 8:94–96

Hooykaas PJJ, Snijdewint FGM, Schilperoort RA (1982b) Identification of the Sym-plasmid of *Rhizobium leguminosarum* strain 1001 and its transfer to and expression in other rhizobia and *Agrobacterium tumefaciens*. Plasmid 8:73–82

Hooykaas PJJ, Peerbolte R, Regensburg-Tuink AJG, Vries P De, Schilperoort RA (1982c) A chromosomal linkage map of *Agrobacterium tumefaciens* and a comparison with the maps of *Rhizobium* spp. Mol Gen Genet 188:12–17

Johnston AWB, Beynon JL, Buchanan-Wollaston AV, Setchell SM, Hirsch PR, Beringer JE (1978) High frequency transfer of nodulating ability between strains and species of *Rhizobium*. Nature 276:634–636

Kerr A, Manigault P, Tempé J (1977) Transfer of virulence in vivo and in vitro in *Agrobacterium*. Nature 265:560–561

Klapwijk PM, Schilperoort RA (1979) Negative control of octopine degradation and transfer genes of octopine Ti-plasmids in *Agrobacterium tumefaciens*. J Bacteriol 139:424–431

Klapwijk PM, Beelen P Van, Schilperoort RA (1979). Isolation of a recombination deficient *Agrobacterium tumefaciens* mutant. Mol Gen Genet 173:171–175

Klapwijk PM, Breukelen J Van, Koorevaar K, Ooms G, Schilperoort RA (1980) Transposition of Tn904 encoding streptomycin resistance into the octopine Ti-plasmid of *Agrobacterium tumefaciens*. J Bacteriol 141:129–136

Klee JH, Gordon MP, Nester EW (1982) Complementation analysis of *Agrobacterium tumefaciens* Ti-plasmid mutations affecting oncogenicity. J Bacteriol 150:327–331

Koekman BP, Ooms G, Klapwijk PM, Schilperoort RA (1979) Genetic map of an octopine Ti-plasmid. Plasmid 2:347–357

Koekman BP, Hooykaas PJJ, Schilperoort RA (1980) Localization of the replication control region on the physical map of the octopine Ti-plasmid. Plasmid 4:184–195

Koekman BP, Hooykaas PJJ, Schilperoort RA (1982) A functional map of the replicator region of the octopine Ti-plasmid. Plasmid 7:119–132

Kondorosi A, Vincze E, Johnston AWB, Beringer JE (1980) A comparison of three *Rhizobium* linkage maps. Mol Gen Genet 178:403–408

Krol AJM, Hontelez JGJ, Kammen A Van (1982) Only one of the large plasmids in *Rhizobium leguminosarum* strain PRE is strongly expressed in the endosymbiotic state. J Gen Microbiol 128:1839–1847

Larebeke N Van, Engler G, Holsters M, Elsacker S Van Den, Zaenen I, Schilperoort RA, Schell J (1974) Large plasmid in *Agrobacterium tumefaciens* essential for crown-gall inducing ability. Nature 252:169–170

Lippincott BB, Lippincott JA (1969) Bacterial attachment to a specific wound site as an essential stage in tumor initiation by *Agrobacterium tumefaciens*. J Bacteriol 97:620–628

Moore L, Warren G, Strobel G (1979) Involvement of a plasmid in the hairy root disease of plants caused by *Agrobacterium rhizogenes*. Plasmid 2:617–626

Murai N, Kemp JD (1982) Octopine synthase messenger RNA isolated from sunflower crown gall callus is homologous to the Ti-plasmid of *Agrobacterium tumefaciens*. Proc Natl Acad Sci USA 79:86–90

Nuti MP, Ledeboer AM, Lepidi AA, Schilperoort RA (1977) Large plasmids in different *Rhizobium* species. J Gen Microbiol 100:241–248

Nuti MP, Lepidi AA, Prakash RK, Schilperoort RA, Cannon FC (1979) Evidence for nitrogen fixation *(nif)* genes on indigenous *Rhizobium* plasmids. Nature 282:533–535

Ooms G, Klapwijk PM, Poulis JA, Schilperoort RA (1980) Characterization of Tn904 insertions in octopine Ti-plasmid mutants of *Agrobacterium tumefaciens*. J Bacteriol 144:82–91

Ooms G, Hooykaas PJJ, Moolenaar G, Schilperoort RA (1981) Crown gall plant tumors of abnormal morphology, induced by *Agrobacterium tumefaciens* carrying mutated octopine Ti-plasmids; analysis of T-DNA functions. Gene 14:33–50

Ooms G, Hooykaas PJJ, Veen RJM Van, Beelen P Van, Regensburg-Tuink AJG, Schilperoort RA (1982) Octopine Ti-plasmid deletion mutants of *Agrobacterium tumefaciens* with emphasis on the right side of the T-region. Plasmid 7:15–29

Panagopoulos CG, Psallidas PG (1973) Characteristics of Greek isolates of *Agrobacterium tumefaciens* (Smith & Townsend) Conn. J Appl Bacteriol 36:233–240

Petit A, Delhaye S, Tempé J, Morel G (1970) Recherches sur les guanidines des tissus de crown gall. Mise en évidence d'une relation biochimique spécifique entre les souches d'*Agrobacterium tumefaciens* et des tumeurs qu'elles induisent. Physiol Veg 8:205–213

Picker A De, Montagu M Van, Schell J (1978) Homologous DNA sequences in different Ti-plasmids are essential for oncogenicity. Nature 275:150–153

Prakash RK, Schilperoort RA (1982) Relationship between nif plasmids of fast-growing *Rhizobium* species and Ti-plasmids of *Agrobacterium tumefaciens*. J Bacteriol 149:1129–1134

Prakash RK, Schilperoort RA, Nuti MP (1981) Large plasmids of fast-growing rhizobia: homology studies and location of structural nitrogen fixation *(nif)* genes. J Bacteriol 145:1129–1136

Prakash RK, Veen RJM Van, Schilperoort RA (1982) Restriction endonuclease mapping of a *Rhizobium leguminosarum* Sym-plasmid. Plasmid 7:271–280

Risuleo G, Battistoni P, Costantino P (1982) Regions of homology between tumorigenic plasmids from *Agrobacterium rhizogenes* and *Agrobacterium tumefaciens*. Plasmid 7:45–51

Rogler CE (1980) Plasmid-dependent temperature-sensitive phase in crown gall tumorigenesis. Proc Natl Acad Sci USA 77:2668–2692

Rosenberg C, Boistard P, Dénarié J, Casse-Delbart F (1981) Genes controlling early and late functions in symbiosis are located on a megaplasmid in *Rhizobium meliloti*. Mol Gen Genet 184:326–333

Schilperoort RA (1969) Investigations on plant tumors – crown gall. On the biochemistry of tumor induction by *Agrobacterium tumefaciens*. Thesis, Univ Leiden, Netherlands

Schröder J, Hillebrand A, Klipp W, Pühler A (1981) Expression of plant tumor-specific proteins in minicells of *Escherichia coli:* a fusion protein of lysopine dehydrogenase with chloramphenicol acetyltransferase. Nucleic Acid Res 9:5187–5202

Skoog F, Miller CO (1957) Chemical regulation of growth and organ formation in plant tissues cultured in vitro. Symp Soc Exp Biol 11:118–131

Spano L, Pomponi M, Costantino P, Slogteren GMS Van, Tempé J (1982) Identification of T-DNA in the root-inducing plasmid of the agropine type *Agrobacterium rhizogenes* 1855. Plant Mol Biol 1:291–300

Tempé J, Goldmann A (1982) Occurrence and biosynthesis of opines. In: Kahl G, Schell J (eds) Molecular biology of plant tumors. Academic Press, London New York, pp 427–449

Thomashow MF, Nutter R, Montoya AL, Gordon MP, Nester EW (1980) Integration and organization of Ti-plasmid sequences in crown gall tumors. Cell 19:729–739

Thomashow MF, Knauf VC, Nester EW (1981) The relationship between the limited and wide host range octopine-type Ti-plasmids of *Agrobacterium tumefaciens*. J Bacteriol 146:484–493

Trinick MJ (1979) Structure of nitrogen-fixing nodules formed by *Rhizobium* on roots of *Parasponia andersonii* Planch. Can J Microbiol 25:565–578

Vincent JM (1980) Factors controlling the legume-*Rhizobium* symbiosis. In: Newton WE, Orme-Johnson WH (eds) Nitrogen fixation, vol II. University Park Press, Baltimore, pp 103–129

Vos G De, Beuckeleer M De, Montagu M Van, Schell J (1981) Restriction endocnuclease mapping of the octopine tumor-inducing plasmid pTiAch5 of *Agrobacterium tumefaciens*. Plasmid 6:249–253

Watson B, Currier TC, Gordon MP, Chilton M-D, Nester EW (1975) Plasmid required for virulence of *Agrobacterium tumefaciens*. J Bacteriol 123:255–264

White FF, Nester EW (1980a) Hairy root: plasmid encodes virulence traits in *Agrobacterium rhizogenes*. J Bacteriol 141:1134–1141

White FF, Nester EW (1980b) Relationship of plasmids responsible for hairy root and crown gall tumorigenicity. J Bacteriol 144:710–720

Willmitzer L, Schmalenbach W, Schell J (1981) Transcription of T-DNA in octopine and nopaline crown gall tumours is inhibited by low concentrations of α-amanitin. Nucleic Acid Res 9: 4801–4812

Willmitzer L, Sanchez-Serrano J, Buschfeld E, Schell J (1982a) DNA from *Agrobacterium rhizogenes* is transferred to and expressed in axenic hairy root plant tissues. Mol Gen Genet 186: 16–22

Willmitzer L, Simons G, Schell J (1982b) The t_1-DNA in octopine crown-gall tumours codes for seven well-defined polyadenylated transcripts. EMBO J 1:139–146

Zaenen I, Larebeke N Van, Teuchy H, Montagu M Van, Schell J (1974) Supercoiled circular DNA in crown-gall inducing *Agrobacterium* strains. J Mol Biol 86:109–127

Genetic Analysis of Host Range Expression by *Agrobacterium*

V.C. KNAUF[1], M.F. YANOFSKY[1], M.P. GORDON[2], and E.W. NESTER[1]

Introduction

Host range is an important feature of plant-bacteria interactions. Specificity, compatibility, recognition, successful pathogenesis, effective resistance – all of these indicate highly-evolved mechanisms in both host and pathogen which are reflected in host range. The expression of host range by *Agrobacterium* is a complex phenomenon. The total host range of *Agrobacterium* probably includes thousands of different plant species (DeCleen and DeLey 1976). The known hosts are predominantly dicotyledonous plants, but some species of gymnosperms have also been reported to be susceptible to crown gall disease. It is questionable whether any monocotyledonous plants form crown gall tumors (DeCleen and DeLey 1976). Although narrow host range strains of *A. tumefaciens* (Panagopoulos and Psallidas 1973, Loper and Kado 1979) have been described, almost all of the molecular biology of *Agrobacterium* tumorigenesis has focused on a small group of similar strains which express very wide host ranges (Nester and Kosuge 1981). In this communication, we will discuss several aspects of host range expression by *Agrobacterium,* some experiments identifying the nature of genetic loci involved in the expression of host range, and the likely basis for a narrow host range of *Agrobacterium* strain Ag162.

Expression of Host Range by Grapevine Isolates of *Agrobacterium tumefaciens*

We have studied a collection of *Agrobacterium* strains isolated from the grapevine, *Vitis vinifera.* Grapevine isolates of *Agrobacterium* are interesting because some exhibit narrow host ranges and because three significantly different kinds of *Agrobacterium* are found on grapevine (Panagopoulos and Psallidas 1973, Süle 1978). For lack of a better taxonomy almost all *Agrobacterium* strains can be assigned to one of three biotypes based on biochemical tests, protein profiles, serology, or DNA homology (Keane et al. 1970, DeLey et al. 1973, Kerr and Panagopoulos 1977).

1 Department of Microbiology and Immunology, SC-42, University of Washington, Seattle, WA 98195, USA

2 Department of Biochemistry, SJ-70, University of Washington, Seattle, WA 98195, USA

Molecular Genetics of the Bacteria-Plant Interaction
(ed. by A. Pühler)

Biotype 3 *Agrobacterium* strains are found almost exclusively on grapevine to date, whereas strains of *Agrobacterium* of Biotypes 1 and 2 have been found on many plants including grapevines.

The *Agrobacterium* Ti-plasmid is the primary determinant of host range, as well as virulence in general (Thomashow et al. 1980b, Loper and Kado 1979, Knauf et al. 1982). When the Ti-plasmid is transferred from one strain of *Agrobacterium* to another, the host range properties of the donor strain are transferred. *Agrobacterium* strain C58 (Currier and Nester 1976) expresses a very wide host range. Cured of its Ti-plasmid pTiC58, this strain was completely avirulent. When the Ti-plasmid of the narrow host range strain Ag162 was transformed into the avirulent derivative of C58, the transformed strain expressed the narrow host range of strain Ag162 and not the wide host range of the original C58. We have transformed 25 different octopine or nopaline type plasmids from grapevine isolates of *Agrobacterium* into the avirulent derivative of C58, strain A136 (Knauf et al. 1982). In almost all cases, the transferred plasmid coded for the same host range in the C58 background as it did in the original wild-type background.

Another aspect of host range in *Agrobacterium* is illustrated by the strains 2/6 and PPI-1. These wide host range strains have almost the same host range. However, strain PPI-1 can induce tumors on *Kalanchoë daigremontiana,* whereas strain 2/6 cannot. Other data (Anderson and Moore 1979) also indicate that wide host range strains may express slightly different wide host ranges. This property also appears to be Ti-plasmid-coded since pTiPPI-1 in the C58 background codes for a wide host range including *Kalanchoë;* pTi2/6 in the C58 background codes for a wide host range that does not include *Kalanchoë.*

Two strains, Ag105 and Ag123, caused tumors on a range of plants including tobacco (*(Nicotiana tabacum* var. Xanthi) and tomato *(Lycopersicon esculentum).* However, in the C58-type strain, neither pTiAg105 or pTiAg123 coded for virulence on tobacco or tomato. Since strain Ag123 contains no other detectable plasmids, this indicates that chromosomal factors also play a role in the determination of host range. This conclusion is supported by studies in which transposon mutagenesis of the *Agrobacterium* chromosome resulted in strains with altered host ranges (Garfinkel and Nester 1980).

In collaboration with Christos Panagopoulos (Knauf et al. 1982), we have found that Ti-plasmids like pTiB6 and pTiA6 (Sciaky et al. 1977) code for virulence on some cultivated varieties of grapevine but not on others. This demonstrates that at least one kind of host factor which determines crown gall susceptibility can vary between members of a single plant species. It is noteworthy that the plasmid pTiAg-162 codes for virulence on all grapevine varieties tested.

Complementation of Host Range by Cosmid Clones of pTiA6 DNA

In order to identify the Ti-plasmid-coded factors determining host range, *Agrobacterium* strains A348 and A856 were studied. Both strains have the same C58-type chromosome and large cryptic plasmid (Thomashow et al. 1980b), but strain A348 contains pTiA6 whereas strain A856 contains pTiAg162. pTiA6 is a 181 kb octopine-agropine type plasmid which codes for a wide host range and was originally found in a Biotype 1 strain isolated in the United States. pTiAg162 is a 231 kb octopine type plasmid which codes for a narrow host range and was originally found in a Biotype 3 strain isolated from grapevines grown in the U.S.S.R. Although these two plasmids belong to the same incompatibility group (Knauf and Nester 1982), DNA homology studies indicate that they are quite divergent in terms of DNA sequence (Knauf et al. 1983).

In order to identify host range determinants coded by pTiA6, fragments of pTiA6 were cloned and then introduced into the narrow host range strain A856. pTiA6 DNA was partially digested with the restriction endonuclease *Sal*1 and cloned into the *Sal*1 site of pVK102 using the in vitro packaging system of phage lambda (Enquist and Sternberg 1979). pVK102 is a cosmid cloning vector coding for kanamycin and tetracycline resistance (Fig. 1) that can be mated into *Agrobacterium* and many other Gram-negative bacteria where it is stably maintained (Knauf and Nester 1982). Since restriction enzyme maps of pTiA6 are available (Knauf and Nester 1982), the cloned insert in each recombinant plasmid was characterized as to size, organization, and specific map co-ordinates. Cosmids with inserts covering overlapping portions of pTiA6 were selected so that every locus on pTiA6 could be tested for complementation of host range characters in strain A856.

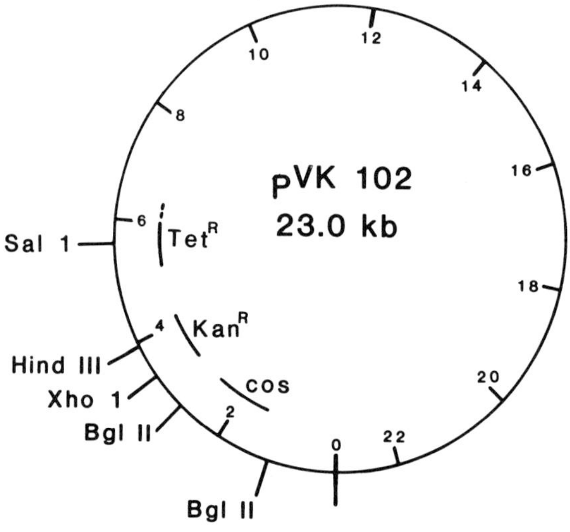

Fig. 1. Physical map of the wide host range cosmid cloning vehicle pVK102 (Knauf and Nester 1982). pVK102 has RK2 replication properties

Cosmid clones containing the replication region of pTiA6 excluded pTiAg162 from strain A856 by plasmid incompatibility (Knauf and Nester 1982), causing avirulence. The only cloned molecules that expanded the host range of strain A856 to tomato, tobacco, and sunflower were those that contained the entire T-DNA region of pTiA6 (Fig. 2). This region is defined by the regions of pTiA6 which are transferred into plant tissue where they are stably maintained as part of the genome of

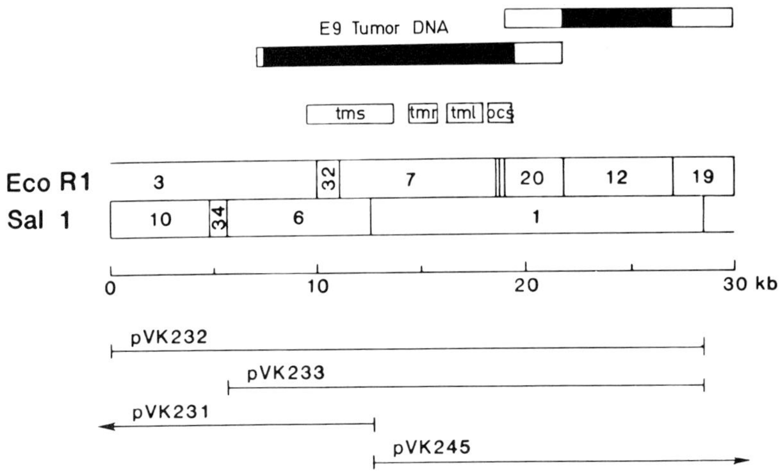

Fig. 2. Recombinant clones which extend the host range of strain A856. The map shows only 30 kb of wide host range Ti-plasmid pTiA6 (181 kb). *Solid bars* flanked by indefinite borders indicate the regions of pTiA6 DNA found in the tobacco tumor line E9 (Thomashow et al. 1980). Regions directly *underneath the solid bar* represent genetic loci defined by transposon insertions (Garfinkel et al. 1981); *tms* shoot phenotype of tumors on tobacco plants; *tmr* root phenotype of tumors on Kalanchoë plants; *tml* large tumors on Kalanchoë plants; *ocs* the octopine synthetase locus. The *Eco*R1 and *Sal*1 restriction enzyme maps are shown (Knauf and Nester 1982). The lines associated with pVK232, pVK233, pVK231, and pVK245 show the extent of pTiA6 DNA cloned into the *Sal*1 site of pVK102. When pVK232 or pVK233 was present in narrow host range strain A856, the host range was increased to include tomato, tobacco, and sunflower

crown gall tissue (Thomashow et al. 1980a). Although the cloned inserts of pVK231 and pVK245 (Fig. 2) are directly adjacent to each other in the T-DNA region of pTiA6, neither clone alone extends the host range of strain A856. Only when *Sal*1 fragments 1 and 6 of pTiA6 were both present, did strain A856 cause tumors on tobacco, tomato, or sunflower. This suggests that a critical locus spans the *Sal*1 site between *Sal*1 fragments 1 and 6. The *tms* locus (Garfinkel et al. 1981) does straddle this site (Fig. 2) and is important in oncogenicity. Since T-DNA functions are expressed in plant tissue (Gelvin et al. 1982, Wilmitzer et al. 1982), it is possible that the entire T-DNA must be present on a single plasmid replicon to be transferred to a plant cell. That is, left and right "ends" must flank the host-range-expanding loci if the DNA is to be transferred. Another possibility is that the coordinate action of two or more loci is necessary (von Leemans et al. 1982, Ream et al. submitted).

If the T-DNA of pTiA6 can extend the narrow host range coded by pTiAg162 in strain A856, what are the implications for the T-DNA of pTiAg162? From earlier studies (Thomashow et al. 1981, Knauf et al. 1983), we know that pTiAg162 is unusual in that its DNA sequences share little homology with probes containing the so-called "common DNA" (Chilton et al. 1978, Depicker et al. 1978), a portion of the T-region conserved among most Ti-plasmids so far examined. Indeed, pTiAg-162 and pTiA6 share little homology. However, wide host range octopine Ti-plasmids such as pTiAg86 exist which are very closely related to pTiAg162, distantly related to pTiA6, but share strong homology to the common DNA probe from pTiA6 (Knauf et al. 1982a, b). In fact, all wide host range Ti-plasmids studied so far possess sequences homologous to the common DNA; naturally-occurring avirulent plasmids and limited host range Ti-plasmids uniformly lack this homology.

Since restriction maps of pTiAg162 based on cosmid clone banks are available, very specific probes from pTiA6 can be used to map homologous DNA sequences in pTiAg162. This has revealed that the T-DNA organization in pTiAg162 differs markedly from pTiA6. No homology was detected between the *tmr* locus of pTiA6 (Garfinkel et al. 1981) and pTiAg162 DNA when fragment *Bam*29 of pTiA6 was used as a hybridization probe. There is distinct homology to the *tms* and octopine synthase *(ocs)* loci. In pTiA6 *tms* and *ocs* are about 3 kb apart and are found in the same T-DNA insert in crown gall tissue. In pTiAg162, the loci defined by DNA homology are about 45 kb apart (Fig. 3).

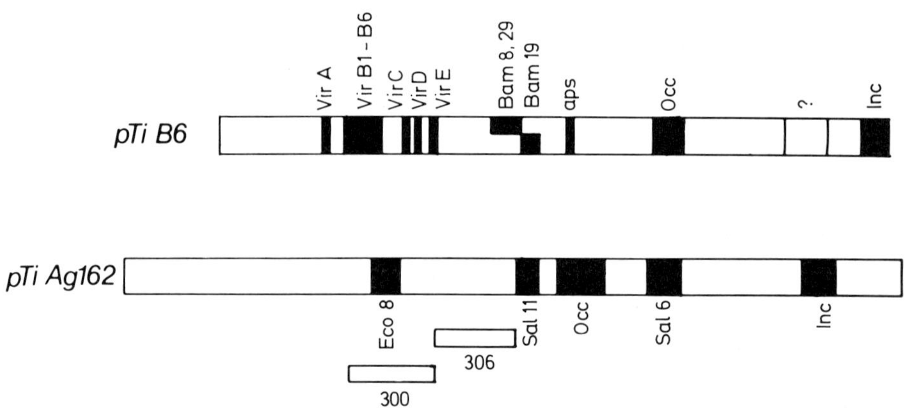

Fig. 3. Organization of a wide host range octopine Ti-plasmid, pTiB6, and a narrow host range octopine Ti-plasmid, pTiAg162. Vir regions contain loci where transposon insertions result in avirulence (Iyer et al., submitted). Occ regions contain loci coding for octopine catabolism (Knauf and Nester 1982). *Inc* regions contain loci determining plasmid incompatibility properties. The *Bam*8, 29 and *Bam*19 regions represent two parts of the core T-DNA of pTiB6 and pTiA6 type plasmids (Thomashow et al. 1980). *Sal*11 of pTiAg162 shares DNA homology with *Bam*19 of pTiB6. *Bam*8, 29 of pTiB6 shares DNA homology with *Sal*6 and a region directly to the left of *Sal*11. *Eco*8 of pTiAg162 shares DNA homology with a region of pTiB6 contained within the *vir*B loci. Moreover, cloned pTiAg162 DNA containing *Eco*8 (Box designated 300) restored wide host range to *vir*B mutants of pTiA6. Cloned DNA of pTiAg162 represented by the box numbered 306 complemented *vir*D and *vir*E mutants of pTiA6 restoring a wide host range as well as virulence

Notably, clones containing the *vir* loci of pTiA6 had no effect on host range expression by strain A856 for the plants tested. *Vir* loci are genetic loci required for the expression of virulence (Garfinkel et al. 1981, Ooms et al. 1980). Since *vir* mutations in pTiA6 can be complemented in *trans* (Klee et al. 1982, Iyer et al submitted), the expression of *vir* gene products most likely takes place in the bacterium. Unless an undescribed *vir* locus exists within the T-DNA region of pTiA6, the failure of known *vir* loci to complement host range expression suggests that the co-ordinate expression of virulence functions in the bacterium is not a major determinant in differentiating susceptible and resistant plant hosts. Using cosmid-cloned fragments of the narrow host range plasmid pTiAg162, we have been able to restore virulence to seven different *vir* mutations of pTiA6 (Fig. 3). In these cases, pTiAg162 DNA restored virulence on tobacco and *Kalanchoë* plant hosts, even though pTiAg162 itself does not code for virulence on these plants. These results indicate that the *vir* loci of pTiAg162 can function in the expression of a wider host range than actually is coded for by pTiAg162.

Implications and Outlook

The host range differences between pTiA6 and pTiAg162 led to a rather surprising explanation. Although one can hypothesize roles for factors present in the bacterium (such as binding of the pathogen to host plant cells), the failure of strain A856 to cause tumors on *Nicotiana tabacum* or *Lyocpersicon esculeritum* can be traced to the T-DNA of pTiAg162 even though that T-DNA of pTiAg162 is obviously functional in *Nicotiana glauca* and *Vitis vinifera.* The idea that pTiAg162 oncogenes are ineffective on certain plant hosts suggests that the lack of a crown gall response on resistant plant hosts should not be considered proof that *Agrobacterium* cannot transfer and establish T-DNA in those plant species.

However, it is unlikely that all differences in host range associated with the plasmid will be due to T-DNA factors. For example, the virulence differences on *Kalanchoë* plants coded by the wide host range nopaline plasmids pTiPPI-1 and pTi2/6 should be analyzed to see if this variation is associated with *vir* loci or T-DNA loci. Other limited host range strains should be examined to see if that phenotype is Ti-plasmid-coded and if that plasmid contains sequences homologous to the portion of T-DNA found so far in all wide host range Ti-plasmids (Knauf et al. 1983). Since transposon mutagenesis of the *Agrobacterium* chromosome resulted in altered host ranges (Garfinkel and Nester 1980), it would be useful to identify the function and cellular localization of the relevant wild-type gene products. Also, the ability to transform certain plant species should prove to be a useful tool in efforts to understand how these genes exert their effects on plant tissues.

The observation that T-DNA loci on a separate DNA replicon can complement host range suggests that a genetic locus on a small cloning vector might be engineered in such a way that the Ti-plasmid will mobilize that locus into the plant genome. This would provide a much simpler method for analyzing gene expression in plant tissue than the marker exchange techniques currently used (Matzke and Chilton 1981, von Leemans et al. 1981).

References

Anderson AR, Moore LW (1979) Host specificity in the genus *Agrobacterium*. Phytopathology 69:320–323

Chilton M-D, Montoya AL, Merlo DJ, Drummond MH, Nutter R, Gordon MP, Nester EW (1978) Restriction endonuclease mapping of a plasmid that confers oncogenicity upon *Agrobacterium tumefaciens* strain B6-806. Plasmid 1:254–269

Currier TC, Nester EW (1976) Evidence for diverse types of large plasmids in tumor-inducing strains of *Agrobacterium*. J Bacteriol 126:157–165

DeCleen M, DeLey J (1976) The host range of crown gall. Bot Rev 42:389–466

DeLey J, Tijtgat R, DeSmedt J, Michiels M (1973) Thermal stability of RNA:DNA hybrids within the genus *Agrobacterium*. J Gen Microbiol 78:241–252

Depicker A, Montagu M Van, Schell J (1978) Homologous DNA sequences in different Ti-plasmids are essential for oncogenicity. Nature 275:150–153

Enquist L, Sternberg N (1979) In vitro packaging of Dam vectors and their use in cloning DNA fragments. In: Wu R (ed) Methods in enzymology, vol 68. Academic Press, London New York, pp 281–298

Garfinkel DJ, Nester EW (1980) Mutants in *Agrobacterium tumefaciens* affected in crown gall tumorigenesis and octopine catabolism. J Bacteriol 144:732–743

Garfinkel DJ, Simpson RB, Ream LW, White FF, Gordon MP, Nester EW (1981) Genetic analysis of crown gall: Fine structure map of the T-DNA by site-directed mutagenesis. Cell 27: 143–153

Gelvin SB, Thomashow MF, McPherson JC, Gordon MP, Nester EW (1982) Sizes and map positions of several plasmid-DNA-coded transcripts in octopine-type crown gall tumors. PNAS 79:76–80

Iyer VN, Klee HJ, Nester EW (1983) Units of genetic expression in the virulence region of a plant tumor-inducing plasmid of *Agrobacterium tumefaciens*. Mol Gen Genet (submitted)

Keane PJ, Kerr A, New PB (1970) Crown gall of stone fruit. II. Identification and nomenclature of *Agrobacterium* isolates. Aust J Biol Sci 23:585–595

Kerr A, Panagopoulos CG (1977) Biotypes of *Agrobacterium raciobacter* var. *tumefaciens* and their biological control. Phytopathol Z 90:172–179

Klee HJ, Gordon MP, Nester EW (1982) Complementation analysis of *Agrobacterium tumefaciens* Ti-plasmid mutations affecting oncogenicity. J Bacteriol 150:327–331

Knauf VC, Nester EW (1982) Wide host range cloning vectors: A cosmid clone bank of an *Agrobacterium* Ti-plasmid. Plasmid 8:45–54

Knauf VC, Panagopoulos CG, Nester EW (1982) Genetic factors controlling host range of *Agrobacterium tumefaciens*. Phytopathology 72:1545–1549

Knauf VC, Panagopoulos CG, Nester EW (1983) Comparison of Ti-plasmids from three different biotypes of *Agrobacterium* isolated from grapevine. J Bacteriol 146:484–493

Leemans J, Shaw C, Deblaere R, DeGreve H, Hernalsteens JP, Maes M, Montagu M Van, Schell J (1981) Site-specific mutagenesis of *Agrobacterium* Ti-plasmids and transfer of genes to plant cells. J Mol Appl Genet 1:149–164

Leemans J, Deblaere R, Willmitzer L, DeGreve H, Hernalsteens JP, Montagu M Van, Schell J (1982) Genetic identification of functions of TL-DNA transcripts in octopine crown galls. EMBO J 1:147–152

Loper JE, Kado C (1979) Host range conferred by the virulence-specifying plasmid of *Agrobacterium tumefaciens*. J Bacteriol 139:591–596

Matzke AJM, Chilton M-D (1981) Site specific insertion of genes into the T-DNA of the *Agrobacterium* tumor-inducing plasmid: An approach to genetic engineering of higher plant cells. J Mol Appl Genet 1:39–49

Nester EW, Kosuge T (1981) Plasmid specifying plant hyperplasias. Annu Rev Microbiol 35: 531–565

Ooms G, Klapwijk PM, Poulis JA, Schilperoort RA (1980) Characterization of Tn904 insertions in octopine Ti-plasmid mutants of *Agrobacterium tumefaciens*. J Bacteriol 144:82–91

Panagopoulos CG, Psallidas PG (1973) Characteristics of Greek isolates of *Agrobacterium tumefaciens* (EF Smith and Townsend). Conn J Appl Bacteriol 36:233–240

Ream LW, Gordon MP, Nester EW (1983) Multiple mutations in the T-region of the *Agrobacterium tumefaciens* tumor-inducing plasmid. PNAS (submitted)

Sciaky D, Montoya AL, Chilton M-D (1977) Fingerprints of *Agrobacterium* Ti-plasmids. Plasmid 1:238–253

Süle S (1978) Biotypes of *Agrobacterium tumefaciens* in Hungary. J Appl Bacteriol 44:207–213

Thomashow MF, Nutter R, Montoya AL, Gordon MP, Nester EW (1980a) Integration and organization of Ti-plasmid sequences in crown gall tumors. Cell 19:729–739

Thomashow MF, Panagopoulos CG, Gordon MP, Nester EW (1980b) Host range of *Agrobacterium tumefaciens* is determined by the Ti-plasmid. Nature 283:794–796

Thomashow MF, Knauf VC, Nester EW (1981) Relationship between the limited and wide host range octopine-type Ti-plasmids *Agrobacterium tumefaciens*. J Bacteriol 146:484–493

Willmitzer L, Simons G, Schell J (1982) The TL-DNA in octopine crown-gall tumors codes for seven well-defined polyadenylated transcripts. EMBO J 1:139–146

The Biology of Genetic Transformation of Higher Plants by *Agrobacterium rhizogenes*

D. TEPFER [1]

> "So what molecular biology has done, you see, is to prove beyond any doubt but in a totally new way the complete independence of the genetic information from events occurring outside or even inside the cell, to prove by the very structure of the genetic code and the way it is transcribed that no information from outside of any kind, can ever penetrate the inheritable genetic message."
>
> One of the founders of molecular biology, quoted in The *Eighth Day of Creation,* by Horace Freeland Judson, Simon and Schuster 1979

Introduction

The eukaryotic genome is normally thought to be isolated from foreign nucleic acid encoded information. The well known exceptions are due to agents which engage in genetic transformation: certain viruses and *Agrobacterium tumefaciens.* The latter carries an episome, the Ti or tumor-inducing plasmid, a region of which, the T-DNA or transferred DNA, becomes integrated into plant DNA in the course of infection by the bacterium. Another T-DNA has recently come to light, originating in the Ri, or root-inducing, plasmid of *Agrobacterium rhizogenes.* This T-DNA is present in the adventitious roots induced by *Agrobacterium rhizogenes* (Chilton et al. 1982, White et al. 1982, Willmitzer et al. 1982). The purpose of this paper is to show that the Ri T-DNA, unlike the Ti T-DNA, is compatible with the processes of regeneration and meiosis, and that its presence in roots, in plants derived from those roots, and in the progeny of those plants confers a phenotype which, rather than appearing harmful, suggests adaptive significance. The possibility that the Ri T-DNA encodes beneficial information capable of passing through the barriers usually assumed to insulate organisms from foreign genetic information is the subject of a discussion of parasexual genetic exchange presented below.

1 Laboratoire de Biologie Cellulaire, C.N.R.A., 78000 Versailles, France

Molecular Genetics of the Bacteria-Plant Interaction
(ed. by A. Pühler)

Results

Acquisition and Transmission of T-DNA: The Ri plasmid appears to behave much like the Ti plasmid in its mode of plant cell transformation. A defined region (the T-DNA) is stably integrated into the plant genome. The essential difference between the Ri and Ti T-DNAs lies in their effects on the plant cell. In the present study *Agrobacterium rhizogenes* was used to induce roots in three species: carrot, tobacco and morning glory *(Convolvulus arvensis).* Plants regenerated spontaneously from cultured roots of tobacco and morning glory, and were obtained artifically through somatic embryogenesis from carrot roots. T-DNA is present in *Agrobacterium rhizogenes* induced roots, in plants regenerated therefrom, and in the progeny of transformed plants (Fig. 1).

The Rhizogenes Phenotype: In axenic, continuous culture, roots induced by *Agrobacterium rhizogenes* differ morphologically and physiologically from normal

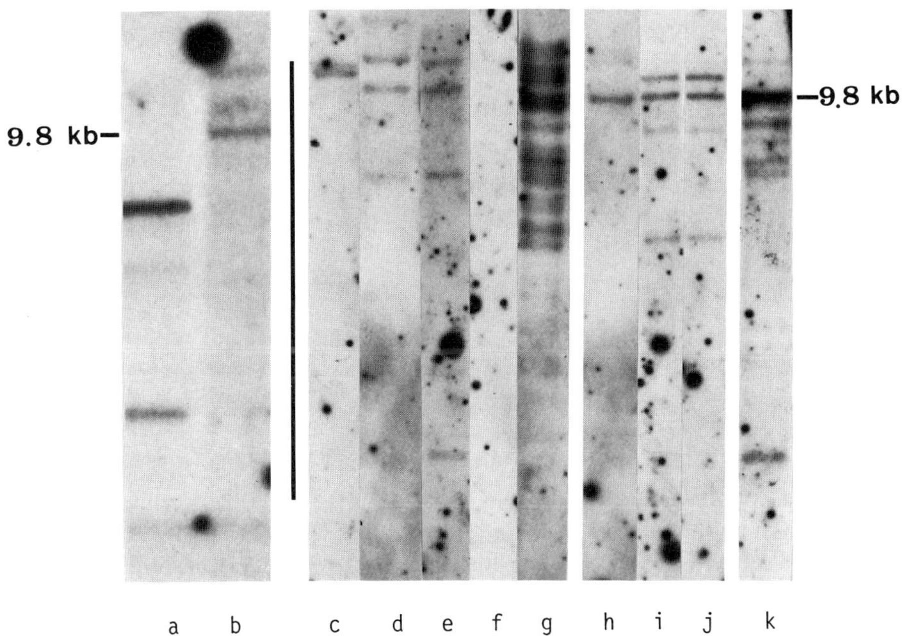

Fig. 1. T-DNA in roots induced by *Agrobacterium rhizogenes* in plants regenerated from transformed roots, and in the progeny of transformed plants. *a* DNA extracted from *Convolvulus sepium* roots induced by *Agrobacterium rhizogenes* strain 8196, probed with nick translated 8196 plasmid, *b* same as lane a, but induced by strain A4, *c* tobacco DNA extracted from plants regenerated from A4 induced root clone number 6, probed with A4 plasmid, *d* same as in lane c, but clone number 4, *e* one of the progeny of the plant in lane d, bearing the morphological markers for transformation, *f* same as in lane 3, but a progeny not exhibiting the *rhizogenes* phenotype, *g* a single copy reconstruction, i.e., one copy of A4 plasmid per tobacco diploid genome, *h Convolvulus arvensis* regenerated from A4 induced root clone number 5, *i* same as lane h, but clone number 7, *j* progeny of plant in lane i, *k* carrot plant, progeny of transformed plant regenerated by somatic cell embryogenesis from roots transformed by strain A4

roots: (1) they grow faster, (2) they are highly branched (reduced apical dominance), and (3) they are plagiotropic, i.e. they tend to grow horizontally instead of downward. Their high growth rate makes them effectively self-disinfecting, since they outgrow contaminating bacteria. Rapid tip elongation combined with a high frequency of lateral root formation results in an elevated accumulation of biomass in culture. Different opines of the agropine family are present, depending on the Ri plasmid involved in the original transformation (Fig. 2).

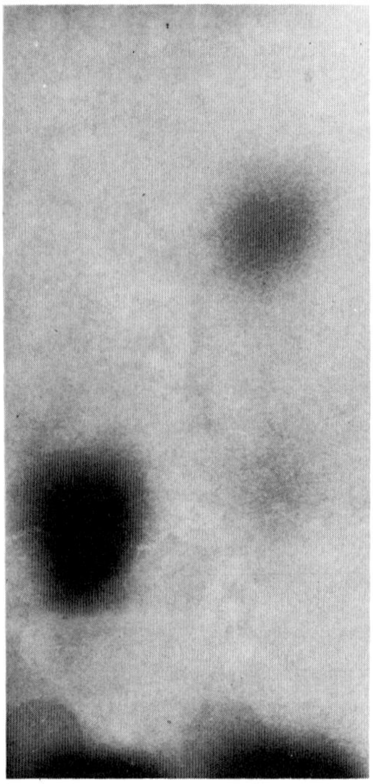

Fig. 2. Opines in *Convolvulus sepium* induced by *a* strain 8196 and *b* strain A4. (See Fig. 1, lanes *a* and *b*, for T-DNA content of the same roots)

The root systems of T-DNA containing plants, regenerated from transformed roots, exhibit the *rhizogenes* phenotype (Fig. 3). The aerial parts of these plants are also modified: (1) the leaves are wrinkled (Fig. 4), a phenomenon first reported in similar plants by Ackerman (1977), (2) apical dominance is reduced, i.e., they are highly branched (Fig. 5), and (3) the biennial species examined, carrot, became an annual. (The finding that Ri T-DNA confers a biennial to annual switch has been confirmed in endive – Touraud and Tepfer, unpubl. results). Some of these transformed plants continue to make opines (Fig. 6). The presence of T-DNA is correlated with essentially the same phenotype in tobacco, carrot, and morning glory.

A number of the progeny of transformed plants exhibit the *rhizogenes* phenotype and contain, in the cases examined, Ri T-DNA. Normal appearing plants do not contain T-DNA (Fig. 1). The *rhizogenes* phenotype was used to determine the pattern of inheritance of Ri T-DNA for plants originating from tobacco root clone number 4.

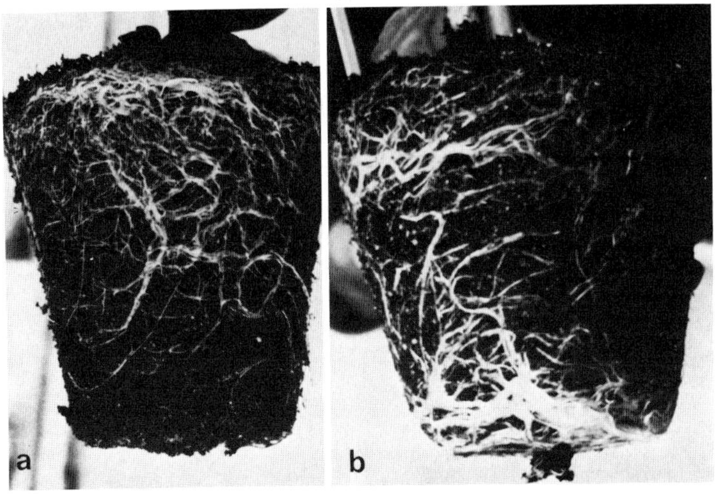

Fig. 3a, b. Root system of **a** a tobacco plant transformed by A4, and **b** a control regenerated from normal roots in organ culture

Fig. 4a, b. Leaves on **a** a transformed (A4) tobacco plant are wrinkled, and on **b** a control plant they are smooth

Fig. 5a, b. Mature tobacco plants: a control and b transformed (A4)

A transformed plant pollinated by a normal plant gave 35 transformed and 31 normal progeny. A normal plant fertilized by a transformed plant resulted in 21 transformed and 37 normal offspring. A transformed plant, self-fertilized, gave 36 transformed and 22 normal daughter plants. These results are consistant with the Ri T-DNA being transmitted as a dominant Mendelian element, but associated with reduced male gamete production (gametic selection), a phenomenon often observed in plant mutants. In fact, seed production is generally reduced in plants bearing the Ri T-DNA (Fig. 7). Pollen from clone number 4 germinates poorly in in vitro germination tests, although other clones of transformed tobacco produce pollen which germinates normally, and do not exhibit gametic selection. Spano and Costantino (1982) have reported partial male sterility in similar plants, derived from *Agrobacterium rhizogenes* induced roots, but not rigorously shown to contain T-DNA.

All components of the *rhizogenes* phenotype segregate together, with the exception of opine synthesis, which is not present in any of the progeny of a transformed, opine-synthesizing *Convolvulus arvensis* parent, although apparently unaltered T-DNA is present (Figs. 1 and 6).

Aside from reduced seed production and, in some cases, partial male sterility, there is no evidence that the *rhizogenes* phenotype is disadventageous. The Ri T-DNA is conserved during regeneration and meiosis. Its presence is correlated with essentially the same phenotype in three species. One cannot help but wonder what the significance of this phenotype might be in nature.

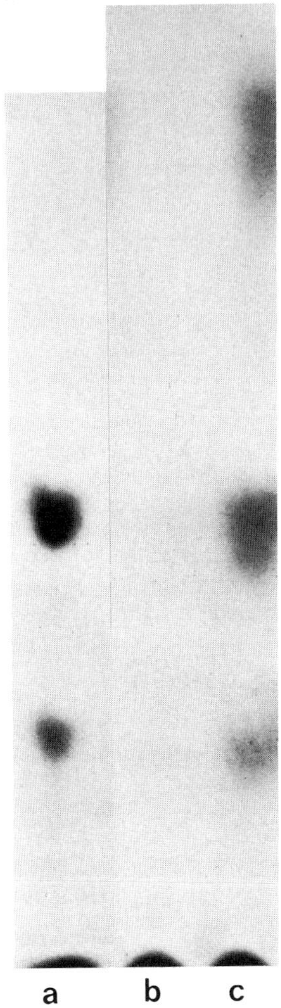

Fig. 6a–c. Opines in transformed *Convolvulus arvensis:* **a** standards *(from top to bottom)* agropine, mannopine, and start of high voltage electrophoresis; **b** opine extract of the progeny of a transformed (A4) *Convolvulus arvensis* containing T-DNA (see Fig. 1, lane *j*); **c** opine extract of parent of plant in lane *b*, containing the same T-DNA (Fig. 1, lane *i*), and also producing mannopine, agropine, and an unknown agropine derivative

Plants which naturally regenerate from roots, such as *Convolvulus arvensis,* presumably would be capable of acquiring an Ri T-DNA without human intervention. Those species which are not known to regenerate may do so at a low frequency, or may be induced to do so when they contain an Ri T-DNA. It is plausible that the Ri T-DNA is capable of directing the formation of a supplementary root system from a primary inoculation, and equally plausible that in some species this secondary root system could give rise to a transformed genotype. The only fact supporting this contention is the ease with which the primary transformation, regeneration, and subsequent sexual transmission take place in the laboratory.

The T-DNA of the Ti plasmid can also be induced to enter a plant genome and be carried through meiosis, but only after extensive laboratory manipulation, accompanied by modification of the T-DNA (Otten et al. 1981). In contrast to the Ri T-DNA it would seem that the oncogenic character of the Ti T-DNA would prevent it

Fig. 7. Seed capsules on transformed (A4) tobacco *(top)* and control *(bottom)*

from becoming integrated naturally into a plant species. The Ri-T-DNA would likely be a better vector for introducing foreign genes into plants.

Discussion

In view of the current discussion of the possibility of horizontal gene transfer, based on the similarities of protein coding sequences for similar genes in widely different species (see Lewin 1982, for a recapitulation), it would be appropriate to discuss the ecological and evolutionary implications of the Ri T-DNA and its associated phenotype. A thesis is offered consisting of a brief statement, followed by a series of at least partially testable hypotheses, beginning with the most probable and ending with the most speculative.

A thesis: Sexually isolated species can exchange genetic information encoded in specialized structures, T-DNAs, capable of moving in and out of eukaryotic genomes under appropriate selective pressure, using microorganisms as intermediaries.

Hypothesis 1: A Selective Advantage is Conferred by the Ri T-DNA

The entry of *Agrobacterium rhizogenes* into a root, and the subsequent formation of transformed roots bearing the *rhizogenes* phenotype (reduced apical dominance and plagiotropism) would result in the production of a secondary root system, composed of tightly interlaced rootlets placed near the surface of the soil. The adaptive significance of supplementary roots might be to improve anchorage, augment the possibility of interactions with other microorganisms, protect against anoxia due to flooding, or (conversely) protect against drought. Moore et al. (1979) reported that apple trees infected with *Agrobacterium rhizogenes* better withstood drought than uninfected trees.

As one can imagine a selective advantage resulting from a primary inoculation, it is possible to conceive of the selective advantage of the Ri phenotype in the aerial parts of the plant as well. Transformed tobacco plants exhibit reduced apical dominance, and are thus more highly branched and produce more leaves (Fig. 5). The photosynthetic capacity of transformed plants may therefore be improved. The switch from biennialism to annualism is likewise a possibly beneficial aspect of the *rhizogenes* phenotype under certain environmental conditions. This hypothesis can be directly tested by subjecting transformed and normal plants to various environmental stresses, which might select for this phenotype. A preliminary test shows that conditions can be established in vitro (Fig. 8), which in fact favor the growth of tobacco seedlings having the *rhizogenes* phenotype.

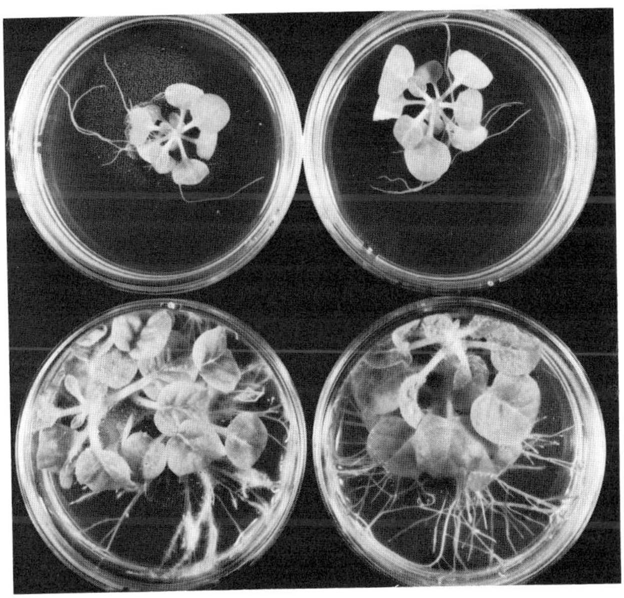

Fig. 8. Growth of progeny of transformed tobacco in sealed petri dishes: plants of normal phenotype, not containing T-DNA (see Fig. 1, lane *f*) on *top*, and plants of *rhizogenes* phenotype, containing T-DNA (Fig. 1, lane *e*) on *bottom*

Hypothesis 2: Other T-DNAs Exist in Other Microorganisms

Other microorganisms engage in interactions with plants. Some of these interactions (whether beneficial or deleterious) might be mediated by parasexual genetic exchange. *Rhizobium* is a likely candidate, due to its taxonomic similarity to the *Agrobacteria* and its specific morphogenic effect, nodulation. Another candidate is the RNA virus which causes the disease "rhizomania" (Putz 1981). The virus is harbored by a soil fungus and causes a syndrome akin to the *rhizogenes* phenotype: numerous, highly branched roots and wrinkled leaves.

A given microorganism-plant relationship can be tested for T-DNA mediation by the biochemical means used for the *Agrobacteria:* demonstrating the presence of T-DNA and/or opines. In addition, among the sequences which are presumably shared by Ti and Ri T-DNAs, those which are necessary for T-DNA integration into plant DNA may turn out to be diagnostic of other T-DNAs in other systems.

Hypothesis 3: Ri T-DNA is, at Least in Part, of Plant Origin. DNA Can Be Transferred from a Plant to a Microorganism

Plant DNA contains sequences complementary to the Ri plasmid (White et al. 1982, Tepfer 1982), and more specifically to the Ri T-DNA (White et al., in press). The significance of these homologies has not been determined, but one might suppose that they represent genetic information similar to that encoded by the Ri T-DNA. One might call these endogenous T-DNAs, as opposed to the exogenous Ri T-DNA carried by the bacterial plasmid. It would be logical to assume that the exogenous T-DNA is constitutively expressed while similar information carried by the endogenous T-DNA is regulated, and only expressed as part of a developmental program, e.g., root morphogenesis, flowering physiology, etc. In fact the various components of the *rhizogenes* phenotype appear in nature, e.g., wrinkled leaves in response to viral infection (Tepfer and Chessin 1959), proliferation of rootlets in the upper layers of the soil as a function of drought stress (Vartanian 1981). The functional relationship between endogenous and exogenous T-DNAs might be investigated using the exogenous Ri T-DNA as a probe for the presence and expression of its endogenous analogue. This hypothesis would predict that both T-DNAs contain structurally similar genetic information and code for similar proteins, while the structures and mechanisms controlling their expression would be different.

Since exogenous and endogenous T-DNAs are similar, one could assume that either they (1) share a common origin, (2) evolved independently in parallel, (3) that the endogenous T-DNA originated in the bacterium and has become stabilized in the plant, or (4), conversely, that the exogenous T-DNA originated in the plant and was at some time transferred to the bacterial plasmid. The latter explanation is particularly interesting and may be partially verifiable by inserting an antibiotic resistance into the Ri T-DNA in the bacterium, transferring the T-DNA plus antibiotic resistance to a plant, and then attempting to clone, in planta, the Ri T-DNA back into a suitable bacterium, relying on the antibiotic resistance as a selectable marker.

Hypothesis 4: Parasexual Genetic Exchange Plays a Role in Evolution

Parasexual transfer of various T-DNAs from microorganisms to plants, under conditions which select for the T-DNA associated phenotype(s), would lead to the simultaneous appearance and maintenance of specific T-DNAs in a variety of species according to enviromental conditions. Furthermore, if T-DNA can originate in a plant and spread via a microorganism intermediate to other species, the evolutionary significance of parasexual genetic exchange is extended. A community could draw upon T-DNA encoded characters carried by another species. Both species must be capable of interacting with the same microorganism intermediary. The taxonomy of parasexual evolution would therefore group species according to their sensitivity to a given intermediary.

In the absence of positive selective pressure the propagation of Ri T-DNA within a species would, if the effects of Ri T-DNA on tobacco are a good example, be restricted by gametic selection and a general reduction in gametic production. Thus acquisition of the *rhizogenes* phenotype would only take place on a large scale under conditions of positive selection.

Hypothesis 5: Pathogenic T-DNAs May Be Instruments of Interspecies Conflict

Just as one can imagine a beneficial role for T-DNAs, it is possible to conceive of a species releasing a set of genes which are pathogenic when inserted into the DNA of another species. A plant might conserve in its genome a T-DNA which when released would infect an insect which is feeding on that plant. Such T-DNAs would become viruses when activated by the insect pathogen and would constitute a defensive tactic. Likewise, plant pathogens (viruses and Ti T-DNA) may have originated as an offensive tactic against a competing species, on the part of a species resistant to self-infection. This hypothesis would predict sequence similarity between T-DNAs and viruses, and the existence of sequences in plant DNA complementary to viruses which attack plant pathogens.

Acknowledgements. The author wishes to thank his colleagues in the Laboratoire de Biologie Cellulaire for their generous assistance, in particular J.P. Bourgin, M. Tepfer, and H. Bannerot for critically reading the manuscript, and in addition, J. Tempé, in whose laboratory the author performed some of the experiments.

References

Ackermann C (1977) Pflanzen aus *Agrobacterium rhizogenes*-Tumoren aus *Nicotiana tabacum*. Plant Sci Lett 8:23–30

Chilton M, Tepfer D, Petit A, David C, Casse-Delbart F, Tempé J (1982) *Agrobacterium rhizogenes* inserts T-DNA into the genomes of the host plant root cells. Nature 295:432–434

Lewin R (1982) Can genes jump between eukarytic species? Science 217:42–43

Moore L, Warren G, Strobel G (1979) Involvement of a plasmid in the hairy root disease of plants caused by *Agrobacterium rhizogenes*. Plasmid 2:617–626

Otten L, De Greve H, Hernalstiens JP, Van Montagu M, Schieder O, Straub J, Schell J (1981) Mendelian transmission of genes introduced into plants by the Ti plasmids of *Agrobacterium tumefaciens*. Mol Gen Genet 183:209–213

Putz C (1981) Etude du virus des nervures jaunes et nécrotiques de la betterave (Beet Necrotic Yellow Virus) agent de la Rhizomanie. Thesis, Univ Louis Pasteur, Strasbourg

Spano L, Costantino P (1982) Regeneration of plants from callus cultures of roots induced by *Agrobacterium rhizogenes* on tobacco. Z Pflanzenphysiol 106:87–92

Tepfer D (1982) La transformation génétique de plantes supérieures par *Agrobacterium rhizogenes*. 2e Coll Rech Fruit, Bordeaux, pp 47–59

Tepfer S, Chessin M (1959) Effects of tobacco mosaic virus on early leaf development in tobacco. Am J Bot 46:496–509

Vartanian N (1981) Some aspects of structural and functional modifications induced by drought in root systems. Plant Soil 63:83–92

White F, Ghidossi G, Gordon M, Nester E (1982) Tumor induction by *Agrobacterium rhizogenes* involves the transfer of plasmid DNA to the plant genome. Proc Natl Acad Sci USA 79: 3193–3197

Willmitzer L, Sanchez-Serrano J, Buschfield E, Schell J (1982) DNA from *Agrobacterium rhizogenes* is transferred to and expressed in axenic hairy root plant tissues. Mol Gen Genet 18: 16–22

Expression of the T-Region of Octopine Plasmid pTi Ach5 into Protein in Bacterial Systems

J. SCHRÖDER [1], W. KLIPP [2], A. HILLEBRAND [1], R. EHRING [3], C. KONCZ [1], and G. SCHRÖDER [1]

Introduction

The T-DNA in crown galls is responsible for opine production and hormone-independent growth of plant cells. To understand the underlying mechanisms, it is necessary to analyze the chain of events from genes through transcripts to proteins and their function. Genes on the T-DNA have been mapped by functional analysis and their transcripts in plant cells have been described (this vol.), but identification of the proteins and analysis of their actions have proved to be difficult. One of the reasons for the problems is the very low concentration of the transcripts. Attempts to identify their protein products have shown that at least some of the transcripts are translated in vitro into defined proteins (McPherson et al. 1980, Schröder et al. 1981a, Murai and Kemp 1982, Schröder and Schröder 1982). In octopine tumors it has been demonstrated that one of the proteins is the octopine-synthesizing enzyme (Schröder et al. 1981a, Murai and Kemp 1982), and it seems likely that other opine-synthesizing enzymes are also T-DNA encoded. The functions of the other proteins detected by in vitro translation are unknown, and with several transcripts it is not yet possible to identify reliably their protein products.

Another approach to T-DNA encoded proteins is the isolation from transformed plant cells. However, functions like induction and maintenance of hormone-independent, undifferentiated growth do not describe single enzymatic reactions but complex morphogenetic events. The enzymatic reactions in these processes, and more importantly in their control, are not understood and therefore it has not been possible to design a functional assay which could be used conveniently to identify and to isolate these proteins.

For these reasons we have developed a different approach to obtain T-DNA encoded peptides or proteins. We investigated whether parts of the T-region could be expressed into protein in bacterial systems, either after fusion with bacterial genes or as complete proteins (Schröder et al. 1981b). Here we report that the T-region of octopine plasmid pTi Ach5 can express in *E. coli* a set of proteins with promoter sequences from the T-region. The protein coding regions are in that part which is responsible for hormone-independent growth of transformed plant cells. Since the same proteins are expressed in a cell-free system prepared from *Agrobacterium*

1 Max-Planck-Institut für Züchtungsforschung, Abt. Schell, 5000 Köln 30, FRG
2 Lehrstuhl für Genetik, Universität Bielefeld, Postfach 8640, 4800 Bielefeld 1, FRG
3 Institut für Genetik, Universität Köln, Weyertal 121, 5000 Köln 41, FRG

Molecular Genetics of the Bacteria-Plant Interaction
(ed. by A. Pühler)

tumefaciens it seems possible that they are functional in Agrobacteria, and the possibility is discussed that expression in the bacteria plays a role in the early events of crown gall tumorigenesis.

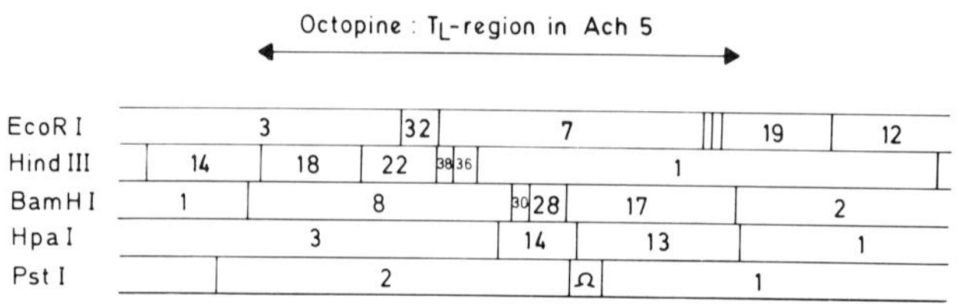

Fig. 1. Restriction map of the T-region in octopine plasmid pTi Ach5 (De Vos et al. 1981). *Line with arrows* indicates the position of the T_L-region which is most often found in transformed plant cells (De Beuckeleer et al. 1981)

Table 1. List of recombinant plasmids with fragments from octopine plasmid pTi Ach5. When orientations of inserts are given (orientation I = *.1;* orientation II = *.2*), they are defined by the presence of a promoter on the vector plasmid which reads from right to left *(.1)* or from left to right *(.2)* into the Ti-plasmid fragment as drawn in Fig. 1

Recombinant plasmid	Ti-plasmid fragments	Vector plasmid	Reference
pGV0219	HindIII (14-18-22-38)	pBR322	De Vos et al. (1981)
pGV0153	BamHI 8	pBR322	De Vos et al. (1981)
pGV0201	HindIII 1	pBR322	De Vos et al. (1981)
pHD106.*1* pHD106.*2*	HindIII 22	pUR51	This vol.
pWK500.*1* pWK500.*2*	EcoRI 7	pACYC184	Schröder et al. (1981b)
pWK500.*2*PstI	EcoRI (ΔPstIΩ)	pACYC184	This vol.
pWK500.*2*HpaI 14	EcoRI 7 (ΔHpaI 14)	pACYC184	This vol.

Results

74 KD Protein

Fragment EcoRI 7 (see Fig. 1) inserted into the chloramphenicol resistance gene of pACYC184 produced in *E. coli* minicells a 74 KD protein (Fig. 2, lane 2). Expression was found only with orientation II of the fragment where transcription from the vector reads from left to right into the Ti-plasmid fragment. This suggested that expression was dependent on the promotor of the chloramphenicol resistance gene and that the protein possibly represented a fusion between the truncated part of the vector-encoded chloramphenicol acetyltransferase produced by insertion of fragment EcoRI 7 into pACYC184 and a protein encoded in the left part of EcoRI 7. Previous experiments have shown that such fusion proteins react with antibodies against

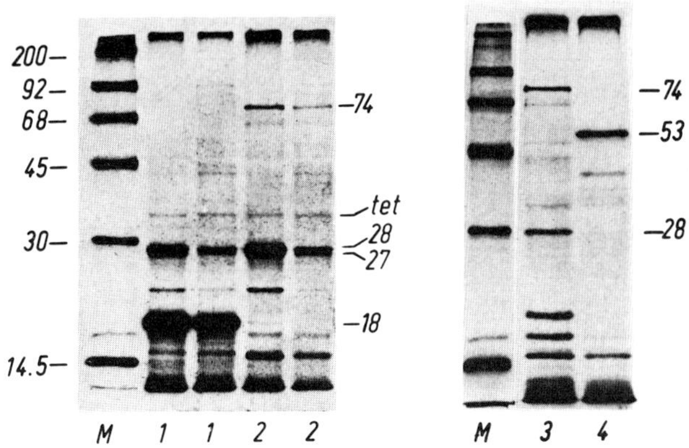

Fig. 2. Gelelectrophoretic analysis of proteins synthesized in *E. coli* minicells with fragments cloned in pACYC184 (see Table 1 and Fig. 1 for identification). *Lanes 1* EcoRI 7 (pWK500.*1;* different exposures); *2* EcoRI 7 (pWK500.*2;* different exposures); *3* EcoRI 7 (ΔPstIΩ); *4* EcoRI 7 (ΔHpaI 14). *M* Marker proteins with their size in KD (= $M_r \times 10^{-3}$) at the *left*. *Numbers* at the *right* indicate the size of proteins synthesized from the Ti-plasmid fragments. *Tet* A 34 KD protein encoded by the tetracycline resistance on pACYC184. Minicells were isolated, labeled and analyzed as described (Schröder et al. 1981b)

this bacterial enzyme. The 74 KD protein did not react with the antibodies and thus was probably not a fusion product. The simplest explanation was then that transcription started on the vector plasmid, but that the ribosome binding site and the protein coding region were on the Ti-plasmid fragment. This was confirmed by transposon mutagenesis with Tn5 (data not shown), and the position of the protein coding region was localized in EcoRI 7 as drawn in Fig. 5. The location was also confirmed by the following experiment: The HpaI fragment in EcoRI 7 was removed by restriction and religation of the recombinant plasmid, and the new plasmid was tested in the minicells. Fig. 2, lane 4, shows that the modified plasmid produced a 53 KD protein, as would be predicted from the position of the left HpaI site in EcoRI 7.

The question whether the 74 KD protein could also be expressed with a promoter activity from the T-region was investigated with fragment BamHI 8. Fig. 5 shows that it contains about 85% of the coding region of the 74 KD protein and also a long stretch of DNA extending to the left of EcoRI 7. When pBR322 containing BamHI 8 was expressed in minicells it synthesized a 49 KD protein and a 66 KD protein (Fig. 3, lane 3). A plasmid containing a composite fragment with Hind III (14-18-22-38) which covers the entire left two thirds of BamHI 8 (see Fig. 1) produced a 49 KD protein but not a 66 KD protein or one of comparable size. This suggests that the 66 KD protein was encoded at the right end of BamHI 8, and that it represents a shortened form of the 74 KD protein. The size of the truncated protein is consistent with the position of the BamHI site in the coding region (Fig. 5). This result therefore indicates that the gene can be expressed in *E. coli* with a promoter activity from the T-region located to the left of EcoRI 7.

49 KD Protein

Plasmid pBR322 containing Ach5 Hind III (14-18-22-38) expressed a protein of 49 KD in *E. coli* minicells (Fig. 3, lane 2) which was also produced with BamHI 8 (Fig. 3, lane 3). The protein was not synthesized with plasmids containing HindIII 14 or 18 when cloned separately or together, suggesting that the gene was in HindIII 22 or close to it. Recloning of HindIII 22 followed by expression in minicells showed that the protein was synthesized in its entire length (Fig. 3, lane 4), indicating that the complete coding region was in the fragment. Expression was independent of the orientation of the insert in the vector plasmid, suggesting transcription with promoter activities within HindIII 22. The finding that BamHI 8 and HindIII (14-18-22-38) expressed the same protein although they contain long stretches of DNA from the T-region either to right or left of HindIII 22 supports the conclusion that promoter activities from the vector are not necessary for expression of the 49 KD protein.

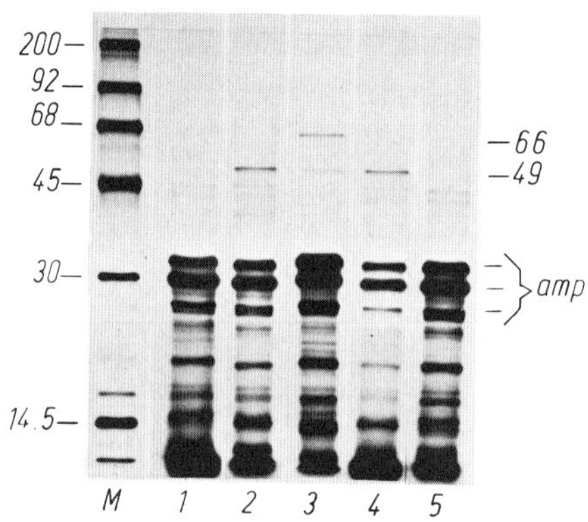

Fig. 3. Gelelectrophoretic analysis of proteins synthesized in *E. coli* minicells. *Lanes 1* Control (pBR322); *2* HindIII (14-18-22-38) (pGV0219); *3* BamHI 8 (pGV0153); *4* HindIII 22 (pHD 106.1); *5* control (pUR51; Rüther 1980). *Numbers* at the *right* indicate the size in KD of proteins synthesized from the Ti-plasmid fragments. *Amp* Proteins produced from the ampicillin resistance genes in pBR322 and pUR51. See legend to Fig. 1 for further details

28 KD and 27 KD Proteins

These two proteins were synthesized with EcoRI 7 cloned in pACYC184 (Fig. 2, lanes 1, 2). Expression was independent of the orientation of the insert in the vector plasmid, suggesting that transcription started within the T-region fragment. Since the left part of EcoRI 7 codes for the 74 KD protein and the right end for the carboxy-terminal part of the octopine-synthesizing enzyme (Schröder et al. 1981b), it seemed likely that the two proteins are encoded in the middle of the fragment. This was confirmed by the following experiments: The small PstI fragment was excised from Eco RI 7 by restriction and religation of the plasmid, and the new plasmid was tested in the minicell system. Fig. 2, lane 3 shows that the 28 KD protein was still synthesized but not the complete 27 KD protein. Although this was not further investigated, the smaller proteins found instead may represent truncated forms of this protein, suggesting that only part of the gene is on the excised PstI fragment. The position was confirmed by removing the HpaI fragment by restriction and religation.

Fig. 2, lane 4, shows that this led to a truncated 74 KD protein and to disappearance of both the 27 KD and the 28 KD proteins. This indicates that part of the gene for the smaller protein is localized in the HpaI/PstI overlap, as drawn in Fig. 5. The gene for the 28 KD protein is entirely to the left of the PstI fragment since excision did not abolish its formation. The results seem to exclude that the two proteins are different forms of the same protein encoded by one gene, but it cannot be excluded that the two genes overlap, and further experiments are necessary to map the coding region for the 28 KD protein more precisely.

Expression in a Cell-Free System from *Agrobacterium tumefaciens*

The finding that *E. coli* can express several well-defined proteins from the T-region without apparent participation of promoter activities from the vector plasmids was somewhat unexpected. It was therefore important to investigate whether Agrobacteria can express the same proteins. This was tested with a cell-free system for coupled transcription/translation prepared from *Agrobacterium tumefaciens.* Fig. 4 shows the results obtained with a number of recombinant plasmids containing T-region fragments which produced the proteins in *E. coli* minicells. The cell-free system from Agrobacteria synthesized the 74 KD, 28 KD and 27 KD proteins, but the 49 KD protein was in most cases visible only after very long exposure times. Since this protein was also very poorly expressed in an extract from *E. coli* (not shown), this seems to reflect a property of the cell-free systems, but the reason is not known. Interestingly, the results also show that the *Agrobacterium* extract synthesized the 74 KD protein with both orientations of EcoRI 7 in pACYC184 (Fig. 4, lanes 6, 7). This is in contrast to *E. coli,* where it was produced only with orientation II and where expression was strictly dependent on the promoter of the vector plasmid (Fig. 2, lane 2). This indicates that EcoRI 7 contains a promoter activity for expression of the 74 KD protein which is recognized in Agrobacteria, but not in *E. coli.* With the possible exception of the 49 KD protein, the results suggest that the proteins expressed in *E. coli* can also be expressed in *Agrobacterium tumefaciens,* although the two systems do not necessarily use the same promoter activities.

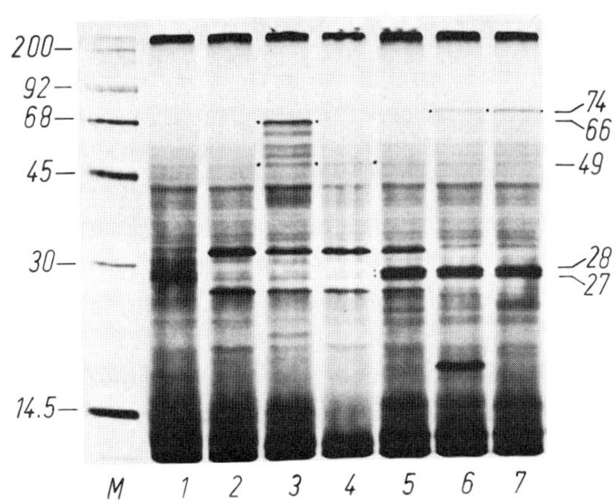

Fig. 4. Expression of T-region encoded proteins in a cell-free system for coupled transcription/translation from *Agrobacterium tumefaciens* (manuscript in preparation). See Table 1 and Fig. 1 for identification of fragments. *Lanes 1* No DNA added; *2* pBR322; *3* BamHI 8 (pGV0153); *4* HindIII 22 (pHD106.*1*); *5* HindIII 1 (pGV0201); *6* EcoRI 7 (pWK500.*1*); *7* EcoRI 7 (pWK500.*2*). *M* Marker proteins. *Numbers* at the *right* indicate the size in KD of the proteins synthesized with promoter activities from the T-region

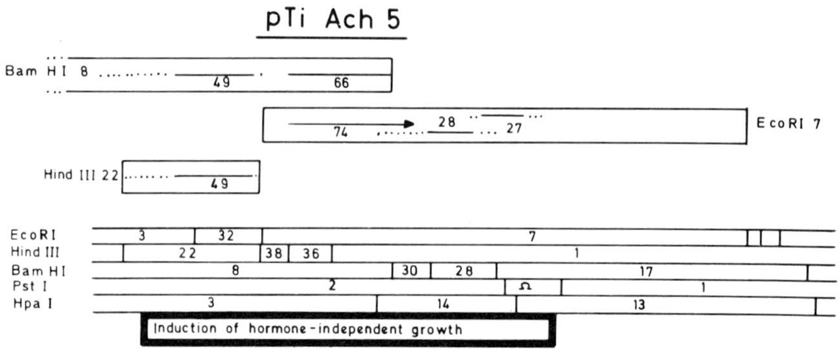

Fig. 5. Protein coding regions expressed in bacteria. The *boxes* show the restriction fragments most often used in the experiments; the *solid lines* within indicate the most likely position of the coding regions, and the *dotted lines* represent the limits of mapping accuracy. *Numbers at the solid lines* give the size of the proteins (KD). *Length of the solid lines* indicates the minimum size of the coding region, as calculated from the size of the protein with the assumption of an average mass of 120 for amino acids

Discussion

The finding that the T-region of octopine plasmid Ach5 contains genes which are expressed in bacteria into defined proteins possibly adds a new aspect in understanding crown gall tumorigenesis. According to the genetic identification of functions encoded in the T-region (this vol.), the 74 KD and 49 KD proteins are encoded in the part responsible for shoot suppression and the 27 KD protein in the part causing root suppression effects on plant morphogenesis. The assignment of the 28 KD protein is more difficult since the coding region was not determined very precisely, but the genetical data of Garfinkel et al. (1981) suggests that there may be a second locus for root suppression which is located to the left of the part coding for the 27 KD protein, and the 28 KD protein possibly represents expression from this second locus. It is important, however, to point out that all of these functions were not defined by expression in Agrobacteria, but in axenic plant tissue cultures. It is therefore an interesting question whether the proteins expressed in the bacteria are functional and possibly participate in hormone metabolism or its control. Agrobacteria do produce cytokinins and auxin, and Ti-plasmids influence their formation (Claeys et al. 1978, Liu and Kado 1979, McCloskey et al. 1980, Weiler and Spanier 1981, Liu et al. 1982, Regier and Morris 1982). Although in most cases the role of Ti-plasmids is not clear, recent experiments indicate that at least one locus in the T-region is involved in trans-zeatin production in Agrobacteria (Regier et al, submitted). More experiments are necessary to define a possible role of the T-region in hormone-metabolism.

The part of the T-region expressed in the bacteria is also transcribed in transformed plant cells (see this vol.), but the protein products have not yet been identified. It is an interesting question whether the same or different genes are expressed in bacteria and in plants. Fig. 6 indicates that three of the protein coding regions in bacteria

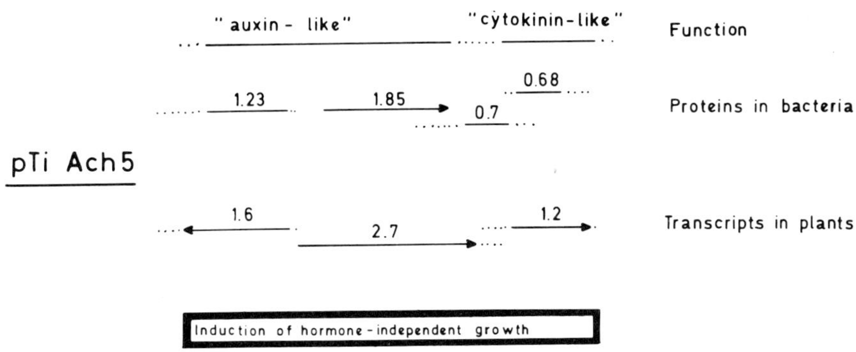

Fig. 6. Location of coding regions expressed in bacteria and in plants. Proteins: The *numbers* indicate the minimum length of the coding regions (see legend to Fig. 5). Transcripts: The data were taken from Willmitzer et al. (1982) and drawn with the assumption that the genes are not spliced. *Top line* Functions identified in this part of the T-region by genetic analysis (Leemans et al. 1982)

correlate with three transcript coding regions in plants in several points: The locations correspond within the limits of mapping accuracy, the sizes correspond, and in each case the transcript is larger than required for the protein. The comparison is particularly striking with the 74 KD protein (calculated minimum size of coding region 1.85 KB): The protein coding region is within the transcript coding region, and the coding strand is the same in bacteria and in plants. This suggests a working hypothesis that the same genes may be expressed in procaryotic and eucaryotic cells, but further evidence is necessary to support this hypothesis. Fig. 6 shows also that the fourth protein expressed in bacteria (28 KD; calculated minimum size of coding region 0.7 kb) appears to have no corresponding transcript in plant cells. The significance is not clear, but this may be a gene which is only expressed in bacteria.

The part of the T-region discussed here controls plant differentiation when expressed in plant cells, and it has therefore been said to exert auxin-like and cytokinin-like effects (Ooms et al. 1981, Garfinkel et al. 1981). A careful genetical analysis of the effect of the different T-DNA transcripts in plants (Leemans et al. 1982, Joos et al., submitted) has indicated that at least two T-DNA genes are involved in the suppression of shoot formation (auxin-like) and one gene is involved in the suppression of root formation (cytokinin-like). Since the same part, and possibly the same genes, are expressed in Agrobacteria, it is tempting to speculate that the bacterial gene products may fulfill a similar function. Such a role could be envisioned in the following model: Efficient tumor induction requires a wound in plant tissue which induces cell divisions, and transformation is thought to occur in this period (see review by Braun 1978). In non-infected tissue cell divisions cease after a few rounds while they continue in infected tissue. It seems possible that the bacterial gene products participate in mediating hormone-effects from the bacteria to plant cells to the effect that wound-induced cell division is maintained until the T-region is transferred and becomes operative in the plant cells. Certain plant pathogenic bacteria do exert hormone effects on plants, and with *Pseudomonas savastanoi* it has been demonstrated that plasmid-coded genes are involved (Comai and Kosuge 1980, 1982, Comai et al. 1982), but the genes are not transferred to plants. It seems possible that Agrobacteria use similar mechanisms in the early stages of infection, and in addition succeed in transferring the genes into the plant cells.

Acknowledgements. This work is the result of a close interaction with several people contributing to various aspects of the experiments. We especially wish to express our sincere gratefulness to Jeff Schell and the crown gall research groups in Gent and Brussels (Belgium). They provided cloned T-region fragments, and, together with our colleagues from the Max-Planck-Institut in Köln, shared results prior to publication and contributed with stimulating discussions.

References

Beuckeleer M, Lemmers M, Vos G De, Willmitzer L, Montagu M Van, Schell J (1981) Further insight on the transferred DNA of octopine crown gall. Mol Gen Genet 193:283–288

Braun AC (1978) Plant tumors. Biochim Biophys Acta 516:167–191

Claeys M, Messens E, Montagu M Van, Schell J (1978) Quantitative determination of cytokinins in culture media of *Agrobacterium tumefaciens.* Quant Mass Spectrom Life Sci 2:409–418

Comai L, Kosuge T (1980) Involvement of plasmid deoxyribonucleic acid in indoleacetic acid synthesis in *Pseudomonas savastanoi.* J Bacteriol 143:950–957

Comai L, Kosuge T (1982) Cloning and characterization of iaaM, a virulence determinant of *Pseudomonas savastanoi.* J Bacteriol 149:40–46

Comai L, Surico G, Kosuge T (1982) Relation of plasmid DNA to indoleacetic acid production in different strains of *Pseudomonas syringae* pv. *savastanoi.* J Gen Microbiol 128:2157–2163

Garfinkel DJ, Simpson RB, Ream LW, White FF, Gordon MP, Nester EW (1981) Genetic analysis of crown gall: fine structure map of the T-DNA by site-directed mutagenesis. Cell 27:143–153

Leemans J, Deblaere R, Willmitzer L, Greve H De, Hernalsteens JP, Montagu M Van, Schell J (1982) Genetic identification of functions of TL-DNA transcripts in octopine crown galls. EMBO J 1:147–152

Liu ST, Kado CI (1979) Indoleacetic acid production: a plasmid function of *Agrobacterium tumefaciens* C58. Biochem Biophys Res Commun 90:171–178

Liu ST, Perry KL, Schardl CL, Kado CI (1982) *Agrobacterium* Ti-plasmid indoleacetic acid gene is required for crown gall tumorigenesis. Proc Natl Acad Sci USA 79:2812–2816

McCloskey JA, Hashizume T, Basile B, Ohno Y, Sonoki S (1980) Occurrence and levels of cis- and trans-zeatin ribosides in the culture medium of a virulent strain of *Agrobacterium tumefaciens.* FEBS Lett 111:181–183

McPherson JC, Nester EW, Gordon MP (1980) Proteins encoded by *Agrobacterium tumefaciens* Ti plasmid DNA (T-DNA) in crown gall tumors. Proc Natl Acad Sci USA 77:2666–2670

Murai N, Kemp JD (1982) Octopine synthase mRNA isolated from sunflower crown gall callus is homologous to the Ti plasmid of *Agrobacterium tumefaciens.* Proc Natl Acad Sci USA 79: 86–90

Ooms G, Hooykas PJ, Moleman G, Schilperoort RA (1981) Crown gall plant tumors of abnormal morphology, induced by *Agrobacterium tumefaciens* carrying mutated octopine Ti plasmids; analysis of T-DNA functions. Gene 14:33–50

Regier DA, Morris RO (1982) Secretion of trans-zeatin by *Agrobacterium tumefaciens:* a function determined by the nopaline Ti plasmid. Biochem Biophys Res Commun 104:1560–1566

Rüther U (1980) Construction and properties of a new cloning vehicle, allowing direct screening for recombinant plasmids. Mol Gen Genet 178:475–477

Schröder G, Schröder J (1982) Hybridization selection and translation of T-DNA encoded mRNAs from octopine tumors. Mol Gen Genet 185:51–55

Schröder J, Schröder G, Huisman H, Schilperoort RA, Schell J (1981a) The mRNA for lysopine dehydrogenase in plant tumor cells is complementary to a Ti plasmid fragment. FEBS Lett 129:166–168

Schröder J, Hillebrand A, Klipp W, Pühler A (1981b) Expression of plant tumor-specific proteins in minicells of *Escherichia coli:* A fusion protein of lysopine dehydrogenase with chloramphenicol acetyltransferase. Nucleic Acids Res 9:5187–5202

Vos G De, Beuckeleer M De, Montagu M Van, Schell J (1981) Restriction endonuclease mapping of the octopine tumor-inducing plasmid pTi Ach5 of *Agrobacterium tumefaciens.* Plasmid 6: 249–253

Weiler EW, Spanier K (1981) Phytohormones in the formation of crown gall tumors. Planta 153: 326–337

Willmitzer L, Simons G, Schell J (1982) The TL-DNA in octopine crown-gall tumors codes for seven well-defined polyadenylated transcripts. EMBO J 1:139–146

Liposomes as a Carrier of Ti Plasmid into Protoplasts

T. NAGATA[1]

Introduction

The unified theme of this symposium is, needless to say, the molecular genetics of the bacteria-plant interaction and in the case of the interaction between *Agrobacterium tumefaciens* and plant, crown gall was induced by bacterial infection. On crown gall tumorigenesis, a part of Ti plasmid harbored in *A. tumefaciens* was transferred and integrated into plant DNA. Many approaches to dissolve this process are presented in this symposium by other workers. However, our approach is different from others. We are trying to introduce Ti plasmid directly into plant cells without bacterial infection. The main object is to develop an in vitro transformation system and we expect an increase of the transformation frequency by this procedure. This approach should be important for the development of genetic engineering of plants and for the elucidation of the crown gall tumorigenesis, because the complex process accompanying infection will be avoided. In this context only naked protoplasts can be used as experimental material because of the absence of thick cell walls.

Thus in this paper our experiences on the introduction of genetic materials, inducing Ti plasmid, into protoplasts will be depicted with special reference to the procedures developed by our own hands. At the moment the following methods were presented in relation to this subject; that is to say, the liposome-method (Nagata et al. 1981, Fraley et al. 1982), *A. tumefaciens* spheroplasts mediated transformation (Hasezawa et al. 1981), transformation of protoplasts co-cultured with *A. tumefaciens* (Márton et al. 1979) and PEG mediated transformation of protoplasts with Ti plasmid (Krens et al. 1982). Among these procedures, the liposome method and the spheroplast-mediated transformation were developed by ourselves.

Liposome-Mediated Delivery of the Ti Plasmid into Protoplasts

If one introduces the Ti plasmid into protoplasts, it takes a rather long time to select and identify the transformants. Thus we first tried to establish a system to introduce nucleic acids into protoplasts by using easily detectable nucleic acids such as tobacco mosaic virus (TMV) RNA as a model material. If TMV-RNA retained its function

1 Department of Biology, School of Science, Nagoya University, Furo-cho, Chikusa-ku, Nagoya 464, Japan

Molecular Genetics of the Bacteria-Plant Interaction
(ed. by A. Pühler)

inside cells, its functional expression could be easily monitored by the fluorescent antibody technique. The problem accompanying the use of TMV-RNA instead of the Ti plasmid or another DNA will also be considered.

Although we had some success in introducing TMV-RNA into protoplasts by using liposomes according to the Ca-EDTA chelation method (Fukunaga et al. 1981), we had much more success by using reverse-phase evaporation liposomes (REV) (Nagata et al. 1981). Thus the results obtained by the REV method will be described here. As the author described the whole detailed procedure elsewhere (Nagata 1983a), only the essential process will be described here.

10 μmol of phosphatidylserine (PS) and cholesterol (Chol) (1:1) were dissolved in 1 ml of redistilled diethyl ether. To this was added 100 μg TMV RNA dissolved in TES/his/NaCl buffer (pH 7.4). The resulting two-phase was sonicated for 15 s and then the organic solvent was removed by rotary evaporation. The prepared liposomes were washed twice by ultracentrifugation at 20,000 rpm for 20 min. The liposomes, in an amount corresponding to 2 μmol of phospholipids, were added to 3×10^6 protoplasts. Protoplasts were isolated from *Nicotiana tabacum* L. cv. BY 2 or *Vinca rosea* L. by treatment at 30°C for 1 h with a mixture of 0.1% Pectolyase Y23 and 1% Cellulase Onozuka RS dissolved in 0.4 M D-mannitol, pH adjusted to 5.5 with 1 N HCl. The mixture of protoplasts and liposomes was treated with 12.5% polyethylene glycol (PEG) 1540 or 10% polyvinyl alcohol (PVA) 203 or 202 (degree of polymerization 300 or 200, respectively, and 88% hydrolyzed) at final concentration. These PVA 202 and 203 were prepared specially for us by Kuraray Co. (Kurashiki, Okayama, Japan) and have a lower degree of polymerization than usual, so they had lower viscosity and were easy for handling in protoplast works. PEG and PVA treatment was carried out for 10 and 20 min, respectively, and these polymers were removed by washing with the high pH-high Ca medium of Keller and Melchers (1973). Then the protoplasts were cultured for 24 h and the infection percentage by TMV RNA was determined by staining with fluorescent antibody against TMV particles. Actually 60%–90% of the protoplasts were infected and 1 ng of RNA could infect ca. 300 protoplasts. According to ultrastructural observation the infected protoplasts were filled with TMV particles after 24 h of culture. The most conspicuous feature of this liposome-mediated introduction of TMV RNA into protoplasts was that TMV RNA encapsulated in liposomes was almost resistant to RNase treatment. Now this method was applied to the introduction of Ti plasmid into protoplasts.

When octopine-type Ti plasmid was introduced into protoplasts, the putative transformants were selected on the medium without plant growth regulators and examined for octopine synthesis. The analysis of the T-DNA of these putative transformants is underway. According to a rough calculation this transformation frequency was ca. 10^{-6}.

The problem which was encountered in the case of Ti plasmid was the encapsulation of Ti plasmid in liposomes. According to the procedure developed by Kuroiwa (1982) the percentage of liposomes filled with DNA was examined by staining with 4,6-diamidino-2-phenylindole (DAPI) which is specific fluorescent dye for DNA. Only 1%–5% of the liposomes were filled with Ti plasmid (unpubl. results). Therefore, efforts to increase the encapsulation frequency are necessary, and alternatively sorting-out liposomes encapsulating Ti plasmid should be tried.

However, we applied this procedure to the encapsulation of 2 μ DNA from yeast harboring leu 2 gene and these liposomes were added to leu$^-$ yeast protoplasts and treated with PEG. According to these experiments we obtained at least 10 times higher transformation frequency than the Hicks-Hinnen method using naked DNA in combination with PEG treatment. When this DNA was encapsulated in liposomes, it was resistant to DNase digestion (unpubl. results of A. Toh-e and T. Nagata).

Thus if we can accomplish the encapsulation of Ti plasmid at higher efficiency we could obtain higher transformation frequencies than those of the present experiment.

Agrobacterium Spheroplasts-Mediated Delivery of Ti Plasmid into Protoplasts

In parallel to the work using liposome we tried also to introduce *A. tumefaciens* spheroplasts into plant protoplasts. The spheroplasts were prepared by the treatment of lysozyme in combination with Carbenicillin pretreatment. The uptake of bacterial spheroplasts by plant protoplasts was carried out by treatment with PEG or PVA in combination with high pH-high Ca washing. After the treatment of *Vinca rosea* protoplasts by the spheroplasts, the spheroplasts were killed by heat treatment at 40°C for 4 days. Although *V. rosea* cells were resistant to this heat treatment, other cells including tobacco, were not so resistant. Thus antibiotic treatment was utilized for bacterial removal in order to be able to apply this method to other plant protoplasts (Hasezawa et al. 1981).

According to electron microscopic observation, each protoplast took up ca. 10 spheroplasts. The treated protoplasts were cultured in the medium without plant growth regulators and the obtained offspring after the selection procedure were examined for the genetic traits of Ti plasmid. Actually 10^{-4}–10^{-3} of these protoplasts were transformed to be able to produce opine. However, after long term culture, some of these clones tended to lose the opine producing ability. The advantage of this procedure is that one can apply this method to various types of bacterial strain without the isolation of Ti plasmid. In this context it is very interesting to apply this method to avirulent strains of *A. tumefaciens* harboring Ti plasmid, whose oncogenicity locus outside of T-DNA was mutagenized by transposon. This experiment will bring the clarification of this oncogenicity region of Ti plasmid (work in progress).

Other Procedures

Although we had also some experience in the transformation of protoplasts by co-culture of regenerating protoplasts with *A. tumefaciens* and with micro-injection, the direct comparison between these procedures has not yet been carried out. As each procedure has some advantages, I think that it is necessary to compare them in detail.

In the case of micro-injection we encountered some difficulty in fixing plant protoplasts for injection. Even if they were embedded in the agar medium, they tended to move away on injection with a glass needle. This might be due to the fact that plant protoplasts were suspended in the higher osmotic condition, which brought the surface tension of protoplasts to a high value. As one of the advantages of micro-injection is the introduction of DNA into the nucleus, the experimental procedure must be improved. Probably the pricking method developed by Yamamoto et al. (1981) should also be important for plant protoplasts. If one can overcome this difficulty, micro-injection might become a very suitable method for transformation, because micro-injection into the nucleus will bring a very high transformation frequency (Capecchi 1980).

Uptake of Liposomes and Bacterial Spheroplasts by Plant Protoplasts

As shown above, PEG or PVA treatment in combination with high pH-high Ca treatment was necessary to introduce liposomes or *Agrobacterium* spheroplasts into protoplasts. In this context it should be remembered that PEG, PVA and high pH-high Ca treatment were originally developed for protoplast fusion. Thus we supposed in the beginning that these treeatments might cause fusion between the protoplast membrane and liposomes or bacterial spheroplasts. However, electrom microscopic observation showed that liposomes and bacterial spheroplasts did not fuse with protoplast membranes, but were simply taken up by protoplasts via the process of endocytosis. Subsequently materials inside the liposomes or spheroplasts were released into the cytoplasm of the protoplasts. On the contrary, several authors working with this subject stated repeatedly that this process was fusion without any ultrastructural examination (Schaffner 1980). Thus I would like to stress that this process was endocytosis and not fusion between protoplast membranes and liposomes or bacterial spheroplasts (Hasezawa et al. 1983, Fukunaga et al.1983).

Importance of the Physiological State of Recipient Cells for Transformation

So far all of the published experiments in transformation were carried out by using random population in terms of cell cycle and there was no systematic work on this subject. However, the physiological condition of recipient cells is also important for transformation. Actually according to the old experiments of Braun (cf. Braun and Stonier 1958), the most vigorously growing crown gall was obtained during the period of DNA synthesis which was induced by wounding. Therefore, utilization of S-phase protoplasts, if available, should be very important for transformation. On the other hand, if one can use M-phase protoplasts it might also be interesting, because M-phase cells do not retain the nuclear membrane and may be more susceptible for the introduction of genetic materials into the nucleus.

The prerequisite for the use of the specific stage of cells in cell cycle is to obtain synchronized cells and to prepare protoplasts from them. So far the highest reported mitotic index was 35% (see review of Komamine et al. 1978), although there were many papers concerned with the synchronization of plant cells. Just recently we could resolve this prerequisite by using aphidicolin which is a specific inhibitor for eukaryotic DNA polymerase α.

Actually after tobacco BY 2 cells in liquid culture were treated with 5 μg/ml of aphidicolin for 24 h and then aphidicolin was removed from the medium by washing, a high degree of synchronization was obtained and 70% of the mitotic index was observed 9 h after the removal of aphidocolin. In this process it was also observed that DNA synthesis started immediately after removal of aphidocolin and ended within 4 h. More than 90% of cells completed their DNA synthesis during this period, which was examined by radioautography and by fluorescence microscopy using propidium iodide (unpubl. results).

Furthermore, enzyme treatment with a combination of Pectolyase Y23 and Cellulase Onozuka RS enabled us to prepare protoplasts from various stages of cell cycle. Now protoplasts were obtained 2 h and 9 h after aphidicolin treatment, which corresponded to S-phase and M-phase, respectively. Actually these S-phase and M-phase protoplasts were infected with TMV-RNA encapsulated in liposomes and the significant infection was obtained in both materials (Nagata et al 1982).

Further, as protoplasts from these synchronized cells might be highly susceptible for transformation as is cited above, we are trying to introduce Ti plasmid into these S-phase and M-phase protoplasts. On the other hand, these synchronized cells should also be useful for the transformation by co-culture with *A. tumefaciens* and by *A. tumefaciens* spheroplasts. The S-phase protoplasts of cells may be also suitable for micro-injection to obtain a high degree of transformation.

Conclusions

Although there is still room for improvement of the liposome-mediated and *A. tumefaciens* spheroplast-mediated transformation of protoplasts, these methods as well as micro-injection and bacterial co-culture will become very important for genetic engineering of plant cells. Furthermore, the increase of the susceptibility of transformation by using protoplasts from synchronized cells in cell cycle will also become an important technique.

References

Braun AC, Stonier T (1958) Morphology and physiology of plant tumors. Protoplasmalogia 5a: 1–93

Capecchi MR (1980) High efficiency transformation by direct microinjection into cultured mammalian cells. Cell 22:479–488

Fraley RT, Dellaporta SL, Papahadjopoulos D (1982) Liposome-mediated delivery of tobacco mosaic virus RNA into tobacco protoplasts: A sensitive assay for monitoring liposome-protoplast interactions. Proc Natl Acad Sci USA 79:1859–1863

Fukunaga Y, Nagata T, Takebe I (1981) Liposome-mediated infection of plant protoplasts with tobacco mosaic virus RNA. Virology 113:752–760

Fukunaga Y, Nagata T, Takebe I, Kakehi T, Matsui C (1983) An ultrastructural study of the interaction of liposomes with plant protoplasts. Exptl Cell Res 144:181–189

Hasezawa S, Nagata T, Syono K (1981) Transformation of *Vinca* protoplasts mediated by *Agrobacterium* spheroplasts. Mol Gen Genet 182:206–210

Hasezawa S, Matsui C, Nagata T, Syono K (1983) Cytological study of the introduction of *Agrobacterium tumefaciens* spheroplasts into *Vinca rosea* protoplasts. Can J Bot 61:1052–1057

Keller WA, Melchers G (1973) The effect of high pH and calcium on tobacco leaf protoplast fusion. Z Naturforsch 28c:737–741

Komamine A, Morigaki T, Fujimura T (1978) Metabolism in synchronous growth and differentiation in plant tissue and cell culture. In: Thorpe TA (ed) Frontiers of plant tissue culture 1978. University of Calgary Press, Calgary, Alberta, Canada, pp 159–168

Krens FA, Molendijk L, Wullems GJ, Schilperoort RA (1982) In vitro transformation of plant protoplasts with Ti plasmid DNA. Nature 296:72–74

Kuroiwa T (1982) Mitochondrial nuclei. Int Rev Cytol 75:1–60

Márton L, Wullems GJ, Molendijk L, Schilperoort RA (1979) In vitro transformation of cultured cells from *Nicotiana tabacum* by *Agrobacterium tumefaciens*. Nature 277:129–131

Nagata T (1983a) Entrapement of genetic materials into liposomes and their delivery to plant protoplasts. In: Gregoriadis G (ed) Liposome technology. CRC Press, New York (in press)

Nagata T (1983b) Interaction of liposomes and protoplasts as a model system of protoplast fusion. In: Beers RF Jr, Bassett EG (eds) Cell fusion. Raven Press, New York (in press)

Nagata T, Okada K, Takebe I, Matsui C (1981) Delivery of tobacco mosaic virus RNA into plant protoplasts mediated by reverse-phase evaporation vesicles (liposomes). Mol Gen Genet 184: 161–165

Nagata T, Okada K, Takebe I (1982) Mitotic protoplasts and their infection with tobacco mosaic virus RNA encapsulated in liposomes. Plant Cell Rep 1:250–252

Schaffner W (1980) Direct transfer of cloned genes from bacteria to mammalian cells. Proc Natl Acad Sci USA 77:2163–2167

Yamamoto F, Furusawa M, Takamatsu K, Muiura N, Uchida T (1981) Intracellular introduction of a fixed quantity of substances by pricking cells using a modified microscope. Exptl Cell Res 135:341–345

Plant Protoplast Transformation by *Agrobacterium tumefaciens* and its Ti Plasmid DNA

G.J. WULLEMS, F.A. KRENS, and R.A. SCHILPEROORT [1]

Introduction

The present progress in molecular genetics and biology makes it possible to perform genetic manipulation not only with micro-organisms and cultured animal cells but also with plants and cultured plant cells. This latter development is mainly due to the recent knowledge of the plant disease crown gall which is caused by the soil bacterium *Agrobacterium tumefaciens.* In addition, the potential of plant tissues to regenerate into complete plants, starting from one single plant cell, is also of great importance to the performance of genetic engineering experiments with plants. However, regeneration from single cells is at present restricted to a relatively few genera and still not well accomplished for e.g. monocotyledones, a group to which all cereal crops belong.

Various approaches may be followed to achieve genetic manipulation. Mixing of more or less complete genomes can be obtained by somatic cell fusion, whereas well defined genes may be introduced by transformation of plant protoplast with isolated DNA. The recombinant DNA technology might be helpful in recombining DNA sequences of interest in such a way that they will be inserted into the host genome.

Besides direct application of these techniques in plant breeding they also offer possibilities for a better understanding of genome organisation and regulation of gene expression in higher plants.

Methods for plant cell transformation are based on naturally occurring or constructed DNA vectors that either are capable to integrate into the host genome or show autonomous replication. Such plant DNA vectors should carry: sites in which genes can be inserted stably, selectable markers, DNA sequences for replication or DNA sequences for integration into the plant genome. This last property is necessary for stable maintenance and transmission of inserted genes into offspring plants. Obviously all types of sequences that are needed for the expression in plant cells should also be present. The Ti plasmid present in *Agrobacterium tumefaciens* is at present the most promising plant DNA vector. This plasmid induces on dicotyledonous host plants the so called crown gall tumours. Monocotyledonous plants are not susceptible. The induction of the tumours is due to the insertion in the plant genome of a specific part of the Ti plasmid called T-DNA. This T-DNA becomes integrated into the nuclear DNA and is expressed into mRNA and protein.

1 Department of Biochemistry, State University of Leiden, Wassenaarseweg 64, 2333 AL, Leiden, The Netherlands

Molecular Genetics of the Bacteria-Plant Interaction
(ed. by A. Pühler)

The Crown Gall System

The pathogenic *Agrobacterium tumefaciens*, belonging to the family of *Rhizobiaceae*, like *A. rhizogenes* which causes the hairy root disease, needs a wound on a plant in order to induce a tumour. This wounding is essential for "conditioned" cells to develop. The bacteria seem to need a "bacterial adjustment phase" before they can accomplish tumour induction. In order to transform the plant cells agrobacteria must attach to the cell wall. This attachment presumably is caused by the lipopolysaccharide (LPS) component of the bacterial cell wall and the polygalacturonic acid fraction of the plant cell walls. The insusceptibility of monocotyledonous plants for tumour induction presumably is due to the absence of bacterial attachment sites on the cell walls of these plants. At present almost nothing is known about the molecular processes by which DNA is transferred from virulent agrobacteria into plant cells. However, studies with plants and with plant cells have shown that this process is accomplished in about 32 h, whereafter the bacteria can be killed without affecting tumour development.

The agents responsible for tumour induction are plasmids called tumour-inducing or Ti plasmids. Three classes of Ti plasmids have been described. These are the octopine-, nopaline- and agropine Ti plasmids. Besides tumour induction, the synthesis of tumour specific amino acid derivatives in the tumour tissue named opines (e.g., octopine, nopaline, agropine) is also coded for by Ti plasmid sequences. Octopine and nopaline are synthesized by the plasmid encoded enzymes lysopine dehydrogenase (LpDH) and nopaline dehydrogenase (NpDH) respectively. The Ti plasmids also code for catabolic enzymes that convert these compounds into products that may serve as a carbon and nitrogen source for the bacterium. Therefore, the interaction between the bacterium and the plant cells results in a plant tumour which is for the benefit of the bacterium.

The main properties of crown gall tumours are phytohormone independent growth (hormone autotrophy) of the tissue when brought in tissue culture; and the presence of opines in the tumour cells. These properties can be regarded as phenotypic traits of the transformed nature of the tumour cells. This transformed nature is due to the integration of a segment of the Ti plasmid in the genome of the plant cells (Chilton et al. 1977). This segment is called the T-region in the bacterium and T-DNA in the plant. Nopaline tobacco tumours can carry from one to at least four different T-DNA copies, each with a molecular weight of about 16,000,000. These copies can be arranged in a head to tail tandem configuration (Zambrisky et al. 1980). The T-DNA in octopine tumours can be composed of either one, the TL-DNA, or two Ti plasmid segments, TL-DNA and TR-DNA (Thomashow et al. 1980). In most callus tissues of tobacco the TL-DNA is present and occasionally also the TR-DNA is observed. Studies with octopine Ti plasmid deletion mutants have shown that deletion of the TR-region does not affect oncogenicity, but that deletion of the TL-region results in loss of oncogenicity (Ooms et al. 1982). This means that only TL-DNA carries genes for tumour development. A considerable part of the TL-DNA has common DNA sequences with a nopaline T-DNA segment (Engler et al. 1981), which indicates that they share several gene functions that apparently are essential in tumour formation and maintenance (Willmitzer et al. 1982). Studies on the expression of the octopine T-DNA showed that in total 7 transcripts are made from the TL-DNA (Willmitzer et al. 1982, Gelvin et al. 1982). That the transcription of T-DNA results in functional mRNA's was shown by in vitro translation studies by which three proteins were identified. One of them was characterized as LpDH (Schröder et al. 1981). With respect to this it can be concluded that the prokaryotic T-region appears to carry all the required signals to be functional in the plant cells.

Since T-DNA must play a role in the phytohormone independent growth of crown gall cells it is important to see whether T-DNA genes could be identified with a function that could be correlated to auxin and cytokinin activities. Therefore, octopin Ti plasmid mutants were investigated (Ooms et al. 1982). A number of them appeared to induce tumours with an aberrant morphology. Two mutants (LBA 1501 and LBA 4060) induced tumours on tobacco with a strong shooting capacity, whereas another mutant (LBA 4210) gave tumours with excessive root formation. These results indicate that, contrary to what is observed for normal tumours induced by wild type strains, growth factors, stimulating either shoot or root formation are active in the mutant tumours. The regeneration of shoots might be regarded as a cytokinin-like activity, whereas the root formation can be explained in terms of an auxin-like activity. Inoculation of a mixture of the two types of mutants (LBA 4060 + LBA 4210) resulted in the formation of tumours that were identical to those induced by wild-type strains. From these data and from T-DNA analysis of the mutant tumours we postulated that TL-DNA contains loci with an auxin- and cytokinin-like phenotype (Fig. 1). Inactivation of the "auxin" locus (LBA 1501 and 4060) would then result in a cytokinin-like effect, because this activity of the T-DNA is then no longer counterbalanced by the auxin activity of the T-DNA. This will result in shoot formation. On the other hand, inactivation of the "cytokinin" locus (LBA 4210) will result in root formation on the tumour. Recently, many more mutants in the T-region of the octopine Ti plasmid have been isolated, resulting in a more accurate mapping of these loci, called by these authors tms (tumour morphology shooting) and tmr (tumour morphology rooting). In addition, a new locus was identified called

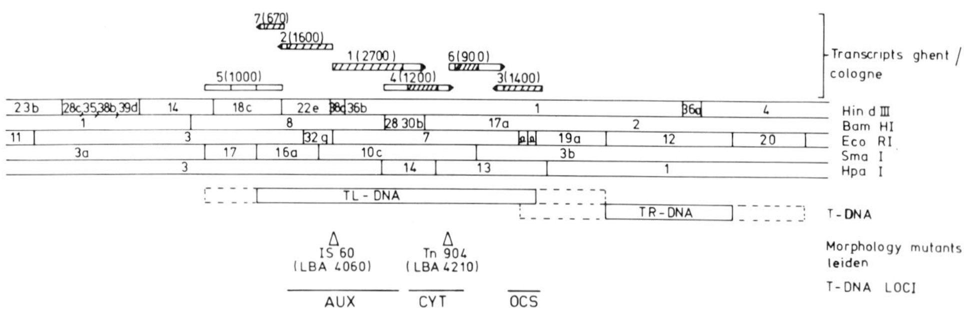

Fig. 1. Summary of the restriction map of part of the octopine pTi Ach5 (De Beuckeleer et al. 1981) and the map position of the T-DNA, and T-DNA derived transcripts. The extent of the TL- and TR-DNA is according to Thomashow et al. (1980). The position of the transcripts is according to Willmitzer et al. (1982). *Below*, the position of mutations in the Ti plasmid leading to a shooting (LBA 4060) or rooting (LBA 4210) phenotype of tumours is indicated. The corresponding T-DNA loci are indicated

tml (tumour morphology large) (Garfinkel et al. 1981, Leemans et al. 1982). An insertion in this locus gave rise to tumours which were much larger than wild type tumours (Fig. 1).

Mutations in the T-DNA described above can also be explained in terms of the presence or absence of T-DNA transcripts. A shooting phenotype can occur when transcripts no. 1 and/or 2 are affected, whereas a rooting phenotype occurs when transcript no. 4 is absent in the tumour (Leemans et al. 1982). The size of the tumour might be correlated with transcript no. 6 (Garfinkel et al. 1981). Experiments with mutant plasmids have indicated that it might be possible to mutate the Ti plasmid in such a way that the oncogenes presumably located in the "auxin" and "cytokinin" loci, are removed. Transformation of plant (cells) with such disarmed plasmids might lead to transformed tissues with retention of the morphogenic potential, leading to the regeneration of complete plants.

Plant Cell Transformation

The properties of transformed crown gall cells, in particular the phytohormone independent growth, allow the selection for transformed cells present in a population of normal cells. With the development of methods to isolate single plant cells as cell wall free protoplasts and to culture these protoplasts, resulting in callus tissues from which plants are regenerated that can be used in breeding programs, different strategies have been developed, based on the crown gall system, to obtain transformation of single cells. The phytohormone autotrophy of plant cells might fill the gap that still exists by the sparse occurrence of other stable selectable auxotrophs in higher plants. On the other hand, described mutant plant tissues like chlorophyll-deficient mutants, drug resistance mutants and mutants carrying metabolic deficiencies could be useful in showing the applicability of transformation systems.

The Co-Cultivation Method

This approach of transforming isolated plant cells is based on the natural process of infection by which *A. tumefaciens* introduces Ti-plasmid DNA into the plant genome. Presumably for acquiring the proper conditions for bacterial attachment and infectivity, protoplasts to be transformed through infection with *A. tumefaciens,* must have regenerated a new cell wall (Marton et al. 1979, Wullems et al. 1981a). For tobacco protoplasts, most of which divide 3–5 days after isolation, the susceptible stage is at day 3. Co-cultivation of protoplasts of this age with *A. tumefaciens,* in the presence of phytohormones results in a heavy aggregation of the cells. This presumably is due to attachment of bacteria to newly formed cell walls of the protoplasts. The co-cultivation time needed for maximum transformation to occur is about 32 h. Thereafter, the bacteria can be killed by the addition of antibiotics and the protoplasts further be cultured for 4–6 weeks in the presence of hormones to form small colonies. After another period of 4–6 weeks they are plated in solid medium without phytohormones. Continued selective pressure allows only transformed cells to survive. According to this procedure transformed tobacco tissue lines have been obtained with octopine- and with nopaline agrobacteria as well as with agrobacteria harbouring a co-integrate between the octopine and nopaline plasmids (Wullems et al. 1981a). The freqeuncy in which such transformants can be obtained lies between 0.1% and 1% of the starting number of protoplasts. A difference between tumour tissues obtained from co-cultivation and those obtained after infection of a plant stem is that the former tissues consist of a homogenous population of transformed cells, whereas tumours induced on a plant in general are a mixture of transformed and normal cells. Another difference between the co-cultivation method and tumour induction in vivo is the fact that within one experiment a high number of independent transformants can be isolated that show segregation of tumour markers. Some of them might be octopine plus, whereas others might be negative. In addition, a number of them show a high shooting capacity, also when the experiment was carried out with a wild type octopine strain. Such a strain does not give regenerating tumours when infected on a plant stem.

The transformants that have been obtained by the co-cultivation method contain T-DNA of a rather definite size. In general, the T-DNA architecture in such tissues does not differ much from T-DNA found in crown gall tumours obtained from infection of a plant (Ooms et al. 1982) (Fig. 2).

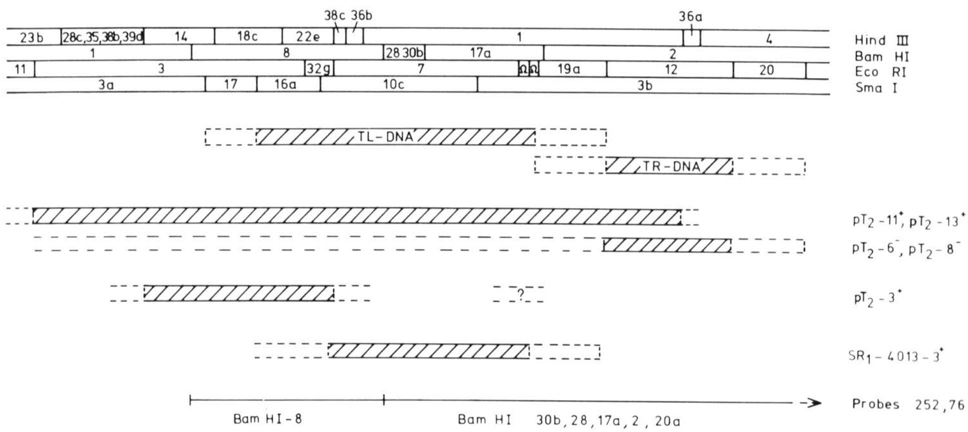

Fig. 2. T-DNA organisation in some octopine tumour tissues derived from co-cultivation (SR1-4013-3), from somatic hybridization (B + S-42-3) and from DNA transformation (pT2-1b1-11 and pT2-1b1-6). SR1-4013-3 and B + S-42-3 are examples for shooting tumour tissues with a deletion of the left part of the TL-DNA. The pTi-DNA transformants pT2-1b1-11 and 6 are examples of the extremes concerning the extent of T-DNA integration by DNA transformation. The presence or absence of octopine is indicated by + or −. The extend of TL- and TR-DNA is according to Thomashow et al. (1980)

The DNA-Transformation Method

The development of a method to transform plant protoplasts with isolated Ti plasmid DNA was based on a method that was used in our lab to achieve somatic fusion of tobacco protoplasts (Wullems et al. 1980, Krens et al. 1982). The procedure involves an incubation of freshly isolated protoplasts with Ti plasmid DNA in the presence of polyethylene glycol (PEG) and herring sperm DNA as carrier DNA. After the incubation the PEG was gradually deluted by addition of a post incubation mixture containing a high Ca^{++} concentration at pH 7. After the treatment the protoplasts were cultured in medium in the presence of phytohormones. When the colonies have a sufficient size they were transferred to hormone free medium in order to select transformed cells. Following this procedure transformants could be isolated in a frequency of 10^{-4} to 10^{-3} of the original number of protoplasts. Characterization of transformants based on the tumour markers hormone autotrophy and octopine synthesis showed that, as was found after the co-cultivation experiments, segregation of these markers occurred between independent transformants. From a total of 12 transformants 6 lines showed both properties, 5 lines were hormone independent but did not synthesize octopine whereas one transformant was hormone dependent and octopine negative (Table 1). These tissue lines have now been analyzed for the presence of T-DNA. The Southern blots have shown in each of the lines the presence of Ti plasmid derived DNA sequences. The most remarkable difference between the T-DNA architecture in these lines compared with the T-DNA in tumour lines obtained from bacterial infection was the great variety in T-DNA among the various pTi DNA transformants (Fig. 2). Some of the transformants harbour a T-DNA fragment that extends more to the left and to the right than the "core" TL-DNA found in crown gall tumours. Other transformants contain both the TL-DNA and the TR-DNA. In that case, the TR-DNA is present in a higher copy number than the TL-DNA. The occurrence of T-DNA of different sizes and sometimes larger than the fixed "core" T-DNA found in tumours induced by the action of the bacterium suggests that a bacterial mechanism exist that is responsible for the integration of a specific fragment

Table 1. Properties of transformants obtained after in vitro DNA transformation of tobacco protoplasts by pTi-DNA

Transformant line	LpDH activity	Hormone autotrophy	Regeneration	T-DNA
pT2-1b1-2	–	+	+	+
3	+	–	–	+
6	–	+	+	+
8	–	+	+	+
9	+	+	–	+
10	+	+	–	+
11	+	+	–	+
13	+	+	–	+
14	+	+	–	+
15	+	+	–	+
20	–	–	+	+
21	–	–	+	+

of fixed size of the Ti plasmid. This mechanism is lacking in the DNA transformation procedure using isolated pTi DNA. The observation of more than two border fragments in some of the transformants and the absence of internal T-DNA fragments in others indicates that following pTi DNA transformation more scrambling and rearrangements of the T-DNA occur after DNA transformation than after bacterial transformation. It is not known whether this is due to degradation of the plasmid DNA before integration or to the absence of a bacterium dependent specific integration mechanism. The results obtained from the DNA transformation experiments clearly demonstrate that the Ti plasmid DNA alone is capable of transforming plant cells. This offers possibilities to develop methods for the transformation of monocotyledonous plant cells, because with this method, specific (attachment) processes that only occur between agrobacteria and dicotyledonous plant cells are not involved in the DNA transformation method.

The Fate of T-DNA in Regenerated Plants

The development in our laboratory of transformation systems at the single cell level was accompanied by the observation that tumour tissues obtained by these methods showed a high shooting capacity. Although shooting of nopaline type tumours is a general feature, octopine type tumours, induced by wild type bacterial strains via infection of plant stems in general have lost their morphogenic potential completely. However, when wild type octopine bacteria are used in co-cultivation experiments, the shoot regenerating potential of the transformants is often restored (Wullems et al. 1981a). Both from nopaline and octopine transformants shoots were separated from the callus tissues and cultured separately. All shoots did not form roots, produced opines and were phytohormone independent when brought back in tissue culture (Table 2). Due to the absence of root formation, shoots were grafted on top of healthy tobacco stems where they developed in mature flowering plants with retention of the opine synthesis. The flowers of such plants were characterized by an outgrowth of the style (heterostyly) and were male sterile (Wullems et al. 1981b). Seeds could be obtained from such plants by cross pollination with pollen from control tobacco plants. Following sawing of seeds on culture medium two classes of seedlings could be distinguished; 56% of the seedlings derived from a nopaline type transformant appeared morphologically normal whereas 44% showed the transformed phenotype. The development of these seedlings was similar as has been described for the

Table 2. Properties of transformed tobacco shoots

No root formation
Teratomatous morphology
Opine positive
Male sterile
Insusceptable for re-infection with agrobacteria
Resistant against plant RNA viruses

parental transformants. To establish DNA from the different nopaline tissue types and seedlings were analysed for the presence of T-DNA (Memelink et al. 1983). From these experiments we could draw the conclusion that the total T-DNA found in the original transformant callus tissue is retained without change during regeneration and development into a mature plant and is transmitted completely to the progeny plants. In this respect the behaviour of these transformants concerning the stability of the T-DNA differs from tumorous shoots described elsewhere. Parental plants and seedlings derived by Yang and Simpson (1981), from BT 37 tumour tissue had lost all tumorous traits and contained only the end fragments of the T-DNA. Seedlings described by Otten et al. (1981), derived from an octopine tumour had retained octopine synthesis but had lost all other tumour markers. In these seedlings, T-DNA has not been detected. In both cases the seedlings were selected on the ability to form roots. The differences between these regenerants and seedlings described by Yang and Simpson and by Otten et al. with our seedlings can be explained in terms of the presumed presence of transcripts coded for by the T-DNA.

It has been postulated that the nopaline T-DNA codes for a number of transcripts involved in morphogenesis of the tumour tissue. Transcript no. 4 is supposed to inhibit root formation, whereas the transcripts no. 1 and 2 should inhibit shooting (Leemans et al. 1982).

Since the first two examples of regenerants were selected on the basis of complete restoration of morphogenesis, root formation will only occur when transcript

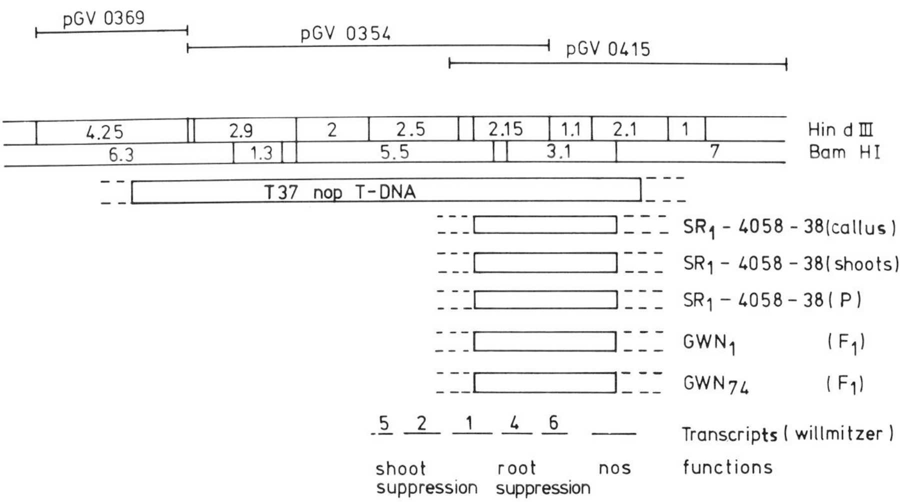

Fig. 3. T-DNA organisation of various tissue types of the nopaline bacterial transformant SR1-4058-38. The restriction map of part of the nopaline pTi-T37 and the extend of T-DNA in T37 crown gall are according to Yang and Simpson (1981). SR1-4058-38 is derived from co-cultivation of tobacco protoplasts with *A. tumefaciens* strain LBA 4058 harbouring pTi-T37. DNA isolated from callus tissue, from shoots, from grafted mature regenerants and from progeny plants is analysed. The progeny plants GWN1 and GWN74 were derived from sexual crosses between SR1-4058-38 and normal SR1 tobacco. All tumour tissues produced nopaline. The positions of transcripts with homologous counterparts in octopine tumours are indicated according to Willmitzer et al. (1982)

no. 4 is absent. This, with the absence of the transcripts no. 1 and 2 that inhibit shooting, implicates that such plants can only occur if most of the T-DNA is absent (Fig. 3). In our case, the regenerants and the seedlings were cultured on culture medium allowing growth without a root system. Therefore, also regenerants that have retained the T-DNA that codes for the transcript that inhibits rooting can survive. This is according with the T-DNA observed in these plants. This T-DNA consists of the right half of the T-DNA that is generally found in nopaline tumours. Passing through meiosis does not change the extent of this T-DNA (Fig. 3). This observation that a large T-DNA fragment is transmitted to progeny plants opens possibilities for introducing genes of interest in plants by using the Ti plasmid as a vector. Since the loci on the T-DNA involved in morphogenesis are known, mutation of these loci on the Ti plasmid, followed by transformation might lead to a complete restoration of the morphogenic potential of the transformed tissues with retention of the (mutated) T-DNA in regenerants and progeny plants. The use of the Ti plasmid would be of less value if the foreign DNA would get lost during meiosis of the plant cells. Also in this respect the Ti plasmid meets the conditions that makes it a good candidate for being used as genetic engineerer of plants.

Resistance Properties of Transformed Plants

When transformed plants and seedlings are wounded and infected with virulent agrobacteria, no tumour formation occurs. The transformed plants are not susceptible to *A. tumefaciens.* Infection with *A. rhizogenes,* which induces on healthy plants hairy roots, results on transformed plants in a strong, tumorous swelling of the stem at the site of the infection. These observations might be explained in terms of auxin and cytokinin activities. The absence of root formation on transformed plants and their thick stems and reduced apical dominance might be the result of a relatively high cytokinin activity. This might also be the reason that *A. tumefaciens* does not induce tumours on these plants. Tumour formation on healthy plants can be inhibited by exogenous supplied Kinetin. The observed swelling after infection with *A. rhizogenes* might be a response to some growth factor with a pronounced auxin-like activity that is excreted by *A. rhizogenes.* This activity together with the presumed high cytokinin activity in the transformed tobacco plants might also trigger cells more outside the wound-site to divide. This high cytokinin theory might also have other implications concerning resistancy against other microorganisms. Certain chemicals, among which cytokinins, applied to several tobacco cultivars gave rise to the induction of the synthesis of soluble proteins which are called pathogenesis-related (PR) proteins. These proteins are also induced by RNA viruses and other pathogens upon infection of plants. The function of these proteins is not yet known, but they seem to be involved in making plant cells resistant towards the pathogens. To test this hypothesis we have infected control plants and T-DNA containing plants with various RNA viruses. Preliminary results of these experiments showed that in contrast to the control plants, no virus specific symptoms were observed on the transformed plants. These results are now studied in more detail to find a molecular basis for the presumed resistancy against virus infection. If T-DNA genes are involved in the protection of plants against various pathogens the Ti plasmids offers a unique system to apply plants with resistancy against virus infection.

Note Added in Proof. Repeated infections of transformed plants with RNA viruses have shown no significant difference in response between control plants and transformed plants.

References

Beuckeleer M De, Lemmers M, Vos G De, Willmitzer L, Montagu M Van, Schell J (1981) Further insight on the transferred DNA of octopine crown galls. Mol Gen Genet 183:283–288

Chilton MD, Drummond MH, Merlo DJ, Sciaky D, Montoya AL, Gordon MP, Nester EW (1977) Stable incorporation of plasmid DNA into higher plant cells: the molecular basis of crown gall tumorigenesis. Cell 11:263–271

Engler G, Depicker A, Maenhaut R, Villarroel R, Montagu M Van, Schell J (1981) Physical mapping of DNA base sequence homologies between octopine and nopaline Ti plasmids of *Agrobacterium tumefaciens.* J Mol Biol 152:183–208

Garfinkel DJ, Simpson RB, Ream LW, White FF, Gordon MP, Nester EW (1981) Genetic analysis of crown gall: fine structure map of the T-DNA by site-directed mutagenesis. Cell 27:143–153

Gelvin SB, Thomashow MF, McPherson IC, Gordon MP, Nester EW (1982) Sizes and map positions of several plasmid-DNA-encoded transcripts in octopine-type crown gall tumors. Proc Natl Acad Sci USA 79:76–80

Krens FA, Molendijk L, Wullems GJ, Schilperoort RA (1982) In vitro transformation of plant protoplasts with Ti plasmid DNA. Nature 296:72–74

Leemans J, Deblaere R, Willmitzer L, Greve H De, Haernalsteens JP, Montagu M Van, Schell J (1982) Genetic identification of functions of TL-DNA transcripts in octopine crown galls. EMBO J 1:147–152

Marton L, Wullems GJ, Molendijk L, Schilperoort RA (1979) In vitro transformation of cultured cells from *N. tabacum* by *A. tumefaciens.* Nature 277:129–131

Memelink J, Schilperoort RA, Wullems GJ (1983) The nopaline T-DNA is maintained during regeneration and generative propagation of transformed tobacco plants. Mol Gen Genet 190:516–522

Ooms G, Hooykaas PJJ, Veen RJM Van, Beelen P, Van, Regensburg-Tuink TJG, Schilperoort RA (1982a) Octopine Ti plasmid deletion mutants of *Agrobacterium tumefaciens* with emphasis on the right side of the T-region. Plasmid 7:15–29

Ooms G, Bakker A, Molendijk L, Wullems GJ, Gordon MP, Nester EW, Schilperoort RA (1982b) T-DNA organisation in homogenous and heterogenous octopine type crown gall tissues of *Nicotiana tabacum.* Cell 30:589–597

Otten L, Greve H De, Hernalsteens JP, Montagu M Van, Schieder O, Straub J, Schell J (1981) Mendelian transmission of genes introduced into plants by the Ti plasmids of *A. tumefaciens.* Mol Gen Genet 183:209–213

Schröder J, Schröder G, Huisman H, Schilperoort RA, Schell J (1981) The mRNA for LpDH in plant tumor cells is complementary to a Ti plasmid fragment. FEBS Lett 129:166–168

Thomashow MF, Nutter R, Montoya AL, Gordon MP, Nester EW (1980) Integration and organisation of Ti plasmid sequences in crown gall. Cell 19:729–739

Willmitzer L, Simons G, Schell J (1982) The Ti-DNA in octopine crown gall tumours codes for seven well-defined polyadenylated transcripts. EMBO J 1:139–146

Wullems GJ, Molendijk L, Schilperoort RA (1980) The expression of tumour markers in intraspecific somatic hybrids of normal and crown gall cells from *N. tabacum.* Theor Appl Genet 56:203–208

Wullems GJ, Molendijk L, Ooms G, Schilperoort RA (1981a) Differential expression of crown gall tumor markers in transformants obtained after in vitro transformation of *N. tabacum* protoplasts by *A. tumefaciens.* Proc Natl Acad Sci USA 78:4344–4348

Wullems GJ, Molendijk L, Ooms G, Schilperoort RA (1981b) Retention of tumor markers in F1 progeny plants from in vitro induced octopine and nopaline tumor tissues. Cell 24:719–727

Yang F, Simpson RB (1981) Revertant seedlings from crown gall tumors retain a portion of the bacterial Ti plasmid DNA sequences. Proc Natl Acad Sci USA 78:4151–4155

Zambrisky P, Holsters M, Kruger K, Depicker A, Schell J, Montagu M Van, Goodman HM (1980) Tumor DNA structure in plant cells transformed by *Agrobacterium tumefaciens.* Science 209:1385–1391

The Study of Integration of Tumor-Inducing DNA from the Nopaline Ti Plasmid of *Agrobacterium tumefaciens*

P. ZAMBRYSKI[1]

Introduction

Crown gall is a neoplastic disease produced in most dicotyledonous plants by infection of the Gram-negative soil bacterium *Agrobacterium tumefaciens.* Crown gall tumors can proliferate autonomously in tissue culture and, in some cases, it is possible to regenerate plants from these cells (Braun and Wood 1976, Otten et al. 1981, Yang and Simpson 1981). The tumorous state of the DNA has been shown to result from the transfer of DNA from *A. tumefaciens* to the plant cell (for recent review, see Kahl and Schell 1982). This natural system for genetic exchange has begun to be studied extensively, since in addition to providing an interesting system of genetic exchange and tumor induction it may be possible to utilize *Agrobacterium* for the introduction of new genetic information into plant cells by combining molecular genetic engineering with plant tissue culture methods.

It has been established by both physical and genetic evidence that the agent responsible for crown gall tumor induction is a large Ti (tumor-inducing) plasmid of roughly 200 kilobase pairs (kbp) which is contained in all oncogenic strains of *A. tumefaciens.* Only a portion of the Ti plasmid is present in transformed plant cells; the transferred segment is called T-DNA (Chilton et al. 1977, Lemmers et al. 1980, Thomashow et al. 1980, De Beuckeleer et al. 1981). Almost nothing is known about the mechanism of transfer of Ti plasmid sequences to plants. Most likely, the transfer involves two stages: (1) the bacterial and plant cell walls interact in a manner which allows the Ti plasmid to enter the plant cell, and (2) part of the Ti plasmid DNA is transferred to plant DNA. One possibility for elucidating this event, is to compare which Ti plasmid sequences from *Agrobacterium* are part of the final T-DNA structure after it has integrated into the genome of the plant. To date, we have more detailed information about T-DNA in tumors induced by the nopaline strains of *Agrobacterium.* However, some of the recent results on the structure of T-DNAs derived from tumors induced by octopine strains suggest that the mechanism of integration may be similar in both types of *Agrobacteria.*

1 Laboratorium voor Genetica, Rijksuniversiteit Gent, K.L. Ledeganckstraat 25, 9000 Gent, Belgium

Molecular Genetics of the Bacteria-Plant Interaction
(ed. by A. Pühler)

Molecular Cloning Analysis of the T-DNA Borders

Hybridization blotting analysis of plant DNA from several independent nopaline tumor cell lines suggested that specific sequences at the ends of the T-DNA region of the Ti plasmid are recognized during the insertion process; the length of the T-DNA segment found in association with the plant DNA is relatively constant (Lemmers et al. 1980). The T-DNA of independent tumors induced by nopaline Ti plasmids has been shown to be a continuous stretch of DNA 23 kbp in length colinear with the T-region of the Ti plasmid.

In order to investigate this process in detail, the boundaries of the T-DNA from several nopaline tumor cell lines have been isolated by molecular cloning techniques (Zambryski et al. 1980, 1982). The nucleotide sequences of these T-DNA border clones have been studied and the precise junctions of the T-DNA have been determined by comparison with the nucleotide sequence of the same region of the Ti plasmid as contained in *Agrobacterium* prior to transfer to the plant cell genome. The clones isolated so far provide information on the overall organization of the T-DNA and suggest that the mechanism of transfer is rather precise.

There are two major classes of T-DNA border clones, one class contains sequences homologous to the left and right ends of the T-DNA. The structure of these clones suggests that the T-DNA may be organized as tandem repeats. Although the data could also be explained if the T-DNA existed as an independent replicon, we have been unable to detect the presence of T-DNA circles in transformed plant cells. The second class of T-DNA border clones contains sequences homologous to the Ti plasmid on one end and sequences contiguous with plant DNA on the other end. These latter clones provide direct evidence that the T-DNA can integrate into the plant cell genome. The border clones which contain plant DNA hybridize to either unique, intermediately repeated or highly repeated plant DNA. Thus, the T-DNA does not exhibit preference for integration into a particular plant sequence.

Six independent T-DNA clones have been isolated from nopaline tumor lines. Three clones, 1, 1′, and 2, were derived from a tumor line induced by the nopaline Ti plasmid T37, and clones 3, 4, and 5, were derived from a tumor line induced by the nopaline Ti plasmid C58. The exact sequences at the borders of all these clones were compared to the sequences found in the same region of the Ti plasmid (Zambryski et al. 1982). Clones 1 and 1′ are derived from tandem junctions so each contain a left and a right end of the T-DNA and both borders were determined for each of these clones. Fig. 1 is a summary diagram of these results which reveal that the site of T-DNA integration is very precise on the right end and less precise on the left.

Only a variation of a single base was found in the analysis of the right borders; this variation was found in clone 1′ which is part of a tandem array. It is especially striking that the T-DNA/plant junction occurs at exactly the same base pair in the T-DNA from two independent tumor lines induced by two different Ti plasmids (clones 2 and 3). The most frequently observed left border varies over 6 bp. We have called this border a secondary border since other borders have been observed to fall further to the left. For example, clone 5 is actually an internal T-DNA fragment implying that the T-DNA border extends further to the left of the Ti plasmid *Eco*RI site which is 31 bp to the left of the secondary left border. In addition, another left

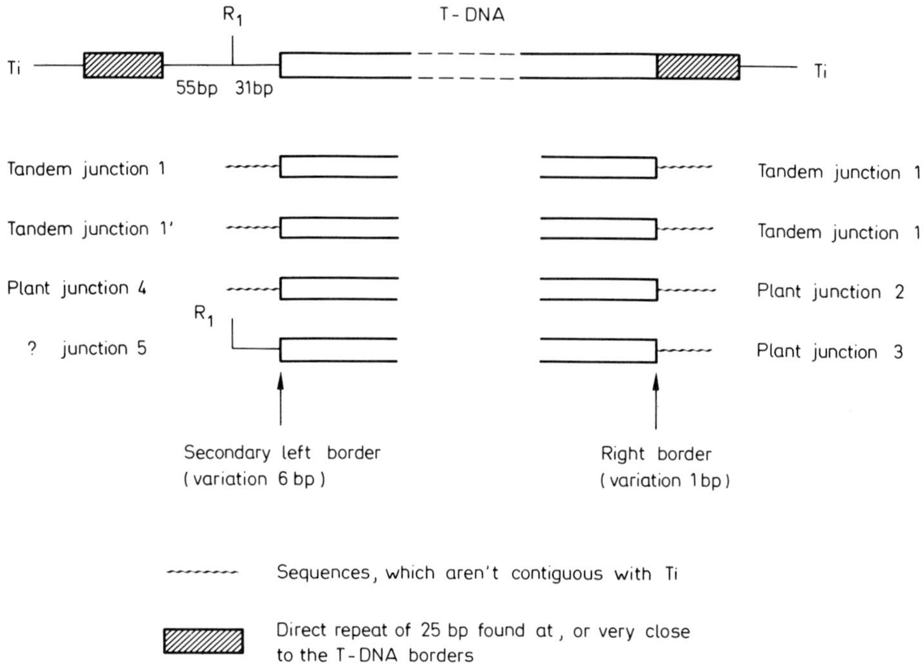

Fig. 1. Summary of nopaline T-DNA borders compared to the nopaline Ti plasmid. The borders of nopaline T-DNAs are diagrammed in comparison to the border regions of the nopaline Ti plasmid; the exact nucleotide sequences of these DNAs have been published (Zambryski et al. 1982). The origin of clones 1, 1′, 2, 3, 4, and 5, is described in the text. One *Eco*RI restriction endonuclease site which is close to the left T-DNA border is shown. In additon, the relative position of the direct repeat of 25 bp shown in Fig. 2 is indicated

border from a T37 tumor has been found approximately 95 bases to the left of the secondary left border (Yadav et al. 1982).

Comparison of the sequences of the Ti plasmid with the T-DNA at the borders did not suggest a mechanism of how T-DNA transfer occurs. For example, there are no repeated sequences found near the ends as is often found in other transposable elements. The one general conclusion from the nucleotide sequence analysis of the T-DNA junctions is that these sequences are enriched for A-T base pairs at the immediate junctional site. This high A-T content of the integration site is a more or less universal observation for most movable DNA; this includes bacteriophage lambda integration into bacterial DNA (Landy and Ross 1977, Pinkham et al. 1980), the sites favored for integration of prokaryotic (Kühn et al. 1979, Van Emmelo et al. 1980, Heffron et al. 1981, Miller et al. 1980, Reed 1981), and eukaryotic transposons (Farabaugh and Fink 1980, Roeder et al. 1980, Dunsmuir et al. 1980), as well as the sites for retrovirus integration (Majors and Varmus 1981, Hishinuma et al. 1981).

To provide further information on DNA sequences which may be important for transfer, we analyzed the sequences immediately around the T-DNA borders in the Ti plasmid. One of the most remarkable characteristics is the presence of a direct repeat of 25 bp. This repeat is found exactly to the right of the right T-DNA border,

G C T G G	T G G C A G G A T A T A T T G	T G	G T G T A A A C	A A A T T	Nopaline L
G T G T T	T G (A) C A G G A T A T A T T G	G C	G (G) G T A A A C	C T A A G	Nopaline R
A G C G G	(C) G G C A G G A T A T A T T C	A A	T T G T A A A T	G G C T T	Octopine L (TL)
C T G A C	T G G C A G G A T A T A T (A) C	C G	T T G T A A (T) T	T G A G C	Octopine R (TL)

Fig. 2. Summary of the 25 bp direct repeat found at the T-DNA borders of the nopaline and octopine Ti plasmids. A direct repeat of 25 bp which is found at or very close to the T-DNA borders in nopaline and octopine Ti plasmids is shown. In the nopaline Ti plasmid (pTiC58 or pTiT37) this repeat occurs exactly at the right T-DNA border and starts 111 bp to the left of the secondary left border. In the octopine Ti plasmid (pTiAch5 or pTiA6S3) this repeat occurs very close to and just outside the right T-DNA border; two T-DNA junctions analyzed so far vary within 6 bp from this sequence (R. Villarroel, M. De Beuckeleer, J. Gielen, H. De Greve, M. Holsters, unpublished results). On the left, in one tumor line this sequence occurs exactly at and within the T-DNA border (Simpson et al. 1982). In two other tumor lines, the left T-DNA border, either occurs within this sequence or 59 bp internal to this sequence (Holsters et al. 1983)

and starts 111 base pairs to the left of the secondary left border. This sequence is shown in Fig. 2, and its relative location is diagrammed in Fig. 1. In addition, a very similar sequence is found as a direct repeat at or very close to the right and left borders of the octopine T-DNA (TL) and is also shown in Fig. 2 (Simpson et al. 1982; Holsters et al. 1983). This repeat has two domains of homology of 15 bp and 8 bp. This repeat seems to be rather significant since the probability of finding even only the 15 bp sequence is greater than one in 10^8. The proof of its relative importance will be determined in future studies where this sequence is specifically altered by genetic mutations. While the analyses so far are limited, the results suggest that the right T-DNA border may be essential for T-DNA transfer; a large deletion in this region drastically reduces tumor formation (Holsters et al. 1980, Joos et al. 1983). However, large deletions of the left T-DNA border have no effect on oncogenicity (Joos et al. 1983).

The structure of the tandem T-DNAs is intriguing. DNA sequence analysis has shown that the junctions are not an exact joining of the left and right ends of the T-DNA. In clone 1 there are 136 bp of junctional sequence and in clone 1′ there are 22 bp. We do not know how these junctional sequences arose; however, they contain sequences which are also found in the Ti plasmid just close to the observed T-DNA borders. For example, the junctional sequence of clone 1 is composed of repeats of 12, 13, 14, 15, 18, and 28 bp; however, these repeats are now scrambled with respect to their orientation and location to the same sequences found at the ends of the T-DNA. Recently, octopine T-DNA (TL) has also been found to exist in a tandem arrangement; in addition, the junctional sequence between the tandem copies contain repeated sequences (Holsters et al. 1983).

Fig. 3 summarizes in a simple manner what is known to date about the nopaline T-DNA structure in plants; however, it also indicates how little is known about the mechanism of transfer. The T-DNA region is transferred to the plant genome and either the T-DNA can remain as a single copy or it sometimes seems to be amplified resulting in a tandem array. It is not known whether this reorganization occurs as

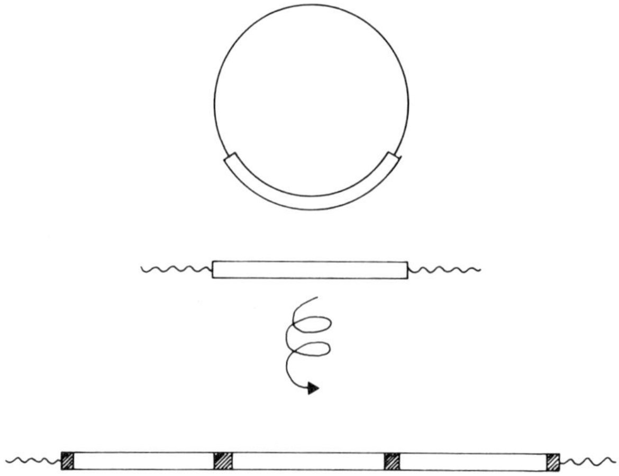

Fig. 3. T-DNA transfer and stabilization. *Circle* represents the Ti plasmid with the T-DNA as a *box*. *Wavy lines* represent plant DNA and the *cross-hatched areas* represent rearranged sequences at the ends of the T-DNA. All T-DNAs are in direct orientation. *Curved arrow* indicates a reaction wherein there is an amplification of the T-DNA

a consequence of the initial integration of T-DNA, or as a result of subsequent recombination events which lead to a stabilized structure. Site-specific amplification of DNA is a property common to prokaryotic and eukaryotic genomes and it is not surprising that the introduction of foreign DNA into the DNA of a susceptible host results in some perturbation of the local environment. Whether or not tandemization occurs may depend on the region of the plant genome which serves as a site for T-DNA integration.

Analysis of the T-DNA Borders when the T-DNA is Used as a Vector for the Insertion of Foreign DNA

A further test of the specificity of T-DNA transfer is the proof that foreign DNA can be inserted within the T-DNA and be transferred to the plant genome as part of the T-DNA structure. This phenomenon has been studied in most detail for a tumor line which was induced by the nopaline plasmid pTiT37::Tn*7nos* (pGV3171) (Hernalsteens et al. 1980, Holsters et al. 1982). Here the transposon Tn*7* of approximately 15 kbp is inserted near the right end of the T-region and this enlarged T-DNA is found to be transferred and stably integrated to the genome of transformed tobacco cells. Thus, the integrative mechanism responsible for T-DNA transfer must recognize specific Ti plasmid sequences as the endpoints of the T-DNA so that the length of the DNA between these endpoints is not critical. Furthermore, the T-DNA-containing Tn7 was reisolated from the tumor cell DNA by molecular cloning; and it was shown that the Tn7 was still functional in transposition and expression of its resistance markers (Holsters et al. 1982). Thus, the Tn7 DNA was not modified during its residence in the plant genome.

Conclusions to Date

The following points summarize what is known to date about T-DNA transfer from nopaline Ti plasmids of *Agrobacterium* to plant cell DNA:

1. T-DNA integration is a precise event. The right T-border varies over a single base pair, and the left T-DNA border varies over 100 base pairs. Furthermore, foreign DNA can be inserted within the T-DNA ends and is transferred utilizing the same borders.
2. There are sequences (25 bp direct repeats) which are found at the ends of the T-DNA region of the Ti plasmid which may be important in the excision of the T-region during the T-DNA transfer.
3. The T-DNA can be found as a single copy or in multiple copies which are organized in a tandem array. The sequences comprising the junction of two tandem T-DNA copies is often composed of repeats. This implies that reorganization of the ends of the T-DNA can occur during the events which lead to tandemization.

These results are rather encouraging since they indicate that specific regions of the Ti plasmid are involved in T-DNA transfer. This is an important finding if the Ti plasmid is to be manipulated in vitro in the hopes of designing a vector which will be useful for the insertion of foreign genes, and which contains the T-DNA border regions to allow transfer without causing tumors. That this approach may succeed is supported by recent studies which analyze the T-DNA in plant regenerates from tumor tissue. Here, normal shoot development results from a selection against tumorous growth such that deletions of the regions of the T-DNA controlling abnormal growth occur. However, the shortened T-DNA is now stable and will even be transmitted through seeds in a Mendelian fashion (Otten et al. 1981, De Greve et al. 1982). Thus, we can systematically create the appropriate T-DNA region deletions while retaining the T-DNA borders to produce a Ti plasmid which can still transfer DNA without altering the normal growth pattern of the recipient plant.

Acknowledgements. The author would like to thank all the members of the laboratories of Profs. Marc Van Montagu and Jeff Schell for their helpful discussions, constant encouragement and support during this work. The research described was supported by USDA grant 5901-0410-9-0374-0, NIH grant CA2742A. The author is currently indebted to EMBO for a long-term fellowship.

References

Beuckeleer M De, Lemmers M, Vos G De, Willmitzer L, Montagu M Van, Schell J (1981) Further insight on the transferred-DNA of octopine crown gall. Mol Gen Genet 183:283–288

Braun AC, Wood HN (1976) Suppression of the neoplastic state with the acquisition of specialized functions in cells, tissues, and organs of crown gall teratomas of tobacco. Proc Natl Acad Sci USA 73:496–500

Chilton M-D, Drummond HJ, Merlo DJ, Sciaky D, Montoya AL, Gordon MP, Nester EW (1977) Stable incorporation of plasmid DNA into higher plant cells: the molecular basis of crown gall tumorigenesis. Cell 11:263–271

Dunsmuir P, Brorein WJ Jr, Simon MA, Rubin GM (1980) Insertion of the *Drosophila* transposable element copia generates a 5 base pair duplication. Cell 21:575–579

Emmelo J Van, Devos R, Ysebart M, Fiers W (1980) Construction and characterization of a plasmid containing a nearly full-size DNA copy of satellite tobacco necrosis virus RNA. J Mol Biol 143:259–271

Farabaugh PJ, Fink GR (1980) Insertion of the eukaryotic transposable element Ty*1* creates a 5-base pair duplication. Nature 286:352–356

Greve H De, Leemans J, Hernalsteens JP, Thia-Toong L, Beuckeleer M De, Willmitzer L, Otten L, Montagu M Van, Schell J (1982) Regeneration of normal and fertile plants that express octopine synthase, from crown galls after deletion of tumour-controlling functions. Nature 300:752–755

Heffron F, Kostriken R, Morita C, Parker R (1981) Tn*3* encodes a site-specific recombination system: identification of essential sequences, genes, and the actual site of recombination. Cold Spring Harbor Symp Quant Biol 45:259–268

Hernalsteens JP, Vliet F Van, Beuckeleer M De, Depicker A, Engler G, Lemmers M, Holsters M, Montagu M Van, Schell J (1980) The *Agrobacterium tumefaciens* Ti plasmid as a host vector system for introducing foreign DNA in plant cells. Nature 287:654–656

Hishinuma F, Bona PJ De, Astrin S, Skalka AM (1981) Nucleotide sequence of acceptor site and termini of integrated avian endogenous provirus *ev*I: integration creates a 6 bp repeat of host DNA. Cell 23:155–164

Holsters M, Silva B, Vliet F Van, Genetello C, Block M De, Dhaese P, Depicker A, Inzé D, Engler G, Villarroel R, Montagu M Van, Schell J (1980) The functional organization of the nopaline *A. tumefaciens* plasmid pTiC58. Plasmid 3:212–230

Holsters M, Villarroel R, Montagu M Van, Schell J (1982) The use of selectable markers for the isolation of plant-DNA/T-DNA junction fragments in a cosmid vector. Mol Gen Genet 185: 283–289

Holsters M, Villarroel R, Gielen J, Seurinck J, De Greve H, Montagu M Van, Schell J (1983) An analysis of the boundaries of the octopine TL-DNA in tumors induced by *Agrobacterium tumefaciens*. Mol Gen Genet 190:35–41

Joos H, Inzé D, Caplan A, Sormann M, Montagu M Van, Schell J (1982) Genetic analysis of T-DNA transcripts in nopaline crown galls. Cell 32:1057–1067

Kahl G, Schell J (1982) Molecular biology of plant tumors. Academic Press, London New York

Kühn S, Fritz HJ, Starlinger P (1979) Close vicinity of IS*1* integration sites in the leader sequence of the gal operon of *E. coli*. Mol Gen Genet 167:235–241

Landy A, Ross W (1977) Viral integration and excision: structure of the lambda *att* sites. Science 197:1147–1160

Lemmers M, Beuckeleer M De, Holsters M, Zambryski P, Depicker A, Hernalsteens JP, Montagu M Van, Schell J (1980) Internal organization, boundaries and integration of Ti-plasmid DNA in nopaline crown gall tumours. J Mol Biol 144:353–376

Majors JE, Varmus HE (1981) Nucleotide sequences at host-proviral junctions for mouse mammary tumour virus. Nature 289:253–257

Miller JH, Calos MP, Galas D, Hofer M, Büchel DE, Müller-Hill B (1980) Genetic analysis of transpositions in the lac region of *Escherichia coli*. J Mol Biol 144:1–18

Otten L, Greve H De, Hernalsteens JP, Montagu M Van, Schieder O, Straub J, Schell J (1981) Mendelian transmission of genes introduced in plant by the Ti plasmids of *Agrobacterium tumefaciens*. Mol Gen Genet 183:209–213

Pinkham JL, Platt T, Enquist L, Weisberg RA (1980) The secondary attachment site for bacteriophage λ in the *pro*A/B gene of *Escherichia coli*. J Mol Biol 144:587–592

Reed RR (1981) Resolution of cointegrates between transposons $\gamma\delta$ and Tn*3* defines the recombination site. Proc Natl Acad Sci USA 78:3428–3432

Roeder GS, Farabaugh PJ, Chaleff DT, Fink GR (1980) The origins of gene instability in yeast. Science 209:1375–1380

Simpson RB, O'Hara PJ, Krook W, Montoya AL, Lichtenstein C, Gordon MP, Nester EW (1982) DNA from the A6S/2 crown gall tumors contains scrambled Ti-plasmid sequences near its junctions with the plant DNA. Cell 29:1005–1014

Thomashow MF, Nutter R, Montoya AL, Gordon MP, Nester EW (1980) Integration and organisation of Ti-plasmid sequences in crown gall tumors. Cell 19:729–739

Yadav NS, Vanderleyden J, Bennett DR, Barnes WM, Chilton M-D (1982) Short direct repeats flank the T-DNA on a nopaline Ti plasmid. Proc Natl Acad Sci USA 79:6322–6326

Yang F, Simpson RB (1981) Revertant seedlings from crown gall tumors retain a portion of the bacterial Ti plasmid DNA sequences. Proc Natl Acad Sci USA 78:4151–4155

Zambryski P, Holsters M, Kruger K, Depicker A, Schell J, Montagu M Van, Goodman HM (1980) Tumor DNA structure in plant cells transformed by *A. tumefaciens*. Science 209:1385–1391

Zambryski P, Depicker A, Kruger K, Goodman H (1982) Tumor induction by *Agrobacterium tumefaciens:* analysis of the boundaries of T-DNA. J Mol Appl Genet 1:361–370

Transcription of the Ti-Plasmid in Crown Gall Tumors

S.B. GELVIN, S.J. KARCHER, V.J. DI RITA, and E.W. TALIERCIO [1]

Introduction

The T-region of the *Agrobacterium tumefaciens* Ti-plasmid is a unique piece of DNA in that it is currently the only known nucleic acid which can naturally exchange between organisms of different phylogenetic kingdoms (Chilton et al. 1977, Thomashow et al. 1980a, 1980b, Yadav et al. 1980, Zambryski et al. 1980, Lemmers et al. 1980). Evidence has recently accrued indicating that the T-DNA can be transcribed both in its eukaryotic (Gurley et al. 1979, Gelvin et al. 1981, 1982, Willmitzer et al. 1981a) as well as in its prokaryotic (Gelvin et al. 1981) hosts. Although the T-DNA-encoded transcripts found in crown gall tumors share many similarities with other eukaryotic mRNAs, including transcription by RNA polymerase II (Willmitzer et al. 1981b), polyadenylation at the 3′ end (Gelvin et al. 1981, 1982, Willmitzer et al. 1981a), and probably the presence of a "cap" at the 5′ terminus (Schröder and Schröder 1982), the details of the mechanisms and regulation of this transcription have yet to be elucidated.

Early studies examining both octopine and nopaline-type crown gall tumors identified regions of the T-DNA which were expressed as mRNA (Gurley et al. 1979, Gelvin et al. 1981). These studies also demonstrated that the steady-state levels of RNA encoded by different regions of the T-DNA often varied considerably. More recent studies using Northern blot analysis have determined more precisely the number and relative abundance of mRNAs which map to discrete regions of the T_L region of several octopine-type crown gall tumors (Gelvin et al. 1982, Willmitzer et al. 1981a). This region, which has been found in all octopine-type tumors examined to date (Thomashow et al. 1980a), contains information essential for tumor maintenance and octopine production. Genetic analyses, employing both transposon (Koekman et al. 1979, Klapwijk et al. 1980, Ooms et al. 1981, Garfinkel and Nester 1980, De Greve et al. 1981, Garfinkel et al. 1981) and deletion (Ooms et al. 1982, Leemans et al. 1982) mutagenesis, have defined four distinct regions of T_L according to the phenotype of the tumors containing the mutagenized T-DNA: (1) a region which may be involved with the synthesis or regulation of auxin by the plants (the shooting or TMS locus); (2) a region which may play a role in the synthesis or regulation of cytokinins by the plant (the rooting or TMR locus); (3) a region which may regulate the growth rate of the tumors (TML); (4) the locus encoding octopine synthase, the

1 Department of Biological Sciences, Purdue University, West Lafayette, IN 47907, USA

Molecular Genetics of the Bacteria-Plant Interaction
(ed. by A. Pühler)

enzyme responsible for the production of octopine by the tumors (Garfinkel et al. 1981). Each of these loci codes for discrete mRNAs (Gelvin et al. 1982, Willmitzer et al. 1981a). Additional mRNAs are encoded by other regions of T_L which, when mutagenized, direct the synthesis of tumors which are phenotypically identical to tumors containing wild-type T-DNA (Gelvin et al. 1982, Willmitzer et al. 1981a, Garfinkel et al. 1981).

In addition to the T_L region, many octopine-type crown gall tumors contain one or more copies of an adjacent region of the Ti-plasmid termed T_R (Thomashow et al. 1980a). Although the precise function or functions of T_R in the tumor is yet unknown, this region of the T-DNA has been implicated in the synthesis by some tumors of the recently discovered compounds mannopine and agropine (Firmin and Fenwick 1978). In this article, we report the sizes and map positions of four transcripts encoded by T_R. We also show that the relative abundance of these RNAs can vary between tumor lines and even within a given line when grown under different conditions. We have also begun an investigation of two aspects of DNA structure, the chromatin conformation and the extent of DNA methylation, which may affect the transcription of the T-DNA in crown gall tumors.

Results

Mapping of Transcripts Within the T_R Region in the T-DNA

The E9 tobacco crown gall tumor, as well as containing 1–2 copies of T_L, contains 15–30 copies of T_R, a region including Eco R1 fragment 13 and extending to the adjacent Eco R1 fragments 22 and 24. Although T_L and T_R are contiguous in the Ti-plasmid in *Agrobacterium tumefaciens,* they are not in the E9 tumor (Thomashow et al. 1980a). Restriction endonuclease fragments derived from Bam H1 fragment 2 from the Ti-plasmid pTi$B_6$806 were cloned in *E. coli.* These fragments, denoted 2_{a-e_3}, encompass all of T_R (see Fig. 1).

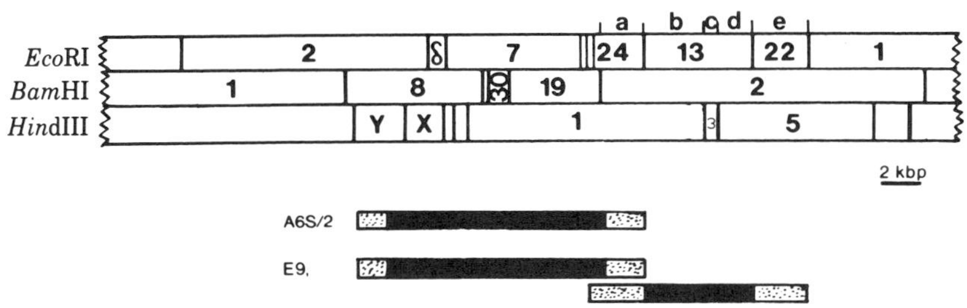

Fig. 1. Restriction endonuclease map of the T-DNA of a typical octopine-type Ti-plasmid. *Letters* above the map indicate the subcloned regions of Bam H1 fragment 2. *Bars* beneath the map indicate the extent of the T-DNA in each of the tumor lines. *Stippled areas* of the bars indicate regions in which a T-DNA-plant DNA junction occurs

Total RNA or polyadenylated RNA was isolated from E9 suspension or callus cultures and subjected to electrophoresis on denaturing 2% agarose, 6% formaldehyde gels (Gelvin et al. 1982). The RNA was transferred to nitrocellulose (Thomas 1980) and hybridized with the radioactive T_R subclones. The results are shown in Fig. 2. Four discrete transcripts were detected: A relatively abundant 0.78 kb RNA and a relatively rare 1.05 kb RNA show hybridization entirely within region 2b. Extending from region 2b, through region 2c, and into region 2d is a transcript 1.45 kb in size. A fourth transcript 1.65 kb in size extends from region 2d, through regions $2e_1$ and $2e_2$, and into region $2e_3$. These last two RNAs, although moderately abundant in E9 tumor cells grown in suspension culture (2_{sus} in Fig. 2), are relatively rare when the same tissue is grown on solid agar as a callus (2_{cal} in Fig. 2). It thus appears that the steady-state levels of the E9 T_R transcripts can vary according to the growth conditions of the cells.

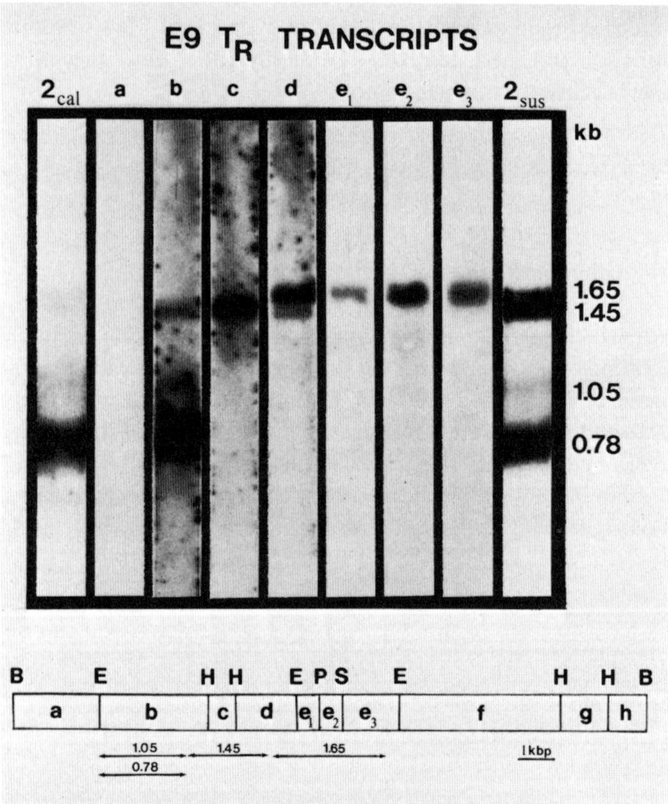

Fig. 2. Northern blot analysis of transcripts from the T_R region of the E9 crown gall tumor. Total RNA was extracted from E9 tumor cells, subjected to electrophoresis on denaturing agarose gels, blotted onto nitrocellulose, and hybridized with radioactive T-DNA probes. *Letters* above the lanes indicate the subclones of Bam H1 fragment 2 used as probes. A map of these regions is shown *below the blots*. 2_{cal} and 2_{sus}, Bam H1 fragment 2 used as a hybridization probe to blots made from E9 callus and suspension cell RNA, respectively. Restriction endonuclease sites: *B* Bam H1; *E* Eco R1; *H* Hind III; *P* Pst I; *S* Sal I

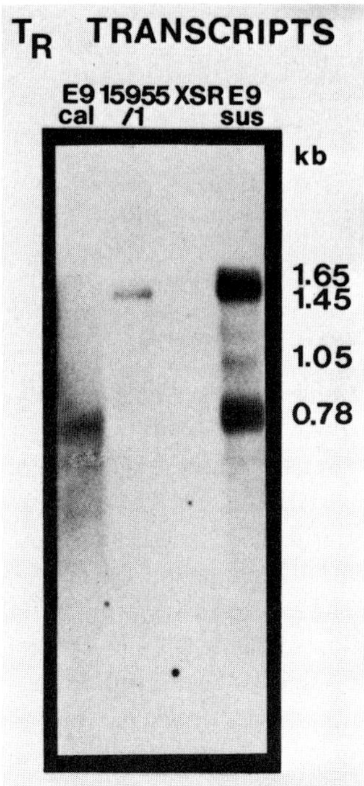

Fig. 3. Northern blot analysis of transcripts from the T_R region of various cell lines. The hybridization probe was Bam H1 fragment 2. Two micrograms of total cellular RNA was used for each lane. *$E9_{cal}$* and *$E9_{sus}$*, RNA from callus cells and suspension cells, respectively. *15955/1,* another octopine-type tobacco tumor line. *XSR,* nontransformed *N. tabacum* cv Xanthi nc

Another tobacco tumor line (15955/1) which contains 1 copy of T_R (Thomashow et al. 1980a) was also analyzed by Northern blot analysis for T_R transcripts (Fig. 3). Only the 1.45 kb transcript was present, despite the fact that this tumor contains a complete copy of region 2b (Thomashow et al. 1980a). Figure 3 also shows that the nontransformed *N. tabacum* cv Xanthi callus XSR does not contain transcripts homologous to Bam fragment 2. Equal amounts of total cellular RNA from each cell line were analyzed in Fig. 3. Although the E9 tumor contains 15–30 more copies of T_R DNA than does the 15955/1 tumor, the steady state level of the 1.45 kb T_R transcript is only 3–5 times more abundant in the E9 suspension culture line. This may indicate that not all copies of T_R are transcribed in this line. Other explanations, such as decreased initiation or RNA stability in the E9 suspension cells, are also possible.

Recent evidence has indicated that many tumors containing T_R and T_L, but not those containing only T_L, can synthesize the opines agropine and/or mannopine (John Kemp, pers. commun.). Table 1 shows that both the E9 suspension and callus cultures produce large quantities of both mannopine and agropine. The 15955/1 callus also produces detectable levels of both these compounds. These results, together with the transcription data discussed above, suggest that a protein encoded by the 1.45 kb transcript is involved with the biosynthesis of these opines.

Table 1. Mannopine and agropine production by crown gall tumor lines

Line	Mannopine production	Agropine production
E9 (callus)	+ + +	+ + +
E9 (suspension)	+ + +	+ + +
15955/1	+	+

T-DNA Structure in Crown Gall Tumors

The chromatin of eukaryotic organisms is arranged in complexes of deoxyribonucleoprotein called nucleosomes. Multimers of the basic 165-220 bp DNA from these nucleosomes can be separated by agarose gel electrophoresis after mild digestion of the DNA in the nuclei with micrococcal nuclease. Because of the role that the chromatin structure may play in the regulation of the transcription of animal genes (Garel and Axel 1976, Gazit et al. 1980, Weintraub et al. 1981, Groudine and Weintraub 1981, Groudine et al. 1981), we have been interested in what effects the structure of crown gall tumor chromatin may have on the expression of the T-DNA. As a first step in this direction, we have isolated DNA from nucleosomes of crown gall tumors. Figure 4 shows a typical nucleosomal DNA ladder from the A6S/2 crown gall tumor. Blots of these gels are currently being hybridized with cloned T-DNA fragments to determine whether the T-DNA, when in the tumor, is packaged in a nucleosomal conformation.

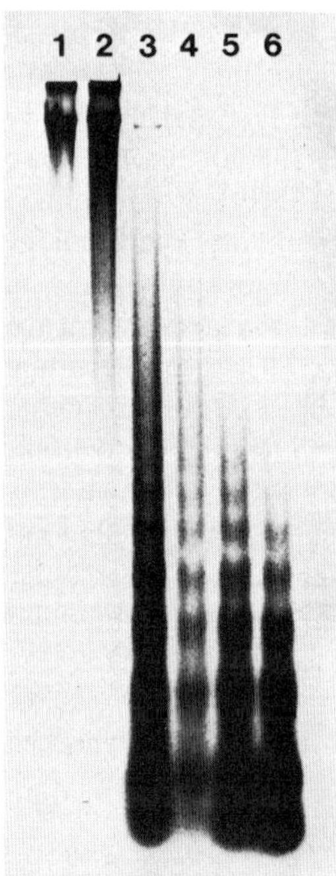

Fig 4 Nucleosome pattern of DNA isolated from A6S/2 nuclei. Nuclei were isolated from the A6S/2 tumor line, incubated with various amounts of DNase I, the DNA extracted, and subjected to electrophoresis on a 1.8% agarose gel. *Lane 1* DNA from unincubated nuclei, *Lanes 2–6* DNA from nuclei digested (pH 7.0) with 0, 50, 100, 250, and 500 units DNase I/ml, respectively

Methylation of the T-DNA in *Agrobacterium* and in Crown Gall Tumors

Methylation of the DNA in eukaryotes has been implicated with the regulation of gene activity. Although a few exceptions have been noted (Bird and Taggart 1980), transcriptionally active genes are in general undermethylated relative to inactive genes in animal systems (Naveh-Many and Cedar 1981, Compere and Palmiter 1981). The methylation of plant nuclear DNA occurs as 5-methyl cytosine, and the recognition sequence for the plant DNA methylase(s) is 5′CXG3′, where X can be any of the four bases. Over 80% of the time X is the base cytosine (Gruenbaum et al. 1981). The restriction endonucleases Msp I and Hpa II contain within their recognition sequence 5′CCGG3′. Msp I digests the DNA except when the external cytosine is methylated. Hpa II digests the DNA except when the internal cytosine is modified. These enzymes are therefore useful diagnostic tools to measure the degree of methylation at particular sites in the DNA.

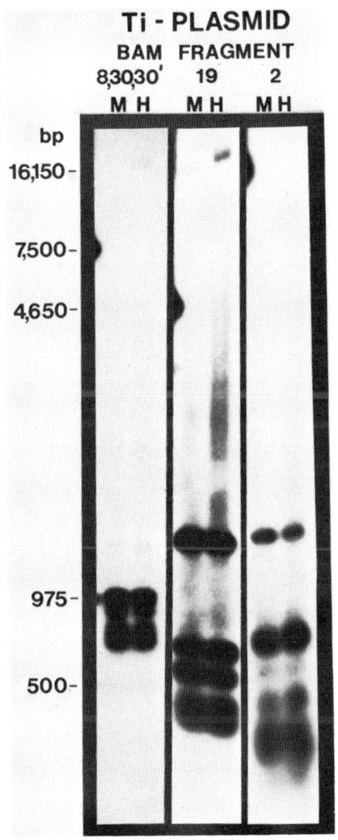

Fig. 5

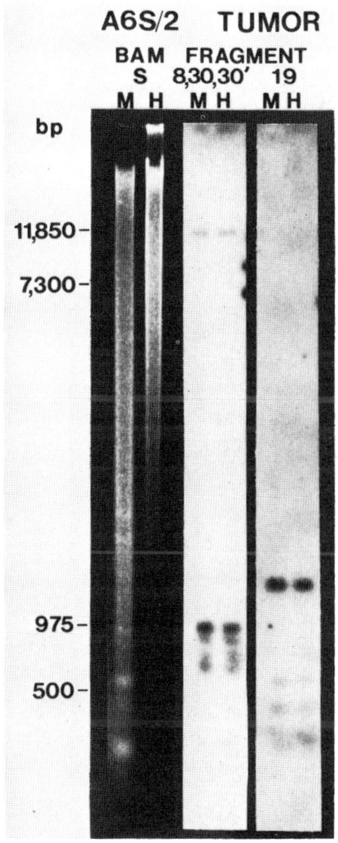

Fig. 6

Fig. 5. Msp I and Hpa II T-DNA fragments of the Ti-plasmid pTiB$_6$806. The DNA was digested with Msp I *(M)* or Hpa II *(H)*, subjected to electrophoresis, blotted, and hybridized with radioactive T-DNA clones. The Bam H1 fragment used as the probe is noted *above each set of lanes*

Fig. 6. Msp I and Hpa II T-DNA fragments from the A6S/2 tumor. Experimental details are the same as Fig. 5

DNA of the octopine-type plasmid $pTiB_6806$, isolated from *A. tumefaciens* strain A277, was digested with Msp I or Hpa II, subjected to electrophoresis on agarose gels, transferrred to nitrocellulose (Southern 1975), and hybridized with cloned Bam H1 restriction endonuclease fragments encompassing the entire T-DNA. Figure 5 shows that the restriction endonuclease patterns of these T-DNA fragments are identical whether they are digested with Msp I or Hpa II. Other data (not shown) indicate that the patterns obtained by digestion of cloned fragments of the T-DNA in *E. coli* (which does not methylate DNA at the sequence 5′CCGG3′) are identical to these patterns. Thus, the T-DNA is not methylated at the sequence 5′CCGG3′ when in *A. tumefaciens.*

We next digested the DNA from the crown gall tumor A6S/2 with Msp I or Hpa II and analyzed the T-DNA for methylation as described above. The A6S/2 tumor contains a single copy of the T_L region of the T-DNA (Thomashow et al. 1980a) and this region is transcribed in the tumor (Gelvin et al. 1981, 1982). Comparison of the hybridization pattern shown in Fig. 6 with that of Fig. 5 indicates that the T-DNA of this tumor line is not methylated at the sequence 5′CCGG3′.

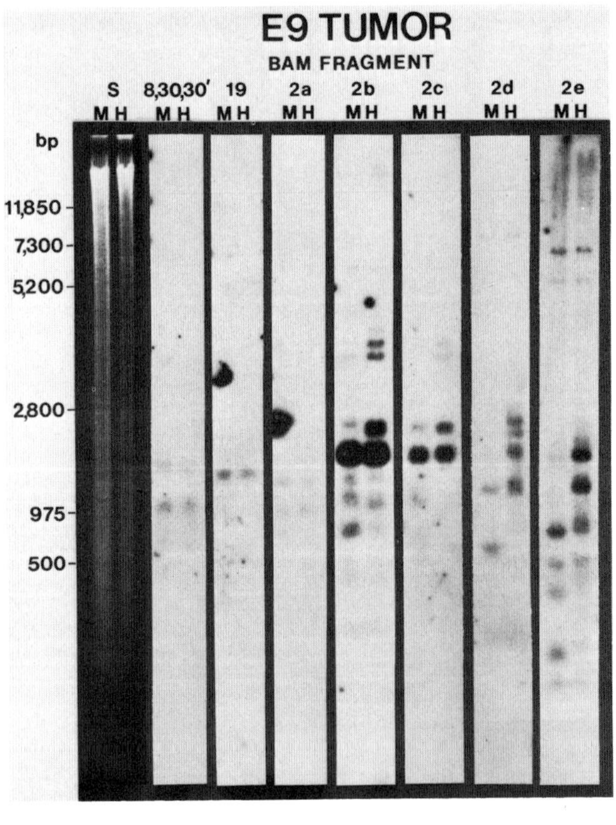

Fig. 7. Msp I and Hpa II T-DNA fragments from the E9 tumor. Experimental details are the same as Fig. 5

As discussed above, the E9 crown gall tumor contains 1–2 copies of T_L and 15–30 copies of T_R (Thomashow et al. 1980a). Both T_L and T_R are transcribed in this tumor (Gelvin et al. 1981, 1982). Restriction endonuclease analysis of the T-DNA of callus cultures of the E9 tumor line, using the enzymes Msp I and Hpa II, indicates that the T_L DNA is not methylated at the sequence 5′CCGG3′ (Fig. 7). One or a few copies of T_R are also not methylated. The majority of copies or T_R are heavily methylated at either the internal cytosine or external cytosine residue, or at both. This is shown by the large number of high molecular weight restriction endonuclease fragments seen in the T_R DNA of the E9 crown gall tumor but not in the unmethylated Ti-plasmid (compare with Fig. 5). By analogy with DNA methylation and gene inactivity in animal systems, these data suggest that the majority of T_R copies are not transcribed in callus cultures of this tumor.

Discussion

The T_R region of octopine-type crown gall tumors possesses many properties which are ideal for genetic and molecular analyses. It can be covalently integrated into the plant nuclear DNA in many tumor lines, sometimes as multiple copies (Thomashow et al. 1980a). The steady-state levels of transcripts from this region are relatively high compared to the levels of RNA encoded by T_L (Gelvin et al. 1981, Karcher and Gelvin, unpubl.), simplifying transcriptional studies. The large number of copies of T_R in some tumor lines presents an interesting system for the study of the potential differential transcription of multiple T-DNA inserts. Finally (and perhaps most important when considering the T-DNA as a vector for delivering foreign genes to plant cells in genetic engineering experiments), T_R does not code for functions necessary for tumor maintenance, as evidenced by the lack of T_R sequences in some octopine-type crown gall tumors (Thomashow et al. 1980a). With these facts in mind, we have recently begun a comprehensive study of the structure and expression of T_R T-DNA sequences in crown gall tumors. These studies will eventually encompass the molecular and genetic elucidation of the functioning of T_R from the sequence to the chromatin structural levels.

As a first step in these studies, we have analyzed the transcripts encoded by T_R in two tobacco crown gall tumors, E9 and 15955/1. Four discrete RNA species were detected in both total and polyadenylated RNA populations from the E9 tumor line. The steady-state levels of these transcripts vary markedly. In all of the approximately 10 preparations of E9 RNA that we have made, the 0.78 kb species was relatively abundant and the 1.05 kb RNA relatively rare. In some of our preparations, the 1.05 kb RNA was not detectable. Two other RNA species, 1.45 and 1.65 kb in size, were moderately abundant in E9 suspension cells, but the steady-state levels of these two transcripts were considerably lower in the two preparations of E9 callus-derived RNA we analyzed. It is not known whether the disparate levels of these transcripts reflect differences in transcriptional initiation, elongation, transport to the cytoplasm, or RNA stability. Similar differences in the levels of T_L transcripts have been described in other tumor lines grown under different conditions (Willmitzer et al. 1981b).

Despite the fact that the 15955/1 crown gall tumor line contains a complete copy of Eco R1 fragment 13 (regions 2_{b-d} in Fig. 1) (Thomashow et al. 1980a, Gelvin, unpubl.), only one of the transcripts (the 1.45 kb RNA species) found in the E9 tumor is detectable in this tumor line. Again, the reason for this is not known. This tobacco tumor line was initiated by a strain of *Agrobacterium tumefaciens* containing the Ti-plasmid pTi15955. This plasmid is highly homologous with the Ti-plasmid, pTiB$_6$806, used to initiate the E9 tumor (Thomashow et al. 1980a). It is interesting to note that a sunflower tumor initiated by pTi15955 also makes only the 1.45 kb RNA (Murai and Kemp 1982). (In that paper the size of this transcript was determined to be 1.6 kb; the transcript maps, however, to the same location as the 1.45 kb transcript we have described). Perhaps the plasmid pTi15955 contains small mutations not detectable by restriction endonuclease analysis which inhibit the transcription of the other T_R RNAs we have detected in the E9 tumor.

Although the functions of the four genes we have detected in the T_R region of the E9 tumor have not yet been determined, we can speculate on the possible functions of the 1.45 kb RNA. This is the only T_R transcript we could detect in the 15955/1 tumor line. This tumor produces both mannopine and agropine. The insertion of a phaseolin gene into the Hind III sites through which the 1.45 kb RNA is transcribed leads to the production of tumors which do not synthesize either of these opines (John Kemp, pers. commun.). Thus, it is likely that the 1.45 kb transcript encodes some function(s) necessary for the production of mannopine and agropine.

We have investigated the T-DNA for methylation at the sequence 5′CCGG3′ both when in the Ti-plasmid and when integrated into the plant nuclear DNA. This was accomplished by digesting the various DNA samples with the methylation-specific isoschizomeric restriction endonucleases Msp I and Hpa II, subjecting the fragments to electrophoresis, blotting onto nitrocellulose, and hybridizing with various radioactive T-DNA probes. The results of those studies indicated that the T-DNA is not methylated at these sequences when in the Ti-plasmid pTiB$_6$806 in *Agrobacterium*. Other experiments (data not shown) suggest that, with two minor exceptions, the T-DNA is also not methylated at this sequence in the nopaline-type Ti-plasmid pTiT37. Methylation of the T-DNA does not occur in the tumor A6S/2 (which contains one copy of T_L) or in the single copy of T_L in the E9 tumor. At least one copy of T_R in the E9 tumor is not methylated; extensive methylation of the T-DNA does occur, however, on most of the 15–30 copies of T_R in this line. Other data (not shown) indicate that in the octopine-type tumors 15955/1 and 15955/01, and the nopaline-type tumor HT37#15, at least one copy of the T-DNA is not methylated at the sequence 5′CCGG3′. DNA methylation can occur, however, when multiple copies of T-DNA are present.

In all tumors investigated, at least one copy of T-DNA was not methylated. In animal systems, DNA undermethylation has been correlated with transcriptional competency (Groudine et al. 1981, Naveh-Many and Cedar 1981, Compere and Palmiter 1981). If this correlation holds for crown gall tumors, the results of the experiments on the E9 tumor line would suggest that most copies of T_R are not transcriptionally competent. The high steady state levels of transcripts from T_R relative to T_L in this line may therefore be attributable to strong promoters in T_R. Other explanations, however, are possible.

When the T-DNA of the Ti-plasmid integrates into the plant nuclear DNA, it may take on specific characteristics found in the DNA of eukaryotes, e.g., extensive methylation at the sequence 5′CCGG3′. Experiments are currently in progress to determine whether the T-DNA can also become associated with nucleosomes in the plant nucleus, and whether actively transcribed T-DNA genes are preferentially DNAse I sensitive.

Acknowledgements. The authors would like to thank Ms. Jeanette Richter for technical assitance, Dr. Peter B. Goldsbrough for critical reading of the manuscript, and Marge Miles for help in preparation of the manuscript. Special thanks go to Dr. John Kemp for performing the agropine and mannopine assays. This work was supported, in part, by grants from the National Science Foundation (PCM-8109626) and the Indiana Chapter, American Cancer Society and the Indiana Elks Club.

References

Bird AP, Taggart MH (1980) Variable patterns of total DNA and rDNA methylation in animals. Nucleic Acids Res 8:1485–1497

Chilton M-D, Drummond MH, Merlo DJ, Sciaky D, Montoya AL, Gordon MP, Nester EW (1977) Stable incorporation of plasmid DNA into higher plant cells: The molecular basis of crown gall tumorigenesis. Cell 11:263–271

Compere SJ, Palmiter RD (1981) DNA methylation controls the inducibility of the mouse metallothionein-1 gene in lymphoid cells. Cell 25:233–240

DeGreve H, Decraemer H, Seurinck J, Montagu M Van, Schell J (1981) The functional organization of the octopine *Agrobacterium tumefaciens* plasmid pTiB6S3. Plasmid 6:235–248

Firmin JL, Fenwick K (1978) Agropine – a major new plasmid-determined metabolite in crown gall tumours. Nature 276:842–844

Garel A, Axel R (1976) Selective digestion of transcriptionally active ovalbumin genes from oviduct nuclei. Proc Natl Acad Sci USA 73:3966–3970

Garfinkel DJ, Nester EW (1980) *Agrobacterium tumefaciens* mutants affected in crown gall tumorigenesis and octopine catabolism. J Bacteriol 144:732–743

Garfinkel DJ, Simpson RB, Ream LW, White FF, Gordon MP, Nester EW (1981) Genetic analysis of crown gall: Fine structure map of the T-DNA by site-directed mutagenesis. Cell 27:143–153

Gazit B, Panet A, Cedar H (1980) Reconstruction of a deoxyribonuclease I-sensitive structure on active genes. Proc Natl Acad Sci USA 77:1787–1790

Gelvin SB, Gordon MP, Nester EW, Aronson AI (1981) Transcription of the *Agrobacterium* Ti plasmid in the bacterium and in crown gall tumors. Plasmid 6:17–29

Gelvin SB, Thomashow MF, McPherson JC, Gordon MP, Nester EW (1982) Sizes and map positions of several plasmid-DNA-encoded transcripts in octopine-type crown gall tumors. Proc Natl Acad Sci USA 79:76–80

Groudine M, Weintraub H (1981) Activation of globin genes during chicken development. Cell 24:393–401

Groudine M, Eisenman R, Weintraub H (1981) Chromatin structure of endogenous retroviral genes and activation by an inhibitor of DNA methylation. Nature 292:311–317

Gruenbaum Y, Naveh-Many T, Cedar H, Razin A (1981) Sequence specificity of methylation of higher plant DNA. Nature 292:860–862

Gurley WB, Kemp JD, Albert MJ, Sutton DW, Callis J (1979) Transcription of Ti plasmid-derived sequences in three octopine-type crown gall tumor lines. Proc Natl Acad Sci USA 2828–2832

Klapwijk PMJ, Breukelen J Van, Korevaar K, Ooms G, Schilperoort RA (1980) Transposition of Tn 904 encoding streptomycin resistance in the octopine Ti plasmid of *Agrobacterium tumefaciens.* J Bacteriol 141:129–136

Koekman BP, Ooms G, Klapwijk PM, Schilperoort RA (1979) Genetic map of an octopine Ti plasmid. Plasmid 2:347–359

Leemans J, Deblaere R, Willmitzer L, DeGreve H, Hernalsteens JP, Montagu M Van, Schell J (1982) Genetic identification of functions of TL-DNA transcripts in octopine crown galls. EMBO J 1:147–152

Lemmers M, Beuckeleer M De, Holsters M, Zambryski P, Depicker A, Hernalsteens JP, Montagu M Van, Schell J (1980) Internal organization, boundaries, and integration of Ti-plasmid DNA in nopaline crown gall tumors. J Mol Biol 144:353–376

Murai N, Kemp JD (1982) T-DNA of pTi-15955 from *Agrobacterium tumefaciens* is transcribed into a minimum of seven polyadenylated RNAs in a sunflower crown gall tumor. Nucleic Acids Res 10:1679–1689

Naveh-Many T, Cedar H (1981) Active gene sequences are undermethylated. Proc Natl Acad Sci USA 78:4246–4250

Ooms G, Hooykaas PJJ, Moolenaar G, Schilperoort RA (1981) Crown gall plant tumors of abnormal morphology induced by *Agrobacterium tumefaciens* carrying mutated octopine Ti plasmids: analysis of T-DNA functions. Gene 14:133–150

Ooms G, Hooykaas PJJ, Vleen RJM Van, Beelan P Van, Regensburg-Tuink TJG, Schilperoort RA (1982) Octopine Ti-plasmid deletion mutants of *Agrobacterium tumefaciens* with emphasis on the right side of the T-region. Plasmid 7:15–29

Schröder G, Schröder J (1982) Hybridization selection and translation of T-DNA encoded mRNAs from octopine tumors. Mol Gen Genet 185:51–55

Southern EM (1975) Detection of specific sequences among DNA fragments separated by gel electrophoresis. J Mol Biol 98:503–518

Thomas PS (1980) Hybridization of denatured RNA and small DNA fragments transferred to nitrocellulose. Proc Natl Acad Sci USA 77:5201–5205

Thomashow MF, Nutter R, Montoya AL, Gordon MP, Nester EW (1980a) Integration and organization of Ti plasmid sequences in crown gall tumors. Cell 19:729–739

Thomashow MF, Nutter R, Postle K, Chilton M-D, Blattner FR, Powell A, Gordon MP, Nester EW (1980b) Recombination between higher plant DNA and the Ti plasmid of *Agrobacterium tumefaciens.* Proc Natl Acad Sci USA 77:6448–6452

Weintraub H, Larsen A, Groudine M (1981) α-globin-gene switching during the development of chicken embryos: expression and chromosome structure. Cell 24:333–344

Willmitzer L, Otten L, Simons G, Schmalenbach W, Schröder J, Schröder G, Montagu M Van, Vos G De, Schell J (1981a) Nuclear and polysomal transcripts of T-DNA in octopine crown gall suspension and callus cultures. Mol Gen Genet 182:255–262

Willmitzer L, Schmalenbach W, Schell J (1981b) Transcription of T-DNA in octopine and nopaline crown gall tumors is inhibited by low concentrations of α-amanitin. Nucleic Acids Res 9:4801–4812

Yadav NS, Postle K, Saiki RK, Thomashow MF, Chilton M-D (1980) T-DNA of a crown gall teratoma is covalently joined to host plant DNA. Nature 287:458–461

Zambryski P, Holsters M, Kruger K, Depicker A, Schell J, Montagu M Van, Goodman HM (1980) Tumor DNA structure in plant cells transformed by *Agrobacterium tumefaciens.* Science 209:1385–1391

T_R Genes Involved in Agropine Production

J. VELTEN[1], L. WILLMITZER[1], J. LEEMANS[2], J. ELLIS[3]*, R. DEBLAERE[2], M. VAN MONTAGU[2,3], and J. SCHELL[1,3]

Introduction

The rightmost portion, or T_R, of the T-DNA of octopine-type Ti plasmids is frequently, but not always, found within plant tissue transformed by appropriate strains of *Agrobacterium tumefaciens* (Thomashow et al. 1980, De Beuckeleer et al. 1981). Tumor tissues lacking T_R show no obvious morphological differences when compared to T_R-containing tumor lines. T_R would, therefore, appear to contain no genetic information required for oncogenesis or tumor maintenance. T_R-DNA is, however, actively transcribed within plant tissue and has been associated with one known phenotype trait; the biosynthesis of the agropine family of opines (Leemans et al. 1982). Agropine biosynthesis has also been observed in tumors induced by agropine type Ti plasmids (Guyon et al. 1980) and root tissue transformed by *Agrobacterium rhizogenes* (Tepfer and Tempé 1981).

We have analyzed T_R expression in: (1) tobacco tumor tissues induced by Ti plasmids containing in vitro mutated T_R regions and (2) spontaneous T_R mutant tissues which have arisen during the propagation of established tumor lines. The results show the production of four T_R specific transcripts within transformed plant tissue. At least three of the five transcripts are involved in opine biosynthesis.

Mapping the T_R Transcripts

Transcripts sizes and approximate positions of corresponding coding regions within T_R were determined by Northern analysis of RNA from tumor line 1GV7 (induced on *Nicotiana tabacum* cv. Wiscons in 38 by *A. tumefaciens* carrying pGV2208; Leemans et al. 1981). Electrophoretically separated poly A^+ selected RNA was transferred to DBM paper and hybridized with various ^{32}P labelled DNA fragments spanning the T_R region. In this way five T_R specific transcripts, three more abundant

1 Max-Planck-Institut für Züchtungsforschung, 5000 Köln 30, FRG
2 Laboratorium voor Genetische Virologie, Vrije Universiteit Brussel, 1640 Dt.-Genesius-Rode, Belgium
3 Laboratorium voor Genetica, Rijksuniversiteit Gent, 9000 Gent, Belgium
* Current Address: CSIRO, Division of Plant Industry, P.O. Box 1600, Canberra City A.C.T. 2601, Australia

Molecular Genetics of the Bacteria-Plant Interaction
(ed. by A. Pühler)

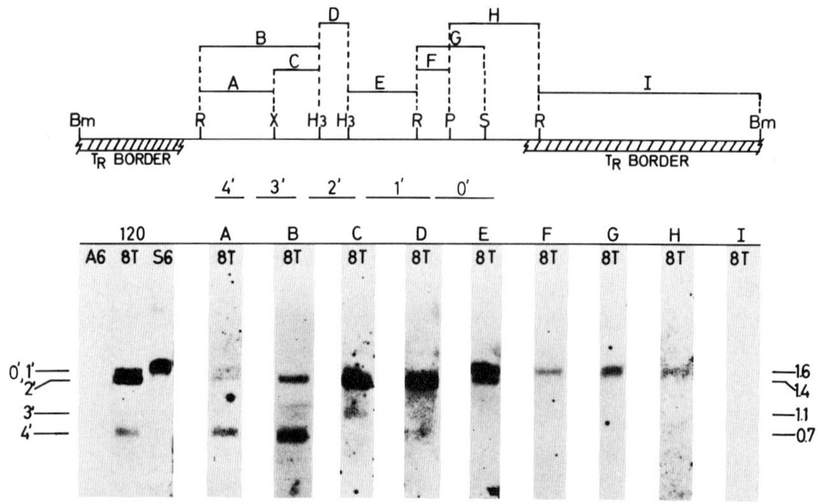

Fig. 1. Map of T_R transcription within tumor tissue specific DNA fragments spanning the T_R region (fragments *A–I*) were labelled with ^{32}P. The labelled fragments were then hybridized to electrophoretically separated poly A-selected RNA bound to DBM paper. The resulting hybridization patterns from each fragment are shown. The sizes and approximate locations of each of the T_R transcripts, *1′–4′*, are indicated. RNA was isolated from several tumor lines: line A6-S1, A6, contains no T_R DNA (De Beuckeleer et al. 1981), line 1GV7, 8T, was induced by Ti plasmid pGV2208, and line 1GV23-S6, S6, with only the 1′ transcript detectable

(1′, 2′ and 4′) and two of less abundance (0′, 3′), were detected within the tumor tissue. The locations of the transcripts within T_R (as diagramed in Fig. 1) were confirmed by the hybridization of ^{32}P labelled cDNA (oligo-dT primed reverse transcription of 1GV7 RNA) against Hinf I digested T_R-DNA (data not shown).

Transcriptional Polarity

The polarities of the T_R transcripts were determined by hybridization of labelled poly A^+ selected RNA to single-stranded DNA fragments covering T_R. Specific single-stranded DNA fragments of known polarity were generated by reannealing denatured double-stranded DNA fragments from the T_R region in the presence of an excess of single-stranded fd bacteriophage DNA containing T_R-DNA of known orientation.

One of the two strands of the double-stranded DNA fragment annealed to the excess bacteriophage DNA and the resulting small T_R-specific single-stranded DNA fragment was electrophoretically separated from the much larger [and now partially double-stranded, fd DNA (see diagram, Fig. 2)]. After cutting the large bacteriophage DNA band from the gel the remaining small fragments were transferred to nitrocellulose paper. The T_R single-stranded DNA fragments were then hybridized with polynucleotide kinase labelled tumor RNA. The results (Fig. 2) indicate that all T_R transcription detected, with the exception of transcripts 2′ and 0′, is directed, 5′ to 3′, rightward or away from T_L.

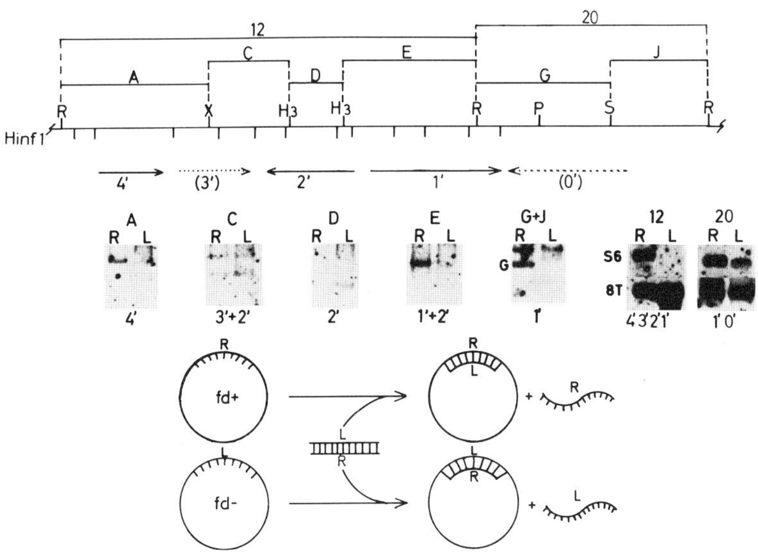

Fig. 2. Transcriptional polarity in T_R single-stranded DNA from the indicated restriction fragments *(A–J)* was generated by annealing denature fragments to single-stranded DNA from fd bacteriophage containing T_R DNA in known orientations (see text and diagram). Poly A selected RNA from tumor line 1GV7 was radioactively labelled with α^{32}P-ATP by T4 polynucleotide kinase and hybridized to nitrocellulose filters containing the electrophoretically separated single-stranded DNA fragments (in those lanes labeled 12 and 20, hybridization was to intact fd viral DNAs carrying each strand of *Eco* RI fragments 12 and 20 respectively, the viral DNAs were also probed with labeled RNA from line IGV23-S6). The polarity assigned for the minor transcript 3′ was based upon the weak rightward hybridization seen to fragment *C*, while that of the 0′ transcript was deduced from the observed hybridization to the leftward strand of fragment *Eco* RI 20. The second set of hybridizing bands seen occasionally in both *R* and *L* lanes result from residual double stranded fragments (lanes *C* and *G*)

T_R and the Agropine Phenotype

Expression of T-DNA within plant tissue is associated with two general phenotypes (1) virulence (or the capacity to induce and maintain tumorous growth) and (2) the production of opines (the particular class, or classes, of opines produced depending upon the Ti plasmid present in infecting *A. tumefaciens*). The T_R region is not required for virulence since: (1) Several established crown gall tumor lines contain no DNA homolgous to T_R region (Thomashaw et al. 1980, De Beuckeleer et al. 1981), and (2) Ti plasmids deleted in T_R allow the induction of tumors which are indistinguishable in growth requirements and morphology from those produced by *Agrobacteria* containing a wild-type octopine Ti plasmid (Ooms et al. 1982, Leemans et al, in prep.). An opine phenotype (agropine biosynthesis) has, however, been assigned to T_R based on the following data: (1) tissues induced by bacteria containing a Ti plasmid in which T_R has been removed by deletion, fail to produce any of the agropine family of opines (Leemans et al., in prep.), and (2) the converse plasmid, in which T_L has been deleted but T_R remains, fails to induce tumors but is able to initiate agropine and mannopine synthesis in infected wound tissue (Leemans et al. 1982).

We have examined the effect of several T_R mutations on transcription and opine biosynthesis within transformed tobacco tissues.

T_R Mutant Analysis

Several mutations, flanking and within the T_R region, have been produced by in vitro manipulation of Ti plasmid DNA (Leemans et al. 1981 and in prep.). The location of these mutations, with respect to the mapped transcripts of T_R, are diagramed in Fig. 3. Tumor tissues induced on *Nicotiana tabacum* cv. Wisconsin 38 by *A. tumefaciens* containing these mutant plasmids were grown as steril callus under axenic conditions and eventually examined for: (1) opine phenotype, (2) the transcription pattern and (3) T_R content within the tumor lines. We have also similarly examined a spontaneous varient cell line discovered during propagation of a suspension culture of tumor line 1GV23 (derived from the Ti plasmid pGV2206; Leemans et al. 1981).

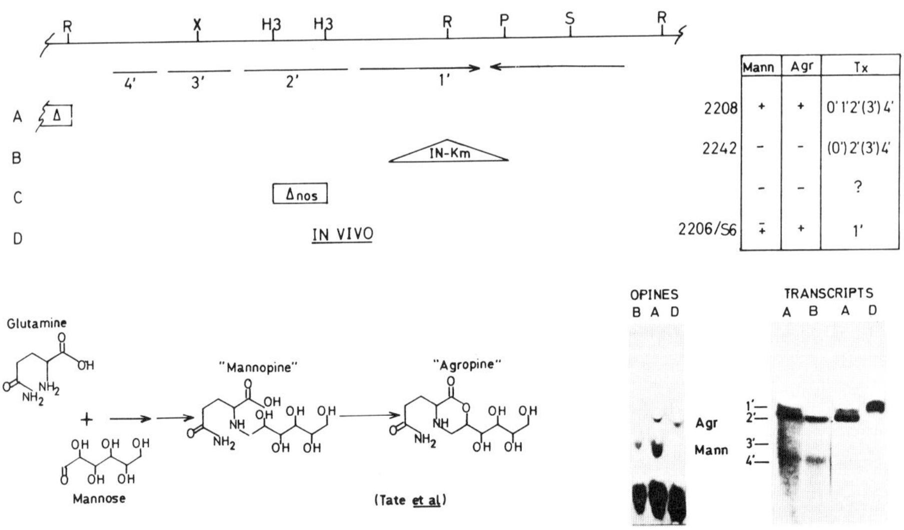

Fig. 3. Mutations within T_R; phenotype and transcription. Mutations within the T_R DNA of pTiB6S3 were produced by in vitro insertion of the kanamycin determination of Tn903, in pGV2242 *(B)*, or deletion and insertion replacing the small *Hind* III fragment with the nopaline synthase gene of pTiC58 in pGV2253 *(C)*. Line *D* (1GV23-S6) occurred spontaneously during propagation of line 1GV23 in suspension culture. The T_R transcripts present in lines *A,B* and *D* were determined by Northern analysis of poly A-selected RNA. The tissues were tested for agropine and mannopine as described by Tepfer and Tempé (1981). The opine structures and proposed biosynthetic pathway is from Tate et al. (1982) and J. Ellis (personal communication)

Tumor Line 1GV7 (Induced by pGV2208)

The tumor line 1GV7, induced by Ti plasmid pGV2208 [deleted from the octopine synthase gene, near the right border of T_L, through the left border of T_R (see Leemans et al. 1981)] produced a full complement of T_R transcripts (Fig. 1) as well as both opines, agropine and mannopine (Fig. 3).

The 1GV7 line, as well as the tumor lines derived from other Ti plasmids mutated in the T_R region, were produced by propagation of primary tumor tissue and therefore are potentially mixtures of independently transformed and untransformed plant cells. The heterogeneity of such uncloned tumor lines makes an exact analysis of the T-DNA content difficult. We have therefore concentrated on determining which sections, as defined by uninterrupted restriction fragments, of the T_R DNA were actually present within the tumor lines. Hybridization analysis of tumor DNA from

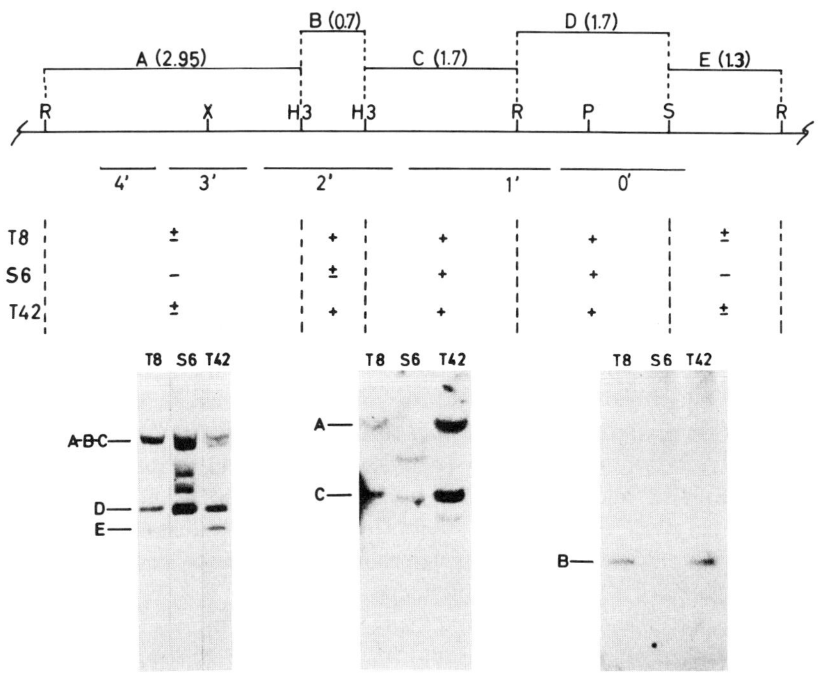

Fig. 4. T_R DNA content of tumor lines. The T_R DNA content of the tumor lines 1GV23-S6 (S6), 1 GV7 (T8) and 1GV69 (T42) was analysed by hybridization of labelled T_R DNA fragments *(A–E)* to various restriction digests of tumor DNA ectrophoretically separated and bound to nitrocellulose. The digestions and labelled probes used were: (1) *left Eco* RI plus *Sal* I digested tumor DNA probed with labelled *Bam* HI fragment *2* (covering all of T_R, see Fig. 1); (2) *middle Eco* RI plus *Hind* III digested tumor DNA probed with labelled *Eco* RI fragment covering fragments *A, B* and *C;* (3) *Hind* III digested tumor DNA probed with labelled fragment *B*. The T_R DNA fragments present in each line are indicated ("+" if present in amounts comparable to the internal standards, fragment *C* or *D;* "±" if present reduced amounts compared to those fragments; "–" if not detected)

the 1GV7 line showed, as expected, the presence of T_R DNA covering the entire region of T_R transcription (Fig. 4). It should be noted, that the restriction fragments flanking T_R (Fig. 4, fragments A and E) were not found at stoichiometric amounts when compared to the internal fragments (Fig. 4, fragments B, C and D) indicating heterogeneous insertion or maintenance of the T_R within these cell lines.

Tumor Line 1GV69 (Induced by pGV2242)

Tumor line 1GV69 was induced by *Agrobacteria* containing the mutant Ti plasmid pGV2242. This plasmid was constructed by the insertion of a restriction fragment, coding for the kanamycin resistance determinate of Tn903, into the *Eco* RI site between *Eco* RI fragments 12 and 20 (Fig. 3, Leemans et al., in prep.). Insertion at this site within TR would be expected to interrupt the coding region for transcript 1′ and as can be seen in Fig. 2, the 1.6 kb RNA band is greatly reduced in this tumor line. The loss of the 1′ transcript was also confirmed by hybridization of labeled cDNA from the 1GV69 line to southern blots of T_R DNA (data not shown). The loss of transcript 1′ was not due to a failure to transfer and maintain the corresponding T_R DNA since the 1GV69 line was found to contain DNA fragments from the entire T_R region (Fig. 4). Two independently produced tumor lines induced by pGV2242 (1GV68 and 1GV69) were both found to produce small amounts of a compound related, but not indentical, to mannopine which comigrates with mannopine in the electrophoresis system used (see Fig. 3).

Tumor Tissue Induced by pGV2253

The primary tumor tissue analysed was induced by bacteria containing the plasmid pGV2253. The T_R mutation within pGV2253 resulted from insertion of a restriction fragment coding for the nopaline synthase *(nos)* gene (from pTiC58, a nopaline type Ti plasmid) in place of the internal *Hind* III of *Eco* RI fragment 12 (see Fig. 3). The insertion served two purposes: (1) To interrupt and delete part of the coding region for transcript 2′ and (2) to provide an assayable marker for the presence of T_R within infected tissue through expression of the nopaline synthase gene. At present only primary, non sterile, tumor tissue has been available for analysis and has therefore been examined only for opine production.

In this tissue nopaline was found (indicating the presence of the *nos* gene, inserted in T_R, within the transformed plant tissue) but no agropine or mannopine were detected. Further analysis of T_R transcription within, and the T-DNA content of this tumor line is required before final conclusions can be made, but it would appear that removal of the gene coding for transcript 2′ results in a loss of both agropine and mannopine synthesis.

Suspension Line 1GV23-S6

The line 1GV23-S6 was produced spontaneously during propagation of a suspension culture of 1GV23. The original tumor line, 1GV23, resulted from infection with bacteria harbouring the plasmid pGV2206 which contains a wild type T_R region (Leemans et al. 1981). The 1GV23-S6 line was found to produce the 1′ transcript (Fig. 3) at a level similar to the other tissues studied, but none of the other T_R specific transcripts were detected.

Correspondingly, the 1GV23-S6 line was found to contain T_R DNA covering the region of the 1′ transcript (fragments C and D, Fig. 4) at levels comparable to the other tumor lines analysed (Fig. 4), while DNA fragments coding for the remaining T_R transcripts (fragments A and B, Fig. 4) were only seen after prolonged exposure and at levels well below that seen for the 1′ region (data not shown). It would appear that, during propagation of 1GV23, cells deleted in a portion of T_R have had a selective growth advantage over these cells containing a normal T_R region. The resulting suspension culture, 1GV23-S6, contains, predominately, the right portion of T_R (see S6, Fig. 5) with the left part of T_R present at only a very low level. The opine phenotype of this culture was found to be agropine positive with no detectable mannopine.

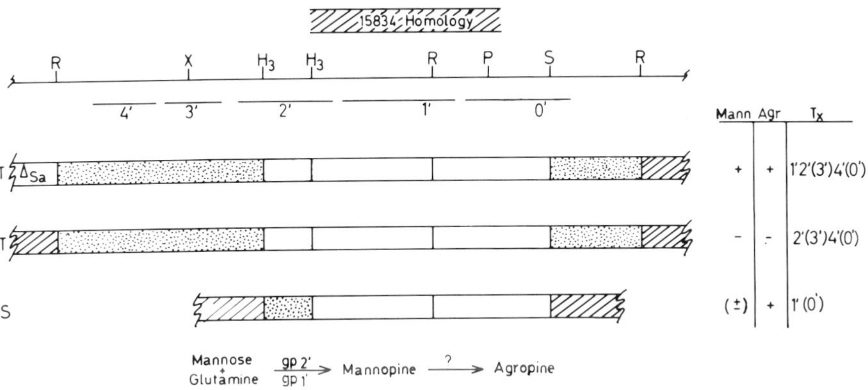

Fig. 5. Comparison of mutant lines analysed. The T_R content of tumor lines 1GV23-S6 (6S), 1GV7 (8T) and 1GV69 (42T) are indicated as well as the opine phenotypes and the T_R transcripts detected. *The stippled fragments* were detected in reduced amounts within the tumor tissue. *Slashed fragments* were not detected as intact restriction fragments. The region of partial homology with the root inducing plasmid of *A. rhizogenes* strain 15834 is indicated *above* the restriction map (Willmitzer et al. 1982). A possible scheme for agropine synthesis by the gene products of the 1′ and 2′ coding regions is indicated *below* the map (see discussion)

Conclusions and Discussion

The results presented in this report address two basic questions: (1) the mapping of transcription from T_R DNA present in tumor tissues induced by *A. tumefaciens* and (2) determination of the gene or genes within T_R responsible for opine production within transformed tissue. The transcriptional mapping data is nearly complete and agrees quite well with an independently determined map (S. Gelvin, this vol.). Specific definition of those genes involved in agropine/mannopine biosynthesis will however require further analysis of the tissues containing mutant T_R DNA (currently in progress). We will thus limit the discussion of the T_R opine genes to those T_R genes so far implicated in agropine production and how they may fit into the presumptive biosynthetic pathway of agropine biosynthesis.

Transcription from T_R in Plant Cells

The deduced transcriptional map of the T_R DNA is summarized in Fig. 2. The five observed transcripts are complementary to DNA comprising most of the T_R region,

We found no evidence for T-DNA specific transcription (other than that from T_L from regions outside the two *Eco* RI fragments (*Eco* RI fragments 12 and 20, De Vos et al. 1981) diagramed in Fig. 2. The active expression of only a portion of the smaller *Eco* RI fragment (transcript 1′ and 0′ in *Eco*20) is consistent with the reported extent of the T_R region. An intact *Eco* 20 fragment does not occur in several Ti induced tumor lines (Thomashaw et al. 1980, De Beuckeleer et al. 1981) indicating that the right most portion of *Eco* 20 is not always integrated or if integrated not always stably maintained. The T_R DNA content of the tumor lines we have analysed confirms this finding since the far right portion of *Eco* 20 (fragment E, Fig. 4) is present in less than stoichiometric amounts within the tumor lines analysed. In considering gene expression from fragment *Eco* 20, it should be noted that the *Arc* locus, involved in arginine catabolism, has been located within the *Eco* 20 fragment (De Greve et al. 1981). The *Arc* function is part of the opine degradative pathway in bacteria (Ellis et al. 1979) and presumably need not be part of the T-DNA carried or expressed within transformed plant tissue. It is likely that the *Arc* locus comprises at least a portion of the untranscribed (within our tumor lines) right-most segment of the *Eco* 20 fragment.

Genes Involved in Agropine Production

The data presented in this report implicate at least two of the T_R transcribed regions, those coding for transcripts 1′ and 2′, in opine production within transformed tissue. Involvement of the product of the 2′ coding region (gp2′) in agropine production was demonstrated by the failure of the plasmid pGV2253, deleted in part of the 2′ region, to induce either agropine or mannopine production in primary tumors. Loss of the agropine-mannopine phenotype was also observed in tumors induced by a Ti plasmid containing a transposon insertion in the 2′ region (unpubl. observ.).

Inactivation of the 1′ coding region by DNA insertion in the Ti plasmid pGV2242 results in a mannopine negative, agropine negative phenotype in transformed tissues. This tissue does, however, produce very low levels of a compound which, although comigrating with mannopine in the electrophoresis system used, is not authentic mannopine (J. Ellis, personal communication). This compound is not likely a precursor to one or more of the agropine type opines.

The 1′ coding region was also found to be partially homologous (see Fig. 5) to the Ri (root inducing) plasmid of the *Agrobacterium rhizogenes* strain 15834, a hairy root disease pathogen. Since root tissues transformed by the Ri 15834 plasmid also produce both agropine and mannopine (Willmitzer et al. 1982), it seems likely that the area of pRi15834 partially homologous with T_R at the 1′ coding region is also involved in agropine production. No other parts of the T_R DNA showed any homology with pRi15834 at the level of stringency used. Finally, in the 1GV23-S6 line, in which the 1′ and 0′ transcripts are the only detected T_R specific transcription, low levels of agropine but no mannopine was found. The 1GV23-S6 line was, however, also found to contain very low amounts of T_R DNA coding for the 2′, 3′ and 4′ transcripts. In view of the finding that inactivation of the 2′ coding region eliminates both mannopine and agropine production it seems likely that in the 1GV32-S6 line the 2′ coding region is transcribed but at levels below detection. More sensitive experiments are in progress to examine this possibility. In any case, the presence of a relative excess of the 1′ and 0′ transcripts is associated with an increased *proportion* of agropine in the 1GV23-S6 line.

We have, as yet, no data pertaining to the products of the 3′ and 4′ coding regions. Mutations affecting this region of T_R are in the process of construction.

The biosynthetic pathway for mannopine and agropine, shown in Fig. 3, was proposed by Tate and coworkers (Tate et al. 1982) based upon: (1) the opine structure and (2) analogy to the anabolic pathways for octopine and nopaline. There is, as yet, no reported biochemical data directly confirming the proposed pathway. Our results are consistent with a two step pathway in that at least two T_R coding regions are involved in agropine biosynthesis; the 2′ region, mutated in plasmid pGV2253, and the 1′ region mutated in plasmid pGV2242. These results can be best fitted to the proposed pathway if the products of the 1′ and 2′ regions are both necessary for the coupling of glutamine and mannose to produce mannopine. An enzymatic function responsible for the cyclization of mannopine to agropine may be coded for by either the plant genome or one of the remaining T_R genes. Of course, independent biosynthetic pathways for each opine, as well as more complicated models involving opine transport functions are not excluded.

Although our data is predominately consistent with the proposed agropine biosynthetic pathway, some results raise questions which require further examination. Results from the analysis of the 1GV23-S6 suspension culture line raise some questions which require further examination. At the level of sensitivity used, this line appears to produce only the 1′ and 0′ transcripts, yet is able to synthesize agropine at respectable levels. This observation is not consistant with the finding that production of the 2′ transcript is required for both mannopine and agropine biosynthesis. The simplest explanation is that in this line transcription of the 2′ region is simply below the level of detection. As was mentioned earlier, 2′ transcription can not be excluded since T_R DNA coding for the 2′ region is present in the line although at a very low level. The possibility of transcription from the 2′ region of T_R shows homology to the pRi15834 plasmid whereas at least the 1′ and 2′ regions are involved in agropine production. Comparison of the agropine biosynthetic systems of *A. rhizogenes* and the octopine Ti plasmids should be very interesting concerning the origin and evolution of opine genes in *Agrobacteria.*

We are currently in the process of completing the analyses of these mutant lines as well as several new T_R mutations. The resulting data should help clarify the functions of T_R genes expressed within plant tissue.

Addendum

The faint opine spot from the 1GV68 and 1GV69 lines (induced by pGV2242), which comigrates with mannopine was subsequently found to be a reaction intermediate of mannopine synthesis (J. Ellis and M. Tate, personal communication). Thus, the T_R-mutant lacking the 1′ transcript would appear to be deficient in a second reaction required for mannopine production. This finding, coupled with preliminary genetic data linking the 0′ region with agropine synthesis, suggests a three step pathway. Thus agropine production appears to require the products of the 0′, 1′ and 2′ transcripts.

Acknowledgements. We would like to thank Barbara Baker for critical comments during preparation of the manuscript and Mrs. E. Schölzel for her accurate and fast typing of the manuscript. J. Velten was the recipient of an EMBO long term fellowhip during this research.

References

Beuckeleer M De, Lemmers M, Vos G De, Willmitzer L, Montagu M Van, Schell J (1981) Further insight on the transferred-DNA of octopine crown gall. Mol Gen Genet 183:283–288

Ellis JG, Kerr A, Tempé J, Petit A (1979) Arginine catabolism: a new function of both octopine and nopaline Ti-plasmids of *Agrobacterium.* Mol Gen Genet 173:263–269

Greve H De, Decraemer H, Seurinck J, Montagu M Van, Schell J (1981) The functional organization of the octopine *Agrobacterium tumefaciens* plasmid pTiB6S3. Plasmid 6:235–248

Guyon P, Chilton M-D, Petit A, Tempé J (1980) Agropine in "null type" crown gall tumors: evidence for the generality of the opine concept. Proc Natl Acad Sci USA 77:2693–2697

Leemans J, Shaw C, Deblaere R, Greve H De, Hernalsteens JP, Maes M, Montagu M Van, Schell J (1981) Site-specific mutagenesis of *Agrobacterium* Ti plasmids and transfer of genes to plant cells. J Mol Appl Genet 1:149–164

Ooms G, Hooykaas PJJ, Veen RJM Van, Beelen P Van, Regensburg-Tuink TJG, Schilperoort RA (1982) Octopine Ti plasmid deletion mutants of *Agrobacterium tumefaciens* with emphasis on the right side of the T-region. Plasmid 7:15–29

Tate ME, Ellis JG, Kerr A, Tempé J, Murray K, Shaw K (1982) Agropine: a revised structure. Carbohyd Res (in press)

Tepfer DA, Tempé J (1981) Production d'agropine par des racines formées sous l'action d'*Agrobacterium rhizogenes,* souche A4. C R Acad Sci (Paris) 292 Ser III:153–156

Thomashaw MF, Nutter R, Montoya AL, Gordon MP, Nester EW (1980) Integration and organisation of Ti-plasmid sequences in crown gall tumors. Cell 19:729–739

Vos G De, Beuckeleer M De, Montagu M Van, Schell J (1981) Restriction endonuclease mapping of the octopine tumor inducing pTiAch5 of *Agrobacterium tumefaciens.* Plasmid 6:249–253

Willmitzer L, Simon G, Schell J (1982) The TL-DNA in octopine crown-gall tumours codes for seven well-defined polyadenylated transcripts. EMBO J 1:139–146

Transcription of T-DNA in Octopine and Nopaline Crown Gall Tumours

L. WILLMITZER [1], P. DHAESE [2], P.H. SCHREIER [1], M. VAN MONTAGU [2], and J. SCHELL [1,2]

Introduction

The molecular mechanism underlying the tumorous transformation of plant cells into crown gall cells is the transfer and stable integration of the T-region, a well-defined segment of the Ti-plasmid of *Agrobacterium tumefaciens,* into the nuclear genome of the plant cell (cf. for a review: Van Montagu and Schell 1982). Octopine and nopaline crown gall tumours are by far the best studied examples of crown gall tumours. Whereas in nopaline tumours the T-DNA was found to consist of one contiguous piece of DNA of about 23 kb size (Lemmers et al. 1981), the T-DNA of octopine tumours can be composed of either one or two fragments from neighbouring sequences of the Ti-plasmid. All octopine tumour lines studied up to now contained at least the fragment located to the left on the Ti-plasmid map (therefore called TL-DNA) whereas the fragment located to the right (and therefore called TR-DNA) can be found as an additional sequence (Thomashow et al. 1980, De Beuckeleer et al. 1981, Velten et al. this vol.). Octopine and nopaline Ti-plasmids share several regions of homology including a segment present in the T-DNA's of all nopaline tumours (Engler et al. 1981). It was therefore assumed that this so-called common DNA plays an essential role in the formation and maintenance of octopine and nopaline crown gall tumours.

To elucidate the mechanism of tumour transformation a double approach was used in our group. On the one hand insertion and deletion mutants were introduced at specific sites of the T-region of octopine and nopaline Ti-plasmids (Leemans et al. 1981, Joos et al. 1982, Leemans et al., this vol.).

On the other hand a careful analysis of the number, sizes and location of the different T-DNA encoded genes was performed by analyzing the different T-DNA derived poly-adenylated transcripts present in octopine and nopaline type tumours, the results of which will be described in this chapter. This double approach has led to a rather defined picture of the location and function of the different T-DNA encoded genes.

1 Max-Planck-Institut für Züchtungsforschung, 5000 Köln 30, FRG
2 Laboratorium voor Genetica, Rijksuniversiteit Gent, 9000 Gent, Belgium

Molecular Genetics of the Bacteria-Plant Interaction
(ed. by A. Pühler)

Results

General Experimental Approach

In one sort of experiments, poly-A$^+$ RNA, isolated from the tumours under conditions preventing degradation of RNA, was separated according to size by electrophoresis and subsequently transferred to diazobenzyloxymethyl paper. These filters were subsequently probed for T-DNA derived transcripts by using a large number of T-DNA fragments labeled to high specific activity by nick translation. In this way a rather precise map of the location of the different T-DNA encoded polyadenylated transcripts was obtained. The polarity of the different transcripts was determined by

a) hybridizing poly-A$^+$ RNA labeled at its 5′-end with polynucleotide kinase towards the separated strands of T-DNA fragments specifically encoding one or a group of transcripts, thus identifying the coding strand
b) determining the chemical polarity of this strand by hybridizing T-DNA fragments specifically labeled at one defined 3′-end with terminal transferase.

TL-DNA of the Octopine Tobacco Tumour Line A6-S1 Codes for Eight Different Poly-Adenylated Transcripts Absent from Nontransformed Plant Cells

The autoradiograms obtained after hybridizing T-region clones containing the total TL-DNA to a DBM-filter containing the separated poly-A$^+$ fraction of RNA isolated from the octopine line A6-S1 (first lane) respectively the untransformed habituated tobacco line NW-S1 (second lane) is shown in Fig. 1.

Seven different RNA species can be readily distinguished by their different size, band 6 being a doublet representing transcripts 6a and 6b. No hybridization is visible

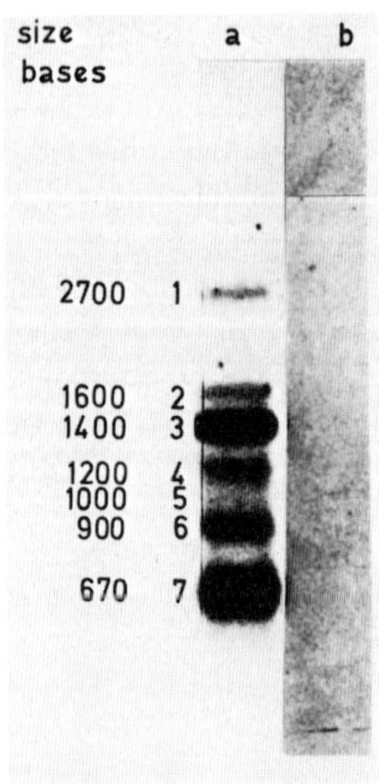

Fig. 1. Hybridization of octopine TL-DNA against poly-A$^+$ RNA isolated from the octopine crown gall tumours A6-S1 *(first lane)* and untransformed tissue *(second lane)*. Nick-translated Hind III fragment 1 of pTi Ach5 and Bam HI fragment 8 both of pTi Ach5 were used as probes. *1–7* transcripts according to diminishing size, the length of the transcripts is indicated in nucleotides without correction for the poly-A tail

to the poly-A$^+$ RNA isolated from untransformed tissue. The sizes and the relative abundance of the different transcripts vary considerably, transcripts 3 and 7 are present at the highest concentrations, transcripts 4 and 6 are less abundant and transcript 1, 2 and 5 have even lower concentrations.

The terms high and low concentrations are used only to compare internal concentrations of the different TL-DNA derived transcripts, compared to other cellular transcripts TL-DNA transcripts all belong to the low abundancy class totally not more than 0.005–0.001% of the poly-A$^+$ RNA population (Willmitzer et al. 1982a). The location of the different transcripts on the T-DNA map was deduced from hybridization experiments using different T-DNA fragments as shown in Fig. 2. The data describing the evidence for the two transcripts named 6a and 6b is described elsewhere (Willmitzer et al. 1982b).

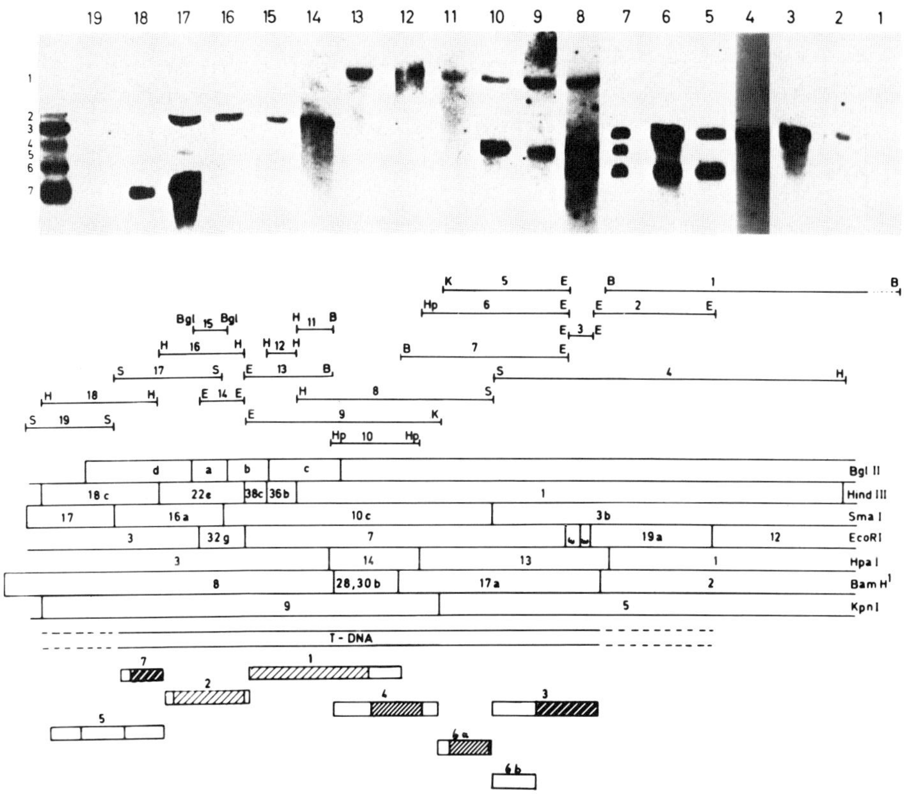

Fig. 2. Hybridization of probes of subfragments of the octopine TL-region against poly-A$^+$ RNA isolated from the octopine crown gall tumour A6-S1. The hybridization patterns are shown *above*, the *numbers above each lane* correspond to the map shown *below* and indicate the T-DNA fragment used as probe. The most probable map position of each transcript is indicated below by the *shaded area* of each box, the *limits of each box* indicate the uncertainty concerning the exact location of each transcript. *Intensity of shading* is proportional to the relative concentrations of each transcript

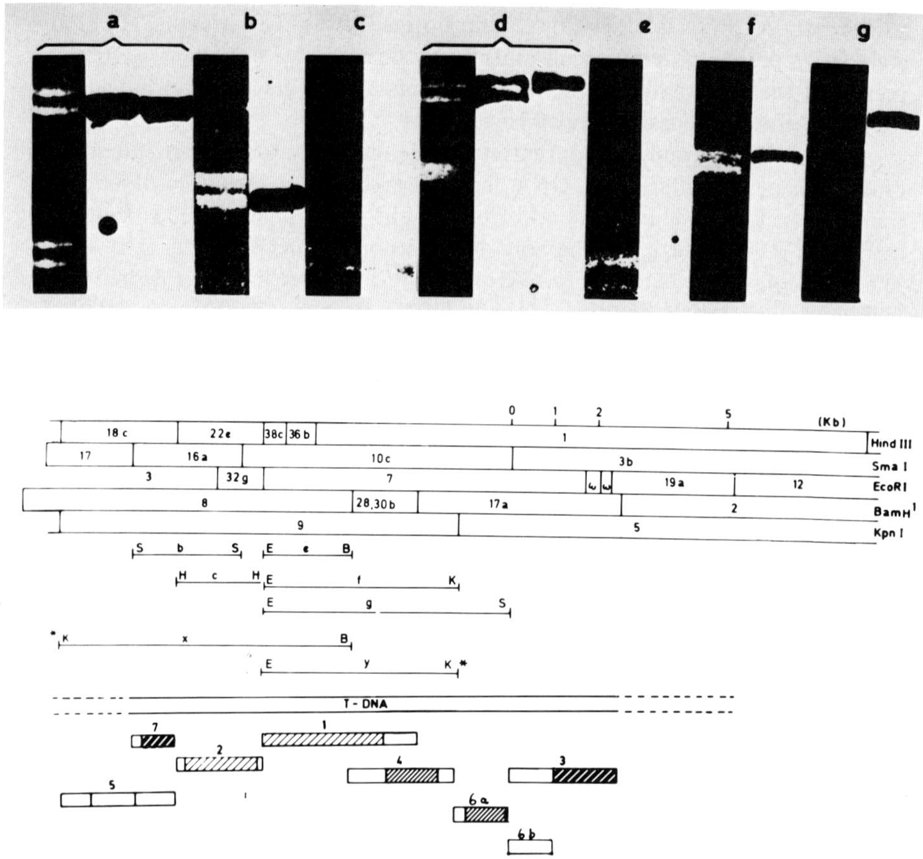

Fig. 3. Hybridization of poly-A$^+$ RNA from the octopine tumour line A6-S1 against separated strands of different T-DNA fragments. Poly-A$^+$ RNA was labeled at its 5′-end with poly-nucleotide kinase. Hybridizations were performed against the following T-region fragments: group *a* Bam HI fragment 8 (the lower bands originate from the vector pBR322), group *b, c* fragments b, c; group *d* EcoRI fragment 7, groups *e, f, g* fragments e, f, g as indicated in the *lower part* of the figure. The hybridization obtained with poly-A$^+$ RNA is always shown in the *second lane*, the ethidium bromide strain of the separated strands is always shown in the *first lane*. In groups *a* and *d*, the *third lane* shows the hybridization of the fragment x respectively a selectively labeled at the protruding 3′-end generated by the KpnI restriction cut *(asterisk)*. The location of the different transcripts as deduced from Fig. 2 is indicated on the *lower part* of Fig. 3

The experimental data determining the polarity of the different transcripts are shown in Fig. 3. Thus e.g. using the EcoRI-BamHI fragment e, specific for transcript 1, hybridization of the 5′-labeled poly-A$^+$ RNA is visible only towards the upper strand. The same, i.e. hybridization to, the upper strand only is also seen in case of using fragments f and g, specifically encoding transcripts 1 and 4 respectively 1, 4 and 6a. Therefore these transcripts must all have the same polarity.

The chemical polarity of the upper strand was determined by hybridizing the EcoRI-KpnI fragment y, specifically labeled at the right 3′-end with terminal transferase, towards the separated strands of the EcoRI fragment 7. Hybridization to the upper strand is visible (cf. Fig. 3, group d, lane 3) thus the 5′–3′ polarity of the transcripts read from this strand must be from left to right.

The data summarizing the size, location and polarity of the eight different TL-DNA encoded transcripts present in octopine crown gall tumours are shown in Fig. 4.

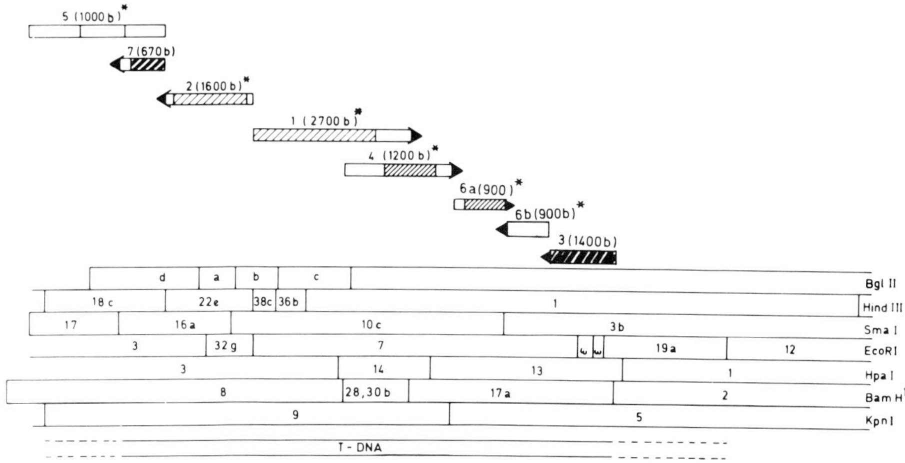

Fig. 4. Sizes, location and direction of transcription of octopine TL-DNA derived transcripts. The location of each transcript is indicated by a separate *box*, the *shaded area* indicates size and most probable map position of transcripts under the untested assumption that no splicing occurs, the *intensity of shading* is proportional to the relative concentration of each transcript. The *size* is given in bases, the *numbers* are not corrected for the poly-A tail. The *arrows* indicate the 5′–3′ polarity. *Asterisks* indicate transcripts from the common DNA

All RNA's map within the T-DNA sequence. This and the observation that transcription is inhibited by low concentrations of α-amanitin (Willmitzer et al. 1981) suggests that each transcript is determined by a specific promoter site on the T-DNA recognized by the plant RNA polymerase II. Considering the groupwise orientation of several transcripts the simplest model would assume one promoter site per group of transcripts, e.g. 1, 4 and 6a. If so, one would expect that the incativation of a 5′-proximal gene of a group would also lead to the disappeareance of the transcripts from the 5′-distal genes. However, analysis of certain tissues containing the T-DNA of Ti-plasmid mutants indicated that groupwise inactivation of genes does not occur (Leemans et al. 1982, Joos et al. 1982, Willmitzer et al. 1982a). The results available so far are therefore consistent with the assumption that each gene on the T-DNA has its own signals for transcription in the eukaryotic plant cell. Current experiments are performed with the aim to use these T-DNA promoters for the expression of foreign DNA in plant cells.

The eight transcripts originating from the TL-DNA of octopine tumours, can be divided into two groups concerning their relationship to genes expressed in nopaline tumours.

The first group comprises six transcripts (i.e. transcript 1, 2, 4, 5, 6a and 6b) encoded by the common DNA which have physically and functionally related counterparts in nopaline crown gall cells. They will be discussed in more detail further below.

The second group containing transcripts 3 and 7, is specific for octopine type tumours. Concerning their function, genetic and biochemical data demonstrated that transcript 3 codes for the octopine-synthesizing enzyme (Leemans et al. 1981, Garfinkel et al. 1981, Schröder et al. 1981, Murai and Kemp 1982); no function has yet been assigned to transcript 7; Ti-plasmids, carrying mutations in this gene do not display any obvious phenotype (Garfinkel et al. 1981).

In this context it should also be mentioned that the four transcripts encoded by the TR-DNA in octopine tumours (cf. Velten et al., this vol.) also belong to the class of octopine-specific tumours.

T-DNA of Nopaline Tumours Codes for at Least Thirteen Different Poly-Adenylated Transcripts

Using rather similar techniques as described for octopine tumours, the polarity, sizes and locations of thirteen polyadenylated transcripts derived from nopaline T-DNA and present in the nopaline crown gall tobacco line BT37 were mapped.

Figure 5 shows the summary of the results obtained (cf. Willmitzer et al. 1982b for detailed description of the experimental results). As already found for octopine TL-DNA encoded transcripts all RNA's map within the T-DNA. Both strands again serve as coding strand. These observation and the absence of any polar effect displayed by the inactivation of one gene on the expression of another gene (Joos et al. 1982) gives strong evidence that also in case of nopaline tumours most if not all transcripts are encoded by their own promoters.

Again the transcripts can be divided into two groups: the first group comprises seven genes (i.e. transcripts a, b, c, d, e as well as the genes acs and nos) which are

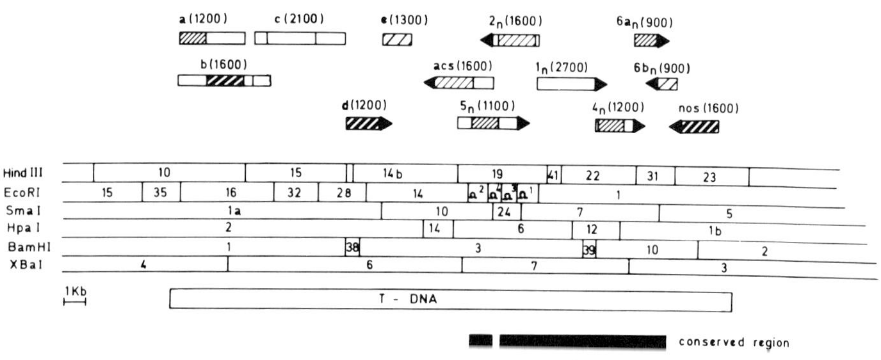

Fig. 5. Sizes, location and polarity of nopaline T-DNA derived polyadenylated transcripts. Further details are as in Fig. 4. *acs* agrocinopine synthesis; *nos* nopaline synthesis. The transcripts encoded by the conserved part of the T-DNA are numbered according to the system used for octopine transcripts, the suffix '*n*' stands for nopaline

specific for nopaline tumours. Up to now the function of only two of transcripts is known: acs codes for a 1.600 nt long transcript located to the left of the common DNA which is responsible for the synthesis of agrocinopine, and opine first described by Ellis and Murphy (1981).

Nos, coding also for a 1.600 nt long transcript and being responsible for the synthesis of nopaline is located immediately to the right of the common DNA (Joos et al. 1982). No function could as yet been assigned to the remaining transcripts a–e.

The second group comprises six transcripts (1n, 2n, 4n, 5n, 6qn, 6bn) originating from the common DNA. They will be discussed below in further detail.

A Subset of Six Transcripts is Common to Both Octopine and Nopaline Type Tumours

As outlined above in both octopine as well as nopaline tumours the RNA's present could be divided in a group of RNA's specific for the respective type of tumour and originating from the nonconserved part of the T-DNA's and into a second group derived from the common DNA conserved between both octopine and nopaline T-DNA's.

In both cases six transcripts were found to originate from the common DNA. If these transcripts possess homologous counterparts in the other tumour type they should be of the same size and polarity, originating from corresponding locations on the T-DNA's and show cross-hybridization with the T-DNA of the counterpart.

The fulfillment of the first three criteria is already obvious from a comparison of Figs. 4 and 5. The last criterium, i.e. cross-hybridization, is fulfilled too (cf. Willmitzer et al. 1982a, b). These data therefore demonstrate that this group of transcripts is strongly conserved between octopine and nopaline type tumours. Genetic data in addition have shown that they also fulfill the same functions in octopine and nopaline crown gall tumours. Thus inactivation of either transcript 1 or 2 or both of them gives rise to tumours which induce shoot formation, whereas inactivation of transcript 4 leads to formation of tumours which induce root formation (Garfinkel et al. 1981, Leemans et al. 1982, Joos et al. 1982, Leemans et al., this vol.). No clear phenotype has as yet been related to the remaining transcripts 5, 6a and 6b.

Conclusion

The detailed physical analysis of the T-DNA derived polyadenylated RNA's present in octopine respectively nopaline tumours has led to a rather defined map of the different genes located on the T-DNA and expressed in the plant tumour cell.

In both octopine and nopaline tumours a set of genes is present which is specific for the type of tumour made. The only function which up to now has been related to this group of genes is their involvement in the synthesis of opines, i.e. synthesis of octopine and agropine/mannopine in octopine type tumours (cf. also Velten et al., this vol.) respectively synthesis of nopaline and agrocinopine in nopaline type tumours.

In both tumour types a set of six transcripts was found which is physically and functionally highly conserved between both tumours. Three of these transcripts have been demonstrated to be involved in suppression of the normal development of the plant cell, i.e. suppression of shoot and root formation. Due to their high conservation between octopine and nopaline type tumours it is reasonable to assume these transcripts to be of high importance for tumour induction and maintenance.

The data shown in Figs. 4 and 5 as well as additional DNA sequence data and S1-mapping experiments (De Greve et al. 1982a, Depicker et al. 1982) demonstrate that most if not all T-DNA derived transcripts are read from independent promoters located on the T-DNA. These promoters, at least some of them, are expressed constitutively (De Greve et al. 1982b), should be of help for the expression of foreign genes in plant cells.

Acknowledgements. The authors wish to thank their colleagues at the Laboratorium voor Genetica, Vrije Universiteit Brussels, and at the Laboratory of Genetics, Gent University, for help and discussion. The excellent technical assistance of W. Schmalenbach and G. Simons during the investigations described in this article is gratefully acknowledged.

References

Beuckeleer M De, Lemmers M, Vos G De, Willmitzer L, Montagu M Van, Schell J (1981) Further insight on the transferred-DNA of octopine crown gall Mol Gen Genet 183:283–288

Depicker A. Stachel S, DHaese P, Zambryski P, Goodman HM (1982) Nopaline synthase: transcript mapping and DNA sequence. J Mol Appl Genet (in press)

Ellis J, Murphy PJ (1981) Four new opines from crown gall tumours – their detection and properties. Mol Gen Genet 181:36–43

Engler G, Depicker A, Maenhaut R, Villaroel-Mandiola R, Montagu M Van, Schell J (1981) Physical mapping of DNA base sequence homologies between an octopine and a nopaline Ti-plasmid of *Agrobacterium tumefaciens*. J Mol Biol 152:183–208

Garfinkel DJ, Simpson RB, Ream L, White F, Gordon M, Nester E (1981) Genetic analysis of crown gall: fine structure map of the T-DNA by site directed mutagenesis. Cell 27:143–153

Greve H De, DHaese P, Seurinck J, Montagu M Van, Schell J (1982a) Nucleotide sequence and transcript map of the *Agrobacterium tumefaciens* Ti-plasmid encoded octopine synthase gene. J Mol Appl Genet (in press)

Greve H De, Leemans J, Hernalsteens J-P, Thia-Toong L, Beuckeleer M De, Willmitzer L, Otten L, Montagu M Van, Schell J (1982b) Normale and fertile plants, expressing octopine synthase, regenerate from tobacco crown galls after deletion of tumour controlling functions. Nature (in press)

Joos H, Inze D, Caplan A, Sormann M, Montagu M Van, Schell J (1982) Genetic analysis of T-DNA transcripts in nopaline crown galls. Cell (submitted)

Leemans J, Shaw C, Blaere R De, Greve H De, Hernalsteens JP, Maes M, Montagu M Van, Schell J (1981) Site-specific mutagenesis of *Agrobacterium* Ti-plasmids and transfer of genes to plant cells. J Mol Appl Genet 1:149–164

Leemans J, Blaere R De, Willmitzer L, Greve H De, Heralsteens JP, Montagu M Van, Schell J (1982) Genetic identification of functions of TL-DNA transcripts in octopine crown gall. EMBO J 1:147–152

Lemmers M, Beuckeleer M De, Holsters M, Zambryski P, Depicker A, Hernalsteeens JP, Montagu M Van, Schell J (1981) Internal organization, boundaries and integration of Ti-plasmid DNA in nopaline crown gall tumours. J Mol Biol 144:353–376

Montagu M Van, Schell J (1982) The Ti-plasmids of *Agrobacterium*. In: Hofschneider PH, Goebel W (eds) Gene cloning in organisms other than *E. coli*. Current topics in microbiology and immunology, vol 96. Springer, Berlin Heidelberg New York, p 237

Murai N, Kemp (1982) Octopine synthase mRNA isolated from sunflower crown gall callus is homologous to the Ti plasmid of *Agrobacterium tumefaciens*. Proc Natl Acad Sci USA 79: 86–90

Schröder J, Schröder G, Hiusman H, Schilperoort RA, Schell J (1981) The mRNA for lysopine dehydrogenase in plant tumor cells is complementary to a Ti-plasmid fragment. FEBS Lett 129:166–168

Thomashow M, Nutter R, Montoya A, Gordon M, Nester E (1980) Integration and organization of Ti-plasmid sequences in crown gall tumors. Cell 19:729–739

Willmitzer L, Schmalenbach W, Schell J (1981) Transcription of T-DNA in octopine and nopaline crown gall tumours is inhibited by low concentrations of α-amanitin. Nucleic Acids Res 9: 4801-4812

Willmitzer L, Simons G, Schell J (1982a) The TL-DNA in octopine crown gall tumours codes for seven well-defined polyadenylated transcripts. EMBO J 1:139–146

Willmitzer L, DHaese P, Schreier PH, Schmalenbach W, Montagu M Van, Schell J (1982b) Size, location and polarity of T-DNA encoded transcripts in nopaline crown gall tumours; evidence for common transcripts present in both octopine and nopaline tumours. Cell (submitted)

Genetic Analysis of T-DNA and Regeneration of Transformed Plants

J. LEEMANS [1]*, J.P. HERNALSTEENS [1], R. DEBLAERE [1], H. DE GREVE [1], L. THIA-TOONG [1], M. VAN MONTAGU [1,2], and J. SCHELL [1,2,3]

Introduction

The evidence presented in a number of chapters in this volume (e.g., by P. Zambryski and by Willmitzer et al.) demonstrates that the tumorous transformation of plant cells into crown gall is the consequence of transfer, integration and expression of T-DNA, a specified region of the Ti plasmids of *Agrobacterium tumefaciens* in these plant cells.

In octopine crown galls (induced by *Agrobacterium* strains such as A6, B6S3 or Ach5), the T-DNA can be composed of either one or two fragments from neighbouring sequences on the Ti plasmid. All tumor lines investigated thus far contain the TL-DNA (13 kb), whereas the TR-DNA (7 kb) can be found as an additional sequence (Thomashow et al. 1980, Lemmers et al. 1980, De Beuckeleer et al. 1981, Velten et al. this vol.). In crown gall cells, T-DNA is transcribed by host RNA polymerase II and results in a defined set of polyadenylated transcripts. The TL-DNA encodes 8 transcripts of different relative abundance. The fact that both DNA strands are coding suggested that independent promoters and poly(A) addition sites on the T-DNA are involved in generating these transcripts. DNA sequence analysis and S1 mapping of T-DNA genes, such as the octopine synthase gene (De Greve et al. 1982a) demonstrated that T-DNA genes indeed carry transcriptional control regions similar to those found in eukaryotic genes.

Common DNA is the DNA homologous between octopine and nopaline T-DNA (Depicker et al. 1978, Chilton et al. 1978, Engler et al. 1981). It encodes transcripts of the same size and polarity in both types of tumors. The transcripts hybridize with corresponding regions of the common DNA of octopine and nopaline T-DNA. It has been suggested that common DNA, and consequently the common transcripts, play a key role in forming and maintaining the tumorous state (Depicker et al. 1978).

In this chapter we describe the genetic identification of the functions of the T-DNA. The methods developed for, and the knowledge obtained from, this analysis have been crucial to the use of the T-DNA as an experimental gene vector for plants.

1 Laboratorium voor Genetische Virologie, Vrije Universiteit Brussel, Paardenstraat 65, 1640 St.-Genesius-Rode, Belgium
2 Laboratorium voor Genetica, Rijksuniversiteit Gent, 9000 Gent, Belgium
3 Max-Planck-Institut für Züchtungsforschung, 5000 Köln 30, FRG
* Present address: Centro de investigaciones sobre fijación de nitrogeno, Ap. Postal 565.A, Cuernavaca, Mexico

Molecular Genetics of the Bacteria-Plant Interaction
(ed. by A. Pühler)

Indeed, they provided efficient methods for cloning into Ti plasmids and they allowed the dissociation of T-DNA transfer and tumorous growth. The latter resulted in the regeneration of normal and fertile plants which still contain T-DNA sequences.

Efficient Gene Cloning into T-DNA and Site-Directed Mutagenesis of Ti Plasmids

The genetic organization and the large size of Ti plasmids do not allow the construction of small Ti plasmids directly suitable as cloning vehicles. Several alternatives have been worked out. They involve cloning or mutagenesis in so-called intermediate vectors, which contain isolated Ti fragments. Site-directed inserts in Ti plasmids are produced by exchange of the modified Ti sequence for the wild-type sequence through in vivo recombination between the intermediate vector and a resident Ti plasmid. The three approaches are outlined in Fig. 1. The first two involve broad-host range intermediate vectors which replicate stably in *Agrobacterium* (Ruvkun and Ausubel 1981, Leemans et al. 1981). The third, most useful method (Van Haute et al. 1983) uses pBR322-derived intermediate vectors. They can be efficiently mobilized into *Agrobacterium* but they cannot replicate in this host. As these vectors contain Ti plasmid sequences, they can be maintained by recombination with a resident Ti plasmid. If two recombinations occur, a pBR322 derivative containing the wild-type Ti sequence is excised and subsequently lost, leaving the desired recombinant Ti plasmid.

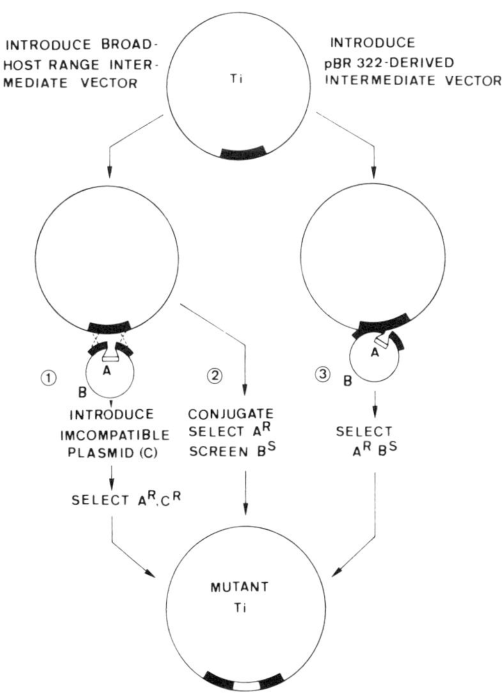

Fig. 1. Site-directed mutagenesis of Ti plasmids. Outline of the 3 different procedures which allow isolation of mutant Ti plasmids by homologous recombination between pTi and intermediate vectors. Homologous DNA is indicated as a *thick line*. Antibiotic resistance markers are indicated as *A* and *B*. *Procedure 1* (Ruvkun and Ausubel 1981) and *2* (Leemans et al. 1981) involve cloning vectors which replicate in *Agrobacterium*. Desired recombinants are obtained either by eliminating the intermediate vector through curing with the incompatible plasmid carrying antibiotic marker C^R *(procedure 1)*, or after conjugation and selection of transconjugants resistant to antibiotic *A*, but sensitive to antibiotic *B (procedure 2)*. The direct mobilization of pBR322 derivatives yields cointegrates (single cross-overs) and mutant Ti plasmids (double cross-overs) in one single step *(procedure 3)*. Immediate screening for B^S cells after selection for A^R colonies provides the mutant Ti plasmid

Functions of T-DNA Genes

Detailed mutation analysis of the T-region of octopine and nopaline Ti plasmids has been performed using the three methods summarized in the previous section. Two different strategies have been followed.

Site-Directed Transposon Mutagenesis

Transposon mutagenesis was used to nearly saturate the octopine (Garfinkel et al. 1981) and nopaline T-DNA (Joos et al. 1983) with Tn*5* and Tn*3* insertions. This allowed the identification of the T-DNA regions involved in the synthesis of octopine, nopaline, and agrocinopine (Fig. 2).

These regions do not form part of the common DNA. Mutations located in the common DNA induced three different phenotypes:

1. tumors which induce shoot formation (mutations in transcripts 1 or 2);
2. tumors which induce root formation (mutations in transcript 4);
3. tumors without aberrant phenotype.

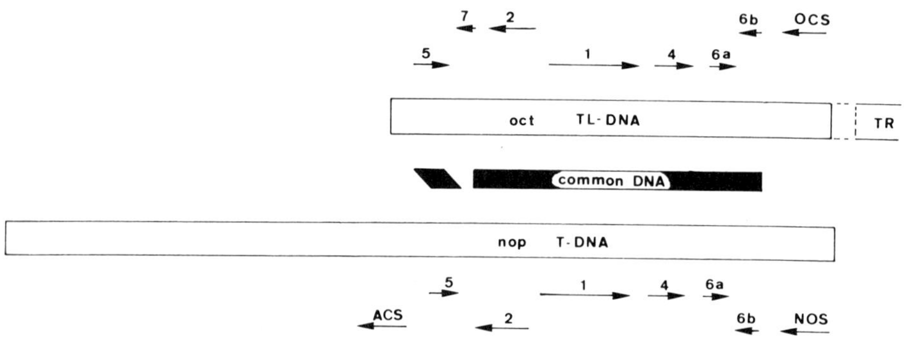

Fig. 2. The T-region of the octopine plasmid pTiAch5 and the nopaline plasmid pTiC58, aligned on their homologous DNA regions, denoted as common DNA (Engler et al. 1981). *Octopine TL-* and *TR-DNA*, and *nopaline T-DNA* indicate the extent of these T-DNAs in most of the tumor lines studied (Thomashow et al. 1980, De Beuckeleer et al. 1981, Lemmers et al. 1980). Direction, size and location of the 8 TL-DNA transcripts are from Willmitzer et al. (1982a, 1983), except for the direction of transcript 5 which is deduced from DNA sequence data. Transcripts are numbered *1–7* according to diminishing size. Six of the 8 transcripts are derived from common DNA sequences. Their homologous counterparts in nopaline tumors are indicated. Ti plasmids mutated in transcript *1* or *2* induce shoot formation, and in transcript *4* root formation. TL-transcript *3* encodes octopine synthase. Nopaline T-DNA-encoded *acs* and *nos* (agrocinopine synthase and nopaline synthase, respectively) are also indicated

Site-Specific Deletion and Substitution Mutagenesis

Since physical maps of the T-DNA for various restriction enzymes were available, it was possible to construct in vitro a series of mutant Ti plasmids carrying well-defined deletions in the T-region which eliminate specifically one or more of the known T-DNA transcripts. This approach allows the combination of nearly any set of T-DNA transcripts. The properties of tissues derived from plants or plant protoplasts infected by these mutant strains and analysis of the T-DNA transcripts in mutant crown gall lines revealed additional data on the genetic organization of T-DNA:

1. The products encoded by transcripts 1, 2, and 4 act by suppressing organ development from both transformed and T-DNA negative cells. The vast majority of normal plants regenerating from tumors induced in vivo by shoot-inducing mutants do not synthesize opines. Most probably, these shoots are derived from untransformed cells present in the tumor or from cells from which T-DNA was largely or totally deleted. Cloned tumor lines, obtained through infection of tobacco protoplasts with shoot-inducing agrobacteria, give rise to transformed teratomata only. These data suggest that transcript 1 and/or 2 suppress(es) the regeneration of both transformed and untransformed cells into shoots by producing diffusible hormone-like substances. Similar arguments indicate that the product of transcript 4 prevents regeneration into roots.

2. Some mutants in the right part of the T-DNA induce tumors of a larger size than the wild-type (Garfinkel et al. 1981). This already suggested that, besides transcripts 1, 2, and 4, other T-DNA genes (6a or 6b) are involved in maintaining the tumorous state. We have constructed pGV2235, a double mutant, in which transcript 6b is inactivated along with transcripts 5, 7, 2, and 1 (Fig. 3). Compared to pGV2219, the corresponding mutant in which transcript 6b is still present, pGV2235 induces smaller tumors with enhanced shoot induction on tobacco.

On decapitated tomato cv "Marmande", pGV2219 induces tumors from which very occasionally shoots appeared. In contrast, pGV2235 induced massive amounts

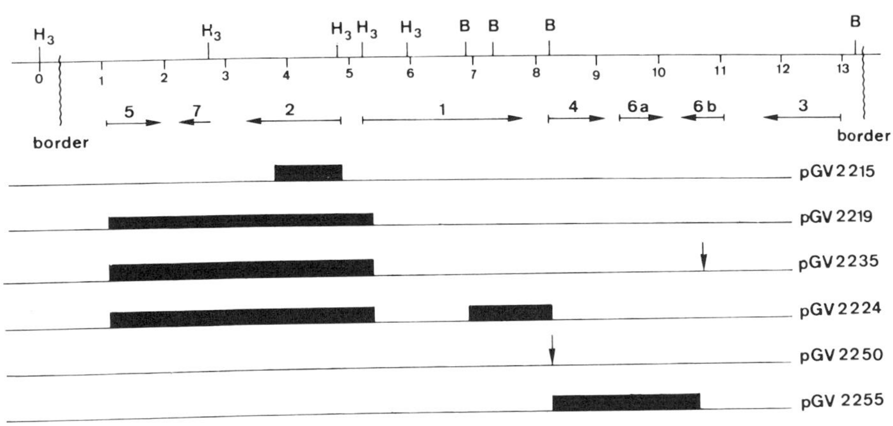

Fig. 3. Location of mutations in the TL-region of the octopine plasmid pTiB6S3 aligned on the *Hind*III (H3) and *Bam*HI (B) physical map of the T-region. Position of the transcripts *(1–7)* is according to Willmitzer et al. 1982, 1983). *Arrows* represent insertions and *thick lines* represent deletion (in pGV2219) or substitutions. pGV2215, pGV2219, and pGV2235 produce shoot-inducing tumors; pGV2224 is avirulent, and pGV2250 and pGV2255 produce root-inducing tumors

Fig. 4. Shoot induction on a pGV2235-induced tomato tumor. *Lycopersicon esculentum* cv. "Marmande" plants were decapitated and the wound site was simultaneously inoculated with *Agrobacterium* C58C1 (pGV2235). Numerous shoots were induced within one month and before growth of tumor tissue was visible. The shoots could be excised, rooted and grown into fertile plants at 100% efficiency

of untransformed shoots on all inoculated tomato plants, long before tumorous tissue was visible (Fig. 4). Without passing through tissue culture, these shoots could efficiently be rooted and grown to maturity. Commercial tomato cultivars, including the one we used, are notoriously difficult to regenerate from tissue culture. From our results, we feel that pGV2235 and similar plasmids might provide a novel and attractive tool for regeneration. For instance, it might extend the selection of plants with altered properties through somaclonal variation (Shepard et al. 1980) to plant species which at present cannot be regenerated from tissue culture.

3. The data obtained through transposon mutagenesis showed that no single insertion in the T-region resulted in complete loss of oncogenicity (Garfinkel et al. 1981, Joos et al. 1983). Using deletion mutants, this result has been confirmed and extended: in tumors induced by pGV2235 (Fig. 3), a shoot-inducing octopine Ti plasmid, only transcripts 4 and 6a are formed, together with transcript 3 encoding octopine synthase. In contrast, transcripts 4 and 6a are absent in root-inducing tumors induced by e.g. pGV2255 and pGV2258 (Fig. 3).

These data prove that:

a) expression of T-DNA transcripts 4 and 6a only is sufficient to allow hormone-independent growth of transformed cells. Experiments are underway to identify whether transcript 4 alone allows tumorous growth.

b) Two complementary mechanisms which prevent organ development by suppressing the formation of either shoots or roots are active in wild-type crown gall tumors. Expression of one of these systems can allow plant cells to grow in the absence of exogenous hormones. The hormone response of the host plant determines whether these transformed cells will proliferate as tumor cells. For instance, the root-inducing mutant pGV2250 induces tumors of wild-type size on carrot slices, while it is nearly avirulent on decapitated tobacco plants. However, when inoculated on the stem of tobacco plants, tumors with profuse root proliferation are formed readily. Similarly, shoot-inducing mutants such as pGV2215 are only slightly attenuated on tobacco, strongly attenuated on carrots and avirulent on most legumes tested. These mutants, when inoculated on carrot slices placed on a medium with 2,4 D (0.1 μg/ml which does not induce callus growth) behave as normally pathogenic.

Non-Oncogenic T-DNAs and Regeneration of Transformed Plants

To use T-DNA as a gene vector for plants, it was important to identify whether suppression or inactivation of the T-DNA functions responsible for tumorous growth (*"onc"* genes) would yield a T-DNA still capable of "transforming" plant cells without preventing them to regenerate into normal plants. Three different approaches can lead to this situation.

1. Exceptionally, normal plants which regenerated from tumors induced by shoot-inducing mutants, synthesized octopine (about 1 in 250). Southern hybridization and the analysis of genomic clones showed that additional mutations had deleted the oncogenic T-DNA functions such that only the octopine synthase gene had remained (De Greve et al. 1982b). In such plants, octopine synthase behaved as a single dominant Mendelian marker (Table 1) (Otten et al. 1981).

2. *Agrobacterium* strains with T-DNA mutations inactivating transcript 4 induce tumors with profuse root proliferations. In axenic culture, some of these roots produce numerous side roots. This abnormal morphology has been correlated with the presence of opines. These cultures, therefore, resemble very much hairy root cultures induced by *Agrobacterium rhizogenes.*

We have induced root cultures from tobacco plants infected with *A. rhizogenes* strain 15834. Reproducibly, the roots form shoots which develop into complete plants. These plants synthesize the same opines as the parental root tissue (we obtained root cultures which produce agropine and mannopine, only mannopine or none of both). The regenerated plants exhibit morphological differences compared to the wild-type Wisconsin 38 in that:

a) the root system is much more developed and roots have reduced geotropism;
b) stems are thicker and have shorter internodes; consequently, plants are much shorter when flowering;
c) the leaves are abnormally elongated;
d) the flowers of some of the plants are abnormal and reduced in fertility. The plants formed variable amounts of seed after selfing.

Table 1. Mendelian transmission of LpDH in transformed tobacco plants. rGV1 originated from a pGV2100-induced tumor line of *Nicotiana tabacum* cv. "Wisconsin 38". rGV4 originated from a pGV2206-induced tumor line of *Nicotiana tabacum* cv. "Wisconsin 38". Untransformed JWB and Wisconsin 38 tobacco was used as LpDH⁻ wild-type (W.T.)

Plantlets from another culture			LpDH$^+$	LpDH$^-$
rGV1			46%	54%
♀		♂	LpDH$^+$	LpDH$^-$
rGV1	X	rGV1	76%	24%
W.T.	X	rGV1	50%	50%
rGV1	X	W.T.	46%	54%
rGV4	X	rGV4	72%	28%
rGV4	X	rGV1	(15)	(1)
F1 (++)	X	W.T.	76%	24%

Conclusions: – LpDH is transmitted as a dominant marker;
– rGV1 and rGV4 are hemizygous for LpDH;
– LpDH$^+$ eggs and pollen are fertile;
– The LpDH genes in rGV1 and rGV4 are located on different loci

( ): Indicative value on 16 plants tested

Since regeneration from hairy roots is relatively efficient, deletions of part of the *rhizogenes* T-DNA are very unlikely to explain the transition from root to plant. More likely, the full size T-DNA is still present but is somehow suppressed. In view of these results, it will be interesting to investigate whether transformed root cultures, induced by mutant Ti plasmids will behave similarly.

3. The low frequency regeneration of transformed plants from shoot-inducing crown gall tissues and the morphological aberrations of plants regenerated from hairy roots are evident drawbacks of the two previous approaches. A third approach would consist in using mutant T-DNAs from which all *onc* genes have been deleted. The first indications that this could be successful were inferred from the fact that agropine and mannopine could be detected in the wound tissue of carrot slices, infected with the avirulent Ti plasmid pGV2217 (Leemans et al. 1982). In pGV2217, the entire TL-DNA has been deleted, but the neighbouring TR sequences are still present. This showed that the genes involved in the synthesis of agropine and mannopine are encoded by TR. Furthermore, assuming opine synthesis in plant tissue can only occur if the relevant genes are introduced by T-DNA transfer, this experiment indicated that the TR-DNA can transfer independently from TL-encoded functions and without transforming the plant cell into a tumor cell. To confirm this hypothesis, regenerating leaf protoplasts of *Nicotiana tabacum* SR1 were infected with *A. tumefaciens* C58C1 pGV2217. Protoplast-derived colonies were cultured on auxin- and cytokinin-containing medium. Among 650 of the resulting calli, one was found that synthesized large amounts of mannopine. We conclude that TR-DNA can be transferred and maintained in the absence of TL-DNA. Therefore, we consider TR-DNA

as a naturally occurring T-DNA vector which might be an alternative to manipulated TL-DNAs from which all *onc* genes have been deleted. Such a TL-DNA is present in the avirulent plasmid pGV2224 (Fig. 3) which also induces octopine synthesis in infected carrot tissues (Leemans et al. 1982). At present, DNA sequence analysis has provided detailed knowledge on the extremities of the T-DNA. This allows to construct T-DNA-derived vectors with appropriate sites for cloning, which are still transferred and integrated and which do not influence the morphogenetic capacities of the plant cells.

Acknowledgements. The authors thank their colleagues of the Laboratory of Genetics (State University Gent, Belgium) and of the Max-Planck-Institut für Züchtungsforschung (Köln, FRG) for help and discussion. This research was supported by grants from the "Fonds voor Kollektief Fundamenteel Onderzoek (F.K.F.O. 2007.77). J.P.H. is a Research Associate of the Belgian National Fund for Scientific Research (N.F.W.O.), and R.D. is indebted for a fellowship as "Aspirant" to the N.F.W.O.

References

Beuckeleer M De, Lemmers M, Vos G De, Willmitzer L, Montagu M Van, Schell J (1981) Further insight on the transferred-DNA of octopine crown gall. Mol Gen Genet 183:283–288

Chilton M-D, Drummond MH, Merlo DJ, Sciaky D (1978) Highly conserved DNA of Ti-plasmids overlaps T-DNA, maintained in plant tumors. Nature 275:147–149

Depicker A, Montagu M Van, Schell J (1978) Homologous DNA sequences in different Ti-plasmids are essential for oncogenicity. Nature 275:150–153

Engler G, Depicker A, Maenhaut R, Villarroel-Mandiola R, Montagu M Van, Schell J (1981) Physical mapping of DNA base sequence homologies between an octopine and a nopaline Ti-plasmid of *Agrobacterium tumefaciens.* J Mol Biol 152:183–208

Garfinkel DJ, Simpson RB, Ream LW, White FF, Gordon MP, Nester EW (1981) Genetic analysis of crown gall: fine structure map of the T-DNA site-directed mutagenesis. Cell 27:143–153

Greve H De, Dhaese P, Seurinck J, Lemmers M, Montagu M Van, Schell J (1982a) Nucleotide sequence and transcript map of the *Agrobacterium tumefaciens* Ti plasmid-encoded octopine synthase gene. J Mol Appl Genet 1:499–512

Greve H De, Leemans J, Hernalsteens JP, Thia-Toong L, Beuckeleer M De, Willmitzer L, Otten L, Montagu M Van, Schell J (1982b) Regeneration of normal and fertile plants that express octopine synthase, from tobacco crown galls after deletion of tumour-controlling functions. Nature 300:752–755

Haute E Van, Joos H, Maes M, Warren G, Montagu M Van, Schell J (1983) Intergeneric transfer and exchange recombination of restriction fragments cloned in pBR322: a novel strategy for the reversed genetics of Ti plasmids of *Agrobacterium tumefaciens.* EMBO J 2:411–417

Joos H, Inzé D, Caplan A, Sormann M, Montagu M Van, Schell J (1983) Genetic analysis of T-DNA transcripts in nopaline crown galls. Cell 32:1057–1067

Leemans J, Shaw C, Deblaere R, Greve H De, Hernalsteens JP, Maes M, Montagu M Van, Schell J (1981) Site-specific mutagenesis of *Agrobacterium* Ti plasmids and transfer of genes to plant cells. J Mol Appl Genet 1:149–164

Leemans J, Deblaere R, Willmitzer L, Greve H De, Hernalsteens JP, Montagu M Van, Schell J (1982) Genetic identification of functions of TL-DNA transcripts in octopine crown galls. EMBO J 1:147–152

Lemmers M, Beuckeleer M De, Holsters M, Zambryski P, Depicker A, Hernalsteens JP, Montagu M Van, Schell J (1980) Internal organization, boundaries and integration of Ti-plasmid DNA in nopaline crown gall tumours. J Mol Biol 144:353–376

Otten L, Greve H De, Hernalsteens JP, Montagu M Van, Schieder O, Straub J, Schell J (1981) Mendelian transmission of genes introduced into plants by the Ti plasmids of *Agrobacterium tumefaciens*. Mol Gen Genet 183:209–213

Ruvkun GB, Ausubel FM (1981) A general method for site-directed mutagenesis in prokaryotes. Nature 289:85–88

Shepard J, Bidney D, Shahin E (1980) Potato protoplasts in crop improvement. Science 208: 17–24

Thomashow MF, Nutter R, Montoya AL, Gordon MP, Nester EW (1980) Integration and organisation of Ti-plasmid sequences in crown gall tumors. Cell 19:729–739

Willmitzer L, Simons G, Schell J (1982) The TL-DNA in octopine crown gall tumours codes for seven well-defined polyadenylated transcripts. EMBO J 1:139–146

Willmitzer L, Dhaese P, Schreier PH, Schmalenbach W, Montagu M Van, Schell J (1983) Size, location and polarity of T-DNA-encoded transcripts in nopaline crown gall tumors: evidence for common transcripts present in both octopine and nopaline tumors. Cell 32:1045–1056

D. Plant Pathogenic Bacteria and Related Aspects

The Genetics of the Cherry Pathogen, *Pseudomonas syringae* pathovar *morsprunorum*

A. VIVIAN [1], J. ERRINGTON [2], and D. VEAL [1]

Introduction

Pseudomonas syringae pathovar *morsprunorum* (hereafter referred to as pv. *morsprunorum*) is the cause of bacterial canker in the stone fruit trees, plum *(Prunus domestica)* and cherry *(Pr. avium),* in England (Crosse 1966). Strains of pv. *morsprunorum* are specific for either plum or cherry, showing little tendency to cross-infection. Cherry-specific strains are further sub-divided into two races, 1 and 2, on the basis of their specificity for particular cultivars (Freigoun and Crosse 1975). Table 1 shows the further sub-divisions of the two races on the basis of laboratory tests (Garrett et al. 1966, 1977). Clearly a spectrum of pv. *morsprunorum* strains is present in cherry orchards, and one aim of the present investigation has been to use genetic analysis as a means of investigating the basis of the differences shown by these organisms.

A race 1, non-fluorescent cherry strain, designated C28 was chosen for genetic analysis. An indigenous gene-transfer system which has been particularly useful for the construction of novel combinations of auxotrophic markers, has been found in C28 (Errington and Vivian 1981). Strains behave as either donors or recipients, and no recombinants can be obtained from crosses involving only recipient strains. Linkage has only been established unequivocally between 3 pairs of markers (Errington 1981) and attempts to extend the mapping to other markers is hampered by low levels of fertility (A.N. Malik and A. Vivian, unpubl. results).

Table 1. Race structure of English cherry strains of *Pseudomonas morsprunorum*

Subdivision		Favoured	Gelatin	Arbutin	Pigment	ØA7	ØB1
Race 1	'Fluorescent'	Napoleon	–	–	+	+	–
Race 1	'Non-fluorescent'	Napoleon	–	–	–	+	–
Race 2	Group 1	Roundel	+	–	–	+	+
Race 2	Group 2	Roundel	+	–	+	+	+
Race 2	Group 3	Roundel	+	+	+	–	+

1 School of Biological Sciences and Environmental Health, Thames Polytechnic, London, SE18 6PF, United Kingdom

2 Present address: Microbiology Unit, Department of Biochemistry, South Parks Road, Oxford, OX1 3QU, United Kingdom

Molecular Genetics of the Bacteria-Plant Interaction
(ed. by A. Pühler)

To complement the latter system, we have attempted to use transmissible plasmids from other bacteria to mobilize the chromosome of C28. Here we describe these experiments, and the discovery of strains with enhanced ability to maintain plasmids (Ipm^-).

All genetic strains derived from strain C28 are designated by a prefix E.

Inheritance of RP4 of Auxotrophic Derivatives of Strain C28

Filter matings performed at 25°C between a strain of *Escherichia coli* carrying RP4 and 3 independently isolated auxotrophic derivatives of the wild type pv. *morsprunorum* strain C28 indicated that inheritance of RP4 in pv.*morsprunorum* was relatively low (Table 2a). Three transconjugants were isolated from each mating and all exhibited both the selected tetracycline resistance (Tet^R) and the unselected kanamycin (Kan^R) and ampicillin (Amp^R) resistances. The presence of RP4 in the transconjugants was confirmed by agarose gel electrophoresis of single colony lysates (Eckhardt 1978).

Table 2. Inheritance of RP4 from *E. coli.* Filter matings were performed with *E. coli* strain J53 *pro met* $RP4^+$ as the donor

			pv. *morsprunorum* recipient	Transfer frequency [a]
a)	Ipm^+	E32	*leu-1, met-3*	1.7×10^{-5}
		E66	*his-1, ilv-2*	1.8×10^{-5}
		E71	*met-5, trp-1*	1.2×10^{-5}
b)	Ipm^-	E39	*leu-1, met-3*	8.9×10^{-2}
		E612	*his-1, ilv-2*	2.0×10^{-1}
		E714	*met-5, trp-1*	1.3×10^{-1}
		E93	*gly-1, thi-3*	9.7×10^{-2}

[a] Tet^R transconjugants per recipient parent

Derivation of Strains with Enhanced Ability to Maintain Plasmids (Ipm^-, Interference with Plasmid Maintenance)

Colonies of strain E71 *met-5, trp-1* were replica plate mated (Hinchliffe and Vivian 1980) to *E. coli* J53 $RP4^+$ and selection was made for Tet^R transconjugants of E71. It was possible to identify in this way a single colony among 1124 tested, which inherited Tet^R at a much higher frequency. This colony was isolated as strain E714, and similar Ipm^- variants of strains E32 *leu-1, met-3* (1/1733 colonies) and E66 *his-1, ilv-2 (1/1990 colonies*) were obtained, and designated E39 and E612 respectively. Filter matings of these Ipm^- strains with *E. coli* $RP4^+$ confirmed that the plasmid was inherited at a much higher frequency (Table 2b). In addition a further strain, E93 *gly-1, thi-3,* was apparently a 'spontaneous' Ipm^- strain, in that it was derived as an auxotroph from C28, but was identical to Ipm^- strains in its inheritance of RP4.

Inheritance of Plasmids from Other Incompatibility Groups by Ipm Strains

The changed ability to inherit plasmids from *E. coli* was not confined to RP4. Other plasmids of incompatibility group P such as R702 and the Inc W plasmid S-a behaved like RP4 in filter matings (Table 3). Replica plate mating was used to investigate a range of plasmids for their ability to transfer to Ipm^+ and Ipm^- strains and these results are shown in Table 4. An interesting observation here was the inheritance of the Inc C plasmid R57b in Ipm^- but not apparently in Ipm^+ strains.

Table 3. Inheritance of plasmids from *E. coli* in pv. *morsprunorum:* filter matings. The donor in each cross was *E. coli* J62 *pro, his, trp, lac* $R702^+$ or J53-1 *pro, met, nal* $S\text{-}a^+$

	Recipient		Inheritance from donor [a]	
			(a) $R702^+$	(b) $S\text{-}a^+$
Ipm^+	E32	*leu-1, met-3*	1.6×10^{-5}	3.8×10^{-6}
	E66	*his-1, ilv-2*	1.8×10^{-6}	1.9×10^{-5}
	E71	*met-5, trp-1*	1.7×10^{-6}	1.4×10^{-5}
Ipm^-	E39	*leu-1, met-3*	4.3×10^{-1}	7.2×10^{-2}
	E612	*his-1, ilv-2*	2.6×10^{-2}	2.2×10^{-2}
	E714	*met-5, trp-1*	2.5×10^{-2}	2.0×10^{-2}

[a] R^+ transconjugants per recipient parent

Table 4. Inheritance of plasmids from *E. coli* in pv. *morsprunorum:* replica plate mating

Plasmid	Inc Group	Selection [a]	Inheritance [b] in Ipm^+	Ipm^-
R57b	C	Kan/Sul	N.D.	Low
R1-16	FII	Kan/Sul	N.D.	N.D.
R64	I	Tet	N.D.	N.D.
R391	J	Kan	N.D.	N.D.
R702	P	Kan/Sul/Tet	Low	High
RP4	P	Kan/Tet	Low	High
R388	W	Sul	N.D.	High
S-a	W	Kan/Sul	Low	High

[a] Resistances: Kan, kanamycin; Sul, sulphonamides; Tet, tetracycline
[b] N.D. = Not detected

The Use of Transposon Tn5

E. coli strain PFB3 carrying a Kan^S derivative of R68.45 was mated with *E. coli* strain UNF510 (both strains kindly provided by Dr. A.W.B. Johnston, John Innes Institute, Norwich, England) harbouring Tn5 which confers Kan^R (Berg et al. 1975) and selection was made for transfer of the plasmid. The resulting Kan^R Tet^R strain was then mated with a further *E. coli,* strain 5K^{--}, which lacks host restriction and modification activity (Colson et al. 1965) and selection made for Kan^R Tet^R transconjugants. One of these transconjugants, PU5K was confirmed as carrying a plasmid of molecular weight corresponding to R68.45 plus Tn5. This strain was crossed with various Ipm$^+$ and Ipm$^-$ strains and selection made alternately for Kan^R and Tet^R transconjugants. The results are shown in Table 5 and again show a difference in plasmid inheritance (Tet^R) between Ipm$^+$ and Ipm$^-$ strains. However Ipm$^+$ strains inherit Tn5 (Kan^R) at a much higher frequency than R68.45 (Tet^R). This implies that R68.45 has difficulty maintaining itself in Ipm$^+$ strains, at least initially following transfer. There is no evidence for instability of R68.45 in Ipm$^+$ strains once these have been recovered as transconjugants. R68.45 and Tn5 are inherited concomitantly in Ipm$^-$, indicating that the interference with plasmid maintenance is absent in Ipm$^-$ strains. Similar results have been obtained using an RP4::Tn5 plasmid in Ipm$^+$ and Ipm$^-$ strains (D. Veal and A. Vivian, data not shown).

Table 5. Inheritance of Tn5 by Ipm$^+$ and Ipm$^-$ strains of pv. *morsprunorum* strain C28. The donor strain in filter matings was *E. coli* PU5K *lac, thr, leu, thi, tonA, hsmK, hsrK.* R68.45::Tn5

Recipient strain			Frequency [a] of transconjugants selected as: Kan^R	Kan^R Tet^R
C28	wild type	Ipm$^+$	1.1×10^{-5}	6.6×10^{-7}
E71	*met-5, trp-1*	Ipm$^+$	7.4×10^{-5}	$<1.0 \times 10^{-8}$
E66	*his-1, ilv-2*	Ipm$^+$	4.5×10^{-6}	$<1.0 \times 10^{-8}$
E714	*met-5, trp-1*	Ipm$^-$	1.9×10^{-3}	2.2×10^{-3}
E612	*his-1, ilv-2*	Ipm$^-$	2.6×10^{-2}	2.2×10^{-2}

[a] transconjugants per recipient parent;
Kan^R, kanamycin resistant; Tet^R, tetracycline resistant

The Ability of Ipm Strains to Donate Plasmids

It was initially considered that the Ipm$^-$ phenotype may have resulted from the loss or inactivation of a host restriction/modification system from C28 and its derivatives. However, if this were the case, rare Ipm$^+$ RP4$^+$ transconjugants should harbour modified RP4, and therefore subsequent transfer to other Ipm$^+$ strains should occur at a higher frequency.

To investigate this possibility, a sample of 100 Tet^R transconjugants from a mating between *E. coli* J53 RP4 and E49 *ilv-4, nic-3, trp-4, str-5* Ipm$^+$ were tested by replica plate mating for their ability to inherit the inc W plasmid, S-a. Two transconjugants were Ipm$^+$ and the remainder were Ipm$^-$. One transconjugant of each type was isolated and these strains, E53 Ipm$^-$ and E54 Ipm$^+$, were filter mated with other Ipm$^+$ and Ipm$^-$ strains. The results are shown in Table 6. Transfer of RP4 was only detected between Ipm$^-$ parents. Subsequent crosses involving transfer of RP4 and

Table 6. Donor ability of Ipm^+ and Ipm^- strains of pv. *morsprunorum* strain C28. All strains prefixed E were derived from C28. Strains $5K^{--}$, J53 and J62.1 were *E. coli*. Transfer frequencies between *E. coli* strains are given for comparison

Donor strain	Recipient strain	Plasmid selected	Frequency of transconjugants [a]
E54 Ipm^+	E32 Ipm^+	RP4	N.D.
E54 Ipm^+	E39 Ipm^-	RP4	N.D.
E54 Ipm^+	$5K^{--}$	RP4	N.D.
E78 Ipm^+	$5K^{--}$	R702	N.D.
E80 Ipm^+	$5K^{--}$	S-a	N.D.
E53 Ipm^-	E32 Ipm^+	RP4	N.D.
E53 Ipm^-	E39 Ipm^-	RP4	1.9×10^{-6}
E53 Ipm^-	$5K^{--}$	RP4	1.4×10^{-4}
E721 Ipm^-	$5K^{--}$	R702	1.1×10^{-6}
E725 Ipm^-	$5K^{--}$	S-a	N.D.
J62-1	$5K^{--}$	R702	2.4×10^{-2}
J53	$5K^{--}$	S-a	6.1×10^{-2}

[a] Tet^R (RP4, R702) or Sul^R (S-a) transconjugants per recipient parent
N.D. = Not detected, $< 1.0 \times 10^{-8}$

R702 (Inc P) from pv. *morsprunorum* strains to *E. coli*, indicated that only Ipm^- strains could donate their plasmid. Furthermore the frequency of plasmid donation was low compared to the frequency of transfer among *E. coli* strains (Table 6). It would appear that Ipm^+ strains are impaired both in their ability to inherit and donate plasmids and consequently it has not been possible to determine whether a host restriction/modification system in C28 plays any role in the behaviour of plasmids.

Stability of Ipm^- Strains

Extensive use of the Eckhardt (1978) technique to examine Ipm^+ and Ipm^- strains on agarose gels has failed to show any consistent change of plasmid status between strains. However, the possiblity was considered that a plasmid not visualized by the above method, may have been lost in Ipm^- strains, and consequently it might be possible to 'reinfect' an Ipm^- strain with the missing plasmid and so convert it to Ipm^+. Six crosses were made between all possible combinations of Ipm^- strains with appropriate Ipm^+ strains, but no Ipm^+ 'transconjugants' were detected amongst a total of 18,654 parental Ipm^- colonies tested.

During some three years since the initial isolation of Ipm^- strains, no spontaneous reversion to Ipm^+ has been detected.

Conclusions

Several workers have shown that various Inc P plasmids can be transferred readily from *E. coli* to a variety of plant pathogenic Pseudomonads (Lacy and Leary 1975, Panopoulos et al. 1975). Transfer of RP4 to pv. *morsprunorum* C28 occurred at a similar frequency, but subsequent re-transfer to other strains of C28 and to *E. coli* was not detected. The results presented here indicate that Ipm^- variants of C28 can be obtained, which show (1) an enhanced ability to receive certain plasmids from *E. coli,* (2) an enhanced ability to maintain those plasmids – at least initially following transfer, and (3) an ability to act subsequently as plasmid donors to *E. coli* and other strains of C28. However the genetic basis ofthe Ipm^- phenotype remains obscure.

Agarose gel electrophoresis of single colony lysates has failed to detect a consistent change in the plasmid status between Ipm^+ and Ipm^- strains. However the possibility remains that this technique does not permit us to vizualize all the plasmids present in C28. An alternative hypothesis is that some transposition or excision event within the genome may be responsible for the creation of Ipm^- variants.

In spite of our lack of a clear understanding of their nature, Ipm^- variants are potentially useful in the development of an alternative mapping system in C28. Although the frequencies of transfer of RP4 and R68.45 are relatively low between Ipm^- strains, and do not permit the detection of chromosome mobilization, the isolation of strains with increased donor potential (J. Errington and A. Vivian, unpubl. results) offer wide possibilities for in vivo and in vitro manipulation of the genome.

Acknowledgements. This work was supported by a CASE award to A.V. from the Science and Engineering Research Council in collaboration with East Malling Research Station, Kent, England and also by project grant No. AG 101/6 from the Agricultural Research Council to A.V. The authors would especially like to thank Dr. Connie Garrett for her interest and advice during the course of this work.

References

Berg D, Davies JE, Allet B, Rochaix J-D (1975) Transposition of R factor genes to bacteriophage lambda. Proc Natl Acad Sci USA 72:3628–3632

Colson C, Glover SW, Symonds N, Stacey KA (1965) The location of the genes for host-controlled modification and restriction in *Escherichia coli* K12. Genetics 52:1043–1050

Crosse JE (1966) Epidemiological relations of the pseudomonad pathogens of deciduous fruit trees. Annu Rev Phytopathol 4:291–310

Eckhardt T (1978) A rapid method for the identification of plasmid DNA in bacteria. Plasmid 1: 584–588

Errington J (1981) The genetics of *Pseudomonas morsprunorum.* Ph D Thesis CNAA

Errington J, Vivian A (1981) An indigenous system of gene transfer in the plant pathogen *Pseudomonas morsprunorum.* J Gen Microbiol 124:439–442

Freigoun SO, Crosse JE (1975) Host relations and distribution of a physiological and pathological variant of *Pseudomonas morsprunorum.* Ann Appl Biol 81:317–330

Garrett CME, Panagopoulos CG, Crosse JE (1966) Comparison of plant pathogenic Pseudomonads from fruit trees. J Appl Bacteriol 29:342–356

Garrett CME, Prunier J-P, Crosse JE (1977) Bacterial canker of stone fruits: virulence of typical and variant froms of *Pseudomonas morsprunorum* in wound inoculations. Rep East Malling Res St 1976:127–128

Hinchliffe E, Vivian A (1980) Naturally occurring plasmids in *Acinetobacter calcoaceticus:* a P class R factor of restricted host range. J Gen Microbiol 116:75–80

Lacy GH, Leary JV (1975) Transfer of antibiotic resistance plasmid RP1 into *Pseudomonas glycinea* and *Pseudomonas phaseolicola* in vitro and in planta. J Gen Microbiol 88:49–57

Panopoulos NJ, Guimaraes WV, Cho JJ, Schroth MN (1975) Conjugative transfer of *Pseudomonas aeruginosa* R factors to plant pathogenic *Pseudomonas* spp. Phytopathology 65:380–388

Towards the Genetical Analysis of Pathogenicity of *Xanthomonas campestris*

M.J. DANIELS, P.C. TURNER, C.E. BARBER, W.G. CLEARY, and G. REED [1]

Introduction

Despite extensive physiological experimentation the basis of bacterial pathogenicity to plants remains poorly understood. A priori one would expect the interaction of the host and parasite to be a complex process; specific reactions occur at many levels. While it is possible to investigate the role of suspected virulence determinants such as toxins, degradative enzymes or polysaccharides, it is difficult even to guess at the mechanisms controlling pathogen attributes such as invasiveness, organ and tissue specificity, evasion of defence mechanisms and aspects of symptom production. A logical approach to the problem consists of isolating non-pathogenic mutants by screening directly on plants, without prejudice to the biochemical nature of the lesions, and then using the powerful methods of molecular genetics to investigate the function of the genes concerned. Isolation of mutants demands the ability to screen large numbers of mutagenised bacteria individually on plants, but with many plant pathogens the inoculation procedures are too complex to make this practicable. However we have developed techniques for *Xanthomonas campestris,* the causal agent of black rot of crucifers, which permit more than 1000 brassica seedlings to be easily and reliably inoculated per day. In this paper we describe the use of this test to isolate non-pathogenic mutants. We have chosen to concentrate on mutations induced by the transposon Tn5 because of their many advantages in subsequent manipulations (Kleckner et al. 1977).

Methods

Bacteria

Xanthomonas campestris pv. *campestris* NCPPB 1145 was chosen as the standard wild-type strain because of its reliable growth and pathogenicity characteristics and convenient antibiotic sensitivity. For most experiments a spontaneous mutant, 8004, resistant to 100 μg/ml rifampicin and retaining full pathogenicity, was used. *Escherichia coli* SM10 (harbouring the plasmid p1011) was obtained from R. Simon via G. Hombrecher, and 1843 (harbouring pJB3JI) from A.W.B. Johnston. Complete

1 John Innes Institute, Colney Lane, Norwich NR4 7UH, England

Molecular Genetics of the Bacteria-Plant Interaction
(ed. by A. Pühler)

medium (NYG) contained peptone (0.5%), yeast extract (0.3%) and glycerol (2%), and minimal medium MMX was essentially the same as Minimal A medium (Miller 1972). *X. campestris* cultures were incubated at 25°C. Matings were carried out by mixing overnight cultures of donor and recipient strains and depositing the cells on membrane filters which were incubated on NYG plates overnight at 25°C. The bacteria were then resuspended in water and plated on appropriate media.

Pathogenicity Test

Seeds of turnip (*Brassica rapa* cv. Just Right) were surface-sterilised with 10% commercial bleach for 30 min, washed several times with sterile water and placed on water agar at 25°C overnight. Germinated seeds were transferred individually to wells of "Replidishes" (Sterilin Ltd., Teddington, England) containing 3 ml of agar-solidified, quarter-strength plant tissue culture medium (Murashige and Skoog 1962). After 3–4 days at 25°C seedlings were inoculated by touching a sharpened nichrome wire on a bacterial colony and then pricking into the stem. Plants were incubated at 25°C in a humid atmosphere and observed daily.

Results

Properties of the Pathogenicity Test

Inoculated seedlings developed symptoms over a period of 3–4 days. After 12–24 h stems appeared constricted and collapsed around the inoculation point. The regions of constriction spread progressively up and down the stems and gradually rotted and liquefied. The cotyledons turned yellow and after 4 days the seedlings were totally collapsed and rotted. Seedlings inoculated with a sterile wire, with $CHCl_3$-killed *X. campestris,* or with living saprophytic bacteria (e.g. *E. coli* or unidentified contaminants) showed no symptoms apart from occasional 0.2 mm brown borders around the punctures, and continued to develop normally. Development of the symptoms appeared to be unaffected by light intensity and followed the same course at 15° or 20°C. When individual colonies of wild-type *X. campestris* were tested on single seedlings 90%–95% gave typical symptoms. Those apparently negative were re-tested and invariably then gave normal symptoms. The number of bacteria in (or associated with) seedlings was determined by homogenising the plant tissue in water and plating serial dilutions of the homogenate. The number of bacteria recovered immediately after inoculation was 10^5–10^6 per seedling, and as symptoms developed the number increased to 10^8–10^9. During the period of maximum rate of increase the apparent doubling time was about 4 h: in comparison *X. campestris* doubles in 3 h at 25°C in shaken broth cultures. By loading the inoculating needle with serially-diluted suspensions of bacteria it was possible to determine that the minimum number of cells required to initiate a successful infection was about 10^3.

Tn5 Mutagenesis

Plasmid p1011 is a derivative of pACYC184 (Chang and Cohen 1978) with RP4-derived *mob* genes inserted into the Tc^r gene. In addition to Tn5-encoded kanamycin resistance the plasmid confers resistance to chloramphenicol. The *mob* genes permit the plasmid to be mobilised by *trans*-acting products from a suitable tra^+ plasmid, in this case RP4::Mu integrated into the host (SM10) chromosome. We anticipated that p1011 would be unable to establish itself in *X. campestris* after conjugal transfer so that kanamycin-resistant colonies should only arise by transposition to Tn5 to *X. campestris* replicons. Resistant colonies arose at a frequency of 10^{-6}–10^{-7} per recipient following overnight matings with *E. coli* SM10, and were picked on to master plates (NYG + rifampicin + kanamycin) and subsequently replica-plated on to MMX, MMX + casamino acids, and again to a further NYG plate. Amino acid auxotrophs (i.e. colonies growing on NYG and MMX + casamino acids but not on unsupplemented MMX) comprised about 0.3%. Apart from one mutant requiring arginine and another requiring threonine, the lesions have not yet been characterised. Colonies on the NYG replica plates were tested for pathogenicity. Those which gave either no symptoms or unusual responses were re-tested and, if giving similar results, were taken for further study. These possible non-pathogenic mutants represented about 0.5% of the kanamycin-resistant colonies. In comparison control tests of about 1000 unmutagenised colonies gave no non-pathogenic variants. The mutants were sensitive to chloramphenicol and carbenicillin, and resistant to the bacteriophage PRD1, which only infects cells harbouring P-group plasmids (Olsen et al. 1974), and kanamycin resistance was not transferred in crosses with other *X. campestris* strains, suggesting that the kanamycin-resistant phenotype was due to transposition of Tn5 rather than stable establishment of p1011 or transfer of RP4::Mu. A selection of mutants grown on membrane filters and lysed in situ (Humphries et al. 1978) showed hybridisation with a ^{32}P-labelled Tn5 DNA probe (a 3.3 kb *Hin*dIII fragment). Finally, a derivative of R68:45 which no longer confers kanamycin resistance but which retains chromosome-mobilising ability (pJB3JI, Brewin et al. 1980) was used to mobilise kanamycin resistance from the presumptive Tn5 auxotrophs into a spectinomycin-resistant *X. campestris* strain. Approximately 250 colonies from each cross were tested for auxotrophy and coinheritance of antibiotic resistance with the amino acid requirement was complete (98.5%–100%). Although similar studies have not yet been completed with all the mutants these results strongly suggest that the non-pathogenic phenotype results from Tn5 insertion.

Properties of Non-Pathogenic Mutants

The non-pathogenic mutants were not contaminants: they retained the rifampicin resistance of the parent strain and gave identical band patterns on SDS-polyacrylamide gel electrophoresis of total cell proteins, which is a sensitive indicator of identify of bacterial strains (Daniels et al. 1980). These experiments also showed that none of the mutations had resulted in the loss of any of the fifty or so major polypeptide bands revealed on one-dimensional protein gels.

None of the pathogenicity mutants showed perceptibly different behaviour ex planta. All were prototrophic and appeared to grow as fast as the parent strain, and had similar colonial morphology. Colonial variants resulting presumably from altered polysaccharide or pigment synthesis were detected among the kanamycin-resistant population, but all tested so far have retained pathogenicity.

The symptoms produced by the mutants on seedlings varied from no visible damage or a faint brown ring around the inoculation point to limited stem collapse around the puncture which failed to advance into healthy tissue. However some mutants appeared to be leaky in that if larger numbers of bacteria were introduced into the seedlings more severe symptoms resulted (Table 1).

Table 1. Properties of Tn5-induced non-pathogenic mutants. Groups of five seedlings were inoculated with serial dilutions of *X. campestris* suspensions. After 3 days mean symptom ratings per seedling were assessed (–, no visible symtpoms; ±, localised brown ring around inoculation point; +, partial stem collapse and rotting; ++, full symptoms – see text). "Growth" means the factor by which the mean number of viable bacteria per seedling increased in 2 days following inoculation

Strain	Symptom rating Relative inoculum concentration					Growth
	10^0	10^{-1}	10^{-2}	10^{-3}	10^{-4}	
8004 (wild-type)	++	++	+	+	–	160
8208-1	+	+	–	–	–	2
8208-2	±	–	–	–	–	10
8208-3	+	–	+	±	+	7
8208-4	–	–	±	–	–	4
8208-6	+	–	–	–	–	10
8208-7	–	+	–	–	–	2
8208-8	+	+	–	–	–	2

All the mutants tested so far show impaired ability to grow in plants. Groups of ten seedlings were inoculated, five were sampled immediately and the remainder two days later. Viable counts of the seedling homogenates showed that whereas the wild-type increased in number by a factor of 160 in the two-day period, the mutants increased by factors of 2–10 (Table 1).

Discussion

The most powerful and economical way of initiating a study of the properties of a complex biological system is to isolate mutants altered specifically in the process. In the case of plant disease this approach has been used to a limited extent (e.g. in studying the role of certain enzymes) but the full potential has not been exploited because of the technical difficulty of testing large numbers of mutants for pathogenicity. We have developed a test which enables us to screen 10^3 mutants in a day's work so that, provided the transposon does not show extreme site specificity for insertion, it should be possible to uncover many "pathogenicity" genes whose function we cannot yet even guess.

It should be stressed that our pathogenicity test is regarded as a screening assay and not as a definitive model system for studying the entire disease cycle. In field infections *X. campestris* often gains entry to plants via natural openings, particularly hydathodes (Smith 1897) and there is evidence that some resistant cultivars may be less susceptible to this mode of invasion (Staub and Williams 1970). Mutations in any bacterial genes which contribute to this mode of attack would not be detected by our screening test. Nevertheless study of mutants such as those we have described should contribute much information about the pathways of disease.

References

Brewin NJ, Beringer JE, Johnston AWB (1980) Plasmid-mediated transfer of host-range specificity between two strains of *Rhizobium leguminosarum*. J Gen Microbiol 120:413–420

Chang ACY, Cohen SN (1978) Construction and characterisation of amplifiable DNA cloning vehicles derived from the p15A cryptic mini plasmid. J Bacteriol 134:1141–1156

Daniels MJ, Archer DB, Stephens MA, Townsend R, Longland JM, Best J (1980) Comparison of spiroplasmas by polyacrylamide gel electrophoresis of cell proteins. Curr Microbiol 4:377–380

Humphries P, Old R, Coggins LW, McShane T, Watson C, Paul J (1978) Recombinant plasmids containing *Xenopus laevis* globin structural genes derived from complementary DNA. Nucleic Acids Res 5:905–924

Kleckner N, Roth J, Botstein D (1977) Genetic engineering in vivo using translocatable drug-resistance elements. J Mol Biol 116:125–159

Miller J (1972) Experiments in molecular genetics. Cold Spring Harbor Laboratory Press, Cold Spring Harbor, New York

Murashige T, Skoog F (1962) A revised medium for rapid growth and bioassays with tobacco tissue cultures. Physiol Plant 15:473–497

Olsen RH, Sang-Siak J, Gray RH (1974) Characteristics of PRD1, a plasmid-dependent broad host range DNA bacteriophage. J Virol 14:689–699

Smith EF (1897) *Pseudomonas campestris* (Pammel). The cause of a brown rot in cruciferous plants. Zentralbl Bakteriol Abt II 3:284–291, 408–415, 478–486

Staub T, Williams PH (1970) The resistant reaction in cabbage to *Xanthomonas campestris*. Phytopathology 60:1315

Molecular Analysis of Virulence Genes in *Pseudomonas solanacearum*

B.J. STASKAWICZ, D. DAHLBECK, J. MILLER, and D. DAMM [1]

Introduction

Pseudomonas solanacearum, the causal agent of bacterial wilt, is a destructive pathogen on many Solanaceous crops (Buddenhagen and Kelman 1964). The ability of this bacterium to produce extracellular polysaccharide (EPS) has been shown to be correlated with virulence (Kelman 1954). The virulent phenotype can be easily assayed by colony appearance on a complex medium containing tetrazolium (Kelman 1954). Virulent colonies are slightly pink and produce copious amounts of EPS, while spontaneous avirulent mutants appear bright red and do not produce EPS. Currier and Morgan (1981) and Zischek et al. (1981) have demonstrated that avirulent mutants contain the same indigenous plasmids as the wild-type strain. Thus, the genetic mechanisms responsible for this change in phenotype are still unknown. To identify and clone genes involved in virulence, we have employed the methods of transposon mutagenesis (Kleckner 1977, Boucher et al. 1981). We herein report the successful identification and characterization of a virulence locus involved in EPS production in *P. solanacearum.*

Materials and Methods

Bacterial Strains, Plasmids and Transposons (see Table 1)

Growth Media and Antibiotic Levels. *Pseudomonas solanacearum* was routinely cultured in a complex medium CPG (Kelman 1954) and non-fluidal mutants were detected in a TZC-CPG medium containing peptone, casamino acids, glucose, 0.05% tetrazolium chloride. *E. coli* strains were grown in Luria agar. When employed, the following level of antibiotics were as used: Kanamycin (Km) 50 μg ml^{-1}, Gentamycin (Gm) 25 μg ml^{-1}, Chloramphenicol (Cm) 25 μg ml^{-1}, Streptomycin (Sm) 25 μg ml^{-1}, Rifampicin (Rif) 100 μg ml^{-1}. AB minimal media as described in Chilton et al. (1974).

Filter Matings. All matings were performed on CPG for 16 h on 25 mm Millipore filters. Donor and recipients were used in a 1:1 ratio in the case of HB101 (pJB4J1)

1 International Plant Research Institute, San Carlos, CA 94070, USA

Molecular Genetics of the Bacteria-Plant Interaction
(ed. by A. Pühler)

Table 1. Bacterial strains, plasmids and transposons

Strains		Relevant characters	Source
Pseudomonas solanacearum	S-82	Wild type	L. Sequiera
P. solanacearum	Ps82Rif	Rifamipicin resistant mutant	This paper
P. solanacearum	Ps82-2NR	Km^R, Gm^S, Tn5, IS50, nonfluidal, avirulent	This paper
P. solanacearum	Ps82-18	Km^R, Gm^S, Tn5, IS50, non-fluidal, avirulent	This paper
P. solanacearum	Ps82-293	Km^R, Gm^S, Tn5, IS50, non-fluidal, avirulent	This paper
P. solanacearum	Ps82-303	Km^R, Gm^S, Tn5, IS50, non-fluidal, avirulent	This paper
P. solanacearum	Ps82-305	Km^R, Gm^S, Tn5, IS50, non-fluidal, avirulent	This paper
P. solanacearum	Ps82-316	Km^R, Gm^S, Tn5, IS50, non-fluidal, avirulent	This paper
P. solanacearum	Ps82-316 FR	Km^R, Gm^S, Tn5, fluidal virulent	This paper
P. solanacearum	Ps82-497	Km^R, Gm^S, Tn5, IS50, non-fluidal, avirulent	This paper
Escherichia coli	HB101	recA Sm^R	N. Panapoulos
E. coli	C600	Sm^S	N. Panopoulos
Plasmids			
pJB4J1		Km^R, Gm^R, Tn5, lysogenic for Mu	J. Beringer
pRZ102		COLE1::Tn5	R. Jorgensen
RSF1010		Su^R, Sm^R	N. Panopoulos
pBR325		Amp^R, Cm^R, Tc^R	N. Panopoulos
pKR2013		Km^R, Tra^+, helper plasmid	G. Ditta
PKT210		Su^R, Sm^R, Cm^R	K. Timmis
pPsDDMS1		Amp^R, Cm^S, Tc^R	This paper
pPsDDMS18		Amp^R, Cm^S, Tc^R, IS50	This paper
pPsDDMS316		Amp^R, Cm^S, Tc^R, IS50	This paper

× Ps82 Rif. Late logarithmic cultures 10^9 cell ml^{-1} were mixed together and placed on millipore filters. Triparental matings were also performed as above with an equal ratio of all parents. After the mating, the bacteria were removed from the filters in distilled water and placed on appropriate selective media.

Plasmid Isolations and *E. coli* Transformations. These procedures were performed as described in Maniatis et al. (1982).

Agarose Gel Electrophoresis. 0.7% agarose gels were run in Tris-acetate, EDTA buffer.

Southern Blots. Southern blots were performed according to the method of Southern (1975).

Nick-Translations. Nick-translations were carried out as described by Rigby (1977).

Colony Hybridization. Colony hybridizations were as described by Grunstein and Hogness (1975).

Restriction Enzyme Analysis and Ligations. Reactions were performed as recommended by the manufacturer.

Electro-Elution of DNA from Agarose Gels. DNA was recovered from agarose gels by electro-elution in dialysis bags as described in Maniatis et al. (1982).

Results

Introduction of Transposon Tn5 into *P. solanacearum* and Isolation of Mutants

Selection for Km^R exconjugants after mating with HB101 (pJB4JI) in *P. solanacearum* Ps82 Rif^R occurred at a frequency of 5×10^{-7}. Over 800 colonies were purified and scored for growth on CPG Gm medium, TZC-CPG, and AB minimal medium. One percent of the Km^R exconjugants retained the ability to stably maintain the pJB4JI plasmid. Miniscreens were performed on these strains and were shown to contain the intact pJB4JI. These strains were also able to transfer pJB4JI back to *E. coli* (data not shown). One half percent of the Km^R exconjugants were unable to grow on minimal medium and were not further analyzed. Approximately one percent of the Km^R exconjugants were unable to produce EPS as was evident by their growth on the TZC medium. These colonies were further tested for virulence by inoculation into the susceptible potato cv. Russett Burbank. All non-fluidal colonies tested were avirulent.

Physical Characterization of Tn5 Insertions

To verify that Tn5 was present in the genome of *P. solanacearum,* Southern blots were performed on several randomly chosen Km^R Gm^S exconjugants. Figure 1 shows the hybridizations of the nick-translated pRZ102 probed against the total EcoR1 digested genome of *P. solanacearum.*

The various mobilities of the hybridized bands suggest that Tn5 is being inserted into single sites and are random throughout the genome (lanes B–H). Lane A represents wild-type DNA showing no homology between pRZ102 and *P. solanacearum* DNA.

We next isolated DNA from six independent Km^R Gm^S non-fluidal exconjugants and digested the DNA to completion with EcoR1. Figure 2 depicts the hybridization pattern seen with the pRZ102 probe. Each isolate contained two bands that hybridized with the pRZ102 probe. In addition, each mutant contained a band larger than 5.7 kb (the size predicted if an entire Tn5 element was present) and a common band (smaller than 5.7 kb) at 3 kb. Since it has been shown that Tn5 contains the internal transposons IS50L and IS50R (Berg et al. 1982) we hypothesized that IS50 was independently transposing in these isolates and it was genetically linked to the non-fluidal phenotype.

Cloning IS50 Insertions from *P. solanacearum*

To demonstrate that IS50 was independently inserted into the *P. solanacearum* genome, total DNA (10 μg) from two independent non-fluidal mutants was cut with EcoR1 and electrophoresed in 0.7% agarose. Total DNA was isolated at a mobility of 2.5 kb–3.5 kb and was electro-eluted from the agarose. This DNA was then ligated into the single EcoR1 site of pBR325 and transformed into *E. coli* HB101 by selection on Amp and Tc. The single transformants were purified and scored for Cm^S.

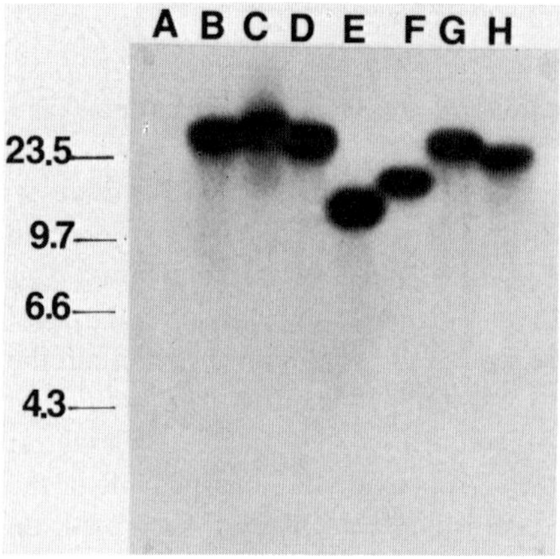

Fig. 1

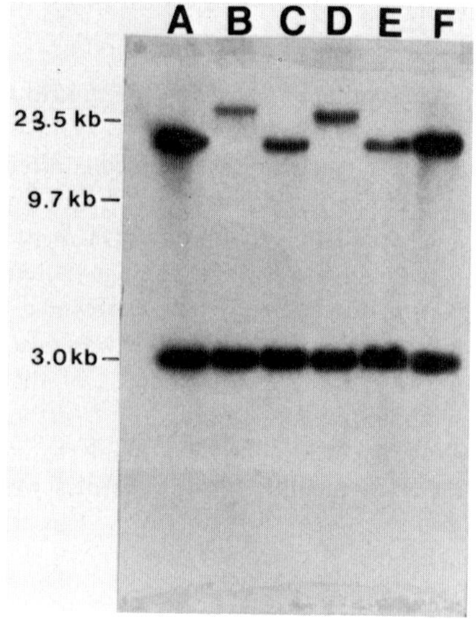

Fig. 2

Fig. 1. Southern blot analysis of random Km^R Gm^S fluidal exconjugants. Total DNA was isolated, cut with the enzyme EcoR1, blotted onto nitrocelluloase and probed with a nick translated pRZ102. Wild-type DNA is represented *(lane A). Lanes B–H* represent seven randomly chosen Km^R Gm^S fluidal exconjugants

Fig. 2. Southern blot analysis of Km^R Gm^S non-fluidal avirulent mutants. Total DNA was isolated from six non-fluidal mutants, cut with EcoR1, blotted onto nitrocellulose, and probed with a nick-translated pRZ102 probe. *Lane A* (Ps8218); *B* (Ps82-316); *C* (Ps82-305); *D* (Ps82-303); *E* (Ps82-293); *F* (Ps82-497)

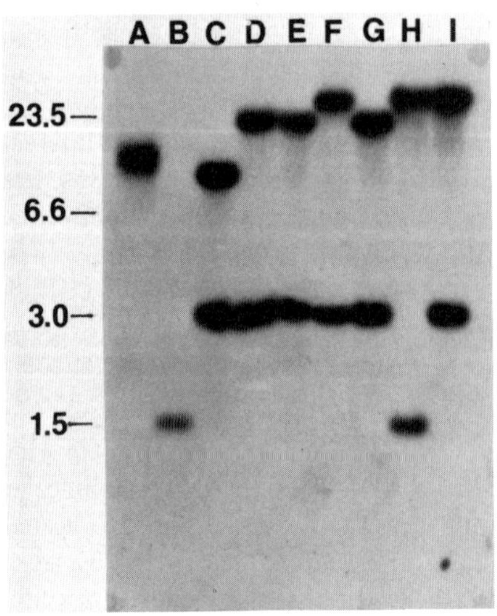

Fig. 3. Confirmation of independent Tn5 and IS50 insertions into the genome of *P. solanacearum*. Total DNA was isolated from non-fluidal mutants, cut with EcoR1, blotted onto nitrocellulose, and probed with pPsDDMS-316. *Lane A* (uncut pPsDDMS-316 plasmid DNA); *B* (wild-type DNA); *C* (Ps82-2NR); *D* (Ps82-18); *E* (Ps82-293); *F* (Ps82-303); *G* (Ps82-305); *H* (Ps82-316FR); *I* (Ps82-316)

Colonies that were Amp^R Tc^R Cm^S were lysed on filters and their DNA transferred to nitrocellulose filters. These filters were probed with nick-translated pRZ102. Hybridizing colonies were isolated and analyzed for plasmid content (Fig. 4). Restriction enzyme mapping confirmed the insertion of IS50 into the *P. solanacearum* genome (data not shown). To verify that these clones were the clones in question, pPsDDMS316 was nick-translated and probed against EcoR1 cut DNA from wild-type *P. solanacearum* (lane B) and six independent mutants (lanes C–G and I) (Fig. 3). This Southern blot demonstrates that the clone pPsDDMS316 consists of a DNA element (IS50 1.5 kb) inserted into a single EcoR1 fragment of *P. solanacearum* DNA (1.5 kb). Furthermore, the probe hybridizes back to the intact Tn5 element and also demonstrates that all IS50 insertions are located within a common 1.5 kb EcoR1 fragment of *P. solanacearum*.

The pPsDDMS316 clone was then used as probe to isolate the corresponding wild-type EcoR1 fragment. Again this was accomplished by the methods described above except DNA corresponding to 1.5 kb was electro-eluted from the agarose gel. A clone (pPsDDMS1) was isolated that hybridized to the probe (pPsDDMS316). Presence of the wild-type fragment was confirmed by agarose gel electrophoresis (Fig. 4).

This clone was then used to probe back to the wild type DNA and DNA from the non-fluidal mutants (Fig. 5). Again, the results were consistent that we had cloned the wild-type fragment and that there was a 1.5 kb insertion into a common fragment that was associated with loss of virulence.

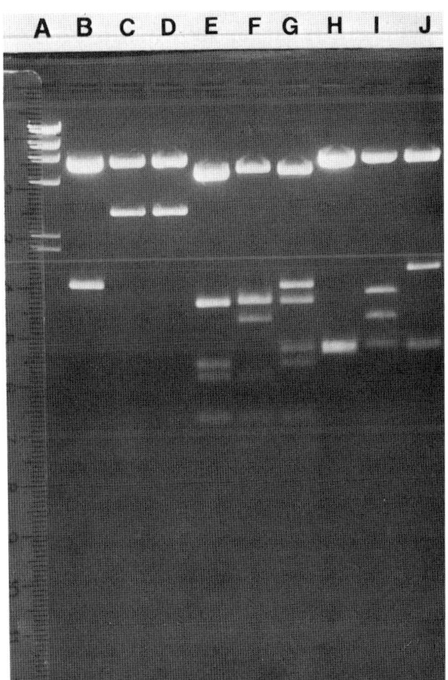

Fig. 4. Agarose gel electrophoresis of IS50 insertions cloned from the genome of *P. solanacearum*. *Lane A* (HindIII digest of phage λ); *B* (pPsDDMS1 cut with EcoR1); *C* (pPsDDMS18 cut with EcoR1); *D* (pPsDDMS316 cut with EcoR1); *E* (pPsDDMS1 cut with EcoR1 and Pst1); *F* (pPsDDMS18 cut with EcoR1 and PstI); *G* (pPsDDMS316 cut with EcoR1 and Pst1); *H* (pPsDDMS1 cut with EcoR1 and XhoI); *I* (pPsDDMS18 cut witn EcoR1 and XhoI); *J* (pPsDDMS316 cut with EcoR1 and XhoI)

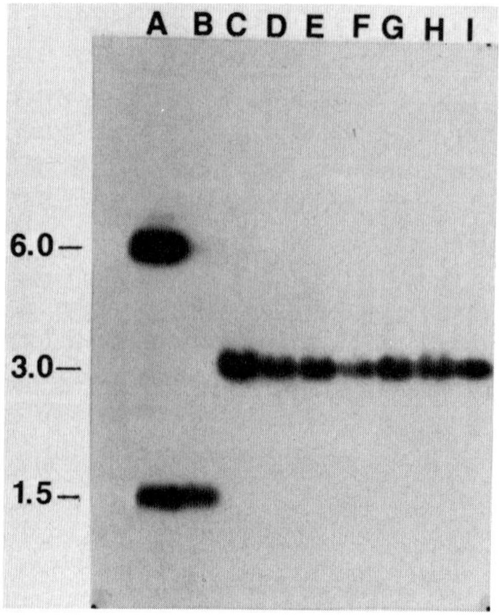

Fig. 5. Southern blot analysis of IS50 insertions in a common 1.5 kb fragment of *P. solanacearum* DNA. *Lane A* (pPsDDMS1 cut with EcoR1); *Lanes B–H* represent total EcoR1 cut DNA of Ps82R, Ps82-2NR, Ps82-18, Ps82-293, Ps82-303, Ps82-305, Ps82-316, Ps82-497. The blot was probed with nick-translated pPsDDMS1

IS50 Excision Restores EPS Production

Further evidence was sought to prove that the IS50 elements were associated with loss of virulence. Spontaneous revertants were observed in one isolate Ps82-316. One such colony was purified, DNA isolated and restricted with EcoR1. Figure 3 shows the results of the hybridization with the pPsDDMS316 probe against total EcoR1 cut DNA of the fluidal revertant Ps82-316FR (lane H). It can be seen that the 3.0 kb band has disappeared and has been restored to the original wild-type size (lane B).

Mobilization of RSF1010-Derived Cloning Vectors into *P. solanacearum* by the Helper Plasmid pRK2013

The wide host range cloning vector RSF1010 and two derivatives were tested from their ability to replicate in *P. solanacearum.* This was accomplished by mobilizing RSF1010 and pKT210 by the helper plasmid pRK2013. Table 2 shows the results of such experiments. DNA miniscreens of these isolates also verified the presence of these vectors in *P. solanacearum* (data not shown).

Table 2. Transmission of RSF1010 and pKT210 to *Pseudomonas solanacearum* by the helper plasmid pRK2013

Donor	Recipient	Selection	Frequency [a]
C600 (RSF1010)	Ps82R	Rif^R, Sm^R	5×10^{-4}
HB101 (pKT210)	Ps82R	Rif^R, Sm^R, Cm^R	1×10^{-5}

[a] Expressed as frequency per donor cells

Discussion

We have shown that both Tn5 and IS50 can be used as insertional elements to isolate mutants in *P. solanacearum.* Most of the Km^R Gm^S exconjugants contained only intact Tn5 elements. Only in the case of non-fluidal avirulent mutants did we see both Tn5 and IS50 insertions. This was the case for over 50 randomly chosen Km^R Gm^S exconjugants. The strong correlation of independent IS50 elements into a common fragment of *P. solanacearum* is suggestive that this region of DNA controls a step in the production of EPS. The additional evidence of an apparent precise excision of IS50 from this fragment is further evidence for this hypothesis. Preliminary experiments have also revealed that other sites in the *P. solanacearum* genome may be important in the determination of the fluidal phenotype as we have detected single Tn5 insertions in other loci (Staskawicz et al., unpubl. observ.). Boucher (pers. commun.) has also observed Tn5 induced mutants in several loci that give rise to the non-fluidal phenotype. Definitive evidence that the particular 1.5 kb EcoR1 locus we have described is involved in virulence comes from site-directed mutagenesis of this locus, as we have proven conclusively its role as one of possible several loci involved in controlling virulence in *P. solanacearum* (Staskawicz et al. 1983, in prep.).

Zischek et al. (1981) have shown that the *P. solanacearum* S-82, Race 3 strain contains a megaplasmid of over 300 megadaltons. Collaborative experiments are underway to determine if the locus we have described is located on this megaplasmid. We are currently constructing cosmid libraries in RSF1010-derived cosmid cloning vehicles to isolate large stretches of DNA that encompass the 1.5 kb fragment in attempts to complement our mutations. This approach should facilitate the isolation of other genes involved in virulence in *P. solanacearum.*

Acknowledgements. Thanks to N.J. Panopoulos, P. Boistard, and A.H. Ellingboe for stimulating ideas and suggestions throughout the course of this work.

References

Berg DE, Johnsrud L, McDivitt L, Ramabhadran R, Hirschel BJ (1982) Inverted repeats of Tn5 are transposable elements. PNAS 79:2632–2635

Boucher C, Message B, Debien D, Zischek C (1981) Use of P-1 incompatibility group plasmids to introduce transposons in *Pseudomonas solanacearum.* Phyto 71:639–642

Buddenhagen I, Kelman A (1964) Biological and physiological aspects of bacterial wilt caused by *Pseudomonas solanacearum.* Annu Rev Phytopathol 2:203–230

Chilton MD, Currier T, Farrand SK, Bendich AJ, Gordon MP, Nester EW (1974) *Agrobacterium tumefaciens* DNA and PS8 bacteriophage DNA not detected in crown gall tumors. PNAS 71:3672–3676

Currier TC, Morgan MK (1981) Plasmids are not associated with formation of noncapsulated variants of *Pseudomonas solanacearum.* Proc 5th Int Conf Plant Pathol Bact, Cali, Colombia, pp 420–425

Grunstein M, Hogness D (1975) Colony hybridization: A method for the isolation of cloned DNA's that contain a specific gene. DNAS 72:3961

Kelman A (1954) The relationship of pathogenicity of *Pseudomonas solanacearum* to colony appearance on a tetrazolium medium. Phytopathology 44:693–695

Kleckner N, Roth JR, Botstein D (1977) Genetic engineering in vivo using translocatable drug-resistance elements. J Mol Biol 116:125

Maniatis T, Fritsch EF, Sambrook J (1982) Molecular cloning – a laboratory manual. Cold Spring Harbor Lab

Rigby PWJ, Dieckmann M, Rhodes C, Berg P (1977) Labeling deoxyribonucleic acid to high specific activity in vitro by nick-translation with DNA polymerase I. J Mol Biol 113:237

Southern E (1975) Detection of specific sequences among DNA fragment separated by gel electrophoresis. J Mol Biol 98:503

Zischek C, Tuyen LTK, Boistard P, Boucher C (1981) Presence of a megaplasmid in strains of *Pseudomonas solanacearum*. Proc 5th Int Conf Plant Pathol Bact, Cali, Columbia, pp 427–437

Cloning of Genes Involved in Bacterial Ice Nucleation and Fluorescent Pigment/Siderophore Production

C. ORSER[1], B.J. STASKAWICZ[2], J. LOPER[1], N.J. PANOPOULOS[1], D. DAHLBECK[2], S.E. LINDOW[1], and M.N. SCHROTH[1]

Introduction

Microorganisms interact with plants in a variety of ways, both direct and indirect. Direct interactions include pathogenicity, of which there are many different types, and symbiotic, of which the best understood example is the *Rhizobium*-legume interaction. Indirect interactions are those in which individual microorganisms, or groups thereof, interact with others which in turn affect plant growth and health. Microbial antagonism and competition in the rhizosphere or phylloplane fall in this category. All of the above types of interactions are important in the agricultural context. Our knowledge of the molecular and genetic components important to the various interactive processes above or to the ecological survival and competitive fitness of the microorganisms participating in the interactions, with few exceptions, are poorly understood. Better understanding of these subjects is essential for the selection, development and use of biological agents and strategies to control plant pathogens and weeds and to modify the microbial populations in the rhizosphere or phylloplane for useful agronomic purposes. Our laboratories are involved in the investigation of the plant-bacteria interactions and have chosen several different model bacteria which individually or collectively offer opportunities for the study of one or more types of interactions. In this report we discuss some recent work regarding the cloning of genes involved in ice nucleation and fluorescent siderophore production by certain plant pathogenic and epiphytic bacteria.

Bacterial Ice Nucleation and Frost Damage to Plants

Pseudomonas syringae Van Hall and *Erwinia herbicola* (Lohnis) Dye cause nucleation of ice crystals in supercooled water at temperature slightly below 0°C (threshold nucleation temperature about −1.8°C). The first species includes strains, currently classified as pathovars (Dye et al. 1980) that are pathogenic on different plants, while *E. herbicola* is not a plant pathogen. Both species are common plant epiphytes found in substantial numbers on above-ground plant surfaces (leaves, twigs, buds, flowers)

1 Department of Plant Pathology, University of California, Berkeley, CA, USA
2 International Plant Research Institute, San Carlos, CA 94070, USA

Molecular Genetics of the Bacteria-Plant Interaction
(ed. by A. Pühler)

with seasonal fluctuations from undetectable levels up to 10^7 cells g^{-1} tissue fresh weight. A causal relationship between frost damage on frost-sensitive crop plants at relatively warm subzero temperatures (down to $-5°C$) has been established. For instance, the degree of damage after exposure to low temperatures either in the field or in controlled environments (growth chambers) is directly related to the logarithm of the populations of ice nucleation active (INA^+) bacteria present in or on the surfaces of above-ground parts. Also, plants grown aseptically can tolerate temperature of several (ca. 6) degrees C below zero without apparent damage and they are rendered sensitive to such temperatures by spraying with suspensions of INA^+ bacteria prior to the low temperature exposure (Linddow 1982, Lindow et al. 1978). Ice nucleation activity, in addition to causing frost damage, is considered to be a conditional frost-dependent virulence factor in *P. syringae* pv. *syringae.* For example, in spite of the fact that significant populations are present on plant surfaces, disease often occurs only after the ambient temperature drops to or below freezing (Panagopoulos and Crosse 1964). Bacterial ice nucleation is also believed to be important in atmospheric precipitation processes and is of some commercial interest, for example in artificial snow making processes and in cloud seeding.

The nature of the substance(s) responsible for ice nucleation activity (INA) in these bacteria is not known. However, nucleation activity is associated with the bacterial cell (i.e., non-diffusible), evidently localized on the outer membrane. One or more proteins appear to be involved, directly or indirectly, in its expression. Mutants deficient in ice nucleation activity (INA^-) have been previously isolated but have not been extensively characterized (Lindow and Staskawicz 1981). We recently succeeded in cloning the ice nucleation genes from *P. syringae* and *E. herbicola* (Orser et al. 1982, in prep.).

Pseudomonas Fluorescent Pigments/Siderophores

Production of gram-yellow UV-fluorescent water-soluble pigments is a characteristic taxonomic property of fluorescent pseudomonads including *P. syringae* pv. *syringae* (Leisinger and Margraff 1979). Many different environmental factors affect the synthesis of these pigments, among them low iron stress. Regulation by iron levels and the observation that a fluorescent pigment of *P. fluorescens* forms very stable complexes with Fe^{3+} and enhance ^{59}Fe uptake by cells (Meyer and Nornsperger 1978) suggest that these substances function as siderophores. The structure of pseudobactin and yellow-green fluorescent pigment from *P. fluorescens* B-10, and its role in iron transport, was recently established (Teintze et al. 1981). Pseudobactin, *P. fluorescence* B-10 and several other strains of the *P. fluorescens-putida* group enhace growth of certain plants and control certain phytopathogenic microorganisms (Kloepper et al. 1980, Misaghi et al. 1982, Scher and Baker 1982, Schroth and Hancock 1982).

The siderophores of *P. syringae* have not been chemically characterized. Our observation (unpublished) that non-fluorescent mutants of *P. syringae* pv. *syringae* are conditionally iron-dependent suggested that a fluorescent pigment(s) function as siderophore(s) in this bacterium. The involvement of fluorescent siderophores in microbial interactions in the rhizosphere and the interest in *P. syringae* as a biocontrol agent (Lindow 1982, Strobel 1981) prompted us to investigate the genetics of fluorescent pigment production by *P. s. syringae*.

Methods

Construction of genomic libraries is described in detail elsewhere (Orser et al. 1982). Briefly, total DNA from *Pseudomonas syringae* pv. *syringae* strains 31R1 and Cit7 and *Erwinia herbicola* strain 26SR6-2 was purified by two cycles of CsCl-ethidium bromide density gradient centrifugation and dialyzed against appropriate buffers. The DNA was partially digested with *Eco*R1 (0.3 units μg^{-1} DNA for 30 min) and fractionated by sucrose gradient centrifugation in 5–25% neutral sucrose. Fractions were analyzed by agarose gel electrophoresis and those rich in fragments 18–25 kb in size were pooled, ethanol precipitated and ligated (with T4 DNA ligase) to *Eco*R1-linearized plasmid vector, pLAFR1 (Friedman et al. 1982). This vector is a cosmid derivative of the broad host range vector pRK290 (Tc^r) and has a single *Eco*R1 site outside the tetracycline gene. To minimize packaging of vector dimers the ratio of *P. syringae* DNA fragments to linearized pLAFR1 during the ligation rection was 3–4:1. The ligated DNA mixture was packaged into phage particles in vitro and was used to transduce maltose-grown *E. coli* HB101 cells, to tetracycline resistance. Several hundred Tc^r transductants were individually inoculated into the wells of sterile microtitre plates containing 100 μl $well^{-1}$ King's B broth supplemented with 15 μg ml^{-1} tetracycline (to select for presence/maintenance of the vector) and 100 μg ml^{-1} streptomycin and cycloheximide to prevent bacterial and fungal contamination. The microtitre plates were covered with their lids and incubated overnight at 37°C. Glycerol was then added to each well to a final concentration of 30% to permit long-term low-temperature storage of the bacteria. Duplicate copies of the libraries were similarly stored in DMSO. To generate working copies or partial "pools" of these libraries a transfer block was constructed which consisted of 48 4-cm-long stainless steel wires inserted in regularly placed holes of a modular aluminum block. The positions of the wires corresponded exactly to those of individual wells in the microtitre plates. Sterilization of this unit was accomplished readily by dipping in alcohol and passing it over a flame. The sterilized unit was first placed over one-half of a microtitre plate so that the wires were immersed into the wells. After a brief agitation the unit was removed and placed in a smimilar way over the corresponding half of another microtitre plate containing 100 μl of antibiotic-supplemented broth as above. The above steps were repeated until a desired number of colonies was transferred to a set of "copy" plates. All transfer operations were performed under a laminar flow biosafety cabinet.

Recombinant clones carrying specific genes of interest were identified either by direct examination of *E. coli* colonies for expression of the genes of interest or by complementation tests with appropriate mutants of the DNA source strains. For complementation tests, recombinant plasmids were transferred by pRK2013-assisted mobilization either from the total original mixture of transductants, from individually stored recombinant colonies, or from "sub-pools", to appropriately marked mutants of the recipient strains.

For this purpose, 50 µl of freshly grown HB101 (pRK2013) and an equal volume of recombinant plasmid donor and of recipient cell suspension were mixed in sterile microtitre plates and incubated overnight. Alternatively, 50 µl of the triparental cell mixture in each well was transferred to another set of microtitre plates containing 200 µl well^{-1} King B agar to simulate solid agar matings which give considerably greater transmission for P-group plasmids. After overnight incubation 50–100 µl fresh King B broth was added to each well to resuspend the cells, and serial dilutions were prepared as necessary. A small volume (5–10 µl) of these cell suspensions was transferred to appropriate agar plates. To facilitate absorption of the liquid the plates were allowed to solidify with their lids off and partially dried under a biosafety hood or were prepared 2 days before use. These and other similar transfers to and from microtitre plates were facilitated through the use of an 8-prong Eppendorf multipet which performed eight transfers (one microtitre plate column) simultaneously.

Cloning of Ice Nucleation Genes from *P. syringae* pv. *syringae* and *E. herbicola*

The genetic and biochemical complexity of the ice nucleation property of *P. syringae* and *E. herbicola* were not known. It was therefore significant that ice nucleation activity could be detected in the *E. coli* host used to construct the genomic libraries of these bacteria. Approximately 0.3% of several hundred Tc^r transductants, tested by the replica freezing technique (Lindow et al. 1978), were INA-positive. Further examination showed in each such clone the presence of recombinant plasmids each having a different set of fragments but all sharing in common a 21 kb (pLAFRR1) and one or more additional fragments. Some plasmids were subsequently transferred to *E. coli* HB101 and SK1592 as well as into INA$^-$ mutants of the DNA source strains *(P. syringae* and *E. herbicola)* obtained earlier (Lindow and Staskawicz 1981). In all cases, INA$^+$ progeny were obtained, all harboring plasmids identical by restriction endonuclease criteria to those present in the donor recombinant clones. The above results, therefore, show that the cloned inserts determine the expression of the INA$^+$ phenotype both in a homologous and heterologous *(E. coli)* cellular environment.

The expression of INA in *P. syringae* and *E. herbicola* is under temperature as well as temporal control (Lindow et al. 1982). For optimal INA expression cells must be grown at an optimum temperature and assayed at late stages of their growth cycle. Even under the best growth conditions known, most cultures contain only a few percent cells with active ice nucleation sites at any given time. The same was true with respect to both growth temperature (Figs. 1 and 2) and growth cycle for *E. coli* carrying INA$^+$ recombinant plasmids. Furthermore, the ice nucleation activity spectra of several *E. coli* clones carrying INA$^+$ recombinant plasmids were generally similar to those of the DNA source strains (Figs. 3 and 4). These findings indicate that the

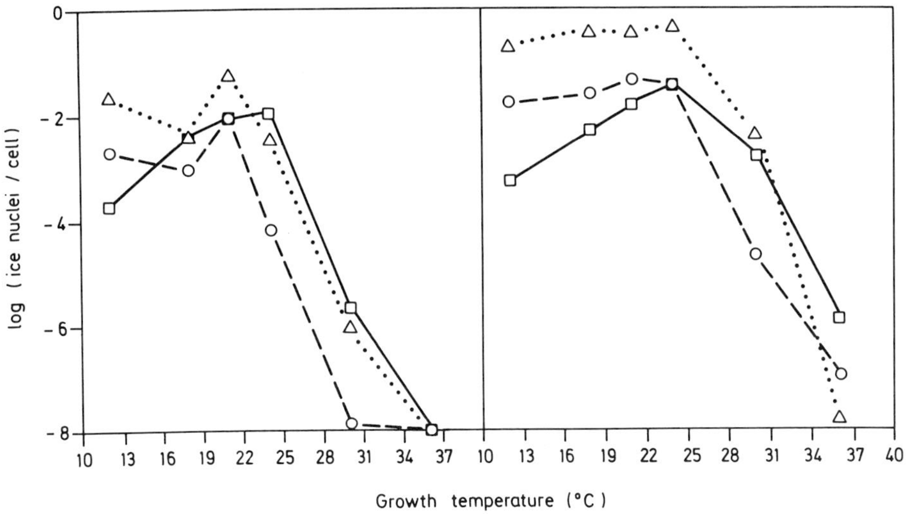

Fig. 1. Frequency of ice nucleation (logarithm of ice nuclei $cell^{-1}$) of *P. s. syringae* cit7 *(squares), E. coli* HB101 (pC-1) *(circles),* and HB101 (pC1BR1) *(triangles)* grown and assayed at different temperatures. Assays were performed at – 5°C *(left panel)* or – 9°C *(right panel)*

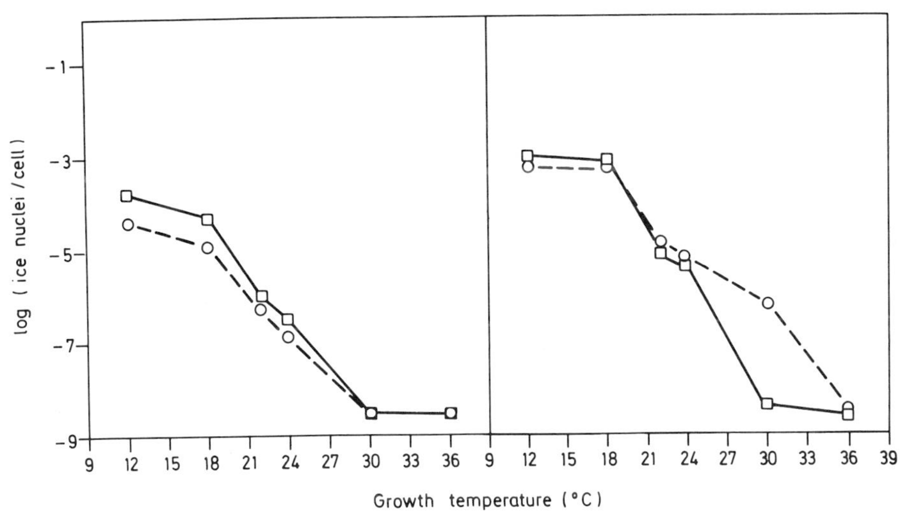

Fig. 2. Frequency of ice nucleation (logarithm of ice nuclei $cell^{-1}$) of *E. herbicola* 26SR6-2 *(squares)* and *E. coli* HB101 (pE-7) grown and assayed at different temperatures. Assays were performed at – 5°C *(left panel)* or – 9°C *(right panel)*

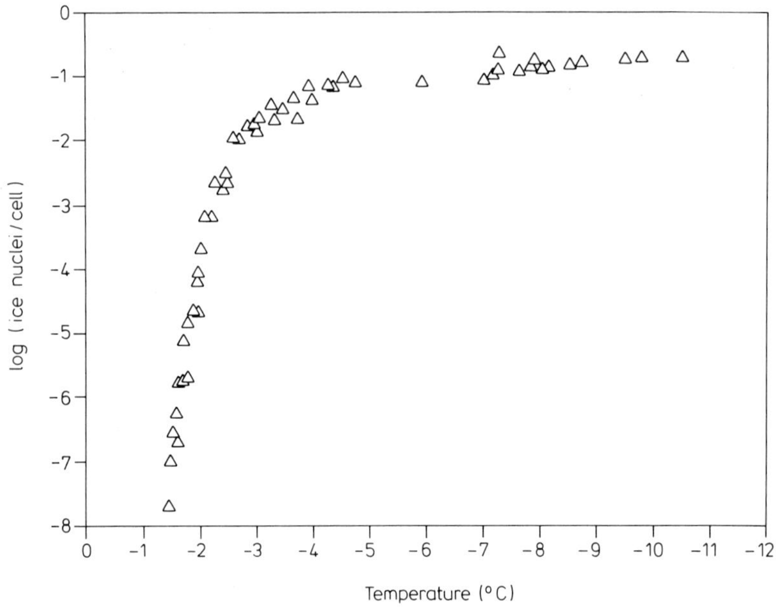

Fig. 3. Ice nucleation activity of *P. s. syringae* cit7 is a function of temperature (ice nuclei cell^{-1}) and is plotted *(ordinate)* against temperature in degrees C *(abscissa).* Methods are described in Lindow (1982)

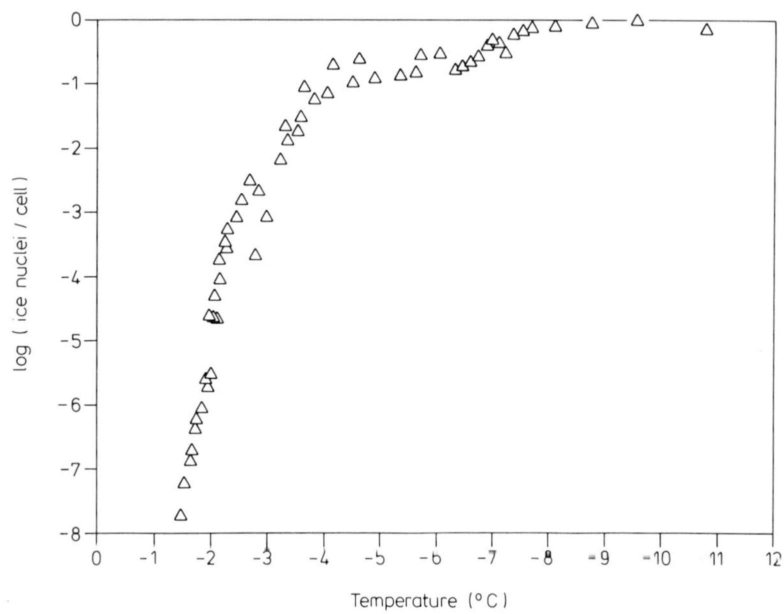

Fig. 4. Ice nucleation activity of *E. coli* HB101 (pC1BR1) on a function of temperature. Parameters and measurements as in Fig. 3

expression of INA in *E. coli* is under similar temporal and temperature control and results in the same threshold nucleation temperature as in the wild type INA^+ strains.

In view of the low expression of INA in the DNA-source strains and in *E. coli*, we explored the possibility of using a high copy number plasmid vector to further increase nucleation frequency. (The copy number of pLAFR1, although not precisely known is believed to be low; S. Long, pers. commun.) Accordingly, we subcloned INA-active *Eco*R1 fragments from one recombinant plasmid, pC-1, onto pBR325. *E. coli*, HB101 transformants were selected on agar plates containing 100 mg ml^{-1} ampicillin plus 100 mg ml^{-1} streptomycin and were screened for sensitivity to chloramphenicol. Several Cm^S INA^+ transformants thus obtained contained plasmids with two *Eco*R1 fragments, one equal in size to pBR325 and the other, about 10 kb, corresponding to one *Eco*R1 fragment of pC-1 and presumably carrying the ice nucleation region. A transformant carrying one such plasmid, designated pC1BR1, was compared with respect to the nucleation parameters with HB101 (pC-1) (Fig. 1). It is clear that *E. coli* (pC1BR1) expressed substantially greater nucleation frequencies relative to *E. coli* (pC-1), as well as to *P. syringae* and *E. herbicola.* When assayed at $-9°C$ almost every *E. coli* (pC1BR1) cell contained an active ice nucleus. It is not known at present whether this reflects solely the greater gene dosage expected of pBR325 recombinant plasmids, transcriptional read-through from the chloramphenicol acetyl-transferase gene in which the 10 kb INA fragment is inserted, or the elimination of negative control elements during subcloning from pC-1.

Isolation of Genes Involved in UV-Fluorescence and Iron-Independent Growth of *P. syringae* pv. *syringae*

Addition of 200 ppm of the iron-chelating compound ethylene diamine dihydroxy phenyl acetic acid (EDDA) to King's B agar results in an iron-deficient medium. *P. syringae* pv. *syringae* 31R1 produces a siderophore(s) which can compete with EDDA for the ferric ion, enabling the strain to grow on this medium. Non-fluorescent mutants of this strain, obtained after EMS mutagenesis, are unable to grow on this medium. Their growth inhibition is reversed upon addition of 10^{-4} M $FeCl_3$ to the medium, indicating that iron starvation is indeed responsible for EDDA inhibition. The specificity of EDDA chelation for the ferric ion is discussed in Archibald and De Voe (1980). The number and location of the genes involved in fluorescent pigment production or iron-dependent growth in *P. syringae* pv. *syringae* is not known. Preliminary screening experiments failed to reveal UV-fluorescent *E. coli* colonies in the two *P. s. syringae* genomic libraries. This failure may be due to one or more of the following reasons: (a) the number of genes involved in fluorescent pigment production in these strains is larger than the coding capacity of the cloned inserts, (b) the genes are not clustered in the *P. syringae* genome, (c) one or more of them fail to express or their products are lethal to, or unstable in, *E. coli.*

En masse matings of the genomic libraries with non-fluorescent mutants yielded fluorescent transconjugants at low frequency. These fluorescent transconjugants were able to grow on media containing 200 μg ml^{-1} EDDA. However, repeated attempts to isolate recombinant plasmid DNA from these transconjugants were unsuccessful. Therefore, individual clones of the HB101 (pLAFR1-*P. syringae*-31R) library stored in microtiter plates were utilized as donors in 718 separate triparental conjugations with HB101 (pRK2013) and non-fluorescent mutants of *P. syringae* as recipients. Two of the 718 separate conjugations yielded a small number of transconjugants which were fluorescent and iron-independent while the majority were not. Again, repeated attempts to isolate recombinant plasmid DNA from the fluorescent transconjugants were unsuccessful. We presume that the DNA sequences present in the two recombinant plasmids above comprised only a portion of the pathway for fluorescent pigment/siderophore production or lacked *cis*-acting expression elements so that integration into the chromosome of the recipient strain by *rec*-mediated single cross-over events was required in order to observe complementation in the mutants. Nevertheless, the recombinant plasmids were isolated from the donor HB101 clones. Restriction endonuclease analysis of these plasmids indicated that they were structurally identical, probably duplicate clones which had been separately stored. One of these plasmids was subsequently shown to complement several additional non-fluorescent mutants but could not complement others, indicating that this plasmid carries only some of the genes involved in fluorescent pigment/siderophore production in *P. syringae* pv. *syringae.* Several additional recombinant plasmids which complement other non-fluorescent mutants of this bacterium have subsequently been obtained.

Summary and Conclusions

We have shown that the DNA sequences responsible for ice nucleation activity in *P. syringae* pv. *syringae* and *E. herbicola* are confined to a small genomic region and, therefore, can be easily cloned. Furthermore, they express well in *E. coli,* normally devoid of ice nucleation activity. Although the actual number of genes needed for INA expression in *E. coli* is not known at present, subcloning, deletion and insertion mutagenesis indicate that the genetic region required for INA in *E. coli* is 4–5 kb. Transfer of the INA^+ inserts to other "expression" vectors is underway to assist in the isolation and characterization of the gene product(s) involved in the expression of INA and the preparation production of highly active cultures and perhaps ice nucleation active cell-free products for commercial purposes.

The isolation of recombinant plasmids involved in the production of the fluorescent pigment/siderophore of *P. syringae* pv. *syringae* is a first step in our continuing study of this property and its importance to the basic and interactive biology of this and other similar bacteria associated with plants.

References

Archibald FS, Voe IW De (1980) Iron acquisition by *Neisseria meningitidis* in vitro. Infect Immunol 27:322–344

Dye DW, Bradbury JF, Goto M, Hayward AC, Lelliott RA, Schroth MN (1980) International standards for naming pathovars of phytopathogenic bacteria and a list of pathovar names and pathotype strains. Rev Plant Pathol 59:153–168

Friedman AM, Long SR, Brown SE, Buikema WJ, Ausubel FM (1982) Construction of a broad host range cosmid cloning vector and its use in the genetic analysis of *Rhizobium* mutants. Gene (in press)

Kloepper JW, Schroth MN (1981) Relationship of in vitro antibiosis of plant growth-promoting rhizobacteria to plant growth promotion and displacement of root microflora. Phytopathology 71:1020–1023

Kloepper JW, Leong J, Teintze M, Schroth MN (1980) Enhanced plant growth by siderophores produced by plant growth-promoting rhizobacteria. Nature 286:885–886

Leisinger T, Margraff R (1979) Secondary metabolites of fluorescent Pseudomonads. Microbiol Rev 43:422–442

Lindow SE (1982a) Population dynamics of epiphytic ice nucleation positive bacteria on frost-sensitive plants and frost control by means of antagonistic bacteria. In: Sakai A, Li PH (eds) Plant cold hardiness. Academic Press, London New York

Lindow SE (1982b) Alternate methods of frost control through control of epiphytic ice nucleation active bacteria. Plant Dis 66 (in press)

Lindow SE, Staskawicz BJ (1981) Isolation of ice nucleation deficient mutants of *Pseudomonas syringae* and *Erwinia herbicola* and their transformation with plasmid DNA. Phytopathology Abstr 71:237

Lindow SE, Arny DC, Upper CD (1978a) The role of bacterial ice nuclei in frost injury to sensitive plants. In: Sakai A, Li PH (eds) Plant cold hardiness. Academic Press London New York

Lindow SE, Arny DC, Upper CD (1978b) Distribution of ice nucleation-active bacteria on plants in nature. Appl Environ Microbiol 20:6446–6457

Lindow SE, Arny DC, Upper CD (1978c) Distribution of ice nucleation-active bacteria on plants in nature. Appl Environ Microbiol 36:831–838

Lindow SE, Hirano SS, Borchet WR, Arny DC, Upper CD (1982) Relationship between ice-nucleation frequency of bacteria and frost injury. Plant Physiol 70:1090–1093

Meyer JM, Hornsperger JM (1978) Role of pyoverdin, the iron-binding fluorescent pigment of *Pseudomonas fluorescens,* in iron transport. J Gen Microbiol 107:329–331

Misaghi IJ, Stowell LS, Grogan RG, Spearman LC (1982) Fungistatic activity of water-soluble fluorescent pigments of fluorescent pseudomonads. Phytopathology 72:33–36

Orser CS, Staskawicz BJ, Panopoulos NJ, Dahlbeck D, Lindow SE (1982) Cloning and expression of ice nucleation genes from *Pseudomonas syringae* and *Erwinia herbicola* in *Escherichia coli.* Phytophathology 72:1000

Panagopoulos CG, Crosse JE (1964) Frost injury as a predisposing factor in blossom blight of pear caused by *P. syringae* van Hall. Nature 202:1352–1355

Scher FM, Baker R (1982) Induction of suppressiveness in soil to Fusarium wilt pathogens with *Pseudomonas putida* and a synthetic iron chelate. Phytopathology (in press)

Schroth MN, Hancock JG (1982) Disease-suppressive soil- and root-colonizing bacteria. Science 216:1376–1381

Strobel GA (1981) *Pseudomonas synringae* as an antagonist: field tests of its effectiveness against Dutch Elm Disease. Phytopathology 71:1007

Teintze M, Hossain MB, Barnes CL, Leong J, Helm D van der (1981) Structure of feric pseudobactin, a siderophore from a plant growth-promoting *Pseudomonas.* Biochemistry 20:6446–6457

The Genetics of Indoleacetic Acid Production and Virulence in *Pseudomonas savastanoi*

L. COMAI and T. KOSUGE [1]

Introduction

Pseudomonas savastanoi (Smith) Stevens is a pathogen of olive *(Olea europea)* and oleander *(Nerium oleander)* plants. The symptoms induced on the host plant consist of tumorous overgrowth called knots or galls, which although similar in appearance to those of the "crown gall" disease, are induced by a different mechanism. No evidence was found to suggest that tumors induced by *P. savastanoi* are the result of an Agrobacterium-type genetic transformation of the host cells (Wilson and Kosuge, unpubl. results); rather all evidence indicates that production of plant hormones by the bacterium is mainly responsible for the development of these symptoms. Indoleacetic acid and cytokinins, both important plant hormones, are produced in culture by *P. savastanoi.* In this paper we summarize some of our results in the molecular organization of the genes for indoleacetic acid production in the *P. savastanoi* genome.

The pathway of indoleacetic acid (IAA) synthesis involves oxidative decarboxylation of L-tryptophan (trp) to indoleacetamide by the enzyme Trp 2-monooxygenase and hydrolysis of the amide to produce IAA by indoleacetamide hydrolase. Smidt and Kosuge (1979) found that selection for α-methyl Trp resistance yielded mutants which had lost production of IAA. Upon inoculation into oleander plants, these mutants multiplied in the plant tissue although they did not induce galls.

The Genetics of IAA Production

The genetics of IAA production was approached by investigating whether the genes coding the biosynthetic enzymes were plasmid borne. This possibility seemed likely since *P. savastanoi* strains harbor two to seven species of plasmid DNA and commonly in the genus *Pseudomonas* unusual catabolic functions are plasmid borne. Such was the case in strain EW2009; by acridine orange curing, loss of IAA production was correlated with loss of a 53 kilobase pair (kb) plasmid called pIAA1. By cotransformation with RSF1010, a small wide host range R factor, pIAA1 was reintroduced in cured derivatives of EW2009 and was shown to reestablish IAA production

1 Department of Plant Pathology, University of California, Davis, CA 95616, USA

Molecular Genetics of the Bacteria-Plant Interaction
(ed. by A. Pühler)

(Comai and Kosuge 1980). Further characterization of pIAA1 permitted precise mapping of the IAA genes. The locus,, *iaaM,* coding Trp 2-monooxygenase, is borne on a 2.7 kb EcoR1 fragment called EcoR1-M, as shown by molecular cloning of that fragment in *Escherichia coli* and by transposon mutagenesis. The locus coding indoleacetamide hydrolase, called *iaaH,* is located promoter distal from *iaaM;* insertion of elements causing polar mutations in the latter prevent expression of the hydrolase gene (Comai and Kosuge 1983). Thus the IAA genes are part of an operon which in strain EW2009 is plasmid borne.

Location of IAA Genes in Different Strains

Many strains of *P. savastanoi* have been isolated from infected oleander and olive plants. On the basis of host specificity a division can be made between oleander-specific and olive-specific isolates. Strains isolated from oleander are capable of causing symptoms both on oleander and on olive plants, while those isolated from olive plants appear to be pathogenic only in olive. Further, the capability of metabolizing sucrose is characteristic of strains isolated from oleander but not of those isolated from olive plants (Wilson et al. 1972).

We wanted to determine if plasmid borne IAA genes were a prerogative of the species *savastanoi.* For this purpose we screened a number of isolates by the two following techniques. First, mutants deficient in IAA production were isolated from each strain by α-methyl Trp selection. Since in EW2009 such mutants usually lose pIAA1, we compared for each strain the plasmid DNA profile in the parent and its mutant progeny. Second, using the cloned gene for Trp 2-monoxygenase as a hybridization probe, we tested plasmid and chromosomal DNA preparations for the presence of homologous sequences. The results from the two approaches correlated perfectly. Whenever deletion of part of, or loss of a plasmid was found in Iaa$^-$ mutants, the probe identified the location of the *iaaM* locus on that plasmid. Conversely, when no change in plasmid DNA composition was observed in Iaa$^-$ mutants, the homologous sequences were located on the chromosome. In addition to pIAA1, two new plasmids pIAA2 and pIAA3, bearing the IAA genes were found in other oleander strains (Comai et al. 1982). Preliminary experiments indicated that these plasmids and pIAA1 show homology in the region of the IAA genes of at least 6 kb. Hybridization studies of the chromosomal IAA genes suggested that they are not the result of integration of the whole pIAA1 (Comai and Kosuge, unpubl. results). The type of mechanism bringing about this varied location of the IAA genes is now being investigated. However, it is interesting that the division revealed by the criterion of host selectivity also holds for the molecular location of the IAA genes.

IAA Production and Virulence

Results of repeated studies have established a linkage between IAA production and virulence and have led to the conclusion that IAA production is necessary for virulence.

First, every Iaa⁻ mutant tested has been avirulent on oleander, whether the lesion preventing IAA production was loss of plasmid DNA, deletions or transpositions of DNA in the IAA region. Second, the capacity to produce symptoms on plants was restored in Iaa⁻ mutants by the introduction of pIAA1, or of a wide host range plasmid carrying *iaaM* (Comai and Kosuge, unpubl. results; Surico et al, unpubl. results).

The Insertion Element IS-51

A 1.3 kb insertion element, called IS-51, was found in the *iaaM* locus in spontaneous Iaa⁻ mutants of EW2009. It has polar effects since it not only inactivated *iaaM* but also prevented expression of *iaaH* in these mutants. The element was cloned in *E. coli* and characterized both physically and genetically. By hybridization studies a family of repeated elements highly homologous to IS-51 was found interspersed throughout the genome of different strains, with a copy number of at least twenty. The polar effect exhibited by this element in *P. savastanoi* was confirmed by studies in *E. coli* in which IS-51 was shown to terminate transcription (Comai and Kosuge 1983).

Evolution of IAA Synthesis

The origin of the IAA genes remains unknown. A likely assumption is that the IAA genes became established in the species *savastanoi* because they confer a selective advantage in nature. The basis of this selective advantage has not been rigorously investigated, but based on a number of observations the following conclusions can be made. *P. savastanoi* does not need the IAA genes to establish a compatible interaction with the host, since Iaa⁻ mutants can grow and survive in the host tissues at least up to 14 days (Smidt and Kosuge 1978). However, wild type strains stimulate the abnormal proliferation of host cells to form the knot. This structure is intensively colonized by *P. savastanoi* allowing the total population per infection site to reach a large number. Thus, while at the early stages following infection growth is comparable for both the wild type and mutants, the mutants fail to stimulate formation of the specialized niche offered by the tumor and their total numbers remain limited. Plants inoculated with Iaa⁻ mutants are indeed almost symptomless. Production of IAA (and probably cytokinins) may be necessary for the construction of a specialized niche in which the pathogen multiplies and occupies during periods of unfavorable climatic conditions.

According to this hypothesis acquisition of an IAA producing system by *P. savastanoi* is a step from a weak and limited parasitic condition to one more virulent and successful. But whence come the IAA genes? The following hypothesis is based on the enzymology and genetics of IAA production and on the extensive information available on the enzymology and genetics of catabolic pathways in saprophytic pseudomonads. A group of catabolic pathways has been characterized which act on

L-amino acids (Takeda and Hayaishi 1966). These pathways are derepressed by the presence of substrate (such as lysine) and consist of a series of enzymatic steps the first two of which are catalyzed by a monooxygenase and a hydrolase functionally similar to those concerned with IAA production from L-Trp. Although a catabolic system acting on L-Trp has not been described, if it existed, it could well be the ancestor for the system concerned with IAA production. Conversion from Trp catabolism to the production of IAA could be achieved by two simple mutations: the first allowing constitutive synthesis and the second causing accumulation of IAA by elimination of a degradation system. This double event is illustrated in Fig. 1.

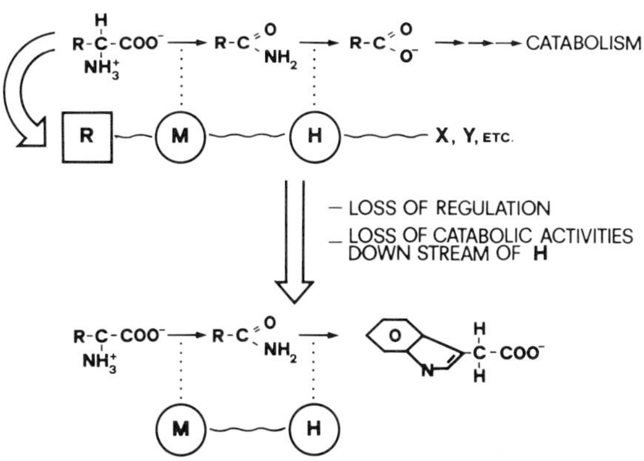

Fig. 1. Conversion by a double mutation of a hypothetical Trp catabolic pathway to the indoleacetic acid (IAA) biosynthetic pathway. The catabolic operon, *(top)* consists of *M* (monooxygenase gene), *H* (hydrolase gene), *X, Y*, etc. (undefined catabolic loci). IAA is produced by this system only when Trp is present in large amounts since *R*, a regulatory locus, codes a repressor inactivated by Trp. In addition, IAA cannot accumulate since it is catabolized further by the product of *X*, a catabolic locus. Two mutations only are needed, resulting in the loss of *R* and *X*, to generate from this hypothetical catabolic system the IAA pathway, shown below

According to this theory, genes for virulence originated from genes for saprophytic growth under the selective pressure of a parasitic condition. Although intriguing and compatible with the limited information this scenario awaits experimental proof.

Acknowledgement. This material is based upon work supported by the National Science Foundation under Grant no. PCM-8011794 and USDA-SEA, Competitive Research Grants Office, Grant no. 59.2063-1-1-643-0.

References

Comai L, Kosuge T (1980) Involvement of plasmid deoxyribonucleic acid in indoleacetic acid synthesis in *Pseudomonas savastanoi*. J Bacteriol 143:950–957

Comai L, Kosuge T (1982) Cloning and characterization of *iaaM*, a virulence determinant of *Pseudomonas savastanoi*. J Bacteriol 149:40–46

Comai L, Kosuge T (1983) Transposable element that causes mutations in a plant pathogenic *Pseudomonas* sp. J Bacteriol 154:1162–1167

Comai L, Surico G, Kosuge T (1982) Relation of plasmid DNA to indoleacetic acid production in different strains of *Pseudomonas synringae* pv. *savastanoi*. J Gen Microbiol 128:2157–2163

Smidt M, Kosuge T (1979) The role of indole-3-acetic acid accumulation by alpha-methyl tryptophan resistant mutants of *Pseudomonas savastanoi* in gall formation on oleanders. Physiol Plant Pathol 13:203–214

Takeda H, Hayaishi O (1966) Crystalline L-lysine oxygenase. J Biol Chem 241:2733–2736

Wilson E, Heskett MG, Johnson ML, Kosuge T (1972) Metabolic behavior of *Pseudomonas savastanoi* isolates from olive and oleander on certain carbohydrate and amino substrates. Phytopathology 62:349–355

Azospirillum Genetics

C. ELMERICH[1]

Introduction

Nitrogen fixing spirillum-like bacteria are commonly associated with the roots of grasses (von Bülow and Döbereiner 1975, Van Berkum and Bolhool 1980). Their taxonomic status changed with time. Designated at the beginning as *Azotobacter spirillum,* they were subsequently termed *Spirillum lipoferum* (see Becking 1982) until a new *Azospirillum* genus was defined (Tarrand et al. 1978). On the basis of GC%, DNA homology and nutritional properties two species were differentiated: *A. brasilense* and *A. lipoferum.*

Nutritional Properties

A. brasilense and *A. lipoferum* are strict aerobes, curved rod shaped, and they contain globules of poly-β-hydroxybutyrate. They can utilize organic acids such as malate or succinate as a sole carbon source. In addition, *A. lipoferum* utilizes a series of carbohydrates, including glucose, and requires biotin (Okon et al. 1976, Tarrand et al. 1978). More recently autotrophy and methylotrophy were reported for the two species (Sampaio et al. 1981). Dinitrogen, NH_4^+ and NO_3^- can be used as nitrogen sources. However, nitrogen fixation requires microaerobic environment (Okon et al. 1976, 1977). Some strains belonging to either species display anaerobic growth using NO_3^- as terminal electron acceptor (Neyra et al. 1977, Nelson and Knowles 1978).

Plasmids

Characterization

Plasmids were characterized by electrophoretic migration, in agarose gel, of cell lysates obtained by various techniques such as those described by Meyers et al. (1976), Eckhardt (1978) and Casse et al. (1979). All *Azospirillum* strains examined contained

1 Unité de Physiologie Cellulaire, Département de Biochimie et Génétique Moléculaire, Institut Pasteur, Paris, France

Molecular Genetics of the Bacteria-Plant Interaction
(ed. by A. Pühler)

at least 1 plasmid and some contained up to 6 molecular species, the size of which ranged from 4×10^6 to over 300×10^6 M.W. (Franche and Elmerich 1981, Heulin et al. 1982, Singh and Wenzel 1982, Wood et al. 1982). Discrepancies between the number and the size of the plasmids carried by a given strain were observed. For example, Singh and Wenzel (1982) found 3 plasmids of 5.4, 12 and 51 kb in strain ATCC 29709 whereas Franche and Elmerich (1981) found a single large plasmid of about 200×10^6 M.W. in the same strain. It is not clear whether this was due to instability of the plasmids, to technical artefacts or to strain misnaming.

Phenotypes Associated with Plasmids

No evidence of self transmissibility of the plasmids was reported. Spontaneous loss (Franche and Elmerich 1981) and temperature or acridine orange induced curing (Heulin et al. 1982, Wood et al. 1982) of some plasmids were described. Phenotypic changes concomittant with plasmid loss were reported (Heulin et al. 1982, Wood et al. 1982). However, no specific feature was demonstrated as plasmid borne. Nevertheless, possible functions of plasmids can be suggested such as: carbon source utilization (Heulin et al. 1982); heavy metals resistance (Wood et al. 1982), plant hormone production, plant host specificity or recognition, nitrogen fixation etc.

Phages

Phage-Like Particle Production

Using mitomycin C as an inducing agent, lysis of most *A. brasilense* and *A. lipoferum* strains examined was observed (Franche and Elmerich 1981, Elmerich et al. 1982). No plaque formation was detected with the lysates on any strain. A defective phage was purified from *A. brasilense* strain Sp7 (ATCC 29145). The purified particles were devoid of DNA (Franche and Elmerich 1981).

Plaque Forming Bacteriophages

Several groups tried to isolate phages and, to my knowledge, phages were only isolated in Brazil. A phage able to form plaques on *A. brasilense* Sp7 was described but not characterized (Franco-Lemos and Maringolo 1980). A temperate phage forming plaques on *A. lipoferum* Br17 (ATCC 29709) was recently characterized (Elmerich et al. 1982). The phage, isolated in J. Döbereiner laboratory, was named Al-1. Its morphology and its size are similar to that of the coliphage λ and it contains a double stranded linear DNA molecule of 36 kb with cohesive ends. Stable lysogens were obtained and the prophage DNA was found to be maintained as a plasmid. Up to now the phage host range is limited to a few strains of *A. brasilense* and *A. lipoferum* of Brasilian origin. No transdruction property has been detected as yet.

Isolation of Mutants and Genetic Exchange

Isolation of Mutants

Classical techniques of mutagenesis previously described in *Escherichia coli* (Miller 1972) such as U.V. irradiation or treatment by chemicals (ethylmethane sulfonate, N-methyl-N′-nitro-N-nitrosoguanidine) followed by penicillin or D-cycloserine enrichment can be successfully applied to *Azospirillum.*

Isolation of mutants impaired in various functions was reported, e.g.: auxotrophy (Polsinelli et al. 1980, Franche et al. 1981, Wood et al. 1982), antibiotic and antimetabolite resistances (Mishra et al. 1979, Franche et al. 1981, Hartmann 1982, Barberio et al. 1982, Hartmann et al. 1983); nitrogen metabolism (Gauthier and Elmerich 1977, Magalhaes et al. 1978, Bani et al. 1980, Barberio et al. 1982), nitrogen fixation (Jara et al. 1983).

Prophage Mu (Bukhari 1977) and transposon Tn*5* (Berg 1977) were also introduced in *Azospirillum* with the aid of IncP plasmids (Datta and Hedges 1971) as vehicles. No Mu production was detected in *Azospirillum* (Franche 1981, Singh 1982). The suizide plasmid pJB4JI (IncP, Mu. Gm, Tn*5*) used by Beringer et al. (1978) to introduce Tn*5* in the chromosome of *Rhizobium* was found to be very stable in *A. brasilense* and no spontaneous Tn*5* transposition was detected in strains ATCC 29710 (Singh 1982) and ATCC 29145 (Elmerich and Franche 1982). However, Elmerich and Franche (1982) briefly reported a method of isolation of Tn*5* containing mutants that was based on the incompatibility between pJB4JI and pJB3JI (IncP, Tc, Cb).

Genetic Exchange

Phages and plasmids of *Azospirillum* cannot be used, as yet, for genetic exchange. Mishra et al. (1979) reported DNA mediated transformation in *A. brasilense* strain Sp7. Until now this technique has not been used to establish genetic linkage.

Plasmid R68-45 (Haas and Holloway 1978) has the property of promoting chromosome mobilization in various Gram-negative bacteria including *Rhizobium* (Kondorosi et al. 1980). Similar properties were observed in *A. brasilense* strain Sp7 and linkage between couples of several markers was defined (Franche et al. 1981).

Genetics of Nitrogen Fixation

Homology Between Total DNA of *Azospirillum* and the *nif*HDK Genes of *Klebsiella pneumoniae*

The *K. pneumoniae nif*HDK operon which codes for the nitrogenase complex polypeptides is carried on a single 6.2 kb *Eco*R1 fragment (Cannon et al. 1980). Using this fragment as a probe, homology was detected with total DNA of a large number of diazotrophs (Ruvkun and Ausubel 1980) including several strains of *Azospirillum*

(Elmerich and Franche 1982, Quiviger et al. 1982). Plasmid localization of the *nif* DNA was not demonstrated (Duggan, Robson, and author's laboratory, unpubl. results).

Cloning of a *nif* Cluster from *Azospirillum*

The 6.7 kb *Eco*R1 fragment of *A. brasilense* strain Sp7, homologous to the *K. pneumoniae nif* HDK probe, was cloned in a bacteriophage λ vector (Quiviger et al. 1982). By heteroduplex analysis the extent of the homology between the two DNAs was found to be approximately 5 kb long. This was consistent with the existence of a complete *nif*HDK gene cluster in *Azospirillum* (Quiviger et al. 1982).

Proteins encoded by the 6.7 kb *Eco*R1 *Azospirillum* fragment were examined in *E. coli* minicells by the technique described by Reeve (1977) Five polypeptides were detected of approximate molecular weight of 54,000; 52,000; 36,000; 31,000 and 27,000 (Quiviger et al. 1982). Some of them might correspond to the nitrogenase component polypeptides.

Genetic Complementation of *nif* Mutants

Using the 6.7 kb *Eco*R1 cloned fragment which carries the *nif*HDK *Azospirillum nif* cluster it was possible to initiate genetic analysis of nitrogen fixation by complementation. Derivatives of pRK290 plasmid (Ditta et al. 1980) which contained *Azospirillum nif* DNA were constructed and introduced by conjugation in *A. brasilense* strain Sp7 Nif^- mutants (Jara et al. 1983). Restoration of nitrogen fixation was observed in the case of mutant 7571, which might be impaired in a structural gene for the nitrogenase complex (Jara et al. 1983). This is the first genetically characterized *nif* mutant of *Azospirillum*.

Conclusion

Bacteria belonging to the *Azospirillum* genus, practically unknown or forgotten ten years ago, appear today as diazotrophs which are amenable to genetic analysis and which may be used as biofertilizer in agriculture.

Although genetics of *Azospirillum* is still at its beginning, basic techniques such as chemical or biological mutageneses, chromosome mobilization, and construction of diploids are now available. The first results are rather encouraging, and the methodology already used should logically lead to the construction of a linkage map of the chromosome and to a better knowledge of the organization of the *nif* genes.

Acknowledgements. The author wishes to thank Drs. Duggan, Klingmüller, Hartmann and Robson for the communication of unpublished results. Research in the author's laboratory was supported by funds from Elf-Aquitaine and Entreprise Minière et Chimique.

References

Bani D, Barberio C, Bazzicalupo M, Favelli F, Gallori E, Polsinelli M (1980) Isolation and characterization of glutamate synthase mutants of *Azospirillum brasilense*. J Gen Microbiol 119: 239–244

Barberio C, Bazzicalupo M, Gallori E, Polsinelli M (1982) Regulation of ammonium assimilation and N_2 fixation in *Azospirillum brasilense*. In: Klingmüller W (ed) *Azospirillum* genetics, physiology, ecology. Birkhäuser, Basel, EXS42, p 52

Becking JH (1982) *Azospirillum lipoferum* – a reappraisal. In: Klingmüller W (ed) *Azospirillum* genetics, physiology, ecology. Birkhäuser, Basel, EXS42, p 130

Berg DE (1977) Isolation and excision of the transposable kanamycin resistance determinant Tn*5*. In: Bukhari AI, Shapiro JA, Adhya SL (eds) DNA insertion elements, plasmids and episomes. Cold Spring Harbor Press, Cold Spring Harbor, New York, p 205

Beringer J, Beynon JL, Buchanan-Vollaston AV, Johnston AWB (1978) Transfer of the drug-resistance transposon Tn*5* to *Rhizobium*. Nature 275:633–634

Berkum P van, Bohlool BB (1980) Evaluation of nitrogen fixation by bacteria in association with roots of tropical grasses. Microbiol Rev 44:491–517

Bülow JFM von, Döbereiner J (1975) Potential for nitrogen fixation in maize genotypes in Brazil. Proc Natl Acad Sci USA 72:2389–2393

Bukhari AI (1977) Bacteriophage Mu as a transposon element. Annu Rev Genet 10:389–412

Cannon FC, Riedel GE, Ausubel FM (1980) Overlapping sequences of *K. pneumoniae nif* DNA cloned and characterized. Mol Gen Genet 174:59–66

Casse F, Boucher C, Julliot JS, Michel M, Dénarié J (1979) Identification and characterization of large plasmids in *Rhizobium meliloti* using agarose gel electrophoresis. J Gen Microbiol 113: 229–242

Datta N, Hedges RW (1971) Compatibility groups among fi^- R factors. Nature 234:222

Ditta G, Stanfield S, Corbin D, Helsinski D (1980) Broad host range DNA cloning system for Gram negative bacteria: construction of a gene bank of *Rhizobium meliloti*. Proc Natl Acad Sci USA 77:7347–7351

Eckhardt T (1978) A rapid method for the identification of plasmid DNA in bacteria. Plasmid 1: 584–588

Elmerich C, Franche C (1982) *Azospirillum* genetics: plasmids, bacteriophages and chromosome mobilization. In: Klingmüller W (ed) *Azospirillum* genetics physiology, ecology. Birkhäuser, Basel, EXS42, p 9

Elmerich C, Quiviger B, Rosenberg C, Franche C, Laurent P, Döbereiner J (1982) Characterization of a temperate bacteriophage for *Azospirillum*. Virology 122:29–37

Franche C (1981) Thesis. University Paris 7, France

Franche C, Elmerich C (1981) Physiological properties and plasmid content of several strains of *Azospirillum brasilense* and *A. lipoferum*. Ann Microbiol Inst Pasteur 132A:3–17

Franche C, Canélo E, Gauthier D, Elmerich C (1981) Mobilization of the chromosome of *Azospirillum brasilense* by plasmid R68-45. FEMS Microbiol Lett 10:199–202

Franco-Lemos MV, Maringolo VL (1980) Abstr 4th Symp Nitrogen Fixation. Canberra, Australia

Gauthier D, Elmerich C (1977) Relationship between glutamine synthetase and nitrogenase in *Spirillum lipoferum*. FEMS Microbiol Lett 2:101–104

Haas D, Holloway BW (1978) Chromosome mobilization by the plasmid R68-45: a tool in *Pseudomonas* genetics. Mol Gen Genet 158:229–237

Hartmann A (1982) Antimetabolite effects on nitrogen metabolism of *Azospirillum* and properties of resistant mutants. In: Klingmüller W (ed) *Azospirillum* genetics, physiology, ecology. Birkäuser, Basel, EXS42, p 59

Hartmann A, Singh M, Klingmüller W (1983) Isolation and characterization of *Azospirillum* mutants excreting high amounts of indoleacetic acid. Can J Microbiol (in press)

Heulin T, Bally R, Balandreau J (1982) Isolation of a very efficient N_2-fixing bacteria from the rhizosphere of rice. In: Klingmüller W (ed) *Azospirillum* genetics, physiology, ecology. Birkhäuser, Basel, EXS42, p 92

Jara P, Quiviger B, Laurent P, Elmerich C (1983) Isolation and genetic analysis of *Azospirillum brasilense* Nif$^-$ mutants. Can J Microbiol (in press)

Kondorosi A, Vincze E, Johnston AWB, Beringer JE (1980) A comparison of three *Rhizobium* linkage maps. Mol Gen Genet 178:403–408

Magalhaes LMS, Neyra CA, Döbereiner J (1978) Nitrate and nitrite reductase negative mutants of N_2-fixing *Azospirillum* spp. Arch Microbiol 117:247–252

Meyers JA, Sanchez D, Elwell LP, Falkow S (1976) Simple agarose gel electrophoresis method for the identification and characterization of plasmid deoxyribonucleic acid. J Bacteriol 127:1529–1537

Miller JH (1972) Experiments in molecular genetics, 2nd edn. Cold Spring Harbor Laboratory, Cold Spring Harbor, New York

Mishra AK, Roy P, Bhattacharya S (1979) Deoxyribonucleic acid-mediated transformation of *Spirillum lipoferum*. J Bacteriol 137:1425–1427

Nelson L, Knowles R (1978) Effect of oxygen and nitrate on nitrogen fixation and denitrification by *Azospirillum brasilense* grown in continuous culture. Can J Microbiol 24:1395–1403

Neyra CA, Döbereiner J, Lalande R, Knowles R (1977) Denitrification by nitrogen-fixing *Spirillum lipoferum*. Can J Microbiol 23:300–305

Okon Y, Albrecht SL, Burris RH (1976) Factors affecting growth and nitrogen fixation of *Spirillum lipoferum*. J Bacteriol 127:1248–1254

Okon Y, Houchins JP, Albrecht SL, Burris RH (1977) Growth of *Spirillum lipoferum* at constant partial pressures of oxygen, and the properties of its nitrogenase in cell free extracts. J Gen Microbiol 98:87–93

Polsinelli M, Baldanzi E, Bazzicalupo M, Gallori E (1980) Transfer of plasmid pRD1 from *Escherichia coli* to *Azospirillum brasilense*. Mol Gen Genet 178:709–711

Quiviger B, Franche C, Lutfalla G, Rice D, Haselkorn R, Elmerich C (1982) Cloning of a nitrogen fixation *(nif)* gene cluster of *Azospirillum brasilense*. Biochimie 64:495–502

Reeve JN (1977) Bacteriophage infection of minicells. Mol Gen Genet 158:73–79

Ruvkun GB, Ausubel FM (1980) Interspecies homology of nitrogenase genes. Proc Natl Acad Sci USA 77:191–195

Sampaio M-JAM, Silva EMR da, Döbereiner J, Yates MG, Pedrosa FO (1981) Autotrophy and methylotrophy in *Derxia gumnosa, Azospirillum brasilense* and *A. lipoferum*. In: Gibson AH, Newton WE (eds) Current perspectives in nitrogen fixation. Aust Acad Sci, Canberra City, p 444

Singh M (1982) Transfer of bacteriophage Mu and transposon Tn*5* into *Azospirillum*. In: Klingmüller W (ed) *Azospirillum* genetics, physiology, ecology. Birkhäuser, Basel, EXS42, p 35

Singh M, Wenzel W (1982) Detection and characterization of plasmids in *Azospirillum*. In: Klingmüller W (ed) *Azospirillum* genetics, physiology, ecology. Birkhäuser, Basel, EXS42, p 44

Tarrand JJ, Krieg NR, Döbereiner J (1978) A taxonomic study of the *Spirillum lipoferum* group, with description of a new genus, *Azospirillum* gen. nov. and two species, *Azospirillum lipoferum* (Beijerinck) comb. nov. and *Azospirillum brasilense* sp. nov. Can J Microbiol 24: 967–980

Wood AG, Menezes EM, Dykstra C, Duggan DE (1982) Methods to demonstrate the megaplasmids (or minichromosome) in *Azospirillum*. In: Klingmüller W (ed) *Azospirillum* genetics, physiology, ecology. Birkhäuser, Basel, EXS42, p 18

Genome Rearrangement of the *Rhizobiaceae*

W. HEUMANN, B. MERKT, A. RÖSCH, R. SPRINGER, B. RÜGER, D. HORN, I. BAUER, K. WINKLER, and E. WAGNER[1]

Introduction

In recent years molecular genetic investigations showed that nature has developed different structural DNA characteristics which can cause strongly controlled rearrangements of genetic material, important to evolution, development, and differentiation (Cold Spring Harbor Symp. Quant. Biol. 1980). We observed that in *Rhizobiaceae* stress conditions trigger a complete rearrangement of the gene order. From the wild-type five different classes of other bacteria can be formed, with a distinctive new combination of traits and markers. This genome rearrangement is characteristic for all *Rhizobium* and *Agrobacterium* species. The five classes of new bacteria which are formed by this process can on the one hand be defined by many different biological and genetic characteristics which are class-specific, and on the other hand elicit unifying characteristics documenting their correlation to the *Rhizobiaceae.* The rearrangement in the five-class scheme is reversible.

Stress Conditions

Originally we observed the appearance of new strains from *Rhizobium* cultures by different treatments, for instance after chemical mutagenesis, after regrowth of old water agar deposits, or after heat treatment. We started to investigate this phenomenon more seriously when very characteristic diazotrophic strains appeared repeatedly in our cultures, strains which grow with elemental N_2 as the sole nitrogen source under atmospheric conditions. This observation, made for the first time about ten years ago, seemed to be important enough to start a basic research program concerning the versatility of the rhizobial genome. In the meantime we tested the reaction of many different *Rhizobium* and *Agrobacterium* wild-type strains to many different cultural conditions intending to induce reliably the cellular state which finally elicits the genomic rearrangement. We found generally that all stress conditions applied to growing cells will induce this cellular state more or less regularly. Routinely we grow logarithmic cultures in an otherwise optimal medium in which the carbohydrate or

1 Institut für Mikrobiologie und Biochemie, Lehrstuhl für Mikrobiologie, Universität Erlangen-Nürnberg, 8520 Erlangen, FRG

Molecular Genetics of the Bacteria-Plant Interaction
(ed. by A. Pühler)

Table 1. Liquid stress media used to induce genome rearrangement and diagnostic plates to differentiate the characteristic new strains

Stress treatment	Carbohydrate-nitrogen source	
1. Liquid minimal medium	Without	$(NH_4)_2SO_4$
2. Liquid minimal medium	Glucose	KNO_3
3. Liquid minimal medium	Without	KNO_3
4. Liquid minimal medium	Dextrin	$(NH_4)_2SO_4$
5. Liquid minimal medium	Dextrin	KNO_3
Diagnostic plates		
1. Lab-lemco nutrient agar		
2. Yeast-extract-mannitol agar		
3. Minimal medium agar		
4. Nitrogen-free *Azotobacter* agar		

nitrogen sources are either missing or replaced by poorly metabolizable compounds, e.g., glucose by dextrine or ammonium sulfate by nitrate (Table 1).

The liquid stress media are inoculated with each strain in several parallel samples and incubated in a tube roller over a prolonged period. From time to time samples will be plated on four different diagnostic media (Table 1), which have been proved to reveal the growth characteristics of the five possible new bacterial classes most distinctively. The time of the appearance of the newly rearranged strains cannot be predicted, sometimes they appear a few days after the start of the stress treatment, sometimes it takes several weeks before new bacterial types can be observed. But as a result of the stress treatment of wild-type strains, one or even several of the new bacterial types are sure to appear, the characteristics of which always fit into one or into several of the five classes. The clearcut definition of these five rhizobial rearrangement classes by many biological, biochemical, and molecular characteristics, the regularity of their appearance and the reversibility of the rearrangement process by further stress treatment seem to be solid arguments for the origin and identity of these strains.

The Five Rearrangement Classes of the *Rhizobiaceae*

Table 2 shows the five rearrangement classes which were formed by stress treatment from wild-type strains of *Rhizobium* and *Agrobacterium*. Whenever the wild-type order of the genomes starts to converse, this process always ends in the genome order which is characteristic for one or several of these five classes. It cannot be predicted which class will appear first nor is there any correlation between a specificity of stress treatment and the number of rearrangement classes to appear. Each class seems to have an equal probability of being formed except class III nif-type, which appears less frequently than the other classes. We investigated a large amount of data of these

Table 2. Wild-types and genome rearrangement classes of the Rhizobiaceae

Class		
I	min-type 1,7170 (1,7170)	← -Ti { *Agrobacterium tumefaciens* wild-type
II	pro wild-type 1,7190	⇄ { *Rhizobium* spec. wild-type
III	nif-type 1,7235 (1,7170)	Asymbiotic aerobic nitrogen-fixing
IV	car-type, y 1,7245 (pl.)	Yellow pigmented by carotenoids
V	car-type, r 1,7310 (pl.)	Red pigmented by carotenoids

five bacterial classes for three different purposes: first, to define the class-specific criteria, second, to show the interdependence of the five classes with each other and with the wild-types by analyzing unifying criteria, and finally to find an approach to an understanding of the mechanism of such a fundamental genetic process.

Class-Specific Criteria

Class I, Min-Type (Minimal-Type). Generally the class I bacteria are very similar to *A. tumefaciens* wild-type. They can be characterized as Ti-cured *Agrobacteria* (Bauer 1981). Class I bacteria are fast-growing rods, the generation time is less than 1 h, the colonies can be rough or smooth. Cells are very motile and fimbriated. The buoyant density of chromosomal and plasmid DNA ρ = 1.7170 g/cm^3. The cells inherit two plasmids: the large one a molecular weight of 300–320,000,000, the other 20,000,000 M.W. in size. The restriction pattern with *Xho*I and other restriction endonucleases is class-specific and very similar to the *A. tumefaciens* restriction pattern (Heumann 1981). Min-type strains are sensitive to *A. tumefaciens* phages but not vice versa (Bauer 1981).

Class II, Pro Wild-Type. This class actually comprises two slightly different groups of strains: one group, coming from *R. leguminosarum, R. phaseoli* and *R. trifolii* nodulates soybean effectively and is therefore identical with *R. japonicum;* while the other group, coming from *A. tumefaciens, R. japonicum* and *R. lupini* has lost the tumor inducing and nodulating capacity. Both pro wild-types grow very slowly, the best medium is YMB, the colony is similar to slow growing wild-type strains. The buoyant density of the chromosomalDNA is ρ = 1,7720 g/cm^3, the existence of plasmids cannot be shown. The restriction pattern with *Xho*I is characteristic but diffeent in each one of the two described groups of class II strains. Especially remarkable seems the stepwise degradation of host specificity in this group from fast growing wild-type strains to *R. japonicum* to non-nodulating strains, and the simultaneous loss of plasmid DNA. We tested non-nodulating pro wild-type strains to regain the wild-type specificity by infecting the respective hosts. In these nodulation experiments we received *R. meliloti* and *R. trifolii* in two cases. After re-isolation, these strains showed new DNA chracteristics (restriction pattern and plasmids). More work has to be done to understand the role of the host plant during infection with non-nodulating pro wild-type strains.

Class III, *Nif*-Type. These are very fast growing bacteria, with a generation time of less than 1 h, very variable capsule material and therefore variable colony appearance. The cells are strongly motile and therefore produce a growing margin around each colony on agar media. The most remarkable characteristic of class III is the asymbiotic nitrogen fixation under aerobic conditions. The activity of these strains signifies a completely derepressed nitrogenase on nitrogen-free agar media, the enzyme reduces acetylen to 15–20 nMol ethylen/mg protein/min, its activity continues over more than 8 weeks. The chromosomal DNA buoyant density ρ = 1.7235 g/cm^3. All *nif*-type strains harbor two large plasmids with the molecular weight of 130,000,000 and 220,000,000 respectively, sometimes smaller plasmids of molecular weight 5,000,000 and 14,000,000 are present in addition. The plasmid DNA buoyant density ρ = 1.7170 g/cm^3. The restriction pattern of all class III DNA with *Xho*I and other restriction enzymes is identical and class-specific. Probing the *nif*-type plasmid DNA for *nif* structural genes with cloned *Klebsiella* or *R. meliloti* structural *nif* DNA resulted in the provisional indication that the *nif* structural genes are located in the 220,000,000 M.W. plasmid. From *nif*-type strains *A. tumefaciens* wild-types and class IV and class V carotenoid-type strains can be obtained by further stress treatment, but we never found *Rhizobium* wild-type strains.

Class IV, Car-Type, y (Carotenoid-Type, Yellow). The main characteristic of these strains is their bright yellow carotenoid pigmentation. The biochemistry and structure of these carotenoids has been analyzed by Kleinig et al. (1977) and proved to be β,β-carotene derivatives. Further characteristics of these strains and especially their very fertile conjugational system have been investigated in detail, reviewed in Heumann (1979). These bacteria grow in normal rods but there is a class IV subgroup whose cells grow in the form of bacteroids which are very similar to nodule bacteroids (Heumann 1981). The DNA buoyant density ρ of rod- and bacteroid-shaped class IV bacteria is 1.7245 g/cm^3. These yellow carotenoid-type strains normally do not inherit plasmids, they have a class-specific restriction pattern with *Xho*I and other restriction endonucleases.

Class V, Car-Type, r (Carotenoid-type, Red). These strains are red pigmented by carotenoids. The structure of this pigment has also been investigated by Kleinig et al. (1979). It is a C30-carotenoic acid glucosyl ester. This pigment is rather unique and different from the class IV carotenoid. It can therefore be assumed that the biosynthetic pathway of both carotenoids are coded by different DNA regions. It is not possible to metamorphose one pigment into the other by mutagenesis of the respective bacterial strains. Class V bacteria have a very characteristic shape, they are large motile and fimbriated cells. They grow with a generation time of > 4 h. The DNA buoyant density ρ is between 1.7290 and 1.7310 g/cm^3. Some of the strains inherit 2–5 plasmids of a molecular weight of 12–350,000,000. Some are free of extra-chromosomal DNA. The XhoI restriction pattern is class-specific. The conjugational fertility between different class V mutants is very low.

The Interdependence of the Bacteria of the Five Classes and of the Wild-Types by Unifying Criteria

We isolated bacteriophages from soil, specific for each class. Phage typing tests with this material indicated that all bacterial strains of one class propagate the class-specifc phages with identical multiplicity. Cross infection experiments with bacterial strains of other classes however also showed growth of the phages though under strong restriction. The phages so modified were then restricted from the original host. These experiments indicate that each class can be characterized by its own restriction and modification system, but also that the strains from all classes are able to grow phages from all classes. Control experiments to propagate these phages in *Escherichia coli* C or other taxonomically unrelated bacteria were unsuccessful. In continuation of the characterization of each class by their own R and M system, we isolated and purified restriction endonucleases from class II and class IV bacteria and showed that the DNA of their own strains was not digested, but the DNA of strains of all other classes was. The restriction activity of the endonucleases of the other classes has so far not been purified, but tests with unpurified preparations also confirmed the class specificity of the different R and M systems.

The interrelation of the strains of the five classes is also documented by conjugational cross fertility: mutants of class IV are fertile with class V-, class III- and class II-mutants. Since the conjugational transfer of the *Rhizobiaceae* is regulated by a taxon-specific chromosomal fertility system (Heumann et al. 1971), these transfer experiments document the interrelation of the respective strains. In the description of the five single classes – see Introduction – the genetic and molecular similarity of pro wild-type strains with *Rhizobium* wild-types and of min-type strains with *Agrobacterium* wild-types has been shown. This further documents the relation between the "classic" *Rhizobiaceae* and their genomic rearrangement scheme. It is of special interest that, from all *Rhizobium* wild-types, *Agrobacterium* wild-types can be obtained by stress treatment. This rearrangement is not reversible.

Approach to the Understanding of the Mechanism of the Genome Rearrangement

In Table 2 the five rearrangement classes are ordered according to the increasing DNA buoyant density from class I ρ = 1.7170 g/cm^3 to class V ρ = 1.7310 g/cm^3. An explanation for this density shift could be that strains of all classes are composed of different proportions of light and heavy DNA and during the shift one part is stepwise amplified and the other diminished and vice versa in the shift reversion. The plasmid DNA is obviously exempted from this process since the extra-chromosomal DNA of class I, class III and of the two wild-type classes in all strains, always has the lowest buoyant density of the *Rhizobiaceae* ρ = 1.7170 g/cm^3.

Another characteristic of the *Rhizobiaceae* DNA is the high frequency of IS elements. Burkardt et al. (1979) and Priefer et al. (1981) characterized IS*R1*-as an IS-element of class IV *Rhizobia*. We probed this IS*R1*-DNA in hybridization experiments with the fragmented DNA of many different class IV strains and found that more than twenty copies of IS*R1* are inserted in different fragments of the chromosome. Ruvkun et al. (1981) characterized another IS-element of *R. meliloti* (IS*Rm1*) and observed that this sequence can transpose to several locations in the *R. meliloti* genome. These results are strong indications that IS elements are at least partly responsible for the dynamic character of the *Rhizobiaceae* genome.

From the conjugation system of class IV bacteria further evidence for a possible mechanism participating in the genome rearrangement can be obtained: we have shown that after conjugal transfer the donor DNA is inserted into the recipient chromosome producing a merodiploid zygote. In the offspring of this zygote all types of different recombinants are formed by intrachromosomal recombination (Heumann and Springer 1977, Springer and Heumann 1978). In connection with the many IS sequences providing homologous regions in the chromosome, the process of intrachromosomal recombination could essentially participate in the mechanism of the genome rearrangement discussed in this paper.

The Bacteriological Significance of Genome Rearrangement

The genome rearrangement in the *Rhizobiaceae* seems at first glance to be incompatible with the concept of species specificity, but it only enlarges the species definition. As I have shown: from one bacterial species several different species can be formed reversibly by genome rearrangement. All these species have the same taxonomic rank, because they come from one and the same genome. The mechanism and the degree of this genomic versatility are strongly regulated. Its meaning is therefore an enlargement of the species definition and not its suspension. Each class of bacteria which we have described as originating from the *Rhizobiaceae* is perfectly stable under growth conditions optimal to the respective class. The conversion into other classes needs strong stress conditions. From the point of evolutionary survival and improvement, genome rearrangement gives to the respective bacterial group a powerful tool with which to enlarge its ecological niche up to the genetic interaction with plants, which leads to symbiotic nodules and pathogenic tumors.

References

Bauer I (1981) Genomumstrukturierung bei *Rhizobiaceae.* Dipl Diss, Univ Erlangen

Burkardt HJ et al (1979) Naturally occurring insertion mutants of broad host range plasmids RP4 and R68. In: Timmis KN, Pühler A (eds) Plasmids of medical, environmental and commercial importance. Elsevier/North Holland Biomedical Press, Amsterdam New York, p 494

Heumann W (1979) *Rhizobium lupini* genetics. Curr Top Microbiol Immunol 88:1–23

Heumann W (1981) *Rhizobium* genetics. In: Bothe H, Trebst A (eds) Biology of inorganic nitrogen and sulfur. p 384

Heumann W, Springer R (1977) Formation of merodiploid clones by conjugation in *Rhizobium lupini.* Mol Gen Genet 150:73–79

Heumann W et al (1971) The two transfer regions of the *Rhizobium lupini* conjugation. Fertility factor elimination and one way transfer. Mol Gen Genet 113:308–315

Heumann W et al. (1973) The two transfer regions of the *Rhizobiumlupini* conjugation. Genetic characterizations of the transferred chromosomal segments. Mol Gen Gent 126:267–274

Kleinig H et al (1977) Carotenoids of *Rhizobia.* I. New carotenoids from *Rhizobium lupini.* Helv Chim Acta 60:254–258

Kleinig H et al (1979) New C_{30}-carotenoic acid glucosyl esters from *Pseudomonas rhodos.* Z Naturforsch 34c:181–185

Priefer UB et al (1981) Is*R1:* An insertion element isolated from the soil bacterium *Rhizobium lupini.* Movable genetic elements. Cold Spring Harbor Symp Quant Biol 45:1025

Ruvkun GB et al (1981) Molecular genetics of symbiotic nitrogen fixation. Cold Spring Harbor Symp Quant Biol 45:494–499

Springer R, Heumann W (1978) Genetical analysis of merodiploidy in *Rhizobium lupini.* Mol Gen Genet 165:57–63

Subject Index

Biology of Inorganic Nitrogen and Sulfur

Editors: H. Bothe, A. Trebst

1981. 144 figures. VIII, 384 pages. (Proceedings in Life Sciences)
ISBN 3-540-10486-0

Contents: General Introduction. – Nitrogen Metabolism in General. – Dinitrogen Fixation. – Sulfur Metabolism in General. – Contributed Papers: Nitrogen Metabolism. – Contributed Papers: Sulfur. – Subject Index.

Physiology and Biochemistry of Seeds in Relation to Germination

Volume 1
J. D. Bewley, M. Black

Development, Germination and Growth

Corrected printing. 1983. 122 figures. XII, 306 pages
ISBN 3-540-08274-3

Contents: Introduction. – The Structure of Seeds and Their Food Reserves. – The Legacy of Seed Maturation. – Imbibition, Germination, and Growth. – Biochemistry of Germination and Growth. – Mobilization of Reserves. – Control Processes in the Mobilization of Stored Reserves. – Glossary and Index of English and Botanical Names. – Author Index. – Subject Index.

Volume 2
J. D. Bewley, M. Black

Viability, Dormancy, and Environmental Control

1982. 153 figures. XII, 375 pages
ISBN 3-540-11656-7

Contents: Viability and Longevity. – Dormancy. – The Release from Dormancy. – The Control of Dormancy. – Perspective on Dormancy. – Environmental Control of Germination. – Glossary and Index of English and Botanical Names. – Author Index. – Subject Index.